HIGHER SPECULATIONS

Grand Theories and Failed Revolutions
in Physics and Cosmology

HELGE KRAGH

University of Aarhus, Denmark

OXFORD
UNIVERSITY PRESS

Great Clarendon Street, Oxford, OX2 6DP,
United Kingdom

Oxford University Press is a department of the University of Oxford.
It furthers the University's objective of excellence in research, scholarship,
and education by publishing worldwide. Oxford is a registered trade mark of
Oxford University Press in the UK and in certain other countries

© H. Kragh 2011

The moral rights of the author have been asserted

First Edition published in 2011

All rights reserved. No part of this publication may be reproduced, stored in
a retrieval system, or transmitted, in any form or by any means, without the
prior permission in writing of Oxford University Press, or as expressly permitted
by law, by licence or under terms agreed with the appropriate reprographics
rights organization. Enquiries concerning reproduction outside the scope of the
above should be sent to the Rights Department, Oxford University Press, at the
address above

You must not circulate this work in any other form
and you must impose this same condition on any acquirer

Published in the United States of America by Oxford University Press
198 Madison Avenue, New York, NY 10016, United States of America

British Library Cataloguing in Publication Data

Data available

Library of Congress Cataloging in Publication Data

Data available

ISBN 978–0–19–959988–2

CONTENT

Introduction — 1

PART I: CASES FROM THE PAST

1. Beginnings of Modern Science — 11
2. A Victorian Theory of Everything — 35
3. Electrodynamics as a World View — 59
4. Rationalist Cosmologies — 87
5. Cosmology and Controversy — 117
6. The Rise and Fall of the Bootstrap Programme — 141

PART II: THE PRESENT SCENE

7. Varying Constants of Nature — 167
8. New Cyclic Models of the Universe — 193
9. Anthropic Science — 217
10. The Multiverse Scenario — 255
11. String Theory and Quantum Gravity — 291
12. Astrobiology and Physical Eschatology — 325
13. Summary: Final Theories and Epistemic Shifts — 355

Bibliography — 371
Index — 403

Introduction

Ever since the birth of modern science during the so-called scientific revolution, it has been the dream of natural philosophers to explain the phenomenal world in terms of a single theoretical system that relies only minimally or not at all on contingent empirical data. This system or theory would ideally unify all partial theories in a coherent and mathematically consistent framework. The dream of theoretical physicists in the twenty-first century is basically the same dream that motivated natural philosophers like Descartes and Newton in the seventeenth century. Although it remains a dream, and will most likely always do so, it has led to great progress in the understanding of nature at its most fundamental level. Not only have physicists uncovered the secrets of nature down to the very smallest level of reality, close to the mysterious Planck scale of 10^{-35} m, they have also established connections to the very largest levels, that is, to the universe—or perhaps the multiverse—as a whole. In a very real sense, microphysics and cosmophysics have been unified, if not completely so.

The subject of this book is a certain class of physical theories that have in common that they are ambitious attempts to describe all or most of nature on a unified basis. Many of the theories are or have been controversial, in particular because their connection to empirical reality is problematic. They are not easily testable, to put it mildly. Another reason why theories of this kind tend to be controversial lies in their reductionist nature. Implicitly or explicitly, they claim to contain all knowledge of other sciences and thus, in principle, make them superfluous. Some of the higher speculation theories are perhaps closer to philosophical world views dressed in the language of mathematical physics. They are all speculative, if of course to different degrees, and they typically rely on nearly unrestricted extrapolations of contemporary scientific knowledge. Moreover, they deal with such knowledge on a fundamental level. Whereas some of the theories I deal with are restricted to a relatively local domain of nature, other even more ambitious theories are intended to be entire physical world views that aspire to finality. They are in principle meant to be 'theories of everything', to use a modern phrase that may be more catchy than informative.

Speculations have always been an integrated part of the physical sciences, sometimes hidden under the more palatable term 'hypotheses'. Indeed, fundamental physics without some element of speculation is hardly conceivable. All scientists agree that speculations have a legitimate place in the construction of scientific theories, but there is no agreement as to *how* speculative a theory may be and still be counted as scientific. Purely speculative theories with no links to empirical reality at all can be dismissed as metaphysical, but where is the boundary between metaphysics and physics? The very term 'speculation' is ambiguous and often used in quite different meanings. For example, what scientists in the past called speculative physics did not always carry the pejorative connotation that the term has today.

This book is not just one more work on the fascinating puzzles of modern quantum physics and cosmology, a genre which is covered by a large number of popular and semipopular titles of varying quality. Contrary to most other books, it adopts a historical perspective and aims to elucidate the present by comparing it with the past—and, in some cases, to elucidate the past by comparing it with the present. In spite of all the novelty of many-dimensional string theories, anthropic arguments, multiverse scenarios, and life-in-the-future speculations, these modern approaches are parts of a historical tradition with roots that extend far back in time. Whereas the higher speculations of the past, be it the vortex theory of atoms or Eddington's fundamental theory, can be analysed by means of the established methods of the history of science, this is not the case with contemporary theories that are still undecided and the subject of discussion among both scientists and philosophers. The historical study of this kind of contemporary science is faced with problems of a methodological nature, but these problems do not prevent the use of the historical approach, they only limit its use. The historian cannot look into the future, but he or she can look into the past. And the past goes on to the present.

Higher Speculations deals with several themes and problems that extend through most of the case studies which make up the book. These themes are fundamental insofar that they relate to the very heart of science, the question of what constitutes science and demarcates it from non-science or pseudoscience. This is basically a problem that belongs to the philosophy of science, and is one that philosophers have dealt with at great length, but the approach of the book is neither philosophical nor systematic. I have no intention of entering the intricate philosophical discussions concerning the nature and proper methods of science, or the different meanings of unification, not to mention the discussion concerning unity versus disunity in the sciences. What I do intend is to illustrate and illuminate the topic by means of historical examples, to discuss it through the words and deeds of the scientists engaged in a particular kind of fundamental and epistemically ambitious science.

By its very nature this kind of science comes close to the borderline between science and philosophy, or between physics and metaphysics. It almost always involves philosophical issues, if rarely explicitly stated, such as the role played by non-empirical considerations (from consistency to aesthetics) in the evaluation of scientific theories. First and most importantly, can the established standards of science—which basically means empirical testability—be applied to those areas of theoretical physics and cosmology that endeavour to understand all of nature? If not, can one appeal to alternative epistemic strategies—new paradigms of science? Such questions are familiar to philosophers, who have long questioned the dominant position of empirical testability relative to other modes of evaluation, but it is only recently that the questions have turned up with some force in the physical sciences. The rallying cry of epistemic shifts or new paradigms is increasingly heard from some quarters of theoretical physics and cosmology. Although somewhat similar proposals were voiced in the past, it is my impression that something new and potentially disturbing—and/or exciting—is happening in areas of contemporary fundamental physics. As this study shows, the claim of a paradigmatic shift in such areas as string theory and multiverse physics is resisted by a considerable part of the scientific community. Undoubtedly, an even greater part of the scientific community ignores the new claims or is even unaware of them. What we experience and what this book is in part about is not only new and amazing theories of physics, it is also a series of connected controversies that for some time have raged within the community of physicists and cosmologists.

The book is mainly composed of 12 separate chapters, each of which can be considered a case study of a theory or an episode in the history of the physical sciences. The chapters follow in a roughly chronological order, from the first half of the seventeenth century to the present. The focus is on the twentieth century. All of the chapters deal with historically important attempts to extend and unify physical knowledge or to question orthodox notions of physics. While some of the cases are limited to the traditional domain of physics, others are concerned either with the universe at large or with interconnections between microphysics and cosmology. I have made no attempt to cover all fundamental theories of this kind, but included case studies that I find to be interesting, instructive, and relevant for the topic of the book. I am well aware that some of the fundamental theories, such as the unified theories of the interwar years based on general relativity, are missing, and also that some of my cases (say the one on the vortex atom theory) may appear curious and of limited interest to many modern readers. Yet I believe that when they read about them, they will realize that they are in fact interesting and more than just curious episodes in the annals of science. They may realize that many of the problems and challenges that face modern fundamental physics are surprisingly similar to those that faced physicists in the past.

The book starts with an introductory chapter in which I briefly deal with three systems of natural philosophy from the early period. These are Descartes' mechanical system of the world, Boscovich's matter theory based on dynamical point atoms, and the Romantic version of natural philosophy dating from the early nineteenth century. The three systems were quite different in scope and details, yet together they may serve to introduce some of the themes that will appear in the later chapters. For example, the highly ambitious theories of Descartes and Boscovich aspired to finality and represent early attempts to establish a theory of everything.

My first elaborated example of a theory belonging to the category of higher speculations is the vortex theory of atoms, an ambitious and beautifully conceived attempt to establish a final framework for all of physics, which was made in the Victorian era. The basic idea of the vortex theorists, namely to picture atoms and other discrete structures as dynamic configurations of an all-pervading continuous medium, is also to be found in the so-called electromagnetic world view of the early twentieth century, which is the subject of Chapter 3. The medium was identified with the electromagnetic ether and the discrete structures were the electrons that were thought to be the building blocks of all matter and fields. The theories associated with this system or world view were no less ambitious and no less unified than those of the vortex physicists. Moreover, the two kinds of theory had in common that they were mathematically complex and only related somewhat peripherally to concrete problems of physics. Mathematically impressive and conceptually attractive as they were, their empirical records were poor. Some of the characteristic features of these two cases will also appear in theories dealt with in later chapters, such as the theory of superstrings.

The focus of Chapters 4 and 5 is pre-1970 cosmology and its connections to microphysics. That there is such a scientifically meaningful micro–macro connection is basically a discovery of the twentieth century. What is perhaps the most radical (and also the most idiosyncratic) modern attempt to establish fundamental physics on a new epistemic basis, namely Eddington's theory of the 1930s, is described in Chapter 4 in some detail, together with Milne's almost equally radical 'world physics'. The nature of these theories of 'cosmophysics' and the controversies that followed in their wake are highly instructive and of more than just historical interest. Chapter 5 is devoted to a somewhat related controversy, this time dealing with a specific cosmological theory that aroused a great deal of antagonism because of its unusual character. The steady-state theory of the universe was widely seen as provocative and speculative, but it was hardly more speculative than other contemporary conceptions of the universe based on the cosmological equations of general relativity. What makes this case of interest is, among other things, that it involved explicit discussions about the foundation and defining features of science in which philosophers took part alongside scientists. Important as these discussions were, they had almost no effect on the outcome of the cosmological controversy: it was settled by observations, not by arguments of a philosophical nature.

The case of the S-matrix or bootstrap theory of strong interactions, which is the subject of Chapter 6, belongs to roughly the same period as the steady-state model, but otherwise the two cases have very little in common. The conception of hadrons (such as protons and neutrons) and their interactions promoted by Geoffrey Chew in particular was not a unified theory of the physical forces, and neither did it relate to the universe. It was an ambitious attempt to create a new framework of particle physics based on the notion of self-consistency alone, as given by the properties of a mathematical quantity known as the S-matrix. In this respect the bootstrap programme shared some of the characteristics of the later string theory, which in its historical origin was in fact indebted to the bootstrap model of hadrons. As will be explained, although bootstrap physics was largely restricted to the hadronic realm, there were also a few attempts to extend it to a more broadly conceived philosophy of nature.

The second half of the book mostly deals with post-1980 developments, and in this part I have made no efforts to organize the case studies chronologically. They all belong to the modern era. The subject of Chapter 7 is not a particular theory or theoretical framework, but the concept of constants of nature and various hypotheses that they might vary in time. This idea, which on the face of it may appear paradoxical or even nonsensical, is interesting in its own right and serves as yet another bridge between microphysics and cosmology. It is also a controversial idea, in particular insofar that it includes a varying speed of light, a hypothesis which has been criticized for violating established scientific norms. Although the idea of varying constants has so far received no solid support in the form of experimental evidence, the idea is still alive and not definitely ruled out by experiment. It belongs to the resources that cosmologists sometimes draw upon. The following chapter examines a class of cosmological models that until fairly recently had a dubious reputation and were often thought of as speculations guided more by emotional desires than scientific facts. However, cyclic or oscillating models have experienced a remarkable revival and are today taken seriously by a not insignificant minority of cosmologists. Some of these models, and especially the so-called 'new cyclic theory', are based on notions of superstring theory and marketed as alternatives to the standard inflation model of the universe. In this case, as in some of the other cases I describe, the crucial question is the standards on which the model should be judged. What is the balance between theoretical foundation and empirical testability?

If cyclic models of the universe are controversial, the anthropic principle that emerged in the 1970s is even more so. Very much has been written about this notorious principle, conjecture, or mode of reasoning, but only a little about its historical context and early uses. Chapter 9 offers a history and summary account of the anthropic principle, with an emphasis on its development and problematic status as a tool of science. Does the 'weak' version of the anthropic principle have genuine predictive and explanatory power? Does the principle belong to science at all, or is it essentially philosophical and therefore non-scientific? Questions like these were asked and debated in the 1970s, when the principle was introduced in its modern form, and they are still being asked and debated. More than 30 years of discussion has only led to greater polarization among physicists and cosmologists with regard to the scientific respectability of the 'A-word'. It is impossible to know how future historians of the physical sciences will look at the anthropic principle. Will they find it to be a grand and fascinating mistake or the beginning of a new scientific revolution? Today the historian of physics can only describe how the principle emerged and try to evaluate what its current status is. By definition, the future is outside historical analysis.

The present status of the anthropic principle is closely connected with ideas of multiple universes, which, insofar that these are motivated in string theory, have produced a trinity of controversial ideas in the form of the 'landscape multiverse'. Chapter 10 introduces and discusses modern multiverse cosmology and relates it to earlier ideas of many universes. Apart from the anthropic principle and string theory, the multiverse is also connected to the inflationary scenario of the early universe that dates from the early 1980s and is widely accepted by cosmologists. During the last decade or so a major controversy has arisen over the scientific nature of the multiverse proposal and the changed style of physics it seems to represent. It is a style radically different from the traditional one, in the sense that it weakens or reinterprets methodological virtues such as explanatory and predictive power. Many multiverse antagonists are willing to admit anthropic multiverse reasoning as an interesting speculation, but not more than that. They tend to see the multiverse as more closely connected to philosophy or science fiction than to science. The ongoing debate over this issue is highly illuminating, among other reasons because it highlights the uneasy relationship between physics and philosophy in the realm of cosmology and thus replays a theme from earlier history. The similarity between the current debate and the one that raged in the 1930s is instructive and I shall return to it in later chapters.

Although modern string theory is part of the multiverse controversy, it is much more than that. It is a topic of great interest in itself, and one than can indeed be characterized as a higher speculation. The class of theories that over a period of 40 years has led to modern superstring and M-theory is the main subject of Chapter 11. After a brief account of the historical development of many-dimensional physics and the string concept, I focus on the problems of string physics and its claim to be a candidate for a final theory (if not necessarily a theory of everything). How do physicists justify a theory that in spite of decades of intense efforts still fails to make connections to the world of experiments? Do string physicists operate with new standards of physics, such as replacing empirical theory evaluation with tests of mathematical consistency? In spite of some rhetoric about string theory being a 'new paradigm' of physics, it turns out that the break with traditional particle physics is not truly revolutionary. Most string physicists do not aim at a new science, but rather at the completion of an old vision of fundamental science. While string theory is the best-known theory of quantum gravity, it is not the only candidate for a unified theory, and, according to some experts, it may not even be the most promising one. The chapter includes a brief account of an alternative theory that unites quantum mechanics and general relativity, the theory known as loop quantum gravity. Whereas the latter theory has not received a great deal of public attention, string theory is highly controversial, both among physicists and in the public arena. A major part of the chapter deals with this much discussed string controversy.

The last case study is concerned with a topic that is as new as it is old, namely, the place and significance of advanced life in the universe. It is a topic with a great past and perhaps also with a great future—but this is probably too early to say. Astrobiology is a new and exciting scientific discipline, yet it has roots that extend far back in time. While Chapter 12 mentions the search for extraterrestrials (SETI) in the Milky Way, it is primarily concerned with speculative scenarios of the existence of life forms in the distant future. This branch of modern cosmological physics, relying on extrapolations in the extreme, is sometimes known as 'physical eschatology'. It is in many ways a fascinating branch of theoretical science, but it shares with the anthropic multiverse a status as a science that has been seriously questioned. Some of the approaches and claims of modern physical eschatologists are fantastic indeed and hard to reconcile with what is traditionally known as science. Eschatology is squarely a theological concept, and it may be tempting to see physical eschatology as closer to unorthodox theology than to physics. Still, these higher speculations seem to be accepted as science—even 'respectable science'—in both an intellectual and sociological sense, at least in parts of the scientific community. No wonder that some commentators have spoken of post-modern or ironic science.

The very name 'physical eschatology' may remind us of the religious dimension that is sometimes claimed to be part, not only of this branch of speculative science, but also of other branches such as string theory and multiverse cosmology. While I mention the associations with religion and some of the debates they have occasioned, I do not deal systematically with the religious dimensions or give them a higher priority than they have in the scientific discussions. When it comes to the scientific publications, these dimensions only enter in a very limited way. In any case, a serious historical examination of the relationship between religious thought and modern concepts such as the multiverse and physical eschatology would demand a book of its own.

The final part of the book, Chapter 13, summarizes some of the main points of the previous chapters and adds some further considerations. I review the idea of a theory of everything in the light of historical examples and also take a second look at the claims of radical epistemic shifts in areas of modern cosmology and theoretical cosmology.

Some of the material presented in this book relies on earlier publications, in the form of articles, books, or book chapters. For example, I have previously examined the vortex theory of atoms (Chapter 2), the cosmophysics of Eddington and Milne (Chapter 4), and the steady-state theory of the universe (Chapter 5). More recently I have published papers dealing with varying-speed-of-light cosmologies (Chapter 7), cyclic cosmological models (Chapter 8), and multiverse cosmology (Chapter 10). Although I draw on these and other publications, the material in the book is presented in a new context and extended with more recent findings and references to the recent literature. The book grew out of a planned research project at the Department of Science Studies at the University of Aarhus, dealing with epistemic shifts in various fields of contemporary science. The discussions I had with my colleague Matthias Heymann on this issue helped to convince me that a book on higher speculations in physics and cosmology might be a worthwhile enterprise. I am grateful to Brandon Carter for having provided me with information about his role in the early history of the anthropic principle, and also to Hubert Goenner for critical and helpful comments on parts of the manuscript.

Helge Kragh

PART I
Cases From the Past

1
Beginnings of Modern Science

To explain all nature is too difficult a task for any one man or even for any one age. 'Tis much better to do with a little certainty, & leave the rest for others that come after you, than to explain all things by conjecture without making sure of any thing.

Isaac Newton, unpublished preface to *Opticks*, 1704.

It has been said that Thales of Miletus, who lived in the sixth century before Christ, was the first unificationist and the first thinker who proposed a theory of everything, in his case expressed by the deceptively simple world equation 'H_2O = TOE'. If we are to believe Aristotle, Thales said that the principle beneath all substances, whether dead or alive, is water. Yet, however simple, attractive and unifying Thales' hypothesis was, it lacked one crucial quality: it was not scientific.

A historically informed account of unified *scientific* theories aiming to explain all or most of nature cannot go arbitrarily far back in time. Science first had to be invented, which means that theories of this kind (as opposed to mere philosophical theories) are effectively limited by the scientific revolution that occurred in Western Europe in the decades around 1600. Historians sometimes limit this period by the publication of Copernicus' *De Revolutionibus* in 1543 and Newton's *Principia* in 1687, but the extension and meaning of the scientific revolution is of less importance. What is perhaps the best example of a scientific higher speculation from this early period is the grand theory of all natural phenomena produced by René Descartes, one of the giants of this revolutionary era. Descartes' mechanical conception of the world, truly a theory of everything, operated with only matter and motion, for which reason it was sometimes suspected of being materialistic or even covertly atheistic. It was a theory that not only provided in-principle explanations of all of inorganic nature, but also of all varieties of organic nature, humans included.

A century after the death of Descartes, the scientific project had stabilized and the science of mechanics greatly advanced, now as part of the Newtonian paradigm that was an important element of the world view of the age of reason. Physics had become deeper and more mathematical, and at the same time the new sciences of heat, electricity, and magnetism broadened its scope. The piecemeal progress and gradual specialization that occurred during the age of the Enlightenment did not imply that ambitious and all-encompassing theories were abandoned. Far from it: we have a remarkable example of a scientific speculator of an almost Cartesian magnitude in the Croatian–Italian astronomer and natural philosopher Roger Boscovich, who developed an ingenious system of matter without positing matter as a primary and independent entity. And Boscovich was not the only enlightenment natural philosopher who claimed to be able to understand the physical world on the basis of first principles. Indeed, some of the developments in the twentieth century can be seen as continuations of ideas that were originally suggested in the era of the Enlightenment.

While the spirits of the great projects of Descartes and Boscovich were related, my third example of a higher speculation belongs to a very different tradition. Romantic natural philosophy, or *Naturphilosophie*, emerged about 1800 as a reaction against the science and mentality of the Enlightenment. Grandiose thoughts were certainly not foreign to Romantic thinkers, whether they were scientists, poets, or philosophers, but they had a different character to earlier thoughts. No less grand, they were markedly different. The system developed by the German philosopher Friedrich Schelling was not scientific in the traditional sense, nor was it meant to be so. On the contrary, rejecting the traditional empirical mode of science, he and his allies wanted to establish a framework of an alternative conception of science. According to this conception, the 'world spirit' was an integral element and imagination was given a much higher priority than in traditional science. With Schelling and the *Naturphilosophen* inspired by his system, 'speculative physics' got a new and positive meaning.

1.1 DESCARTES' DREAM

A pioneer philosopher and mathematician, René Descartes undertook to do what his great contemporaries Kepler and Galileo had neither accomplished nor aimed at, namely to explain, rationally and scientifically, the entire world. He laid out his thoughts on a rationally based natural philosophy principally in three important works, of which only two appeared in his lifetime, the *Discours de la Méthode* of 1637 and the *Principia Philosophiae* of 1644. The third and most controversial of his major works, *Le Monde*, was published posthumously in 1664, but he had started to write it as early as 1629. Educated at the Jesuit College of La Flèche in France, Descartes spent much of his intellectual life in the more liberal Netherlands, where he produced most of his philosophical and scientific works. In 1649 he accepted an invitation from the Swedish court to become the private teacher of the young queen Christina. Tragically, he died of pneumonia in Stockholm the following year.

While still in his thirties, Descartes reached the conclusion that the world could be grasped in terms of rationally conceived ideas with only a minimum of input of empirical data. This astonishing insight he reported to his friend and mentor, the French priest and philosopher Marin Mersenne:

Now I have become bold enough to seek the cause of the position of each fixed star. For although their distribution seems irregular, in various parts of the universe, I have no doubt that there is between them a natural order which is regular and determinate. The grasp of this order is the key and foundation of the highest and most perfect science of material things that men can ever attain, for if we possessed it we could discover *a priori* all the different forms and essences of terrestrial bodies, whereas without it we have to be satisfied with guessing them *a posteriori* and from their effects.[1]

Ultimately Descartes based his philosophical system on the realization that 'I am thinking, therefore I exist', as he first phrased it in 1637. This was the first, undeniable principle from which other sure knowledge could be derived by rational reasoning. The kind of precision and certainty he aimed at was not restricted to existential statements of the *cogito ergo sum* type; it also covered knowledge of the material universe. Such knowledge of natural philosophy, he claimed, could be obtained as a 'mathematical demonstration'. The *a priori* principles on which he based his natural philosophy included the law of inertia, the non-existence of atoms and voids, and also that all action between bodies takes place by means of impact. He deemed action at a distance to be impossible in principle.

Descartes' aim was none other than to give a rational explanation of *all* phenomena in the physical world, and to do so solely on the basis of his conception of matter as a primitive quantity possessing only extension and motion. In *Le Monde* he started by considering a hypothetical world rather than the real one, but argued that his postulated concepts and laws would necessarily make the hypothetical world evolve into one indistinguishable from the world we know. Having presented the basic rules 'that comprise summarily all the effects of nature', he stated that 'we cannot but judge them infallible when we conceive them distinctly, nor doubt that, if God had created many worlds, the laws would be as true in all of them as in this one.'[2] This theme he also developed in some of his other works.

Teleology was popular at the time, and in some form or other accepted by almost all natural philosophers. But there was no room for it in Descartes' cosmogony, which strictly followed mechanical processes governed by inviolable laws. 'It is in no way likely', he said in *Principia*, 'that all things were made for us in the sense that God had no other purpose in creating them.'[3] Descartes dutifully admitted that it was a 'pious thought' to believe that God had created all things for the benefit of man; but he also pointed out that there are or have been many things in the world that have no connection to humans and of which they are not even aware. Pious Catholic as he was, he did not really believe in the teleological idea of intelligent design. As if it were meant to be a comment on the anthropic principle of the late twentieth century, he said: 'It would be clearly ridiculous to use such an opinion to support reasonings about Physics.'[4]

Although Descarte's universe was of course created by God, it was not created in accordance with a literalist reading of Genesis. It gradually *evolved* from an initial chaos into a structured system of stars, planets, and comets. The French philosopher-scientist proudly declared that he was able to show 'how the greatest part of the matter of which this chaos is constituted, must, in accordance with these laws, dispose and arrange itself in a fashion as to render it similar to our heavens; and how meantime some of its parts must form an earth, some planets and comets, and some others a sun and fixed stars'.[5] But it was a claim rather than a demonstration.

As to the cause of motion, 'it seems obvious to me that this is none other than God Himself, who in the beginning created matter with both movement and rest; and now maintains in the sum total of matter, by His normal participation, the same quantity of motion and rest as He placed in it at that time.... He also always maintains in [the world] an equal quantity of motion.'[6] What Descartes called the quantity of motion should be understood as the product of size or volume and speed (a scalar quantity) and not as a kind of momentum in our sense of the term. Only later did the Cartesian philosopher Nicolas Malebranche interpret Descartes' 'quantity of motion' as 'momentum'. According to Descartes, the conservation of the quantity of motion was a direct consequence of God's immutability. God put a certain quantity of motion into the universe at the beginning, but did not leave it at that. He continued to be involved in the universe, namely by securing the conservation of its quantity of motion. However, God was only responsible for motions in straight lines, not for the arrangements of matter that make motions irregular or curved. He was not responsible for the fact that some particular patterns of motion have occurred from the creation rather than some other patterns that might conceivably have been formed.

What would happen, Descartes asked in *Discours*, if God created the matter of a new world 'somewhere in an imaginary space,... and if He agitated in diverse ways, and without any order, the diverse portions of this matter, so that there resulted a chaos as confused as the poets ever feigned, and concluded His work by merely lending His concurrence to Nature in the usual way, leaving her to act in accordance with the Laws which He had established'. Descartes' answer was that the mechanical laws would inevitably lead to the very same world that we inhabit:

> It is certain, and it is an opinion commonly received by the theologians, that the action by which He now preserves it [the world] is just the same as that by which He at first created it. In this way, although He had not, to begin with, given this world any other form than that of chaos, provided that the laws of nature had once been established and that He had lent His aid in order that its action should be according to its wont, we may well believe, without doing outrage to the miracle of creation, that by this means alone all things which are purely material might in course of time have become such as we observe them to be at present.[7]

But if God could have created other worlds, as Descartes granted was within the power of the almighty, couldn't he have provided some of the worlds with laws of nature entirely different from ours? This was not a new question, for it had been raised and discussed by medieval philosophers, but Descartes answered it differently to earlier thinkers. According to him, God's cosmic creativity was effectively limited to cover only the matter of the worlds. The laws of nature, he said, 'are of such a nature that even if God had created other worlds, He could not have created any in which these laws would fail to be observed'. That is, any kind of original chaos would do, the mechanical laws guaranteeing that it will evolve into our world or one indistinguishable from it. This 'indifference principle' continues to play a role in modern cosmology, except that the laws are no longer seen as mechanical only.[8]

Isaac Newton and Gottfried Wilhelm Leibniz, Descartes' two great successors, did not agree that God was constrained by the laws of nature. On the contrary, Newton emphasized God's absolute freedom to create whatever he pleased, including different laws and initial conditions. In the fourth edition of his *Opticks*, Newton indulged in a remarkable multiverse speculation, to use a much later term: 'Since Space is divisible *in infinitum*, and Matter is not necessarily in all places [contrary to what Descartes held], it may also be allow'd that God is able to create Particles of Matter of several Sizes and Figures, and in several Proportions to Space, and perhaps of different Densities and Forces, and thereby to vary the Laws of Nature, and make Worlds of several Parts of the Universe.'[9] Leibniz went a step further, arguing from God as a necessary being that the laws of nature and the matter it contains might have been completely different from what they are in fact: 'It being plain that time, space, and matter, united and uniform in themselves and indifferent to everything, might have received entirely other notions and shapes, and in another order.'[10] But God actualized only this world, the best of all possible worlds because it combines simplicity with fecundity. Leibniz suggested that because our laws of nature are maximally simple, the universe contains the greatest richness of objects and phenomena.

The cosmic machinery of the Cartesian world was driven by mechanical actions of imperceptible matter particles on other particles, giving rise to vortical motions of any size and kind (Fig. 1.1). Although Descartes' matter was made up of corpuscular entities, these were indefinitely divisible. He rejected the idea of the smallest particles, atoms, which at the time had been revived from ancient Greek natural philosophy and had become popular among many advocates of the new mechanical physics. As to another and related controversial question, the existence of a vacuum, he sided with the Aristotelian philosophers, if for reasons entirely different from them. According to Descartes, 'extension' was a primitive property that defined physical reality, and this implied that a body must necessarily be extended. Contrariwise, extension devoid of matter was held to be a contradiction, from which followed the impossibility of void space. The Cartesian world was a plenum.

The physics that Descartes established was kinematical rather than dynamical. He did use the concept of force, but conceived it as the capacity of a body in motion to act on another body by means of impact. He did not attribute force to matter as an active causal agent, nor did he consider mass a quality of matter. For him, matter was completely passive. For example, he did not regard gravity as a real quality of matter but only as a secondary quality of the same kind as the colour of a body.

Descartes distinguished between three elements or kinds of matter, which he sometimes referred to as luminous, transparent, and opaque. In the originally created

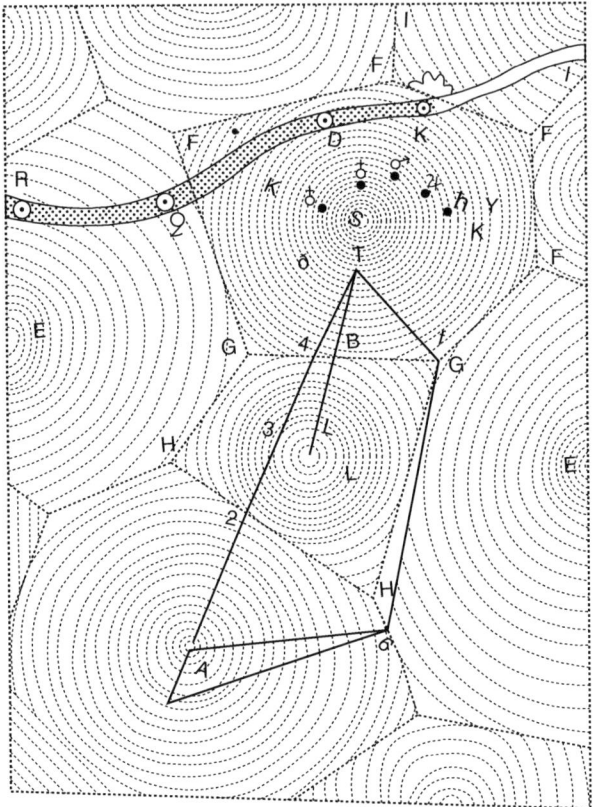

Fig. 1.1. Descartes' mechanical universe, as described in *Le Monde, ou Traité de la Lumiére*, a work published posthumously in 1664. In this picture some of the contiguous vortices are shown, each with its own sun and planets. The symbol S marks our Sun with its six planets revolving around it. The path above the solar system represents a comet.

universe, non-spherical matter particles were distributed uniformly and homogeneously until God endowed them with rotational motion. As a result of the friction caused by the motion, new kinds of elemental matter appeared. The particles of the second element, the globules, were spherical and made up the heavenly vortices. Since spherical particles cannot be packed without leaving a space between them, and since this space cannot be a void, something had to fill up the space. This was done by tiny scrapings of the primary matter, produced when particles of this matter collided with each other. Descartes' first element, filling the entire universe, was composed of these very small and swiftly moving scraping-particles. The extremely active and fine particles of the 'subtle matter' had a large surface area relative to their volume. Finally, the third element was composed of larger particles of primary matter of irregular shapes.

Whereas planets and comets were made up of the third element, the first element constituted the substance of the Sun and the fixed stars. Because the particles of the second element were bigger than those of the first element, they would recede from the centres of the vortices and leave the central parts to the particles of the first element. This was basically Descartes' scenario of how the stars were formed. He likewise provided mechanisms, held to be governed by laws of motion, for the formation of comets and planets. To account for sunspots, he invented a special mechanism based on particles of both the first and the second element.

In the way sketched here, Descartes undertook to explain all natural phenomena and bodies found in the cosmos. His theory was extremely ambitious, covering not only astronomical, physical, chemical, and geological phenomena, but also the realms of life. For example, he provided a fairly detailed theory of the tides in terms of the vortical motion of celestial matter around the Earth. He further explained magnetism by means of his vortex theory and paid much attention to phenomena of fire and light. Light was considered to be a kind of pressure wave caused by a tendency to centrifugal motion; it was thought to be transmitted instantaneously. The transparency of glass, explosion of gunpowder, lightning, the rainbow, the heat generated in stored hay, the nature of smoke, and the poorly understood frictional electricity produced by amber—these were just some of the phenomena that Descartes' theory of everything was capable of explaining. The *Principia* includes sections on such particularities as 'why spirits of wine ignite very easily', 'why steel is better suited for receiving magnetic force than baser iron', and 'why some stars disappear or unexpectedly appear'. I look forward to the day when the theory of superstrings can deliver explanations of such phenomena!

In the *Principia Philosophiae*, Descartes not only proved that atoms cannot exist, but also that the universe is indefinitely extended and that 'the matter of the heaven and the earth is one and the same; and that there cannot be a plurality of worlds'.[11] Whereas the first claim of the unity of matter in the universe stood in sharp contrast to Aristotle's cosmology, the second happened to agree with the Greek philosophical authority. Because Descartes identified space with matter, and there were only the same forms of matter, he found it inconceivable that there could be space to house other worlds. But, as mentioned, he admitted that God *could* have created them. If space were infinite—and from Descartes' perspective this seemed hard to deny—it follows from the space-matter identification that the same must be the case with the material universe. Because only God is truly infinite, Descartes preferred to speak of an indefinite rather than an infinite world, as he pointed out in the introductory chapter of *Principia*. But this was merely a tactical manoeuvre. He certainly believed that the world is infinite in the sense that it is impossible to conceive any limits to the matter of which it is made: 'Because it is not possible to imagine such a great number of stars that we do not believe that God could not have created still more, we shall suppose their number to also be indefinite.'[12]

The framework of physics that Descartes erected was rationalistic and modelled on the clarity of mathematics. It rested on a few concepts and laws, which were claimed to be almost self-evident, and mathematics was supposed to take care of the rest. 'I do not accept or desire in physics any other principles than in geometry or abstract mathematics', he wrote, 'because all the phenomena of nature are explained thereby'.[13] Mathematical deductions had a much higher epistemic authority than sense perceptions, and for this reason empirical investigations held no great significance in Descartes' dream of a rational physics. This physics was based on certain basic principles, and 'there can be no doubt about the truth of these principles, since we sought them by the light of reason and not through the prejudices of the senses'.[14] However, Descartes' mathematical physics was essentially a research programme that remained at the rhetorical level. In fact, mathematics played no real role in his physical arguments, which rarely went beyond crude analogies. His physics was basically qualitative, scarcely more mathematical than Aristotle's. A true mathematical physics had to wait for Newton and his successors. Likewise, although Descartes' theory of the cosmos was of great importance for the acceptance of heliocentric cosmology, it was impotent when it came to quantitative and predictive astronomy.

In spite of the undeniable deductivist and metaphysical flavour of Descartes' system of natural philosophy, it was not really based on metaphysical principles in any strict sense. Descartes realized that deductions from axioms or fundamental principles cannot lead to knowledge about the phenomenal world, but only to laws of a general kind. These laws of motion did not lead uniquely to the course of physical processes, for many different processes were consistent with them. To deduce a particular phenomenon or process, it would be necessary to include information about the circumstances in question. These circumstances were empirical facts of a contingent nature rather than strict consequences of the general laws. It was in this connection that observation and experiment entered Descartes' otherwise closed and non-empirical system after all. To the confusion of many readers, both then and now, he did not state this appeal to observations very clearly. While admitting the necessity of observations, he denied (much as Einstein would do three hundred years later) that the laws of nature could be inferred from collation and comparison of observed instances. He tended to see experiments as an aid in formulating explanations rather than tests that decided the truth or adequacy of an explanation.

In the opinion of the philosopher Patrick Suppes, 'Cartesian physical theory is an example of reductionism at its worst.' It exemplifies a particular kind of theory, which, although impressive and appealing, fails to contribute positively to knowledge of nature. 'A physics based on a very few clear ideas is perennially appealing, but it can be empirically sound and technically interesting only if provided with a powerful mathematical framework, which is precisely what Descartes did not provide for his theory.'[15] We shall later meet other physical theories that are no less reductionistic than Descartes' but do rest on a powerful mathematical framework. And we shall see that it takes more than a strong mathematical foundation to make such theories empirically sound.

1.2 A WORLD OF POINTS AND FORCES

The Croatian born polymath, astronomer, and natural philosopher Roger Joseph Boscovich was one of the Enlightenment era's highly reputed Jesuit scientists. After having entered the Jesuit order in 1725 he studied at the Collegio Romano in Rome, where he was appointed a professor of mathematics in 1740. In the area of mathematics, his main contributions were in probability theory. Boscovich was heavily involved in the preparations to observe the Venus transits of 1761 and 1769, and in connection with this and other astronomical duties he studied the theory of telescopes and other optical instruments. In 1767 he published a dissertation on optics in which he explained the alleged observations of a Venus satellite as optical illusions. Several astronomers, among them the famous director of the Paris Observatory, Jean Dominique Cassini, had seen the mysterious companion of Venus, but Boscovich dismissed the observations as erroneous. In another of his many works, *De lunae atmosphaera* of 1753, he made a careful investigation of the claims that the Moon possessed an atmosphere. He concluded that this was not the case.

In spite of his important contributions to mathematics, astronomy, and optics, today Boscovich is best known for a unified theory of matter based solely on the hypothesis of point atoms governed by a universal law of force. The basic features of this remarkable theory go back to a treatise of 1745, *De Viribus Vivis*, and were described in more detail in a dissertation of 1748 entitled *De Lumine*. His fame is primarily based on the much more comprehensive exposition he published in 1758, the *Theoria philosophiae naturalis*. This important work of Enlightenment theoretical physics appeared in a revised edition in 1763 and was translated into English as late as 1922.

Some of Boscovich's ideas were independently suggested in 1754 by an English librarian and experimental natural philosopher by the name of Gowin Knight. The lengthy title of Knight's book leaves no question about the author's ambitions: *An Attempt to Demonstrate that all the Phænomena in Nature may be Explained by Two Simple Active Principles, Attraction and Repulsion*. Knight reasoned that the reduction of matter to a function of force required two distinct types of primary matter.[16] In addition to the ordinary gravitating matter, he postulated repellent matter particles to account for phenomena as diverse as heat, light, elasticity, cohesion, electricity, and magnetism. These phenomena, as well as gravitation, he claimed to be able to explain in qualitative terms. Knight's system of the world was basically philosophical and it included neither mathematics nor quantitative arguments. Both of the two mid-eighteenth-century works were speculative, but whereas Boscovich's *Theoria* was comprehensive and lucid, Knight's *Attempt* was obscure and attracted very little attention.

The original aim of Boscovich was to develop a unified theory of matter and space that incorporated aspects of both Newtonian action at a distance and Leibniz' rival law of continuity. In his *magnum opus*, he made his ambitious aim clear from the very beginning. 'Dear reader', he said in the preface, 'you have before you a Theory of Natural Philosophy deduced from a single law of Forces.'[17] The result was a grandiose attempt to understand the universe in terms of one fundamental idea, the existence of point atoms interacting with one another according to a complex force law. Actually Boscovich never referred to his material units as atoms, but mostly called them *puncta*, points. (According to the traditional notion of atoms, the smallest parts of matter were extended and could have different forms.) He held that these atoms or points are all that exist in the world. Physical phenomena, whether on the microscopic or macroscopic level, are manifestations of structural changes in the configurations of immutable point atoms in accordance with a single law of force. 'My elements are really such that neither themselves, nor the law of forces can be changed', he emphasized; 'for they are simple, indivisible & non-extended.'[18] Boscovich's theory of matter and forces qualifies as an early example of a theory of everything.

Whereas Newtonian physics operated with matter made up of spatially extended corpuscles located in space, Boscovich's monistic system of 'Pythagorean atomism' did not distinguish between matter and space.[19] Not only did his point atoms have no extension at all, they were also, as points must be, identical. The corpuscles or atoms of Newtonian physics could be ascribed volume as well as mass, as John Dalton later did when he associated the chemical elements with particular atomic weights. In Boscovich's system an individual point atom had neither mass, extension, electrical charge, or any other physical parameter except that it was endowed with the property of inertia. It was the ultimate primitive particle. No more than Newton did he speculate on the origin of inertia, which he was satisfied with ascribing to 'some arbitrary law of the Supreme Architect'.[20] Still, matter does have mass, a property Boscovich interpreted as the number of point atoms in a body. It may seem strange that a large number of massless particles can add up to a massive body, but in Boscovich's theory the concept of mass was expressed by the forces between the particles rather than being a primitive and additive property.

The same is the case with other physical phenomena, which are fully determined in Boscovich's theory by the oscillatory force law that expresses the interaction between any pair of point atoms (Fig. 1.2). This force is alternately attractive and repulsive depending on the distance separating the pair of atoms. For large distances the force is attractive and varies as Newton's law of gravitation, that is, inversely with the square of the distance. At smaller distances the force oscillates between attraction and repulsion, to become steeply repulsive at very small distances. Because the repulsive force approaches infinity as the distance approaches zero, two point atoms cannot occupy the same point in space. Boscovich summarized his theory of matter as follows:

> The primary elements of matter are in my opinion perfectly indivisible & non-extended points; they are so scattered in an immense vacuum that every two of them are separated from one another by a definite interval, ... [which] can never vanish altogether without compenetration of the points themselves; for I do not admit as possible any immediate

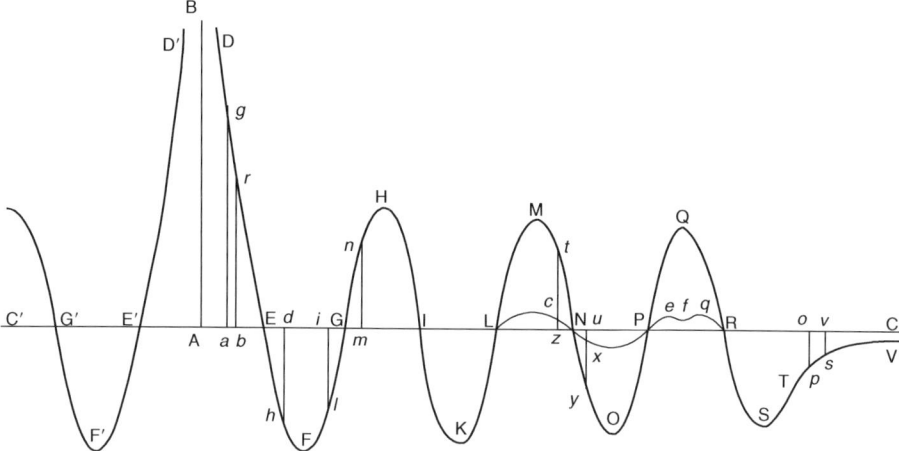

Fig. 1.2. Boscovich's law of force between two point atoms as presented in *Theoria Philosophiae Naturalis*. When the two atoms are very close together (near A), the force is repulsive and tends to become infinite as the distance reduces to zero. For this reason two atoms cannot occupy the same point in space. As the distance increases, the force oscillates between attraction (below the line AC) and repulsion (above the line). At large distances it is attractive only. In accordance with Newton's law of gravity, in this range the force varies inversely with the square of the distance.

Source: Boscovich 1966, p. 22.

contact between them. On the contrary I consider that it is a certainty that, if the distance between two points of matter should become absolutely nothing, then the very same indivisible point of space, according to the usual idea of it, must be occupied by both together, & we have true compenetration in every way. Therefore indeed I do not admit the idea of vacuum interspersed amongst matter, but I consider it interspersed in a vacuum & floats in it.[21]

A real body consists of a very large but finite number of *puncta* arranged in some grid-like configuration, its physical properties being determined by all the involved forces. Like most natural philosophers of his time, Boscovich subscribed to the Newtonian dictum of finding the forces from the motions, and deriving the phenomena from the forces. But the nearly infinite possibilities of his curve of force made it useless to realize the dictum and to derive specific phenomena. This he could do only in a very general way that had little to do with a proper derivation. Boscovich was simply unable to calculate the properties of matter and phenomena from his first principles. Not only did he not specify the details of the force law, or write it down as a definite mathematical function, he was also unable to say anything about the number and configurations of point atoms in, say, a lump of sugar. However, he discussed the problem mathematically for very simple systems and maintained that it was theoretically possible to find a solution for more complex systems too. Yet, in spite of all his efforts and ingenuity he did not come up with a single quantitative explanation or prediction. His grand theory had an air of unreality when confronted with specific problems from the physical world.

This does not mean that Boscovich was uninterested in the applications and consequences of his theory. Referring to the 'bountiful harvest [which] is to be gathered throughout the wide field of general physics', he immodestly claimed that, 'from this one theory we obtain explanations of all the chief properties of bodies, & of the phenomena of Nature'.[22] In fact, the major part of *Theoria* deals with applications, either mechanical or what at the time were called physical applications. Although astronomical and cosmological applications were within the framework of the theory, he had little to say about these areas. Among the applications he did consider were mechanical properties such as collisions and the pressure of fluids, and he also dealt with physical phenomena including gravity, cohesion, viscosity, elasticity, light, electricity, magnetism and heat.[23] Like Descartes, he argued that chemical processes also could be understood on the basis of his unified theory. It has been suggested that Boscovich's theory predicted some phenomena discovered only much later, such as states of matter of extremely high density (as in white dwarf stars), but it did this only implicitly and if seen with hindsight. At any rate, such unhistorical suggestions add nothing to an understanding of what his system of physics was about.

Boscovich was an ardent advocate of mechanical determinism. Although he assumed matter to be endowed with 'a necessary & essential force of inertia & a law of active forces', yet, 'in order that at any subsequent time it may have the determinate state, which it actually has, it must be determined to that state, from the state just preceding; & if this preceding state had been different, the subsequent state would also have been different'.[24] Elsewhere in *Theoria* he expressed his view of determinism in much the same way as Pierre-Simon Laplace would famously do in his *Exposition du Système du Monde* of 1796. Here is Boscovich's version of Laplace's demon or intelligence:

> A mind which had the powers requisite to deal with such a problem in a proper manner & was brilliant enough to perceive the solutions of it, ... could, from a continuous arc described in an interval of time, no matter how small, by all points of matter, derive the law of forces itself. ... Now, if the law of forces were known, & the position, velocity & direction of all the points at any given instant, it would be possible for a mind of this type to foresee all the necessary subsequent motions & states, & to predict all the phenomena that necessarily followed from them. It would be possible from a single arc described by any point in an interval of continuous time, no matter how small, which was sufficient for a mind to grasp, to determine the whole of the remainder of such a continuous curve, continued to infinity on either side. We cannot aspire to this, not only because our human intellect is not equal to the task, but also because we do not know the number, or the position & motion of each of these points.[25]

Incidentally, Boscovich was not the first to formulate this kind of Laplacean determinism, so foreign to the mind of Newton (if not to later Newtonians). Leibniz expressed the view much earlier and more pointedly. Convinced that 'everything proceeds mathematically—that is, infallibly—in the whole wide world', he wrote: 'If someone could have sufficient insight into the inner parts of things, and in addition has remembrance and intelligence enough to consider all the circumstances and to take them into account, he would be a prophet and would see the future in the present as in a mirror.'[26]

Theoria contains a wealth of interesting claims and ideas, and it is understandable that these have appealed to modern readers. For example, Boscovich argued that future eternity is different from past eternity and that only the first kind of infinite time can be real. This was not only a philosophical argument, but also one in agreement with the assumption of a divinely created universe. It is an argument that can still be found in modern cosmology (see Chapter 10). At some places in *Theoria* Boscovich speculated about the possibility of worlds different from the one we experience. Imagining different kinds of substances kept together by different force laws, he wrote:

> It would be possible for one kind to be bound up with another by means of a law of forces, which they have with a third, without any mutual law of forces between themselves, or these two kinds might have no connection with any third. In this latter case there might be a large number of material & sensible universes existing in the same space, separated one from the other in such a way that one was perfectly independent of the other, & the one could never acquire any indication of the existence of the other.[27]

That is, Boscovich suggested, if only as a theoretical possibility, a many-universe scenario in which different universes co-exist here and now. In a supplement to *Theoria* dealing with the nature of space and time, he returned to the scenario:

> What if there are other kinds of things, either different from those about us, or even exactly similar to ours, which have, so to speak, another infinite space, which is distant from this our infinite space by no interval either finite or infinite, but is so foreign to it, situated, so to speak, elsewhere in such a way that it has no communication with this space of ours; & thus will induce no relation of distance. The same remark can be made with regard to a time situated outside the whole of our eternity.[28]

These thoughts about a multiverse of separate worlds are indeed fascinating, but there is no reason to assume that Boscovich entertained them as more than uncommitted speculations.

Several scientists and philosophers have interpreted Boscovich's dynamical theory of matter as an anticipation of modern knowledge. For example, it has been suggested that it anticipated important features in Niels Bohr's quantum theory of atoms of 1913 and also of Erwin Schrödinger's later wave mechanics. During an international symposium in Dubrovnik celebrating the bicentenary of *Theoria*, Werner Heisenberg added to the myth of Boscovich as a precursor of modern physics. According to the founder of quantum mechanics, Boscovich's great insight was that he 'considered the key to an understanding of matter to be the law of nature that determines the forces between elementary particles, [which] view is very close to our present understanding'.[29] Another of the participants at the symposium concluded that 'Boscovich was a forerunner of the modern theory of elemental particles.... The potential [of his force law] has been discovered by the quantum mechanics when calculating the molecules.'[30] Boscovich's unified theory of force and matter was an original contribution to eighteenth-century natural philosophy and is interesting in its own right. To consider it as a precursor of modern physics is to ignore the historical context and only misrepresents the true meaning of the theory.

Not only has Boscovich been regarded a forerunner of atomic theory, his work has also been seen as anticipating some of the crucial insights of the theory of relativity, including relativistic covariance and the space-time continuum. These claims are principally based on a slightly earlier work, the *De Spatio et Tempore* from 1755, which was appended as a supplement to the 1763 edition of *Theoria*. In this work the Jesuit natural philosopher argued that the state of motion of an observer relative to the observed world constitutes the primary reality. The observer can never observe the world as it is in itself, but only describe the relationship between himself and the world. In *De Spatio et Tempore*, Boscovich wrote as follows:

> If the whole Universe within our sight were moved by a parallel motion in any direction, & at the same time rotated through an angle, we could never be aware of the motion of the rotation.... Moreover, it might be the case that the whole Universe within our sight should daily contract or expand, while the scale of forces contracted or expanded in the same ratio; if such a thing did happen, there would be no change of ideas in our mind, & so we would have no feeling that such a change was taking place.[31]

This is certainly an interesting speculation, but it was not quite new. In *Le Livre du Ciel et du Monde*, a work written nearly three hundred years before Boscovich, the medieval Paris philosopher and theologian Nicolas Oresme said about the same. Oresme imagined that if the world were suddenly made one thousand times larger or smaller than it is now, with 'all of its parts being enlarged or diminished proportionally', everything would appear 'exactly as now, just as though nothing had been changed'.[32]

Whereas the connection between Boscovich and modern physics is a mythical reconstruction, his theory did exert considerable impact on physicists and chemists in the period up to the 1840s. Among those who took an interest in the theory were Joseph Priestley, John Michell, Humphry Davy, Michael Faraday, and William Rowan Hamilton.[33] As interpreted by Faraday, more than anyone the founder of the modern concept of fields, Boscovichean atomism implied that one could establish a physics completely without matter. In Faraday's version, matter was an epiphenomenon of forces acting between mathematical points, and forces or fields were all that existed in the world. In considering a mass, he wrote, 'we have not to suppose a distinction between its atoms and any intervening space'. Because, 'each atom extends, so to speak, throughout the whole of the solar system, yet always retaining its own centre of force'.[34] As late as 1902, Lord Kelvin (William Thomson) appealed to Boscovich's conception of matter in order to understand how atoms made up of electrons can have different chemical properties.[35]

Another of the ambitious and speculative theories of the Enlightenment era was a system proposed by Georges-Louis Le Sage, a Geneva-based Swiss natural philosopher. Le Sage drew on earlier ideas of Nicolas Fatio de Duillier, a close friend of Newton who in 1690 presented to the Royal Society a corpuscular theory of gravity, but without publishing his ideas. The basis of Le Sage's theory was the assumption that all of space is filled with tiny *particules ultramondaines* ('otherworldly particles'), so-called because they were supposed to come from outside the visible universe. The hypothetical particles were extremely light, completely inelastic, and moved with great speed in all directions. Moreover, they did not interact gravitationally. On this basis he was able to explain universal gravitation as an effect of the pressure exerted by the ultramundane particles and derive the correct inverse-square law. The main advantage of Le Sage's theory was that it provided a mechanical explanation for gravity as an alternative to the problematic postulate of action at a distance accepted by conventional Newtonian physics.[36] The main disadvantage was that the theory built upon purely hypothetical entities for which there was no independent evidence at all.

Although Le Sage's idea went back to 1748, he only published it 10 years later, in a prize essay entitled *Essai de Chymie Méchanique*. As indicated by the title, it was an attempt to explain chemical reactions and properties in mechanical terms and thereby establish a link between chemistry and gravitational physics. This sounds odd to a modern ear, but it did not to Le Sage and his contemporaries. Although Le Sage is today remembered only for his heroic but ill-fated theory of gravitation, the theory was of a more universal nature and intended to explain macroscopic as well as microscopic phenomena. Thus, he used it in an attempt to explain in mechanical terms the problem of chemical affinities, that is, why some substances react together easily while others do it more slowly or not at all. According to Le Sage, cohesion, chemical affinity, and gravitation were all aspects of the same general phenomenon.

In 1784 Le Sage published a more comprehensive and succinct version of his gravitational mechanism. It was this treatise, with the title *Lucrèce Neutonien*, that made his theory known among the natural philosophers of the period.[37] His kinetic explanation of gravitation also attracted considerable interest in the nineteenth century, when it was reconsidered by Samuel T. Preston, Lord Kelvin, and a few other physicists. However, from the perspective of later physics the original theory appeared hopeless. Thus, Henri Poincaré found that Le Sage's theory would require particles moving with a velocity of at least 2.4×10^{18} times the velocity of light (!) and that the quantity of heat produced by them would raise the temperature of the Earth enormously.[38] Nonetheless, revised gravitation theories of the Le Sage type continue to attract attention among a minority of scientists even today. Such theories have been suggested by Tom Van Flandern, Toivo Jaakkola, and others, but they are not taken seriously by mainstream physicists and astronomers.

Boscovich was well acquainted with the system of Le Sage, with whom he corresponded. However, although he granted that it did provide an explanation of gravity of sorts, he found the theory to be unacceptable for metaphysical and methodological reasons. For one thing, he could not accept the postulate of an enormous amount of ultramundane particles, most of which were of no use. The Scottish mathematician and natural philosopher John Playfair later paraphrased Boscovich's objection to the multitude of 'gravific atoms':

> The actions of these atoms supposes a vast superfluity of matter, and an infinity of corpuscles... An immense multitude of atoms, thus destined to pursue their never ending journey through the infinity of space, without changing their direction, or returning to the place from which they came, is a supposition very little countenanced by the usual economy of nature. Whence is the supply of these innumerable torrents; must it not involve a perpetual exertion of creative power, infinite both in extent and duration? The means here employed seem greater than the end, great as it is, can justify.[39]

That is, Le Sage's theory violated the principle of explanatory simplicity, also known as Ockham's razor.

1.3 ROMANTICISM AND SPECULATIVE PHYSICS

My third example of an early all-comprehensive system of physics is entirely different from the two other cases. Whereas the systems erected by Descartes and Boscovich were forwarded as scientific theories, intended to explain nature in all her colourful diversity from a few basic principles, the system of Friedrich Schelling and his followers was primarily philosophical, although with important scientific implications. Inspired by Kant and Schelling, in the early part of the nineteenth century a movement of Romantic natural philosophy, also known as *Naturphilosophie*, spread over northern Europe. Its stronghold was Germany, but it also made an impact on science in Scandinavia, England, Austria-Hungary and The Netherlands. Among the scientists who were more or less influenced by the new way of conceiving nature were several leading physicists and chemists, including Johann Wilhelm Ritter, Hans Christian Ørsted, Christian Samuel Weiss, Thomas Johann Seebeck, Humphry Davy, and Michael Faraday. Their ideas of science differed as much from those that dominated the age of the Enlightenment as they differ from those accepted today. Moreover, there were significant differences between their ideas—neither Davy nor Faraday were influenced by German *Naturphilosophie*. What makes them of interest in the present context is, not least, that they admitted that speculations and aesthetic sentiments were a legitimate part of science, indeed a necessary part.

Romanticism was to a large extent a counter-movement to the Enlightenment and its emphasis on standardization and 'cold' rational thought. Instead of obtaining knowledge about nature by controlling and dissecting her, scientists affiliated with the Romantic movement wanted to establish a peaceful and harmonious relationship with her. In this process, not only would nature be understood on a deeper and more true level, the self of man would also be understood. According to Davy and his like, true understanding of nature required an attitude of respect and admiration, even of love:

> Oh, most magnificent and noble Nature!
> Have I not worshipped thee with such a love
> As never mortal man before displayed!
> Adored thee in thy majesty of visible creation,
> And searched into thy hidden and mysterious ways
> As Poet, as Philosopher, as Sage?[40]

Schelling and other prophets of the Romantic revelation paid as much attention to man as to nature, for the two were seen as intimately connected. In the mythical golden age, man and nature had been united, only later to be separated. Now the time had come to reunite man with nature.

Although there was little unity among the Romantic scientists, many of them shared a set of values and ideas that may be taken as characteristic of the *Naturphilosophie* that formed the core of the Romantic movement.[41] First of all, materialism and mechanicism were anathema. Matter was regarded as inherently dynamic, in the sense of possessing or even consisting of forces. Natural substances and physical forces—say water and magnetism—were appearances that ultimately had to be understood as manifestations of a single underlying primeval force, an *Urkraft*. The mechanistic view based on the twin concepts of vacuum and material atoms was contemptuously rejected as ideologically incorrect, conceptually inconsistent, and methodologically inadequate. The true objects of nature were not fixed and permanent, but, on the contrary, dynamic and eternally changing manifestations of oppositely polarized forces in equilibrium.

Another important feature was the belief in a fundamental unity, not only between the various forces of nature, but also between mind and nature. Just as mind, or sometimes reason, was invisible nature, so nature was invisible mind—the one was unable to exist without the other. Properly understood, nature constituted an active and creative whole, an organism that could not be grasped by means of the senses alone. Schelling's *Naturphilosophie* involved searching for unity among natural phenomena on the basis of Kant's dynamical theory of matter, the main phenomena of interest being electricity, magnetism, chemistry, heat, and light. Physico-chemical effects were explained and described by the interaction of opposite forces. Ideally, all natural phenomena were to be regarded in relation to a whole and dynamic system, which was again dynamical and constantly active. Matter—even nature as a whole—was believed to be constituted of force or activity.

The Romantic movement involved a strong exchange of ideas between philosophy, science, art, poetry, theology, and politics—indeed between all aspects of life, nature, and society. Rooted in their belief in a unity between nature and human spirit, the Romantics found speculation to be a valid method of scientific inquiry on equal terms with, or sometimes even superior to, empirical investigation. The inner character of nature was inseparable from the human mind or spirit, and for this reason it could be intuitively—or speculatively—recognized by particularly gifted thinkers. Hence, a purely *speculative physics* was a possibility, indeed the only possible path for those seeking to attain insights into nature at the most profound level. The sort of nature that could be objectively sensed by means of observations and experiments was regarded as a somewhat dull wrapping, which contained and also obscured the real, non-objective nature. To get insight into the latter, there was no route other than the one of speculative physics. H. C. Ørsted, the discoverer of electromagnetism and a convinced follower of and contributor to the Romantic movement, expressed the doctrine as follows:

> The investigator of nature seeks to take command of the idea of the whole, so that from there he may take a larger view across the parts, the more perfectly to behold them in their context. This is what occurs in the *Naturphilosophie*. This science attempts to ascertain the nature of the World based on the necessary characteristics of things. It is the highest form of speculation and borrows nothing from experience, just as, conversely, experiential science should not be admixed with speculative sentences.[42]

The most important of the philosophical godfathers of Romantic science was Friedrich Schelling, a German philosopher who resided in the university town of Jena. After studies in philosophy, theology, and science, he developed in about 1800 what he called a transcendental system of natural philosophy. One concrete result was the founding of a journal with the fascinating title *Zeitschrift für spekulative Physik*, intended to include scientific investigations in the spirit of Romantic natural philosophy. The journal, which only survived for two years, included a mixture of reviews and general research articles characterized by lengthy qualitative arguments and an almost complete absence of mathematics and quantitative experiments. Most of the content of the journal was written by Schelling himself. Contrasting speculative physics and empirical physics, Schelling explained that the former 'occupies itself solely and entirely with the original causes of motion in Nature, that is, solely with the dynamical phenomena'. Empirical physics, on the other hand, 'deals only with the secondary motions, and even with the original ones only as mechanical'. Moreover: 'The former, in fact, aims generally at the inner clockwork and what is non-objective in Nature; the latter, on the contrary, only at the surface of Nature, and what is objective and, so to speak, outside in it.'[43]

Schelling's speculative physics built on concepts and principles that could be discovered *a priori* but nonetheless were claimed to apply to the experienced or physical world. Rooted in intellectual rather than sensible intuition (to use his terms), it stood in direct contrast to the empirical physics concerned with only the phenomenal world. His *Naturphilosophie* was an attempt to relate the enquiries of speculative physics to investigations of nature in so far it was accessible to experiments and other sensible intuition. Schelling viewed nature as a dynamic and self-organizing structure where non-equilibrium forces explained natural processes and phenomena on a deeper level than the traditional one associated with empirical physics. For this reason he has sometimes been regarded a precursor of the modern research programme based on complexity and self-organization.[44]

Although his was a philosophical system, Schelling endeavoured to apply it to concrete scientific problems. He knew that philosophy alone would fail to convince chemists and physicists of his ideas. Schelling's *Ideen zu einer Philosophie der Natur*, a book published in 1797, contained chapters on the composition of water, the nature of electricity, theories of light, combustion processes, mechanical physics, the atomic hypothesis, and much more. Not a thinker of modesty, Schelling believed he had provided the basis of a new 'absolute science of nature' that would lead to a true understanding not only of everything physical, but also of everything organic, metaphysical, mental, and spiritual. This was the framework of a new science, a genuine revolution. It was only necessary to fill in the detail.

According to Schelling, *Naturphilosophie* differed in an essential way from the ordinary kind of science that tried to establish laws from phenomena by means of experiments and inductive reasoning. The 'fruitless endeavours' of traditional science could at best establish that certain things are possible, but never that they are necessary. And necessary truths were what Schelling ambitiously aimed at. He sometimes compared his system with a mathematical demonstration, as when he wrote:

> In the Philosophy of Nature, explanations take place as little as they do in mathematics; it proceeds from principles certain in themselves, without any direction prescribed to it, as it were, by the phenomena. Its direction lies in itself, and the more faithful it remains to this, the more certainly do the phenomena step of their own accord into that place in which alone they can be seen as necessary, and this place in the system is the only explanation of them that there is.... For one who has but grasped the coherence as such, and has himself reached the standpoint of the whole, all doubt is likewise removed; he perceives that the phenomena can only be thus, and so must also exist as they are presented in this context: In a word, he possesses the objects through their form.[45]

Although Schelling's *Naturphilosophie* was in many respects the very opposite of Descartes' rationalistic system of nature, the two had some features in common. Given that Schelling's natural philosophy, or speculative physics, was completely non-mathematical, it is remarkable that he nonetheless likened it to mathematical demonstrations. Many later physicists would with greater justification make a similar analogy, or even define physics as mathematical in nature, but their motivations were very different from those of Schelling. In spite of all the differences, there is some similarity between the rhetoric of Schelling's *Naturphilosophie* and the ideas of fundamental physics that Arthur Eddington espoused more than a century later and which we shall look at in Chapter 4. In a work of 1799, *Erster Entwurf eines Systems der Naturphilosophie*, Schelling repeated the message that our knowledge of nature is basically *a priori* and that it follows from a grand law that includes and connects all phenomena:

> We suggest that all phenomena are correlated in one absolute and necessary law, from which they can all be deduced; in short, that in natural science all that we know, we know absolutely

a priori. Now, that experimentation never leads to such a knowing is plainly manifest from the fact that it can never get beyond the forces of Nature, of which it makes use as means.[46]

To most natural philosophers associated with Romanticism, the traditional atomic theory was unpalatable, and especially so in the concrete version introduced by Dalton. The idea of indivisible and immutable particles of matter as the ultimate building stones was scientifically fruitful but philosophically unacceptable. It was often seen as an expression of pure materialism and, for this reason alone, it was abhorrent to the true supporters of *Naturphilosophie*. On the other hand, they could not deny that Dalton's chemical atomism worked very well and had a great deal of explanatory power. For example, it provided a natural explanation of the weight ratios of elements in compounds consisting of two elements only (such a NO, NO_2, and N_2O). Consequently some of the Romantic scientists sought to formulate alternative atomic theories, for example by conceiving the discrete units as localized centres of force, an idea which had more than a little similarity to the one developed by Boscovich. Indeed, Ørsted referred with some sympathy to the ideas of Boscovich and Knight, which he much preferred over those entertained by Dalton. Ørsted and Weiss were among those who favoured a purely dynamical atomic theory, but they were unable to develop their ideas (which they discussed in their correspondence) into an actual theory. Schelling, too, could only accept the atomic theory in the version of a dynamical atomism freed from its associations with materialism and mechanistic physics.

While Schelling was a philosopher and not a practitioner of science, Ørsted was a reputed physicist and chemist and, at the same time, an outspoken supporter of the new Romantic natural philosophy. On the other hand, he received most of his inspiration from Kant rather than Schelling and eventually distanced himself from Schelling's version of *Naturphilosophie*. He insisted that experiments had an important role in science and that there was no contradiction between the Romantic view of nature and accurate experimental work done in the laboratory. Indeed, if only interpreted properly experiments would support the dynamical and holistic view of nature that he saw as the core of the Romantic programme.

Contrary to some of the other *Naturphilosophen*, Ørsted was fully aware that progress in science depended on a proper balance between the traditional and the new mode of doing research. Speculations in the Romantic style were essential, but not sufficient. In an article of 1830 in the *Edinburgh Encyclopædia*, he contrasted two basic attitudes to science: what may be called the speculative and systematic approaches. While the scientists of the first class focused on principles and wholeness, those of the other considered science only as an investigation of facts. They worked methodically in their laboratories, 'but in their laudable zeal they often lose sight of the whole, which is the character of truth'. Ørsted characterized science as a fruitful conflict between the two attitudes, which he saw as complementary rather than mutually exclusive: 'Those who look for the stamp of divinity on every thing around them, consider the opposite pursuits as ignoble and even as irreligious; while those who are engaged in the search after truth, look upon the others as unphilosophical enthusiasts, and perhaps as phantastical contemners of truth.'[47]

According to the view of science defended by Ørsted and several other protagonists of Romantic natural philosophy, the ultimate goal of physics was to find the system of forces that were manifested in and produced material objects. Matter in itself was conceived as an epiphenomenon, hence of less interest. In one of his main works, characteristically entitled *The Soul in Nature*, he wrote: 'Matter itself is only space occupied by the primitive forces of nature, therefore it is the laws according to which a thing is formed from which it derives its invariable peculiarity...[and] peculiar essence.'[48] The laws of nature did not express the limited reason present in man's thinking, but a higher reason that was embodied in nature, which in turn was an expression of God's thinking of which humankind was part and could partake. All of nature was seen as an arrangement of reason, the business of the natural philosopher being to seek for and understand that reason.

There was a distinct rationalistic and even Cartesian element in Ørsted's conception of the laws of nature, which he tended to represent as laws of reason that a few chosen individuals, the geniuses, were able to access. His emphasis on a reason or 'soul' in nature led him to believe that some of the laws of physics were true *a priori*. For example, he thought (wrongly) that the law of inertia was true *a priori*, that it was a 'self-evident necessity of reason'. In *The Soul in Nature* he expressed his view as follows:

> We find that they [the laws of nature] harmonize so perfectly with Reason that we may assert with truth that the harmony of the laws of nature consists in their being adapted to the dictates of Reason, or rather, by the coincidence of the laws of Nature and the laws of Reason. The chain of natural laws, which in their activity constitute the essence of everything, may be viewed either as a natural thought, or more correctly as a natural idea; and since all natural laws together constitute but one unity, the whole world is the expression of an infinite all-comprehensive idea, which is one with an infinite Reason, living and acting in everything. In other words, the world is a revelation of the united power of Creation and Reason in the Godhead.[49]

This passage summarizes much of the philosophy of the Romantic conception of science and nature, including its religious dimensions. It also suggests that the quest of the Romantic scientists was not entirely different from the search for an ultimate theory that other scientists pursued.

Schelling devoted a whole chapter in his *Ideen* to Le Sage's theory, which he used to criticize the foundations of mechanical physics in general. He saw in Le Sage's system a kind of theoretical physics—'hyperphysical' physics, as he called it—that was abstract, hypothetical, and far removed from experience. 'The whole system proceeds from abstract concepts, which cannot be represented in any intuition', he complained. 'If appeal is made to ultimate forces, we thereby admit unreservedly that we stand at the limits of possible explanation.' He further criticized the whole approach of the Geneva physicist, which 'begins with *postulates*, then erects *possibilities* upon these postulates, and finally purports to have constructed a system that is beyond all doubt'.[50] Schelling's objections to the system of Le Sage were not so much that it was speculative—a quality he had nothing against—but rather that it represented an abstract and mechanistic kind of speculation and not the right kind: the Romantic *spekulative Physik*.

Protagonists of Romantic natural philosophy considered it a revolutionary movement, a bold attempt to overthrow the old science and establish a new one based on holistic and anti-materialistic ideals. Although the revolution failed, it was not without consequences. At least some of the progress in the period, in particular in electrochemistry, electrodynamics, and early thermodynamics, was directly influenced by the Romantic approach to science. The most important of these discoveries was probably Ørsted's discovery of electromagnetism, which to a large extent relied on his Romantically based belief in a unity of forces. Another noteworthy result was Ritter's discovery of ultraviolet light and his pioneering works in voltaic electricity, at the time generally known as galvanism.

Yet, in spite of these and other accomplishments, by the 1840s *Naturphilosophie* was in decline and on its way to being replaced by a positivistic conception of science. It would soon be regarded in an unfavourable light, as a fanciful and speculative approach that was more anti-science than alternative science. This was the way the great chemist Justus von Liebig presented it in 1840, violently attacking the remaining followers of the 'false Goddess' of *Naturphilosophie*, which 'promises them light, without troubling them to open their eyes, [and] gives them results without observation or experiment'. According to Liebig, the activities of the Romantic natural philosophers could be characterized as 'the Black Death of the century'.[51]

Notes for Chapter 1

1. Quoted in Gaukroger 1995, p. 249. On Descartes' mathematical-physical theory of the world, see also Gaukroger 2002, Gaukroger, Schuster, and Sutton 2000, and Barbour 2001, pp. 406–450.
2. Descartes 1979, Chapter 7. Quoted from the English translation available on http://www.princeton.edu/~hos/mike/texts/descartes/world/worldfra.htm.
3. Descartes 1983, p. 85.
4. Ibid.
5. Descartes 1996, p. 27.
6. Descartes 1983, p. 58.
7. Descartes 1996, pp. 26–28.
8. The indifference principle is analysed in McMullin 1993. See also Barbour 2001, p. 432.
9. Newton 1952, pp. 403–404. First published 1730.
10. Leibniz 2008, p. 93. First published 1710.
11. Descartes 1983, p. 49.
12. Ibid., p. 14.
13. Ibid., p. 76.

14. Ibid, p. 84.
15. Suppes 1954, p. 152.
16. For an account of Knight's theory in relation to contemporary theories of matter, see Schofield 1970, pp. 175–81.
17. Boscovich 1966, p. 8. Barrow 1992, pp. 17–19, places Boscovich's theory in the historical context of theories of everything.
18. Boscovich 1966, p. 144.
19. For 'Pythagorean atomism', see Whyte 1961. On Boscovich's theory of matter, see also Schofield 1970, pp. 235–42.
20. Boscovich 1966, p. 21.
21. Ibid., p. 20.
22. Ibid., p. 49.
23. On Boscovich's theory of cohesion, see Rowlinson 2002, pp. 49–51.
24. Boscovich 1966, p. 193.
25. Ibid., pp. 141–42. Boscovich's determinism differed from Laplace's in that Boscovich, being a Jesuit, presupposed a transcendent being and free motions produced by spiritual actions.
26. Quoted in Cassirer 1956, p. 12.
27. Boscovich 1966, p. 184.
28. Ibid., p. 199.
29. Markovic 1958, p. 29.
30. Ibid., pp. 71–73.
31. Boscovich 1966, p. 203.
32. Quoted in Grant 2007, pp. 227–28.
33. On Faraday and Boscovich, see Kargon 1964 and Spencer 1967. The connection to Davy is discussed in Siegfried 1967.
34. Faraday 1844, pp. 142–43.
35. Thomson 1902, p. 2: 'When we consider the great and wild variety of quality and affinities manifested by the different substances, . . . we must fall back on Father Boscovich, and require him to explain the difference of quality of different chemical substances by different laws of force between the different atoms.'
36. On Le Sage's theory of gravitation, see Rowlinson 2003 and the contributions in Edwards 2002. There is an informative online article on the theory in *Wikipedia, the Free Encyclopedia.*
37. The title alluded to the Roman poet and philosopher Lucretius, who in *De Rerum Natura* from about 50 BC developed a world system based purely on atoms in motion. Le Sage's theory presented Newtonian gravitation in terms of atomic particles, hence 'Newtonian Lucretius'.
38. Poincaré 1952, p. 247. Originally published 1908.
39. Playfair 1807, p. 148.
40. Poem by Humphry Davy, first published by his son John Davy in 1858. Quoted in Cunningham and Jardine 1990, p. 14.
41. For background, see e.g. Cunningham and Jardine 1990, Gower 1973, Caneva 1997, and Wilson 1993. Sometimes *Naturphilosophie* in the German tradition is distinguished from the broader movement of Romantic natural philosophy.
42. Ørsted 1809, p. 6. On Ørsted's conception of Romantic physics, see Brain, Cohen, and Knudsen 2007.
43. Schelling 2004, p. 196.

44. It has been suggested that Schelling's natural philosophy 'is in many ways astonishingly similar to our present-day picture of nature that lies behind the paradigm of self-organization' (Küppers 1990, p. 54).

45. Schelling 1988, p. 53.
46. Schelling 2004, p. 197.
47. Ørsted 1998, p. 543.
48. Ørsted 1852, p. 450.
49. Ibid., pp. 450–51.
50. Schelling 1988, p. 169 and p. 161.
51. Address of 1840, quoted in Brock 1997, p. 68.

2

A Victorian Theory of Everything

Do not imagine that mathematics is harsh and crabbed, and repulsive to common sense. It is merely the etherealisation of common sense.

William Thomson, *Popular Lectures and Addresses*, 1891, p. 273.

Although one of the grandest and most ambitious attempts to construct a unified theory of all of nature, the vortex theory of matter is almost entirely forgotten today. To call it a Victorian theory of everything is no exaggeration, for this is what it was intended to be. It was just as much a theory of everything as the modern theory of superstrings. Indeed, there is more than a superficial similarity between the vortex theory of the past and the string theory of the present.[1] Of course, there are also many dissimilarities. One of them is that whereas string theory is an international adventure, the vortex atom theory was almost completely confined to the United Kingdom. For reasons that are not very clear, the theory never attracted more than polite attention among scientists on the continent. Among British physicists and mathematicians, on the other hand, the theory of vortex atoms was immensely popular during the 20 years between 1870 and 1890. About 60 scientific papers were written on the subject, by some 25 scientists. Given the modest size of the physics community at the time, this indicates a thriving research field in mathematical physics.

The vortex atom theory has often been seen as part of a Cartesian tradition in European thought.[2] However, in spite of some similarities, there are marked differences between the Victorian theory and Descartes' conception of matter. First and foremost, whereas Descartes' plenum consisted of tiny particles in whirling motion, the vortex theory was strictly a unitary continuum theory. Second, the vortex atoms were described in mathematical detail, contrary to the qualitative nature of Descartes' theory. And third, the vortices of the Victorian physicists were subatomic and did not include the heavenly vortices that were so important to the French philosopher-scientist. What the two theories did have in common, apart from their common aspiration of being the ultimate explanation of nature, was the powerful imagery of vortical motions. Although Newton had demolished Descartes' idea of celestial vortices, he also ventured speculations that have a remarkable similarity to the idea of vortex atoms developed about two hundred years later. In a letter of 1675, Newton wrote:

'Perhaps the whole frame of Nature may be nothing but various Contextures of some certaine ætheriall Spirits or vapours condens'd as it were by præcipitation... Thus perhaps may all things be originated from æther.'[3]

2.1 SMOKE RINGS AND VORTEX ATOMS

From a formal and mathematical point of view, the vortex theory of matter had its foundation in an important analysis of fluid dynamics published in 1858 by the versatile German physicist and medical doctor Hermann von Helmholtz, one of the pioneers of thermodynamics.[4] In this contribution to mathematical physics, Helmholtz defined what he called vortex lines and rings in a hypothetical frictionless fluid, demonstrating among other things that closed vortex rings are permanent structures: they cannot be created or annihilated. He also proved that the 'strength' of a vortex ring formed by lines remains constant during its motion. Helmholtz did not suggest that his abstract theory had any relevance for a physical theory of matter, and it took several years before such a connection was made. When it did happen, in 1867, it was due to one of Helmholtz's friends, the Scottish mathematical physicist William Thomson, another of the giants of nineteenth-century theoretical physics. After he received the title of Baron Kelvin of Largs in 1892, he became better known as Lord Kelvin or just Kelvin. In what follows I shall refer to him as either Thomson or (to avoid confusion with J. J. Thomson) William Thomson.

'I do not believe in atoms.'[5] Thus wrote Thomson in a paper of 1862, expressing an opinion that was not particularly unorthodox at the time, when 'atom' usually referred to the hard and structureless minimal bodies introduced by Dalton. Less than ten years later Thomson had changed his mind and convinced himself that physics required an atomic constitution of matter. His conversion was not caused by any experimental discovery, but by his insight that it was possible to construct a dynamically satisfactory model of atoms in terms of a perfect fluid pervading all space. From early on in his scientific career, Thomson felt attracted to a continuum theory of matter, which he much preferred to the standard Newtonian view of matter as being made up of indivisible atoms interacting at a distance through a vacuum by means of short-range forces. In early 1867 Peter Guthrie Tait, an Edinburgh physicist and close friend of Thomson, demonstrated an apparatus by means of which he could produce smoke rings and illustrate their properties.[6] To Thomson, it was a visual demonstration of Helmholtz's vortex theory and at the same time a model of how matter might be constituted. Shortly after having witnessed the spectacular experiment, he described the connection in a letter to Helmholtz. The letter gives the essence of the idea behind the vortex atomic theory:

> The absolute permanence of the rotation... shows that if there is a perfect fluid all through space, constituting the substance of all matter, a vortex-ring would be as permanent as the

solid hard atoms assumed by Lucretius and his followers (and predecessors) to account for the permanent properties of bodies (as gold, lead, etc.) and the differences of their characters. Thus, if two vortex-rings were once created in a perfect fluid, passing through one another like links of a chain, they never would come into collision, or break one another, they would form an indestructible atom; every variety of combinations might exist.[7]

The idea outlined in the letter was worked out in mathematical detail later the same year. According to Thomson, the theory justified the view that space is continuously filled with a frictionless 'perfect fluid', with vortex structures in the fluid acting as atoms of matter. It was well known that the classical conception of atoms was plagued by conceptual as well as scientific difficulties, but these did not exist in the theory of vortex atoms. For example, by 1867 spectroscopy had proved that each chemical element emits a characteristic spectrum of light, often consisting of a large number of discrete frequencies. This fact was inexplicable according to the classical theory of hard atoms, whereas the many spectral lines could easily be explained, at least in principle, as due to the multitude of modes of vibration in the vortex atom. To actually calculate the frequencies presented 'an intensely interesting problem of pure mathematics', as Thomson put it. To his mind, the mathematical difficulties of calculating the frequencies, and in general the difficulties of deducing phenomena from the theory, were a challenge rather than an obstacle. They only added to the appeal of the vortex atom theory.

William Thomson was the founder of the vortex theory of matter, but far from the only British mathematical physicist who strove to develop the theory and evaluate its consequences. His great contemporary and fellow-Scotsman, James Clerk Maxwell, did not share his enthusiasm and remained unconvinced that the theory could be developed into a realistic theory of all matter. On the other hand, Maxwell had a great deal of sympathy for the structure of the theory, which he praised for its methodological virtues in particular. What appealed to him was the theory's purity, its lack of arbitrary features and *ad hoc* hypotheses. In 1875, in an extensive article on 'Atoms' in the *Encyclopædia Brittanica*, he described eloquently the strengths as well as weaknesses of the new unified continuum theory of matter:

> The greatest recommendation of this theory, from a philosophical point of view, is that its success in explaining phenomena does not depend on the ingenuity with which its contrivers 'save appearances', by introducing first one hypothetical force and then another. When the vortex atom is once set in motion, all its properties are absolutely fixed and determined by the laws of motion of the primitive fluid, which are fully expressed in the fundamental equations.... [Thomson's] primitive fluid has no other properties than inertia, invariable density, and perfect mobility, and the method by which the motion of this fluid is to be traced is pure mathematical analysis. The difficulties of this method are enormous, but the glory of surmounting them would be unique.[8]

The vortex theory as originally proposed by Thomson was an attempt to explain matter in terms of structures in a hypothetical fluid characterized only by inertia and its elastic properties. However, according to most physicists in the second half of the nineteenth century, the world consisted not only of matter in motion but also, and no less importantly, of an ethereal medium. The quasi-hypothetical 'luminiferous ether' was considered necessary to explain the propagation of light and other electromagnetic waves. In addition it served a number of other purposes and was in general regarded as indispensable in physics. When the German physicist Paul Drude in 1894 wrote an advanced textbook on Maxwellian electrodynamics, he chose to entitle it *Physik des Aethers* (Physics of the Ether).

The ether favoured by British physicists was homogeneous and entirely different from atomic matter as usually conceived. But there were numerous other ether models and conceptions of the relationship between ether and matter. For example, a minority of scientists considered the ether to be corpuscular, consisting of extremely small particles that filled the space between atoms and which perhaps were the ultimate constituents of material atoms. In an article on the ether published in the *Philosophical Magazine* of 1885, the American physicist and engineer De Volson Wood found that the 'computed mass of a molecule of the luminiferous æther' was about 2×10^{-45} g or 10^{21} smaller than the mass of a hydrogen atom. For its specific heat he got an equally surprising number, 4.6×10^{12} times the value of water.[9] Dmitrii Mendeleev, the celebrated chemist and father of the periodic system, was among those who conceived the ether in material terms, namely, as a new chemical element of the nature of a highly tenuous gas. In a pamphlet of 1903 he even gave it a name ('newtonium') and estimated from the kinetic theory of gases that it had an atomic weight of about one millionth of that of hydrogen.[10] The kind of ethers considered by Wood and Mendeleev were unorthodox and appreciated only by a minority of scientists.

Although the nature and properties of the ether was a matter of much discussion, few physicists denied its existence. In a textbook on optics published 1902, Albert Michelson, the first American Nobel laureate in physics and famous for his experiments on the ether drift, declared that 'the day seems not far distant when the converging lines from many apparently remote regions of thought will meet on...common ground'. He prophesied that the nature of atoms, the forces of electricity and magnetism, cohesion, elasticity, and gravitation—'all these will be marshalled into a single compact and consistent body of scientific knowledge'. Michelson's confidence that final unification was near was rooted in 'one of the grandest generalizations of modern science...that all the phenomena of the physical universe are only different manifestations of the various modes of motion of one all-pervading substance—the ether'.[11]

The relationship between matter and ether was one of the great problems of theoretical physics and one that the vortex theory had to address. It was generally assumed, if not always stated explicitly, that the perfect and universal fluid was the very same as the ether, in which case material particles would be nothing but structures in the ether. In this way the theory became truly monistic, a feature which added to its appeal. Whatever the different opinions on this matter, from about the mid-1880s many British physicists and mathematicians began to study vortex theories of the ether, hoping in this way to obtain a more fundamental understanding of the transmission of electromagnetic waves and other physical signals. The result of this work, predominantly of a mathematical nature, was a proliferation of vortex theories and vortical objects. Theories of 'vortex sponges', which were unified models of ether and matter, attracted much attention for a time.

The Irish physicist Gerald Francis FitzGerald, professor of natural philosophy in Dublin, undertook a detailed investigation of vortex sponges. These were more complex structures than the simple vortex rings, but also, he thought, more promising because they offered many more possibilities. As he stated in 1888, in an address to the British Association for the Advancement of Science: 'With the innumerable possibilities of fluid motion it seems almost impossible but that an explanation of the properties of the universe will be found in this conception.'[12] A theory that explains nothing less than 'the properties of the universe' may appear to be suspiciously over-ambitious, but FitzGerald was not the only one who spoke in such elevated terms. His compatriot George Johnstone Stoney, professor of natural philosophy at Queen's College in Galway and one of the fathers of the electron, held the vortex theory in similarly high regard. Speaking of the theories of Thomson and FitzGerald, he found it 'pretty certain that either these hypotheses, or something like them, are the true ultimate account of material Nature'.[13]

The favoured view concerning the ether and its relationship to matter in the Victorian period was that the ether was continuous and that it might possibly include atoms as particular structures, as in the vortex atom theory. There were, however, ways to produce matter out of ether other than by means of vortices. One of them, barely known today and not even widely discussed at the time, was investigated by Karl Pearson, the mathematician, biometrician, and philosopher of science.[14] A monist and positivist, Pearson shared the vortex physicists' dislike of the dualistic conception of two primary substances, ether and matter. However, rather than joining the 'extremely beautiful hypothesis' of the vortex atom, in a paper of 1885 he suggested as a possible alternative that the ultimate atom might be a differentiated spherical part of the ether, pulsating with a natural frequency.

In a later theory of 1891 he combined the merits of the extended vortex atom and the Boscovichean point atom by reducing the atomic sphere to a point—an 'ether squirt'— from which ether continuously flows in all directions of space. 'Matter would thus be simply a point at which ether flows into space & mass the rate of the flow', he wrote to one of his colleagues.[15] Elsewhere in the world there were presumably counterparts of the squirts, sinks that absorbed ether and acted like negative matter. As to the question of from where the ether flowed, and to where it returned, he preferred to leave it to the metaphysicians. (He speculated about a 'space of a higher dimensions' as a possibility.) Pearson developed his hydrodynamic and monistic theory in considerable mathematical detail, and endeavoured to turn it into a model that could illuminate concrete problems of physics and chemistry, including gravity, cohesion, and chemical affinity. For example, he deduced that the force between two atoms would vary as the inverse cube of the interatomic distance. Moreover, he introduced a quantitative measure of chemical affinity in terms of the pulsation periods of the ether squirts, arguing that he was in principle able to predict the degree of stability of a molecule. However, like so many other theories of this class, Pearson's did not deliver what it promised.

The original theory of the vortex atom was meant to explain the composition and properties of matter, but it soon mutated into a large number of mathematical models that, more often than not, had little connection to physical reality. What Thomson had called the 'intensely interesting problem of pure mathematics' associated with the theory was an important reason for its popularity among British researchers. For example, the motion and vibrations of various forms of vortex object were eagerly studied by mathematicians with little interest in the physical world, and their research often appeared in mathematical rather than physical journals. Among the most industrious of the vortex mathematical physicists was William M. Hicks, a former student of Maxwell, whose motivation was primarily mathematical. His favoured models of vortex atoms were complicated and conceptually unappealing, as they operated with two ethereal matters rather than one, but this did not worry him too much. So long that the vortex objects could be analyzed mathematically and illuminate, however abstractly, physical phenomena, he was satisfied.

The vortex atom gave impulse not only to advances in mathematical hydrodynamics, but also to a new branch of topology, the theory of knots.[16] This theory can be traced back to a work done by the German mathematician, Johann Listing, in 1843, but it was only with Thomson's and Tait's vortex-related contributions that the field became recognized as an interesting branch of mathematics (Fig. 2.1). Inspired by Helmholtz's great paper of 1858, knot theory was largely established by Tait, who developed the topological ideas associated with the theory of vortex atoms and proposed a classification of various types of knots. Tait thought that knots and vortices were closely related and that the mathematical theory of knots might illuminate physical problems such as the explanation of spectra, the structure of chemical compounds, and the question of why there is only a relatively small number of chemical elements. However, these hopes did not materialize during the Victorian era. Just like the works in mathematical vortex hydrodynamics done by Hicks and others, knot theory had its roots in the vortex atom theory, but it soon developed into an independent mathematical field where Thomson's vision of a unified physical theory dropped out of sight. Only much later did knot theory become important in theoretical physics, with applications that Tait and Thomson could not have dreamt of. Today, knot theory is a major area of physics, with applications ranging from quantum field theory, through biophysics, to chaos theory.[17]

2.2 APPLICATIONS OF THE VORTEX THEORY

In spite of the significance of the mathematics, the vortex theory was intended to be a physical theory and as such it needed to result in explanations and predictions of physical phenomena. It was indeed applied to several such phenomena, sometimes in

Fig. 2.1. Some of Thomson's knots, as shown in his 1869 paper 'On vortex motion' in which he analyzed in mathematical details the properties of vortex atoms.
Source: William Thomson, *Mathematical and Physical Papers*, Vol. 4 (Cambridge: Cambridge University Press, 1910), p. 46.

detail and at other times only sketchily. As mentioned, as early as 1867 Thomson had pointed out that the vortex theory was well suited to explaining the spectra produced by chemical elements and compounds. Spectroscopy emerged as a science at about the same time as the vortex theory, and it attracted a great deal of attention both among experimentalists and theorists. From about 1870 the regularities revealed by the spectroscope were seen as data that an atomic theory needed to confront. Absorption and emission spectral lines seemed to require atoms (or molecules) to possess a very large number of degrees of freedom, which the hard and uniform atoms of the kinetic gas theory could not provide.

Things looked much better from the perspective of the vortex theory, for here the atoms have an almost infinite number of vibrational modes. In his paper of 1867, Thomson not only dealt with the question in general terms, he also suggested that the line spectra could be explained by means of a vortex mechanism. The spectrum of sodium is particularly simple, consisting of a yellow doublet at a wavelength close to 5896×10^{-10} m. Thomson believed he could explain the spectrum on the basis of a long and straight cylindrical vortex model. If the atom was thus composed, 'two approximately equal vortex rings passing through one another like two links of a chain' would reproduce the spectrum. He added that it was 'quite certain that a vapour consisting of such atoms, with proper volumes and angular velocities in the two rings of each atom, would act precisely as incandescent sodium-vapour acts—that is to say, would fulfil the "spectrum test" for sodium'.[18] However, this was at best a rough qualitative explanation or just an illustration. Neither Thomson nor others succeeded in actually calculating the frequency of spectral lines on the basis of their vortex models of matter.

Another new area of the physical sciences where the vortex atom found an application was the kinetic theory of gases, which at the time was developed in mathematical detail by Maxwell and Ludwig Boltzmann in particular. In his paper of 1867, Thomson had indicated that the interaction between vortex rings, when sufficiently worked out, would provide the basis of a new kinetic theory of gases superior to the ordinary one based on collisions between solid molecules. He returned to the anticipated theory in a paper of 1884, and about the same time another and younger Thomson, the Cambridge physicist Joseph John Thomson (of electron fame), developed it into a quantitative vortex gas theory.

Although in many ways successful, the Maxwell–Boltzmann gas theory was somewhat controversial, in part for conceptual reasons and in part because of its problematic relationship to thermodynamics. According to kinetic theory, the ratio between the specific heats at constant pressure and volume was given by

$$\gamma \equiv \frac{c_p}{c_v} = 1 + \frac{2}{n}$$

where n is the number of degrees of freedom. For diatomic molecules it was assumed that $n = 6$, which gave $\gamma = 1.33$, whereas experiments resulted in values close to 1.4.[19] This specific heat anomaly was serious for the standard kinetic theory, but appeared even worse for the vortex interpretation. With its very large number of degrees of freedom, γ would become close to 1, in strong contrast to the measurements. Rather than considering it a refutation of the vortex kinetic theory, J. J. Thomson argued that the anomaly only occurred on the assumption of frequent molecular collisions. This assumption held no validity in the vortex theory, where collisions were unimportant. As a feature in favour of the vortex interpretation, he argued that this theory could explain experiments on air made in the 1840s by the French physicist Henri Victor Regnault. According to Regnault's measurements, the ideal gas laws needed modification in a direction that the ordinary collision theory could not account for.

The vortex theory of gases was proposed as a serious alternative to the kinetic collision theory of Maxwell and Boltzmann, but it attracted curiously little interest in the physics community. J. J. Thomson showed that as the temperature rises, the mean radius of the rings would increase and, as a result of the bigger rings, the mean velocity of the vortex molecules would decrease. This was a result that strikingly diverged from ordinary kinetic theory, where the velocity increases with the temperature of the gas. Consequently it might serve as a means to distinguish between the two theories. Thomson realized as much and spoke of the test as a 'crucial experiment', but without attempting to carry it out or investigating it more closely.

A theory as ambitious and fundamental as the vortex atom theory had to address the old riddle of explaining gravity. Ever since Newton, many scientists were of the opinion that although the theory of gravitation worked wonderfully and was impeccable from an instrumentalist point of view, conceptually it was unsatisfactory because it rested on the postulate of action at a distance. What was needed, they felt, was a proper explanation of the mechanism and transmission of the gravitational force. There was no shortage of such theories in the nineteenth century—most of them mechanical, some based on properties of the ether, and a few were electrodynamical—but they were all short-lived and failed to win general approval.[20]

To mention but one, the Cambridge astronomer and Plumian professor James Challis believed he had explained gravitation in terms of elastic ether waves. His *Essay on the Mathematical Principles of Physics* from 1873 presented an ambitious unified theory in which all the forces of nature were reduced to manifestations of ether pressures or waves. However, after Maxwell pointed out in an anonymous review in *Nature* that it violated the principle of energy conservation, little more was heard of Challis' theory. The level of ambition of the theory may be seen in Maxwell's description, according to which the aim was 'no less than to explain all actions between bodies or parts of bodies, whether in apparent contact or at stellar distances, by the motions of this all-embracing æther, and the pressure thence resulting'.[21] Maxwell found Challis' theory to be plainly wrong, but he was not opposed to explanations of gravitation based on models of the ether. 'It may be hard to say of an infant theory that it is bound to explain gravitation', he said in connection with Thomson's theory of vortex atoms, indicating that the theory was expected to do just that. 'Since the time of Newton, the doctrine of gravitation has been admitted and expounded, till it has gradually acquired the character rather of an ultimate fact than of a fact to be explained.'[22]

Thomson was of course aware of the problem of gravitation, but wavered in his attitude to it. He was greatly interested in Le Sage's old theory of gravitation in terms of 'ultramundane corpuscles', in part for historical reasons but also because he hoped that it might be modified to serve as an acceptable explanation of gravity. The modification he thought of was to replace the hard ultramundane particles with the perfectly elastic vortex atoms. In a lecture delivered to the Royal Institution in 1881, he said of the vortex atom theory that 'this kinetic theory of matter is a dream, and can be nothing else, until it can explain chemical affinity, electricity, magnetism, gravitation, and the inertia of masses (that is, crowds of vortices)'. He continued in a pessimistic mood:

> Le Sage's theory might give an explanation of gravity and of its relation to inertia of masses, on the vortex theory, were it not for the essential æolotropy [anisotropy] of crystals, and the seemingly perfect isotropy of gravity. No finger-post pointing towards a way that can possibly lead to a surmounting of this difficulty, or a turning of its flank, has been discovered, or imagined as discoverable.[23]

From about that time Thomson began to doubt if ordinary ring vortex atoms were eternally stable, as he had originally believed. This may have been the background for an anonymous poem that appeared in *Nature* of 1882:

> The Vortex-Atom was dying
> The last of his shivering race –
> With lessening energy flying
> Through the vanishing realms of Space.
> But as his last knot was dissolving
> Into the absolute nought –
> No more, so sighed he resolving,
> Shall I as atom be caught.
> I've capered and whirled for ages,
> I've danced to the music of spheres,
> I've puzzled the brains of the sages –
> Whose lives were but reckoned by years.[24]

The most developed and whole-hearted attempt to construct a vortex atom theory of gravitation was due to Hicks, who, around 1880, published several papers on the subject. Hicks studied hollow vortex atoms pulsating in the ether and showed that the forces between them must follow a complicated law that could be approximated by an inverse square law. From his mixture of assumptions and mathematical manipulations he concluded that, in addition to ordinary or 'positive' matter, the theory also allowed for 'negative' matter that would repel ordinary matter but attract other negative matter. The suggestion, which Hicks may not have entertained in a realistic sense, had some similarity to Pearson's later speculations based on the ether squirt model.

The ideas of Hicks and Pearson won little support, yet they were independently entertained by Arthur Schuster, professor of physics at the University of Manchester, in what he admitted was a 'holiday dream'. His vision of 1898 included not only 'anti-atoms' but also worlds made up of 'anti-matter'. Schuster lightheartedly suggested that there might exist entire stellar systems of antimatter, indistinguishable from our own except that two stellar systems would be repelled rather than attracted. He even suggested (more than 30 years before Dirac) that colliding matter and antimatter would annihilate each other.[25] Schuster's idea was merely meant as a speculation, and it was received as such. Neither it nor the theories of Hicks and Pearson were given serious attention, nor were other ether-based theories of gravity seriously considered. To summarize, the vortex theory of atoms failed to produce a satisfactory explanation, or even an explanation sketch, of gravitation. The only consolation was that other theories fared no better.

Supporters of the vortex theory could always appeal to the theory's mathematical complexity and consider it a resource rather than a problem. This is what the British physicist Donald MacAlister did. In a review article of 1883, he emphasized, as Maxwell had done earlier, that the consequences of the theory 'flow without subsidiary hypotheses from its initial data'. He compared the new atomic theory to the development of the wave theory of light after it had been introduced in the 1820s: the more the consequences were worked out, the more phenomena it would be able to explain. There was no easy way from theory to phenomena, but with the help of the mathematicians MacAlister was confident that the problems would eventually be solved:

The work of deduction is so difficult and intricate that it will be long before the resources of the theory are exhausted. The mathematician in working it out acquires the feeling that, although there are still some facts like gravitation and inertia to be explained by it, the still unexamined consequences may well include these facts and others still unknown. As Maxwell used to say, it already explains more than any other theory, and that is enough to recommend it. The Vortex-theory is still in its infancy. We must give it a little time.[26]

Applications of a scientific theory usually relate to other branches of science, or to technological uses, but may also be of an ideological nature. The vortex atom theory was an important part of the world picture of late-Victorian Britain, and it entered the societal and ideological struggle of the period. Part of this struggle was concerned with the provocative claims of a new generation of 'scientific naturalists', who argued that the material and spiritual worlds were entirely separate spheres.[27] (A few radicals went further, denying altogether the existence of a spiritual world.) Exponents of scientific naturalism, including Thomas Huxley, William K. Clifford, and John Tyndall, often used the standard atomic theory to argue their more or less materialist cause, but at the time atomism did not necessarily imply materialism. The vortex atom was entirely different from the material atoms of the Newton–Dalton tradition, as it was unconnected with the materialism that since the days of Democritus and Lucretius had been associated with atomism. In fact, one reason why the vortex atom hypothesis was considered attractive by many British scientists was that it resonated with their religious feelings. To Thomson and his kindred spirits, the mathematically proven permanence of vortex atoms implied that they could only have come into existence through a supranatural act, that is, they were created by God. It was through the ether that the vortex atoms sometimes entered as a scientific justification for spiritual thinking and a revived natural theology. Whether the ether was thought to be vortical or not, it became increasingly common to see it as dematerialized.

Among the physicists who held the vortex theory in favour, even though he did not contribute to its scientific development, was Oliver Lodge, a prominent scientific author and specialist in electromagnetism. According to Lodge, not only was all nature emergent from the ether, he also came to see it as nothing less than 'the primary instrument of Mind, the vehicle of Soul, the habitation of Spirit, . . . the living garment of God'.[28] Lodge found the vortex theory, or something like it, to be the best offer of a truly fundamental theory, both scientifically and spiritually.

In *The Unseen Universe*, an important and controversial book published by Tait and fellow physicist Balfour Stewart in 1875, vortex atoms played an important role. It was the major aim of the book to refute extreme philosophies of materialism, and to do so on scientific rather than metaphysical grounds. Its message was that, far from being in conflict, science and religion were in intimate harmony. Stewart and Tait wanted their Christian belief in the immortality of the soul to have a scientific basis, and for this purpose they introduced a kind of parallel universe, a spiritual heaven connected by bonds of energy or ether to the material universe we live in. Their arguments presupposed an ethereal world consistent with the vortex theory, but they did not rely specifically on this hypothesis. Placing the vortex atom theory in a cosmological context, they asked from where the ether vortices had originally come. It must, they answered, be 'an act impressed upon the universe from without,...for if the antecedent of the visible universe be nothing but a perfect fluid, can we imagine it capable of originating such a development in virtue of its own inherent properties, and without some external act implying a breach of continuity?'[29] Tait and Stewart could not. By what they claimed was scientific logic, they were led to the conclusion that 'the visible universe [has been] brought about by an intelligent agency residing in the unseen'.[30]

In spite of the intentions of Stewart and Tait, many readers of *The Unseen Universe* saw it as offering scientific support to spiritual and occult views of nature. Thus, the notorious Russian émigré Helena Petrovka Blavatsky, the founder of the Theosophical Society, used material from Stewart and Tait's work in her influential book, *Isis Unveiled*, published in 1877. According to Blavatsky and her followers, theosophy was scientific in method, a synthesis of science, metaphysics, Eastern wisdom, and ancient religions. In her spiritual and pseudoscientific cosmology, the ether played a similarly important role as in *The Unseen Universe*. In general, the vortex picture and related ether theories greatly appealed to the occult movements that gained popularity in Britain in the late nineteenth century. Of course, in these 'applications' there was no trace of the heavy mathematical machinery that characterized the scientific vortex theory of atoms.[31]

As one might expect, philosophers with more sympathy for materialism than Tait, Stewart, and Thomson did not agree with the semi-spiritual interpretation of the ether, nor did they necessarily accept the ether hypothesis at all. An advocate of rigid atoms in motion, the German philosopher Kurt Lasswitz subjected the vortex atom theory to detailed criticism, one of the few critical reviews that appeared on the Continent. Lasswitz found the theory to be of mathematical interest only and thought that it was doubly obscure because it presupposed *two* miracles: a supranational creation of the eternal fluid and of the vortical motions in it. Somewhat similar critique was heard from some of the British scientific naturalists, including the mathematician William Kingdon Clifford.

Critics of the vortex atom theory sometimes objected that the theory, based as it was on a perfectly homogeneous and incompressible plenum, did not allow macroscopic motion. This was a very old objection to plenum conceptions of the world. It can be found in ancient Greece, when the atomists inferred the existence of a void to make motion possible and thereby avoid the paradoxes of plenum philosophers such as Parmenides and Zeno. In the late seventeenth century the objection was used as an argument against Descartes' vortices. Thus Leibniz pointed out, against the Cartesians, that if rotational motion occurred within a perfectly homogeneous fluid it would produce no observable effects. However, the objection made no impression on the British physicists who realized that elasticity and continuity are not incompatible. This was how Maxwell defended the conceptual consistency of the vortex theory. Many years later, Lodge did the same in plain language. 'There is no real difficulty; fish move freely in the depths of the ocean', he said.[32]

2.3 VORTEX CHEMISTRY?

Fundamental theories of matter were supposed to explain not only physical phenomena, but also to cast light on chemical problems such as affinity and the constitution of molecules. Chemistry was part of the early unified theories of Descartes and Boscovich, and it was also an important component of the holistic conception of natural philosophy that characterized the *Naturphilosophie* in the Romantic era. Likewise, although the vortex atom theory of the late nineteenth century was a physical theory, it naturally included parts of chemistry. As mentioned, young J. J. Thomson was greatly interested in vortex rings, whose properties and interactions he examined in several works of the 1880s. His Adams Prize essay of 1882, published the following year as *A Treatise on the Motion of Vortex Rings*, was not only a mathematical tour de force but also a serious attempt to apply the theory to concrete problems of physics and chemistry. This theory, he wrote, is 'evidently of a very much more fundamental character than any theory hitherto started.... [and] is the only one that attempts to give any account of the mechanism of the intermolecular forces'. The great advantage of the vortex theory was methodological, namely that 'the mutual action of two vortex rings can be found by kinematical principles, whilst the "clash of atoms" in the ordinary theory introduces us to forces which themselves demand a theory to explain them'.[33]

The novelty of J. J. Thomson's work did not so much lie in the complexity of its calculations as in the author's serious attempt to establish the vortex atom hypothesis as an empirically useful theory of matter. Apart from the vortex gas theory, mentioned above, Thomson endeavoured to demonstrate how chemical valency and actions could be understood on a vortex basis, assuming for reasons of simplicity that all vortex rings had the same strength. His theory of combination and dissociation of gases, meant as a contribution to the new physical chemistry that emerged in the 1880s, rested on a quite different basis to the ordinary gas theory and resulted in different predictions. Thus, according to the standard view of thermal dissociation of gases, the cause was the increased number of molecular collisions with temperature; Thomson, on the other hand, explained that dissociation occurred even without collisions. In his analytical treatment of simple cases, he derived expressions that could be compared with experimental data and found in some cases a promising agreement. For example, he deduced a correction to Boyle's law of gases that agreed with recent experiments. However, both in this and other cases it remained unclear to what extent the agreement was explicitly based on the hypothesis of vortex atoms.

The nature of valency, meaning the tendency of elements to combine chemically, was one of the great problems of nineteenth-century chemistry.[34] Why do some elements combine in definite proportions, while some do not combine at all? Taking the valency of an element to be given by the ratio of the number of links in the atom to those in the hydrogen atom, Thomson attempted to establish a theory of valence on the basis of the vortex constitution of atoms. This was not the first proposal of a theory of valence, but it was the first one that aimed to reproduce the valence of elements from a general and fundamental theory of physics. As far as gases were concerned, Thomson found that no single atom would be capable of uniting with more than six atoms of another element so as to form a stable compound. This prediction agreed nicely with chemical knowledge, as there were no examples of gaseous compounds of the type AB_n with $n > 6$. Thomson further applied his notion of valency to other elements and got in some cases, if only with difficulty and by making some arbitrary assumptions, a reasonable agreement with known data. His calculations only appeared as confirmation of the vortex atom theory for those physicists who already accepted the theory. As to the chemists, they tended to ignore them, or, in most cases, were just unaware of them.

In a later review of the vortex explanation of valency and chemical combination, given to the London Chemical Society in 1896, FitzGerald attempted to extend the theory to organic molecules. This he did by sketching a possible vortex-atom explanation of optical asymmetry, and also by applying the theory to one of the classical problems of structural chemistry, the constitution of the benzene molecule. In accordance with the view of the German chemist Friedrich Kekulé, this molecule (C_6H_6) was assumed to be bound together by three single bonds alternating with three double bonds, but it was recognized at the time that the formula did not quite match the properties of the compound. (For example, three double bonds would make the compound highly unsaturated, which benzene is not.) Although FitzGerald was forced to admit that his foray into organic chemistry was not very successful, he maintained that such problems were not outside the reach of a future vortex theory. He adopted the same optimistic attitude to other problems. For example, on the basis of the simple ring vortex hypothesis, the volume of a mercury atom came out as nearly 3000 times as great as that of a hydrogen atom. This figure was much greater than what was known experimentally, and yet the disagreement did not cause FitzGerald to reject the vortex theory. Instead, he proposed studying other forms of vortex motion, say of 'thick rings and of spherical and worm motions', in the hope that the difficulties would then disappear.[35]

Among the most suggestive results of J. J. Thomson's ambitious theory was that it indicated a possible explanation of the periodic system of the elements. Established in 1869 by Mendeleev and, independently, the German chemist Lothar Meyer, by the 1880s the periodic system had become generally accepted as a fundamental and most useful classification of the chemical elements. The system's ability to predict the properties of new elements (such as gallium and germanium) convinced most chemists of its truth. However, apart from a few speculations there was no explanation for the striking periodicity that was the basic message of Mendeleev's system. Mendeleev himself denied that the system reflected the atomic structure of the various elements.

In his work of 1883, Thomson determined the stability of atomic systems by means of elaborate perturbation calculations. However, this was only possible for a small number of interacting vortex rings, and for more complicated systems he relied on the analogy with an experiment with floating magnets that the American physicist Alfred Mayer, professor at the Stevens Institute of Technology in Hoboken, New Jersey, had published in 1878.[36] Subjecting equally magnetized needles floating in water to the attractive force of an electromagnet, Mayer noticed that the needles took up equilibrium positions on concentric circles. As William Thomson was quick to point out, the experimentally determined configurations of equilibrium corresponded to those expected from an atom made up of columnar vortices.

In 1892 J. J. Thomson noted explicitly the suggestive similarity between Mayer's configurations of magnetised needles, the arrangement of interacting vortices, and the periodicity of the properties of chemical elements. 'If we imagine the molecules [atoms] of all elements to be made up of the same primordial atom', he wrote, 'and interpret increasing atomic weight to indicate an increase in the number of such atoms, then, on this view, as the number of [primordial] atoms is continually increased, certain peculiarities will recur.' He made it clear that the atomic arrangement he thought of was 'on the supposition that the atoms are vortex rings'.[37] Mendeleev denied that the periodic system could be explained along the lines indicated by Thomson or, for that matter, in any other way that assumed the chemical atoms to have an internal structure. This was an idea that the great Russian chemist vigorously resisted.

At that time, Thomson no longer identified the primordial particle with an ether vortex, but his thinking about atomic constitution continued to be guided by the vortex atom theory even after the announcement of the true primordial atom, the electron. As late as 1907, a decade after the discovery of the electron, Thomson praised the vortex atom in his comprehensive account of the new electron atom, *The Corpuscular Theory of Matter*. Greatly attracted by fundamental theories of everything, he admitted that the new picture of the atom was 'not nearly so fundamental as the vortex theory of matter'. On the other hand, from a more pragmatic point of view he realized that the electron was to be preferred over the ether vortex: 'The simplicity of the assumptions of the vortex atom theory are, however, somewhat dearly purchased at the cost of the mathematical difficulties which are met with in its development; and for many purposes a theory whose consequences are easily followed is preferable to one which is more fundamental but also more unwieldy.'[38]

In spite of its ambitions and rhetoric, the unwieldy theory of the vortex atom did not make much of an impact on chemistry. Chemists at the time were mostly practical men occupied with experiments (as they still are—except that many are now women). With a few exceptions they had neither interest in nor knowledge of the higher mathematics required to understand the properties of vortex atoms. Still, a few chemists in Britain and the United States responded favourably to Thomson's vision of a vortex chemistry. Such a foundation of chemistry might remain a dream, but if so it was too beautiful a dream to be ignored. According to Francis Venable, a chemistry professor at the

University of North Carolina, the vortex theory of matter was more than just a dream. It was a realistic offer of a future theory of everything, including chemical phenomena. In a book of 1904 he praised it as follows:

> The harmony of the universe is motion, and so at the close of more than twenty centuries we come back to a theory of a universe filled with a continuous matter, and, at the same time, an atomic theory. But the theory is no longer a baseless dream. It would seem to be the culmination of centuries of work, not fancy, and to embody the explanation of all facts known—chemical, physical and mathematical. There is still much to be done and many untrodden paths. The theory must yet stand many exacting tests, but so far at least nothing has been thought out which so satisfies the conditions known to us.[39]

However, by that time the vortex theory of atoms had been abandoned by almost all physicists. The dream of a theory that embodied 'the explanation of all facts known' was still much alive, but by 1904 the favoured theory of this kind was the theory of the electromagnetic electron, which will be considered in the following chapter.

2.4 A THEORY THAT OUGHT TO BE TRUE

By the turn of the century, the theory of vortex atoms was no longer an active field of research. Scientists continued to refer to it and praising it for its methodological virtues, but only an insignificant minority believed that the world really consisted of vortical structures in the ether. The decline and eventual death of the theory was not only rooted in its poor empirical record, but also in the emergence of a vigorous alternative in the form of the electromagnetic world view based on Maxwellian field electrodynamics (see Chapter 3).

The intellectual and scientific environment of the 1890s was increasingly hostile to a mechanical conception of the world, and from this perspective the ether of the vortex programme was no longer appealing. Although not conceived as a kind of ordinary matter, William Thomson's ether was mechanical in the sense that it possessed inertia as an irreducible property, and non-material only in the sense that it was continuous and not derivable from the physics of matter. Moreover, the etherial vortex atoms were different from the surrounding electromagnetic ether, with the latter playing little role in the vortex theory of matter. By contrast, the new electron physics was based on a unified picture of the electromagnetic ether that differed from the more or less mechanical ether on which the vortex atom theory built. Although the contrast is conceptually significant, British physicists saw no great discontinuity. They smoothly went from accepting the vortex atom programme to accepting the research programme of electron physics. J. J. Thomson is one example of the continuity between the vortex theory and the electromagnetic conception of nature; the Cambridge physics professor Joseph Larmor counts as another. There are many more examples.

Like many other British physicists in the 1890s, Larmor found the hydrodynamical vortex atom appealing while at the same time admitting the serious problems with Thomson's old conception of matter. As he pointed out in an important work of 1894, a rise in temperature would increase the energy, the result being that the vortex rings would expand and their translational energy decrease. As a consequence, not only would the velocity–temperature dependence be wrong (as previously recognized), but the changed size of the atom would also imply a change in the frequency of the emitted light, something that had not been observed experimentally. The prediction contradicted 'the fundamental fact that the periods of the radiations corresponding to the spectral lines of any substance are precisely the same whatever be its temperature'.[40] This came close to an experimental refutation of the admired vortex theory of atoms.

The disappearance of the vortex theory from the scene of physics was not directly caused by its conflict with empirical data, nor by insurmountable mathematical problems. It was primarily abandoned because of its lack of progress over a period of about 20 years and also because of the appearance of a more attractive alternative. One might believe that the theory was unambiguously falsified by experiments, but this was not really the case. Even the problem mentioned by Larmor did not amount to a clear-cut falsification. It is instructive to look at the attempts of some physicists to keep the theory alive, if not as a candidate of truth then at least as a research programme of inspirational value.

It is characteristic that even after the vortex theory had been abandoned as a research field, many physicists continued to speak highly of it and defend it as a theory whose possibilities had not *yet* been harvested. Even physicists outside the vortex atom research programme recognized the theory to be, if not true then highly attractive. In 1899 the Birmingham physicist John Henry Poynting gave an address to the British Association on theories of matter, singling out Thomson's vortex atom as a particularly satisfying attempt to avoid the dualism between material atoms and the ethereal plenum: 'Here all space is filled with continuous fluid—shall we say a fluid ether?— and the atoms are mere loci of a particular type of motion of this frictionless fluid. The sole difference in the atoms are differences of position and motion. Where there are whirls, we call the fluid matter; where there are no whirls, we call it ether.' Although Poynting was careful to speak of the vortex theory as merely a 'mental picture', he was impressed by the general idea underlying it: 'So, as we watch the weaving of the garment of Nature, we resolve it in imagination into threads of ether spangled over with beads of matter. We look still closer, and the beads of matter vanish; they are mere knots and loops in the threads of ether.'[41]

One of the last scientists to praise the vortex atom theory as a fundamental theory of matter, the American physicist Silas Holman, emeritus professor at the Massachusetts Institute of Technology, chose to focus on the theory's positive merits, while he tended to disregard the lack of progress for over a decade. Holman summarized his opinion of the vortex theory as follows: 'The theory has not yet, it is true, been found capable of satisfactorily accounting for several important classes of phenomena, ... but this constitutes no disproof... The theory must be judged by what is has accomplished, not by what we have not yet succeeded in doing with it. And when thus tested, the theory still remains preeminent.'[42] Michelson was another American physicist who, about the turn of the century, continued to find the vortex theory attractive—indeed, he believed that the vortex theory 'ought to be true even if it is not'.[43]

From its beginning in 1867 to its end at about 1900, the theory was frequently justified on methodological and aesthetic grounds rather than its ability to explain and predict physical phenomena. In an 1883 review of ether physics, Lodge described the vortex atom theory as 'beautiful' and 'the simplest conception of the material universe which has yet occurred to man'. He added, just as Michelson would do twenty years later, that it was 'a theory about which one may almost dare to say that it deserves to be true'.[44]

The audience listening to William Hicks' address at the 1895 meeting of the British Association of the Advancement of Science would not suspect that the vortex theory of atoms was dying. Without paying much attention to the theory's disappointing record with regard to empirical physics, Hicks reviewed in an optimistic tone the theories of various vortex objects such as rings, spheres, and sponges. He realized that relatively little progress had been made over the years in the mathematical development of the theory, and that progress was even more lacking in the theory's contact with experiments. However, these problems he deftly turned into a defence of the theory, for the undeveloped mathematical framework meant that the theory could not be rigorously tested. Hicks was convinced that the road towards progress would be to develop still more advanced mathematical models. The vortex theory, he said, 'is at present a subject in which the mathematicians must lead the attack'.

As Hicks saw it, a generalized vortex theory of both matter and ether was the best offer of an ultimate theory of nature in the history of science. Without referring to the latterly popular term, 'theory of everything', he reflected at length on the nature of just such a theory:

> The ultimate aim of pure science is to be able to explain the most complicated phenomena of nature as flowing by the fewest possible laws from the simplest fundamental data. A statement of a law is either a confession of ignorance or a mnemonic convenience. It is the latter if it is deducible by logical reasoning from other laws. It is the former when it is only discovered as a fact to be a law. While, on the one hand, the end of scientific investigation is the discovery of laws, on the other, science will have reached its highest goal when it shall have reduced ultimate laws to one or two, the necessity of which lies outside the sphere of our cognition. These ultimate laws—in the domain of physical science at least—will be the dynamical laws of the relations of matter to number, space and time. The ultimate data will be number, matter, space, and time themselves. When these relations shall be known, all physical phenomena will be a branch of pure mathematics.[45]

Hicks considered the vortex atom an evocation of a special dynamical method that enabled the physicist to bridge the gap between his sense experiences and the inner workings of nature. We must, he said in his address, 'make a bridge of communication between the mechanism and our senses by means of hypotheses'. He further explained how to construct the bridge:

> By our imagination, experience, intuition we form theories; we deduce the consequences of these theories on phenomena which come within the range of our senses, and reject or modify and try again. It is a slow and laborious process. The wreckage of rejected theories is appalling; but a knowledge of what actually goes on behind what we can see or feel is surely if slowly being attained. It is the rejected theories which have been the necessary steps towards formulating others nearer the truth.[46]

Hicks was not the only British physicist who used the meetings of the British Association to reflect on the aim and ultimate laws of physics. Four years later, Poynting argued that the aim of physical theory, in so far it was explanatory and not merely descriptive, was 'to reduce the number of laws as far as possible, by showing that laws, at first separated, may be merged in one; to reduce the number of chapters in the book of science by showing that some are truly mere sub-sections of chapters already written'.[47]

To many physicists trained in the British tradition, the value of dynamical models such as the vortex atom did not lie in their truth but in their heuristic and illustrative functions. The vortex atom model might not represent real atoms, but it was nonetheless a fruitful and suggestive way of conceiving the constitution of matter and ether. Five years after Hicks' address, Larmor gave another address before the British Association. The intervening years had seen a minor revolution in physics—involving the spectacular discoveries of X-rays, radioactivity, and the electron—and by 1900 the vortex theory was no longer an active field of research. Although Larmor considered the vortex atom to belong to the past, he maintained that the ether was essential to the atomic theory of matter. 'The material atom', he said, 'must be some kind of permanent nucleus that retain around itself an aetherial field of physical influence.' Larmor chose to defend models of the vortex atom type, focusing on their heuristic and methodological virtues: 'For purposes of instruction such models, properly guarded, do not perhaps ever lose their value: they are just as legitimate aids as geometrical diagrams, and they have the same kind of limitations.' More specifically Larmor praised the model of vortex atoms as follows:

> This vortex-atom theory has been a main source of physical suggestion because it presents, on a simple basis, a dynamical picture of an ideal material system, atomically constituted, which could go on automatically without extraneous support. The value of such a picture may be held to lie, not in any supposition that this is the mechanism of the actual world laid bare, but in the vivid illustration it affords of the fundamental postulate of physical science, that mechanical phenomena are not parts of a scheme too involved for us to explore, but rather present themselves in definite and consistent correlations, which we are able to disentangle and apprehend with continually increasing precision.[48]

The hope that progress would be effected through mathematics, as expressed by Hicks and others, was a persistent theme in the history of the vortex theory, as it would be in several later theories of a similar scope. Given that the theory was immensely complicated from a mathematical point of view, it could be and was in fact argued that it was not *yet* understood sufficiently to be physically useful. Referring to the problems of a complete analysis of the collision of two vortex rings, Tait wrote that such an investigation would 'employ perhaps the lifetimes for the next two or three generations of the best mathematicians in Europe'. He admitted this was a formidable difficulty, but consoled himself by noting that 'it is the business of mathematicians to get over difficulties of that kind'.[49] More than a century later, Tait's remarks would be almost literally repeated by physicists occupied with understanding the theory of superstrings. We shall return to the analogy in Chapter 11.

The point is that for a theory so appealing that it deserved to be true, one could easily find ways to avoid the unwelcome and prosaic conclusion that it was just wrong. As a concrete illustration, consider one of the empirical problems that faced the theory, the one associated with the speed of gas molecules at different temperatures. As mentioned, J. J. Thomson had deduced that the mean velocity of vortex molecules would diminish with the temperature. Osborne Reynolds, professor of engineering at Owens College, Manchester, and a specialist in fluid dynamics, pointed out that it was unnecessary to conduct new experiments to test the prediction, as such experiments already existed. 'It appears to be an almost obvious deduction from the vortex theory', Reynolds said in 1883, 'that the velocity of sound must be limited by the mean velocity of the vortex atoms'. Then, since 'experimentally it is found that the velocity of sound increases as the square root of the temperature it appears that the verdict must be against the vortex atom'.[50] J. J. Thomson apparently ignored Reynolds' criticism. The same problem was considered some years later by FitzGerald, who confirmed that it was a genuine anomaly. However, he did not consider it a fatal one, for it rested on J. J. Thomson's standard assumption of the extra-atomic ether being a simple fluid. FitzGerald saw no reason why the ether should not have a very complicated structure, in which case the anomaly did not need to arise. In this way the appealing theory of vortex atoms could be kept alive.

To put it in a nutshell, there never was a single, concisely defined theory of vortex atoms. The theory was so rich and flexible, and so underdetermined by empirical physics, that in any practical sense it was unfalsifiable. The mathematical richness of the theory might be considered a blessing—it served as a protection against falsification—but it was also a curse. It made FitzGerald believe that it was 'almost impossible' that the universe would not be explicable in vortex terms. The generalized vortex theory that he and other physicists dreamt of (but never succeeded in constructing) could in principle explain everything, including the properties of the one and only universe. (Contrary to later multiverse physicists, the vortex physicists did not speculate on other universes.) But could it also explain why the numerous other conceivable states of the universe, all of them describable within the theory's framework, did not exist? Could it explain why the speed of light was about 300 000 km s^{-1} rather than some other value?

The theory could in principle (not in practice!) explain the mass of an atom of chlorine, but had chlorine had any other atomic weight the theory could account for that as well.

In short, the theory explained too much—and therefore too little. Or, to express it differently, the ratio between the theory's explanatory and predictive force was ridiculously large. In later chapters we shall meet recent theories, such as the idea of a multitude of universes based on string theory, that share this feature of a distinct asymmetry between explanation and definite prediction.

The vortex theory of matter is of course a theory of the past, of interest mainly to historians of science and ideas. Yet it has left a legacy also in modern physics, where vortex imagery and some of the mathematical methods introduced by Thomson, Tait, and their contemporaries can still be found. This is particularly the case in connection with knot theory, where modern work sometimes acknowledges the legacy of the Victorian past. For example, in a 1997 research paper on knot theory, including references to the works of Thomson and Tait, the authors introduce their subject as follows: 'In 1867, Lord Kelvin proposed that atoms—then considered to be elementary particles—could be described as knotted vortex tubes in ether. For almost two decades, this idea motivated an extensive study of the mathematical properties of knots, and the results obtained at that time by Tait remain central to mathematical knot theory today.'[51]

As mentioned above, the role of the vortex atom in the nineteenth century was not restricted to scientific applications but also included attempts to relate it to the spiritual and religious dimensions of life. Strangely, applications such as those proposed in *The Unseen Universe* can still be found today, if mostly in the pseudoscientific literature. In one book of this dubious genre we are told: 'The vortex idea, with all its enormous potential, was thrown away with the moribund billiard ball model. But the baby was chucked out with the bath water. It is time to look again at the vortex. Today, in the light of everything that has been discovered, this forgotten principle could provide a completely new foundation for science.' Not only is the vortex 'the missing element in the present account of the physical universe', it also 'points to a bridge between the physical and the unseen, non-physical worlds'.[52] Stewart and Tait would have rejoiced to see that their ideas were still alive more than a century later. I shall leave it unsaid as to how they would have responded to some other modern references to their work, such as Frank Tipler's (see Chapter 12).

Notes for Chapter 2

1. This chapter builds on Kragh 2002, which gives a comprehensive account of the history of the vortex atom theory. For the similarity to string theory, see Kragh 2002, pp. 93–95. John Barrow notes of the vortex theory that there are 'many striking parallels with the aims and attractions of modern string theory' (Barrow 2007, pp. 102–104). The similarity was also noted by the author and science journalist Martin Gardner, who in a review in *The New Criterion* pointed out that the vortex theory of atoms 'had an uncanny resemblance to string theory' (Gardner 2007).

2. For the 'Cartesian spirit' of the vortex atom theory, see Čapek 1961, pp. 106–16.

3. Newton to Henry Oldenburg, 7 December 1675, in Newton 1961, p. 364.

4. Helmholtz's work originated in an attempt to understand the mechanism of how sound is produced in organ pipes. See Darrigol 2005, pp. 145–66.

5. Thomson 1872, p. 318.

6. On Tait's experiments and the beginning of vortex physics, see Silliman 1963.

7. Thompson 1910, pp. 515.

8. Maxwell 1965, part II, p. 472.

9. Wood 1885.

10. On the confusing variety of ethers in the late nineteenth century and different views on their relationship to matter, see Kragh 1989a. On Mendeleev's ether, which he used in his fight against what he saw as occult tendencies in science, see Gordin 2004.

11. Michelson 1903, p. 163. Three years later, Einstein's special theory of relativity made the ether superfluous, at least as seen in retrospect. In fact, it took at least another decade until the classical ether disappeared from physics.

12. FitzGerald 1888, p. 562.

13. Stoney 1890, p. 476.

14. On Pearson's theories of atoms and ether, see Kragh 2002, pp. 53–54 and Porter 2004, pp. 179–92. Pearson included both the vortex atom and his own ether squirt theory in his influential book on philosophy of science, *The Grammar of Science* (Pearson 1900, pp. 265–68).

15. Pearson to William H. Macaulay, 23 November 1888. Quoted in Porter 2004, p. 189. Pearson's ether squirts may remind readers of the hypothetical white holes of modern cosmology, that is, sources of spontaneous matter formation that are time-symmetric versions of black holes.

16. For a detailed account of the development of knot theory, including its connections to the vortex atom, see Epple 1999. See also Silver 2006.

17. For example Kauffman 2000.

18. Thomson 1867, p. 23.

19. The anomaly received an explanation when Boltzmann argued that a diatomic molecule has five degrees of freedom instead of six. On the specific heats anomaly and other aspects of the kinetic theory of gases, see Brush 1976.

20. Theories of the inverse square law are reviewed in Rosevear 1982, pp. 95–114.

21. Maxwell 1965, part II, p. 339.

22. Ibid., p. 472.

23. Thomson 1891, p. 153.

24. 'The lay of the last vortex-atom', *Nature* **26** (1882), 297. The poem, of which I have only quoted three of its six verses, was submitted by 'K' from Edinburgh University.

25. Schuster 1898. In a subsequent letter to *Nature* (27 October 1898, p. 618), Schuster referred to Pearson's earlier speculation and suggested that antimatter might possibly be found in comet tails and the Sun's corona.

26. MacAlister 1883, p. 279.

27. For British scientific naturalism, see for example Turner 1974.

28. Lodge 1925, p. 39. On Lodge's ether, see Wilson 1971 and Raia 2007. The extra-scientific role of the Victorian ether is discussed in Noakes 2005.

29. Stewart and Tait 1881, p. 223. See also Heimann 1972. Interestingly, and rather strangely, the theoretical physicist Frank Tipler has come to a conclusion somewhat similar to the one of *The Unseen Universe*. Tipler is 'proud to acknowledge Stewart and Tait as my most distinguished predecessors in the endeavor to make life after death scientifically respectable' (Tipler 1994, p. 352). As another predecessor he might have mentioned Oliver Lodge. On Tipler's ideas of physically justified immortality, see Chapter 12.

30. Stewart and Tait 1881, p. 218.

31. On late-nineteenth-century occultism, including alchemy and speculative atomic theory, and its links to the ethereal world view, see Morrison 2007, pp. 71–83.

32. Lodge 1925, p. 155. For a vortex theorist's earlier refutation of the objection, see MacAlister 1883. On the question of motion versus plenum in the history of ideas, see Čapek 1961, pp. 111–16.

33. Thomson 1883, pp. 1–2. More details on Thomson's vortex chemistry are given in Kragh 2002, pp. 60–69, Sinclair 1987, and Chayut 1991.

34. Russell 1971 provides a detailed account of the development of ideas of valency.

35. FitzGerald 1896.

36. On Mayer's experiment and its significance, see Snelders 1976. The early attempts of a subatomic explanation of the periodic system, whether based on electrons or not, are reviewed in Kragh 2001a.

37. Thomson 1892, p. 410.

38. Thomson 1907, p. 2. On the similarities between vortices and electrons, as perceived in the 1890s, see Kragh 2002, pp. 69–73.

39. Venable 1904, pp. 268–69.

40. Larmor 1927, p. 488. Larmor was aware that a temperature increase will broaden the spectral lines because of the Doppler effect, but also that the Doppler broadening does not affect the mean frequency.

41. Poynting 1899, p. 619.

42. Holman 1898, p. 225.

43. Michelson 1903, p. 162.

44. Lodge 1883, p. 329.

45. Hicks 1895, p. 595.

46. Ibid., p. 596.

47. Poynting 1899, p. 617.

48. Address on 'The Methods of Mathematical Physics'. Reprinted in Larmor 1929, pp. 192–216. Quotations from pp. 202, 209 and 211.

49. Tait 1876, p. 298.

50. Reynolds 1883.

51. Faddeev and Niemi 1997, p. 58. For another example of a modern work in knot theory which explicitly makes the connection to the vortex atom theory, see Lomonaco 1995. Also the eminent mathematician Michael Atiyah considered the vortex origin of modern knot theory, recognizing that after the demise of the vortex atom, 'the study of knots became an esoteric branch of pure mathematics' (Atiyah 1990, p. 6).

52. Ash and Hewitt 1994, p. 23 and p. 12.

3

Electrodynamics as a World View

The entire mass of the electrons, or, at least, of negative electrons, proves to be totally and exclusively electrodynamic in its origin. Mass disppears. The foundations of mechanics is undermined.

Vladimir Lenin, *Materialism and Empirio-Criticism*, 1908.

The period 1895–1915 witnessed very drastic changes in physics, both theoretically and experimentally. As far as theory is concerned, the most important developments were the emergence of quantum theory and the theory of relativity, the latter known, for historical reasons, under two different names: the special and the general theory. The discovery of X-rays and radioactivity in the late 1890s were followed by the discovery of the corpuscular nature of cathode rays, soon understood in terms of the electron being an elementary particle residing in all matter. These discoveries eventually gave rise to a new understanding of matter and radiation culminating with the nuclear atom (1911), the discovery of X-ray diffraction (1912), and the first quantum model of atomic constitution (1913). As to fundamental physics, the predominant theoretical view in the period was a unified conception of matter and ether which, however, turned out to be a blind alley. According to the so-called electromagnetic world view, all of nature was made up of an all-pervading continuous ether governed by the laws of electromagnetism. Concentrated structures in the ether were identified with electrical particles, known as ions or electrons. (None of the terms had quite the same meaning as today. Thus an 'ion' typically meant an electrified subatomic particle, not an electrified atom or molecule.)

The ambition of the electromagnetic theory of electrons was to provide a unified and mathematically precise account of all of nature based on the fundamental equations of electromagnetism. Yet, in reality the theory was curiously sterile and lacked connections to the physical discoveries of the period. Although interest in the theory was greatly stimulated by the experimental discovery of the electron, it had almost nothing to say about quantum phenomena, radioactivity, low-temperature physics, or the regularities revealed by spectroscopy. The theory was similarly impotent when it came to chemical affinity, the periodic system, and other aspects of chemistry. Nor was it of any use in understanding astronomical and astrophysical phenomena. As is often the case with general and unified theories, there was a deep divide between the theoretical framework and the more mundane phenomenal physics. The electromagnetic world view never succeeded in bridging the divide.

Like the vortex theory of matter, the electromagnetic view of nature died a natural death, abandoned by almost all physicists but never properly falsified. Yet, although the view was barely alive at the time when Bohr announced his quantum theory of atoms, or Einstein his new theory of gravitation, elements of it lived on for some time. The general idea that ordinary matter consisted of electrical particles with a mass of electromagnetic origin remained popular, but typically without connecting the idea to the ether or thinking of the particles as internally structured. To mention but one example, Ernest Rutherford, a physicist not attracted by theoretical speculations, concluded in 1914 that 'the electron is to be regarded as a condensed charge of negative electricity existing independently of matter as ordinarily understood'. He found it probable that 'the hydrogen nucleus of unit charge may prove to be the positive electron, and that its large mass compared with the negative electron may be due to the minuteness of the volume over which the charge is distributed'.[1] Although Rutherford spoke in the language of the electromagnetic world view, he never seriously endorsed it. This kind of rhetoric remained popular as late as the 1920s, but should not be mistaken for support of the pre-World War I version of a world built up of electromagnetic fields.

3.1 ELECTRONS, MATTER, AND ETHER

The 1890s saw a general trend away from the mechanical world view based upon the idea that nature consists of forms of matter, ultimately atoms, that behave in complete accordance with the laws of Newtonian mechanics. According to this view, the atoms themselves are elementary and, although thought of as extended bodies with a definite volume and mass, without an internal constitution. Apart from problems of a technical and conceptual nature, the mechanical world view, and materialism in particular, also became unfashionable for philosophical and ideological reasons. It increasingly came to be seen as inconsistent with the Zeitgeist of the *fin de siècle* physical sciences. There were largely two ways of getting rid of the unwelcome brute matter, by replacing it with either energy or ether. In the late 1890s both approaches were followed, sometimes in combination.

The first option was advocated by the German physicist Georg Helm, professor at the Technical Institute in Dresden, who in 1890 suggested elevating the energy principle to such a status that it would replace mechanics as the foundation of physics. A generalized theory of energy, what he called *energetics* (or energeticism), would serve as a new unifying principle that would subsume mechanics and other branches of science. Indirectly supported by the influential Austrian physicist and philosopher Ernst Mach, the programme of energetics found a valuable ally in the eminent physical chemist Wilhelm Ostwald, who became the leader and spokesman of what is better characterized as a movement than a scientific theory.[2] The central doctrine of this movement was that energy was more fundamental than matter and thermodynamics more fundamental than mechanics. In what Ostwald and his followers liked to see as a revolt against 'scientific materialism', mechanics was subsumed under the more general laws of energetics, in the sense that the mechanical laws were held to be reducible to energy principles. Moreover, they rejected atomism, arguing that the belief in atoms and molecules was metaphysical and that all empirical phenomena could be explained without the atomic hypothesis.

The ambitious aim of Ostwald, Helm, and their allies was to rid science of visualizable hypotheses and analogies with mechanics, to construct an alternative *hypothesenfreie Wissenschaft*. However, the ambitious programme of the energeticists was controversial and resisted by distinguished physicists such as Boltzmann and Planck. Although initially not unsympathetic to the ideas of energetics, Planck came to the conclusion that it was an unsound and unproductive version of natural philosophy, consisting of nothing but a series of 'dilletantish speculations'. He objected that energetics had nothing positive to offer and was not vulnerable to experimental data, and for this reason could scarcely be called a science. As he said in a paper of 1896, 'a theory that in order to survive is dependent on evading the real problems is no longer rooted in the empire of the natural sciences, but on the metaphysical ground, where it can no longer be harmed by empirical weapons'.[3]

Although the energetics alternative was mostly a German movement, and one that received only modest support even from German physicists and chemists, somewhat similar views held considerable appeal also outside Germany. Many scientists had sympathy for the view that energy, not matter, was the essence of a reality that could be understood only as processes or actions. At any rate, energetics failed to establish itself as a new foundation of science and by 1905 the movement had degenerated into a philosophical system with quasi-religious associations. This system, often referred to as 'monism', aimed at reforming and unifying philosophy, culture, and social thought on a scientific basis. In 1902 Ostwald established a new journal, *Annalen der Naturphilosophie*, to further the cause of a monistic and organicist conception of science. As the title of the journal indicates, the vision of the new kind of science had much in common with the Romantic science movement of the early nineteenth century. But monism, the successor of energetics, was ignored by the large majority of physicists.[4]

One of the weaknesses of energetics was that it had little to say about electromagnetism, a science that became increasingly important during the 1890s. It was only in this decade that the power of Maxwell's field theory was fully appreciated by the majority of physicists. Whatever the reasons of the failure of energetics, its decline in no way implied a revived confidence in the mechanical–materialistic world view. To many physicists at the turn of the century, electricity and ether were the entities on which a unified physics of the future would have to be built. Moreover, these entities were considered to be deeply connected, perhaps even identical in some way.

Perhaps the most basic problem of physics at the time was the relationship between ether and matter. Was the ether the fundamental substratum out of which matter was constructed? Or, on the contrary, was matter a more fundamental ontological category of which the ether was just a special instance? As mentioned in Chapter 2, very similar questions had been discussed in connection with the vortex atom theory and similar non-electrodynamical conceptions of the ether.

The first view, where primacy was given to the structures in the electromagnetic ether, became increasingly popular at the turn of the century, when mechanical ether models (such as those based on the vortex theory) were replaced by electrodynamical models. If electromagnetism was considered more fundamental than mechanics, it made sense to try to derive the mechanical laws from those of electromagnetism, and this was precisely what many theoretical physicists aimed at. Electromagnetism came to be considered a unifying principle of all science, not unlike the role assigned to energy in the energetics approach favoured by Ostwald and Helm. In both cases, materialism was discarded and matter declared an epiphenomenon of a more basic entity, either energy or the electromagnetic field. If matter was not the ultimate reality, but merely some manifestation of the immaterial electromagnetic ether, it would seem reasonable to challenge such established doctrines as the permanence of chemical elements and the law of conservation of matter. Indeed, such speculations were not uncommon in the period around 1900, years before they were scientifically justified.

The replacement of the mechanical by the electromagnetic ether was a very important change in the physics of the 1890s. No less important, and closely related to this change, was the emergence of the electron, the first modern elementary particle. The history of this particle is complex and predates J. J. Thomson's celebrated and 'official' discovery of 1897 by several years. Not only had the name 'electron' as a quantum of electric charge been coined by Stoney as early as 1891, but it was also discussed as an elementary particle before Thomson's experiments with cathode rays.

For example, in an important memoir of 1894, entitled *A Dynamical Theory of the Electric and Luminiferous Medium*, Larmor introduced the hypothetical electrons—'or let us say monads'—to explain both electromagnetic, optical and material phenomena. He conceived the particles as primordial units of all matter, but was careful to picture them as concentrated parcels of the ether rather than as being material by nature. According to Larmor, electrons or monads were 'the sole ultimate and unchanging singularities in the uniform all-pervading medium'. He supposed the atoms of the chemical elements 'to be built up of combinations of a single type of primordial atom [the electron], which itself may represent or be evolved from some homogeneous structural property of the aether'.[5] This notion of discrete ether aggregates was nothing less than 'the master-key to a complete unravelling of the general dynamical and physical relations of matter', as he expressed it his Adams Prize essay some years later.[6] Larmor's model required two kinds of electrically charged electrons, with the one kind being 'simply perversions or optical images of the other'. However, Larmor realized that this picture, theoretically and conceptually appealing as he found it to be, was difficult to reconcile with the known facts of chemistry. For example, according to the new physical chemistry the unit of hydrochloric acid was the ionic form H$^+$Cl$^-$, whereas the charge-symmetric form H$^-$Cl$^+$ did not exist. Why? A few years later, J. J. Thomson would run into the same kind of problem concerning the dissimilarity of the two electric charges.

Larmor aimed at unifying all of physics by linking the older concepts from mechanics and hydrodynamics to the more modern ones emerging from electromagnetism, and also by linking continuous models to discrete models. His world view was thus not purely electromagnetic, but based on a combination of mechanical and electromagnetic concepts. Nevertheless, the physics he aimed at was unified in the sense that at the fundamental level the world consisted solely of dynamical structures in the ether. His electron was not a material particle located in a continuous sea of ether, as was the case with Lorentz's electron, but a rotational strain in the ether.[7] To the mind of Larmor, there was no contradiction between an etherial conception of matter and a foundation of physics in which mechanics remained indispensable.

The electron became a more physical and definite particle in the autumn of 1896 when the Dutch physicist Pieter Zeeman, at the University of Leiden, discovered the effect named after him—the magnetic influence on the frequency of light—and his compatriot Hendrik A. Lorentz explained the phenomenon in terms of electron theory. This and other work at the time led Lorentz and other theorists to consider the electron as a subatomic, negatively charged particle with a mass-to-charge ratio some 1000 times smaller than the electrolytically determined value of hydrogen.[8] That is, $(m/q)_H \cong 1000 \, (m/q)_e$.

Under the impact of the work of Zeeman and Lorentz, Larmor reconstructed his electron and his picture of the atom. In May 1897 he wrote in a letter that he was inclined 'to the view that an atom of 10^{-8} cm is a complicated sort of a solar system of revolving electrons, so that the single electron is very much smaller, 10^{-14} would do very well'.[9] At about the same time, J. J. Thomson communicated his important series of experiments on cathode rays, which he interpreted in the strong sense that all matter consisted of electrons and electrons only. For reasons which are not important in the present context, Thomson did not use the term 'electron' but preferred to call them 'corpuscles'.

In Thomson's original view, the electron differed in important respects from the particle envisaged by theorists such as Lorentz and Larmor. While they conceived the particle as a structure in, or an excitation of, the electromagnetic ether, Thomson's electron was material. In 1897 he was even willing to consider it a kind of chemical element (as did a few later scientists, including the Swedish spectroscopist Johannes Rydberg and the famous English chemist William Ramsay). Even though the electron did not qualify as a chemical element, it did make up the atoms of the elements, and for this reason might seem to justify the possibility of transmutation. FitzGerald suggested that Thomson's discovery implied that 'we are within measurable distance of the dreams of the alchemists'.[10]

Two years after his discovery of the electron, Thomson succeeded in determining its electric charge. Together with the already known value of the specific charge e/m, it led to the finding that the particle was about 1000 times lighter than the hydrogen atom. These very light particles he pictured as tiny constituents of the atom, configured in dynamical equilibrium positions in a massless and frictionless positive fluid. The atomic model he developed in the early years of the new century was ambitious and unitary, a worthy follower of the vortex theory of atoms he had earlier investigated and with more than a passing similarity to it. Assuming only a single elementary particle, the electron, Thomson was faced with the problem of representing positive electricity as an effect of the electrons. If he could not do that, the unitary theory of matter would have to be abandoned and replaced by the less attractive alternative of a dualistic theory. In 1904 he described the problem as follows:

> I have...always tried to keep the physical conception of the positive electrification in the background because I had always had hopes (not yet realised) of being able to do without positive electrification as a separate entity, and to replace it by some property of the corpuscles. When one considers that all the positive electricity does, on the corpuscular theory, is to provide an attractive force to keep the corpuscles together, while all the observable properties of the atom are determined by the corpuscles, one feels, I think, that the positive electrification will ultimately prove superfluous and it will be possible to get the effects we now attribute to it, from some property of the corpuscles.[11]

To jump ahead in time, 26 years later, in the very different context of the quantum electron, Paul Dirac suggested an analogous idea, namely that protons were electrons in disguise, so-called antielectrons described by the linear quantum wave equation that Dirac had constructed in 1928. He found this to be a most attractive idea because it promised an understanding of all matter in terms of just one fundamental entity, the electron. With such an understanding, 'the dream of philosophers' would no longer be a dream but have become a reality.[12] However, the dream remained a dream, both in the case of Thomson and Dirac. The problem of the Thomson electron atom increased drastically when experiments indicated that the number of electrons was of the same order as the atomic weight. This implied that the atom would be unstable and also that the mass of the positive electricity could not be of electromagnetic origin. The appealing unitary picture of atoms made up from a single particle apparently had to be replaced with the not quite as appealing picture of two different elementary particles as the building blocks of matter.

The problem of the dissimilarity between negative and positive charges continued to plague the electromagnetic theory of matter, although it was rarely addressed directly. One of the few who did address it was young James Jeans, who in 1901 suggested that the atom might be a collection of point-like negative and positive electrons (or 'ions') in dynamical equilibrium. He thus substituted Thomson's hypothetical sphere with equally hypothetical positive electrons. Although Jeans' model was short-lived, it merits attention because it included the bold prediction of e^+e^- annihilation: 'It is not hard to see that positive and negative charges would rush together and annihilate one another until there would be nothing left to distinguish the point at which a body ought to be from a point in empty space.'[13] Jeans noted that his picture of atomic constitution did not agree with those theories that attempted to 'place the structure of matter on a purely electrical or ætherial basis'. Because, according to 'the æther-equations of electricity' positive and negative charges could only differ in their sign, and so 'Any attempt to explain matter in terms of æther must therefore face the problem of reducing what appears to be a difference in quality to a difference in sign only.'[14] With the discovery of the proton, the problem only aggravated. No theory of electromagnetism could explain why the proton is nearly 2000 times as heavy as the electron and yet has precisely the same numerical charge.

The idea of electromagnetic mass goes back to a work of 1881 in which J. J. Thomson showed that when a charged sphere moves through the ether, self-induction will result in an effective mass greater than the sphere's mechanical mass. Some years later, Oliver Heaviside deduced from Maxwell's theory that this apparent or 'electromagnetic' mass follows the expression $m = 2e^2/3Rc^2$, where e is the charge and R the radius of the sphere, and c denotes the speed of light. 'It seems', said Heaviside in a paper of 1893, 'not unlikely that in discussing purely electromagnetic speculations, one may be within a stone's throw of the explanation of gravitation all the time'.[15] The concept of electromagnetic mass played an important role in the electron theories around 1900 and was crucial to the electromagnetic world view. A positive sphere of atomic dimension would have an electromagnetic mass that was negligible as compared with that of a single electron. This was not a problem in the original Thomson model, where the mass was made up of thousands of electrons, but after about 1910, when it was realized that there are only few electrons in atoms, it did become a serious problem. A unitary theory building on electromagnetic electrons seemed incompatible with the Thomson model of the atom.

3.2 ELECTROMAGNETISM AS A WORLD VIEW

Other physicists, closer to the electromagnetic world view than Thomson, believed that electrons were partly or wholly of electromagnetic origin and that material or ponderable mass could perhaps be entirely disregarded. This was the opinion of the Königsberg physicist Emil Wiechert, today mostly known for the pioneering contributions to seismology and geophysics he made after being appointed professor in Göttingen. In 1895–96 Wiechert published several papers on ether and electron theory, in which he suggested that all the laws of nature might be reducible to the properties of the electromagnetic ether. Not unlike Larmor, he speculated that electrons or 'electric atoms' might be excitations of the ether and the constituents of the complicated structures traditionally called atoms. However, he did not believe that the new electromagnetic ether physics, or any other theory of physics for that matter, would ever lead to a theory of everything. As he emphasized in a lengthy paper of 1896, no theory could possibly account for all phenomena in a universe inexhaustible in all directions:

> So far as modern science is concerned, we have to abandon completely the idea that by going into the realm of the small we shall reach the ultimate foundations of the universe. I believe we can abandon this idea without any regret. The universe is infinite in all directions, not only above us in the large but also below us in the small. If we start from our human scale of existence and explore the content of the universe further and further, we finally arrive, both in the large and in the small, at misty distances where first our senses and then even our concepts fail us.[16]

Not all physicists in the electromagnetic research tradition that began in the late 1890s shared Wiechert's pessimistic (or realistic?) view of a universe so diverse that it could not be grasped by physical theory. Although foreshadowed by the works of Larmor and Wiechert, a programme of establishing physics fully on the Maxwell–Lorentz theory of electromagnetism only emerged in the early years of the new century, principally in the works of the German physicists Wilhelm Wien, Max Abraham, Walter Kaufmann, Karl Schwarzschild, and Adolf Bucherer.[17]

Wien, a professor at the University of Würzburg, argued in 1900 that matter was to be understood as conglomerates of electrons (positive and negative) whose mass was of electromagnetic nature. According to Wien, electromagnetic energy and electromagnetic mass were related as $4E = 3mc^2$. Moreover, he suggested that the mass would depend on the electron's velocity relative to the world ether and its value would therefore be different to that obtained from Heaviside's expression for very high velocities. This had been shown a few years earlier by George Searle, a physicist at the Cavendish Laboratory, who had found that the total energy of an electrified sphere of charge e moving with velocity $\beta = v/c$ would be

$$E = \frac{e^2}{2R}\left[\frac{1}{\beta} \ln\left(\frac{1+\beta}{1-\beta}\right) - 1\right]$$

The electromagnetic mass of the sphere would increase accordingly. Wien adopted Searle's result, which he considered important for the electromagnetic programme he had in mind. His paper of 1900, significantly titled *Ueber die Möglichkeit einer elektromagnetischen Begründung der Mechanik* ('On the possibility of an electromagnetic foundation of mechanics'), included the basic features of a new research programme in fundamental physics, a new view of nature founded on the theory of electromagnetism. In 1905 Abraham referred to the programme as the *electromagnetic world picture*, a name that indicates the theory's scope and ambitions. It was meant to be more than just a scientific theory covering a limited domain of nature.

In its pure form this theory, research programme, or world view envisaged by Wien, Abraham and others encompassed the following positions:

(1) A rejection of mechanical modelling and ultimate explanations in terms of mechanics.
(2) A belief that physical reality is of an electromagnetic nature.
(3) A belief that the laws of mechanics can be understood electromagnetically.
(4) A commitment to a research programme aiming at a unified physics based on electromagnetic laws and concepts only.

Although only a handful of physicists subscribed to all these positions, in the first decade of the twentieth century many were in favour of the general view that electromagnetism or electron theory was more profound and fundamental than mechanics. They adopted parts of the programme, and even more of them flirted with it by giving it rhetorical support. It was generally recognized that the programme was a promise for the future, and not a fully fledged physical theory, but this did not diminish its appeal. The electromagnetic view of nature was primarily supported by European physicists, and Germans in particular, whereas British physicists hesitated to eliminate mechanical concepts in favour of electromagnetic ones. The British ether, like the one favoured by Larmor and also by J. J. Thomson, was not as completely non-mechanical as the ether of German advocates of the electromagnetic programme.

It was a central aim of the German electron theorists to unify physics by bringing together the apparently disparate sciences of mechanics and electromagnetism. However, whereas the natural inclination since Maxwell had been to derive the laws of electromagnetism from a mechanical foundation, the new generation of physicists insisted on reversing the relationship and giving priority to electromagnetism. In his programmatic paper of 1900, Wien aimed to show that Newton's laws of motion were just special cases of the more general and fundamental laws governing the electromagnetic field. Nor was the law of gravitation to be left in peace. It, too, had to be understood on the basis of electromagnetism, although it was unclear how to do it. In his *Aether and Matter* of 1900, Larmor speculated that a non-linear modification of the Maxwell equations might lead to gravitation, but concluded that the idea was probably not fruitful. One possibility was to follow the approach of Lorentz, who in 1900 had derived, on the basis of his electron theory, a gravitational law that he considered a possible generalization of Newton's. Although there were several attempts of this kind, that is, to reduce gravitation to electromagnetism, none of them were regarded as satisfactory or widely accepted.

At a meeting of the German Association of Natural Scientists and Physicians in 1900, Wien addressed some of the larger questions connected with the new approach to fundamental physics:

> I asked myself if we could not do just with the apparent mass and leave out the inertial mass and replace it by the electromagnetically defined apparent mass to give a uniform representation of the mechanical and electromagnetic phenomena. ... I have tried to pose the

question of whether by starting from Maxwell's theory we could not attempt to encompass mechanics, too. This would provide the opportunity of founding mechanics on electromagnetism now that Lorentz has developed a conception of the law of gravitation according to which gravitation is said to be closely related to electrostatics.[18]

The following year Walter Kaufmann, a physicist at the University of Bonn, reviewed the state of and hopes for the new electron theory. In his view, it was the culmination of an older tradition in non-Maxwellian electrodynamics, building on electromagnetic forces acting directly and instantaneously over distance. In agreement with Wien, he focused on key problems such as the electromagnetic mass of the electron, the full reduction of mechanics to electromagnetism, the possibility of understanding molecular forces on the basis of electrodynamics, and the problem of establishing and verifying electron theories of gravitation. Instead of the 'sterile efforts' to reduce electrical to mechanical phenomena, he suggested following the reverse process. In addition, he speculated about atomic structure in a manner that was unusual among German electron theorists but strikingly similar to the approach of Larmor and J. J. Thomson in England.

According to Kaufmann, electrons were probably the sole constituents of atoms, the legendary primordial matter. If the atoms of the chemical elements consisted of stable configurations of electrons, transmutation of the elements might be possible and the periodic system might be explained on a scientific basis: 'Perhaps a mathematical treatment will one day succeed in presenting the relative frequency of the elements as a function of their atomic weights and perhaps also in solving many other of the puzzles of the periodic system of the elements.' Moreover, the power of electron theory was not limited to the microcosmos: 'If we cast a glance into the world of space we see many phenomena which await the application of the electron theory, notably the solar corona, the tails of comets, and the aurora.'[19] Characteristically, Kaufmann spoke only vaguely of these phenomena and their relationships to electron theory. He had no idea of how to apply the theory to either these or most of the other phenomena that were presumably manifestations of the behaviour of electrons.

The works of the electron theorists were highly mathematical and might seem remote from empirical reality, but they were not unconnected with experiments. Then, as later, physicists were keenly aware that a theory must somehow and at some stage relate to the world of experiment. For example, the electron theories of the early twentieth century resulted in several models for the constitution of the electron that could be and actually were tested experimentally. In a detailed study of the dynamics of the electron in 1903, Abraham developed a model that pictured the electron as a tiny rigid sphere with a uniform surface or volume charge distribution.[20] Only such a model, he argued, was fully based on the theory of electromagnetism. He found that the mass of the rigid electron increased with velocity according to a definite mathematical expression (see below). There were a few other models of the electron, of which the Lorentz model from 1904 was the most important. By that time the Dutch theorist was ready to accept one of the key doctrines of the electromagnetic world view, namely that the mass of the electron must be of electromagnetic origin. However, his model differed from Abraham's by being deformable, that is, the electron would contract in the direction of motion (relative to the world ether) and thus acquire an ellipsoidal shape instead of the spherical shape it had at rest. Another difference was that the mass of the Lorentz electron increased with velocity in a slightly different way to the rigid electron of Abraham.

At the time the relationship between the two rival models was hotly debated. Abraham, Kaufmann, Arnold Sommerfeld, and some other theorists argued that the Lorentz electron needed a non-electromagnetic force and that this went against the spirit of the electromagnetic view of nature. However, there was no agreement among the specialists. For example, Minkowski considered the Lorentz electron complied better with Maxwell's theory than the Abraham electron. All the same, by 1906 the general opinion among German electron theory specialists was in favour of the purer form of the electromagnetic world view advocated by Abraham. Among the supporters of the rigid electron was Sommerfeld, who, at a meeting of German scientists in 1906, made it clear that he considered the Lorentz (or Lorentz–Einstein) theory hopelessly conservative. As he saw it, it was an attempt to save from the revolutionary wave what little could be saved of the old mechanistic world view.

By 1904 the electromagnetic view of the world had taken off and emerged as a highly attractive substitute for the mechanical view that was widely seen as outdated, materialistic, and primitive. *Le roi est mort, vive le roi!* As another expression for this change in view one may turn to the addresses given by some of the leaders of physics at the Congress of Arts and Sciences held in St Louis in September 1904. The general message of many of the addresses was that physics was at a crossroads and that electron physics was on its way to establishing a new paradigm of understanding nature. In a sweeping survey of problems in mathematical physics, Henri Poincaré spoke eloquently of the 'general ruin of the principles' that characterized the period. Himself an important contributor to the dynamical theory of electrons, he was now willing to conclude that 'the mass of the electrons, or, at least, of the negative electrons, is of exclusively electrodynamic origin'. The address of another French physicist, Paul Langevin, was no less eloquent and no less in favour of the electromagnetic world view that was destined to open up a new era of theoretical physics:

> The electrical idea, the last discovered, appears to-day to dominate the whole, as the place of choice where the explorers feels he can found a city before advancing into new territories.... The actual tendency, of making the electromagnetic ideas to occupy the preponderating place, is justified, as I have sought to show, by the solidity of the double base on which rests the idea of the electron [the Maxwell–Lorentz equations and the empirical electron].... Although still very recent, the conceptions of which I have sought to give a collected idea are about to penetrate to the very heart of the entire physics, and to act as a fertile germ in order to crystallize around it, in a new order, facts very far removed from one another.... This idea has taken an immense development in the last few years,

which causes it to break the framework of the old physics to pieces, and to overturn the established order of ideas and laws in order to branch out again in an organization which one foresees to be simple, harmonious, and fruitful.[21]

Although Langevin was clearly enthusiastic with respect to the electromagnetic view of nature, he was not blind to its problems and limitations. He thought that molecular forces might one day be understood electromagnetically, but doubted that the same could be said of gravitation. Whatever these limitations, he was adamant that the supremacy of the laws and concepts of mechanics belonged to the past, and that the future would belong to the electromagnetic ether.

Aiming at establishing a unitary theory of all matter and forces, the methodology behind the electromagnetic research programme was markedly reductionistic, a general feature of theories of this type of higher speculation. The basis of the theory was the electrodynamics of Maxwell and Lorentz, possibly in some modified and generalized version. When this most ambitious programme was completed and turned into a genuine theory, nothing would be left unexplained—in principle, that is. In this sense, it was clearly meant as a theory of everything, no less than the earlier vortex theory was. Elementary particles, atomic and quantum phenomena, and even gravitation were held to be manifestations of that fundamental substratum of everything physical, the electromagnetic ether. The vision of the electron theorists was, in a sense, that they were approaching the end of physics as far as fundamental physics was concerned. They realized that they missed all the details, but thought that at least they had found the ultimate framework of a final theory.

The electron theories that were developed from about 1900 marked an increasing mathematization of physics and the introduction of a style in which mathematical arguments were given higher priority than earlier. The new style would soon lead to ideas of physics being essentially mathematical in nature, as we shall see below. Although many German physicists joined the mathematical trend, there were also those who opposed it. Paul Drude, a recognized expert in optics and electromagnetism at Leipzig University, disliked what he called the 'mathematical–philosophical direction' in theoretical physics. Adopting a phenomenalist view of science, he favoured the 'practical–physical' direction and cautioned that physicists should not place too much emphasis on mathematical methods and standards. In an address of 1894 he warned against the 'real danger in the application of mathematics, or, as I could better say, of rigid formalism'.[22] Although Drude's view was shared by some of his colleagues—including notables like Planck and Boltzmann—during the following decade mathematics became increasingly important in physics, suggesting to some that there was an intrinsic harmony between mathematics and the world studied by the physicists.

The majority view around 1905 was that the ether was an indispensable part of the new electron physics—indeed, another expression for the electromagnetic field. It was what was left after the removal of ponderable matter. One physicist, August Föppl of Munich, compared the possibility of space without ether to the contradictory notion of a forest without trees.[23] The ether survived the attack on the old physics, but it was a highly abstract ether, devoid of material attributes. Lorentz was led to consider the ether as nothing but the frame of reference in which absolute time was to be measured; being just a frame of reference, it was undetectable in principle. Many scientists spoke of the ether as equivalent to the vacuum, or sometimes space, as Max Planck did in 1909. However, contrary to most electron theorists, Planck favoured a dualistic theory, stressing the difference rather than the similarity between electrons and ether. The ether, he wrote in a letter, is 'completely different from the electrons'. Referring to the atomistic structure of matter and electricity, he pointed out that 'the ether is constituted continuously [and] in this respect the electrons, that is, the atoms of electricity, are therefore much closer to the ponderable matter than to the ether'.[24]

The dematerialized ether was far more popular on the continent than among British physicists, who were unwilling to accept a purely non-mechanical vacuum-like ether. In spite of his sympathy for the electromagnetic electron, J. J. Thomson's view of the ether differed considerably from that of the German electron theorists. Revealingly, in 1907 he described the relationship between ether and matter in terms that could have been taken from Tait and Stewart's *The Unseen Universe*, a work he was thoroughly acquainted with. 'We are then led to the conclusion that the invisible universe—the ether—to a large extent is the workshop of the material universe, and that the natural phenomena that we observe are pictures woven on the looms of this invisible universe.'[25]

Two years later, in his presidential address to the British Association for the Advancement of Science, Thomson reminded his audience that the ordinary matter of the chemists 'occupies, however, but an insignificant fraction of the universe, it forms but minute islands in the great ocean of the ether, the substance with which the whole universe is filled'. 'This etherial medium,' he continued, 'is not a fantastic creation of the speculative philosopher; it is as essential to us as the air we breathe.' Thomson's ether resided in an 'invisible universe' and worked as 'a bank in which we may deposit energy and withdraw it at our convenience'.[26] Joseph John Thomson, Nobel laureate and celebrated discoverer of the electron, lived to see the consequences of both relativity theory and the new theory of quantum mechanics. But he never lost his faith in the ether.

It should be noted that the electromagnetic view of nature, all-encompassing as it was, did not include a cosmology in the usual sense of the term. Like the vortex theory, it had nothing to say about the structure and development of the universe. Although mainly a theory of matter and electromagnetic phenomena, it was not isolated from astronomy and astrophysics, although these areas were seen as peripheral. It was sometimes suggested that the new electron theory might be applied to astrophysical problems, such as the nature of the solar corona or the mysterious green line in the spectrum of the aurora borealis. A few physicists used electrogravitational ideas to investigate astronomical phenomena. For example, the anomalous behaviour of Mercury's motion round the Sun, a well-known problem for Newtonian gravitation theory, was studied on the basis of electron theory by the German physicist Richard Gans. Although he obtained a value for the perihelion advance of the right order, neither he nor others concluded that the problem had been solved.

By the beginning of the second decade of the twentieth century it was generally recognized that the electromagnetic view of nature was probably not the answer to the puzzle of a new theory of gravitation. The long-sought solution only came with Einstein's general theory of relativity, an answer that was completely different from the ones associated with the electromagnetic world picture.

The electromagnetic or etherial world view, understood in a broad sense, was not limited to the community of physicists. It was used and misused by a variety of thinkers of nearly all philosophical inclinations, mostly by idealists and occultists but also by some materialists of the dialectical school. The French author, psychologist, and amateur physicist Gustave LeBon may represent the first group. In his very popular book of 1905, *The Evolution of Matter*, LeBon concluded that all matter is epihenomenal, slowly degrading into ether-like radiations. To him and many others, the ether represented 'the final nirvana to which all things return after a more or less ephemeral existence'.[27] As an example of the second group, consider the revolutionary communist Vladimir Lenin, who in his *Materialism and Empirio-Criticism* of 1908 discussed with sympathy and in some detail the new physical world view based on fields, ether and electrons. Far from admitting that the view supported idealism, he considered it solid support for the dialectical materialism that formed the philosophical basis of communism. 'The electron is as *inexhaustible* as the atom, nature is infinite, but it infinitely *exists*', he said.[28] To Lenin, the modern world view of physics was congruent with dialectical materialism and opposed to 'relativist agnosticism and idealism'.

3.3 PROBLEMS OF RELATIVITY AND QUANTA

The early years of the twentieth century have a special place in the history of science because of the emergence of two foundational theories that were to revolutionize physics: of quanta and relativity. However, as seen by theoretical physicists at the time, the ether theories of electrons were far more exciting and important. How did the eventually so-successful ideas of energy quantization and relative motion relate to the fashionable idea of an electromagnetic world?

Einstein's special theory of relativity appeared in 1905, in a paper in the *Annalen der Physik* entitled *Zur Elektrodynamik bewegter Körper* ('Electrodynamics of moving bodies'). The title may have added to the impression, widespread for several years, that the theory belonged to the tradition of Poincaré, Lorentz, and Abraham, although this was not the case at all. Einstein arrived at the same transformation formulae for length and time that Larmor had derived in 1900 and Lorentz in 1904, in both cases based on electron theory. Moreover, for the mass of the electron he obtained the very same velocity dependence that Lorentz had found in his theory of the deformable electron, namely

$$m(v) = \frac{m_0}{\sqrt{1-\beta^2}} \cong m_0\left(1 + \frac{1}{2}\beta^2 + \frac{3}{8}\beta^4\right)$$

where $\beta = v/c$. Nor were electron theorists shocked by Einstein's prediction of an equivalence between energy and mass (as given by $E = mc^2$), for an equivalence of this kind was more or less expected within the framework of the electromagnetic view of matter, as Wien had pointed out in 1900. The literature of physics includes half a dozen proposals of energy–mass relationships before the publication of Einstein's famous formula.[29] No wonder many readers of Einstein's paper mistakenly thought that it was one more contribution to electron theory, a new version of Lorentz's theory. However, not only had the electromagnetic structure of the electron no place in relativity theory at all, Einstein also dismissed the ether as superfluous. Nonetheless, the true nature of what soon became known as the theory of relativity—a name introduced by Planck in 1906—was not recognized immediately, and it was often assumed to be an improved version of Lorentz's electron theory. The name 'Lorentz–Einstein theory' was commonly used and can be found in the literature as late as the 1920s. Even Hermann Minkowski, who in 1907 offered an important reformulation of Einstein's theory in terms of space–time geometry, considered it to be within the framework of the electromagnetic world view. He still hoped for the completion of this world view and thought that the theory of relativity might be of value for the completion.

Until about 1909 the fictitious Lorentz–Einstein theory was more important than the real Einstein theory, in particular because it entered the debate over the electromagnetic structure of the electron through its predicted mass variation. As mentioned above, the main competitors were Abraham's rigid electron and Lorentz's theory of the deformable electron, which resulted in different mass variations at very high velocities. While Lorentz's expression was the same as Einstein's, the one given by Abraham can be written as

$$m(v) \cong m_0\left(1 + \frac{2}{5}\beta^2 + \frac{3}{2\times 5 \times 7}\beta^4\right)$$

A third possibility, proposed by Bucherer and Langevin, was a deformable electron of invariant volume. To settle the question of the electromagnetic mass, Kaufmann made a series of elaborate experiments from which he concluded that the entire mass of the electron was of electromagnetic origin, in agreement with Abraham's view. On the other hand, his experiments contradicted the Lorentz–Einstein theory, or so he concluded. However, the experiments and their interpretation were delicate, so although they were taken to support the Abraham theory, they did not prove it. More precise experiments made by Bucherer in 1908 gave the opposite result, a confirmation of the Einstein–Lorentz theory, and within a couple of years it was accepted that Abraham's rigid electron did not correspond to experimental reality (Fig. 3.1).[30] Because Abraham's model was the very embodiment of the electromagnetic world view, the result was considered a serious blow to the general view of the electromagnetic ether as the basis of all physics.

A well-defined theory can be falsified by an experiment, but a world view cannot. The electromagnetic view of nature was not shot down by experiments alone, although they did contribute to its decline. The electromagnetic research programme in its standard form was incompatible with the theory of relativity, and as Einstein's theory became understood and accepted, the electromagnetic programme lost a great deal of its credibility. Although not falsified in any direct sense, it came to be seen as irrelevant and unimportant, even a mistake.

The founding father of relativity, Einstein was also a pioneer of the other of the period's revolutionary theories, the theory of quanta. With its origin in Planck's explanation of the blackbody spectrum in 1900, quantum theory was initially seen as a theory of electromagnetic radiation and more or less restricted to this area. For the first couple of years after 1900, the postulate of energy quantization was not given serious attention and Planck's radiation law was not seen as implying a break with classical physics, not even by Planck himself. What mattered was that the law agreed beautifully with experiments. At about the time that Lorentz completed his electron

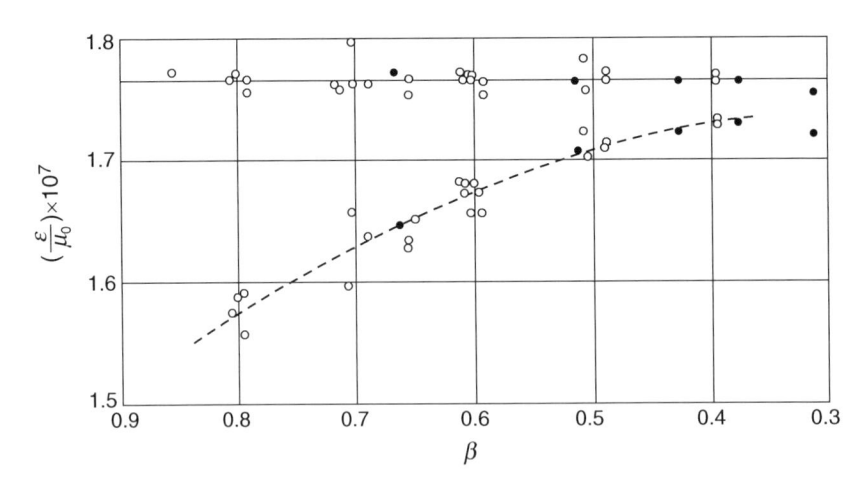

Fig. 3.1. According to Abraham's theory, which many German physicists saw as representing the electromagnetic world view, the mass of a rapidly moving electron should increase with the speed in a way different from the variation prescribed by Einstein's theory of relativity. Both theories assumed that the quantity e/m_0 is constant. In this graph from 1916, experimental values are reduced to e/m_0 as calculated from the two theories and plotted against $\beta = v/c$. The data confirm the Einstein–Lorentz formula while they disagree with Abraham's formula.

Source: Clemens Schaefer, 'Die träge Masse schnell bewegter Elektronen,' *Annalen der Physik* **49** (1916), 934–938.

theory, he began to study Planck's theory of blackbody radiation in the hope that he could derive it on the basis of electron theory, thereby avoiding the unsatisfactory and unclear elements in Planck's derivation.[31] To his surprise and dismay, he could not. Basing his reasoning on ideas from the electron theory of metals, he ended up with what is generally known as the Rayleigh–Jeans law, so named after Lord Rayleigh and James Jeans. According to this law, which reproduces the blackbody spectrum only for long wavelengths, the energy density of electromagnetic radiation varies as the square of the frequency: $\rho \sim \nu^2 T$. The result was theoretically satisfactory, but unfortunately it did not match experiments.

Recognizing the problem, in the years after 1903 Lorentz called attention to the apparent incompatibility of quantum ideas with the electron theory in its classical form. In a lecture of 1908 he sharpened his earlier result, now proving unambiguously that existing electron theory must lead to the Rayleigh–Jeans law or what at the time was sometimes called the Jeans–Lorentz law. The electromagnetic world view appeared to be incompatible with the experimentally confirmed Planck equation. As he formulated it in his *Theory of Electrons*, 'it will be very difficult to arrive at a formula different from that of Rayleigh so long as we adhere to the general principles of the theory of electrons'.[32] Still, he and some other electron physicists, including Sommerfeld, nourished the vague hope that a more complete electromagnetic theory might result in the correct equation. Thus, Wien was troubled over Lorentz's result, which showed that 'Maxwell's theory must be abandoned for the atom'. But he also saw the result as an incitement 'to find an extension of Maxwell's equations within the atom'.[33]

In spite of the vague attempts to keep the electromagnetic programme alive in the case of radiation physics, Lorentz's insight had the effect that confidence in the programme declined. At about the same time the epistemic strength and novelty of quantum theory became clearer, especially after Einstein had shown how the photo-electric effect and the specific heats of solids could be explained on the basis of the theory. It turned out that quantum theory was much more than a theory of heat radiation. With the increasing recognition of the fundamental nature of quantum theory, its incompatibility with the electromagnetic world view became more and more obvious.

The decline of the electromagnetic research programme was a complex process with both scientific causes and causes related to changes in the period's cultural climate. The conflicts with the theory of relativity and quantum theory were the most important of the scientific causes. In neither case was the programme clearly refuted, but it lost its drive and its relevance to problems of mainstream physics. In general, electron theory had to compete with other developments in physics that did not depend on this theory, and after 1910 new developments attracted interest away from the once-so-exciting theory of ether and electrons. So many new and interesting events occurred, so many discoveries were made, so why bother with the complicated and over-ambitious attempt to found all of physics on electromagnetic fields? By 1911, at the time of the first Solvay meeting in physics, few physicists felt a need for the electromagnetic view of nature, which less than a decade previously had been heralded as a revolution in physical thought. About the worst thing that can happen to a proclaimed revolution is that no one cares about it.

3.4 MIE'S FIELD THEORY OF MATTER

Although the theory of relativity was a major reason for the decline of the electromagnetic world view, it was possible to keep to the essence of this view and, at the same time, make use of Einstein's special theory of relativity. This was one way of keeping the dream of an electromagnetic universe alive, as did the German physicist Gustav Mie, professor at the University of Greifswald. Mie published his theory of matter and gravitation in a series of lengthy papers in the *Annalen der Physik* of 1912 and 1913.[34] To the extent that Mie is remembered today it is not for his grandiose but failed unified theory, but for an important work of 1908 in which he calculated from electrodynamics the scattering of light on spherical bodies. (The 'Mie effect' continues to be important in many areas of physics.) Mie was the most productive and influential of the contributors to the late phase of the electromagnetic programme. His *Grundlagen einer Theorie der Materie* ('Foundations of a theory of matter'), as the common title of his sequel of papers read, was based on three assumptions:

(1) Electrons are structures in the electromagnetic ether and inseparable from it.
(2) The principle of relativity holds universally.
(3) All physical phenomena can be characterized by the electromagnetic quantities associated with the ether.

As to the last assumption, he formulated it as follows: 'The presently known states of the ether...suffice completely to describe all phenomena of the material world.'[35]

In spite of adopting Einstein's principle of relativity without qualification, Mie was no less an adherent of the electromagnetic world view than were other physicists of earlier theories such as Abraham and Wien. In a popular book of 1907 he described in qualitative terms the essence of his theory of the world, leaving no doubt that it belonged to the tradition of the electromagnetic world view:

> Elementary material particles...are simply singular places in the ether at which lines of electric stress of the ether converge; briefly, they are 'knots' of the electric field in the ether. It is very noteworthy that these knots are always confined within close limits, namely, at places filled with elementary particles.... The entire diversity of the sensible world, at first glance only a brightly coloured and disordered show, evidently reduces to processes that take place in a single world substance—the ether. And the processes themselves, for all their incredible complexity, satisfy a harmonious system of a few simple and mathematically transparent laws.[36]

Mie believed that the electron, either positively or negatively charged, was a tiny portion of the ether in 'a particularly singular state'. He pictured it as consisting of 'a core that goes over continuously into an atmosphere of electric charge which extends to infinity', adding that outside the core the atmosphere was so rarefied 'that it cannot be detected experimentally in any way'.[37] That is, strictly speaking, his electron did not have a definite radius, but extended throughout the universe. Near the centre of the electron, inside its core, the strength of the electromagnetic fields would be enormous and Mie argued that under such circumstances the Maxwell equations would no longer be valid. He thus adopted the possibility that Wien had indicated in his letter to Sommerfeld, namely that the electrodynamical equations had to be modified at very small distances. This he did by developing a set of generalized non-linear equations that, at relatively large distances from the electron's core, corresponded to the ordinary Maxwell equations. The new equations provided an attractive force of electrical nature that held the etherial structure together, and which was not a mechanical force as in Lorentz's old theory.

Being a unificationist, Mie wanted to include gravitation in his electromagnetic theory, or rather to derive gravitation from his electromagnetic equations. His approach to the problem of gravitation differed drastically from Einstein's, among other reasons because he did not accept the equivalence principle on which Einstein based his theory. At any rate, Mie was soon forced to realize that gravitation could not be derived from his equations. It had to be introduced by means of special quantities and assumptions, and his electrogravitational theory was therefore no more satisfactory than the earlier theories of Lorentz, Wien, Gans, and others. To his dismay, he had to admit that 'gravity . . . shows itself as obstinate as ever'.[38] Nor was he able to make sense of the quantum of action, although he tried to. For example, he noted (as Einstein had done earlier) that the square root of Planck's constant was 'analogous' to the elementary charge—meaning that e^2 was of the same dimension and roughly the same order as hc—from which he suggested an interpretation of Planck's constant in terms of electric lines of force. The idea proved to be of no scientific value.[39]

The strength of Mie's theory was its scope, ontological parsimony, and advanced mathematical structure, not its physical content. It did result in a few predictions, but these were of such a kind that they could not be tested in realistic experiments. For example, Mie found that the ratio between the gravitational and inertial mass would decrease with the internal motion of the particles making up a body. This implied that the ratio, and therefore also the gravitational constant, would decrease with the temperature and also with the average atomic weight of the body. However, Mie estimated that the effect would only be measurable in experiments that could determine the acceleration of gravity to an unrealistic accuracy. He consequently admitted that the prediction, although theoretically interesting, was 'experimentally useless'. He further deduced that an oscillating particle would not only emit transversal electromagnetic waves but also longitudinal gravitational waves in the ether. Alas, since he calculated the intensity ratio to be about $1:10^{-43}$ he again had to conclude that the prediction could not be tested in any real sense.

From Mie's fundamental equations it was possible to calculate the charges and masses of the elementary particles as expressed by a certain 'world function' (Hamiltonian), which was a notable advance from a formal point of view. The advance was limited to the mathematical programme, however, and the grandiose theory was conspicuously barren when it came to real physics. As Einstein pointedly said about Mie's theory, 'it is a fine frame, but one cannot see how it can be filled'.[40]

By 1913, two elementary particles were known, the electron and the 'positive electron' (or proton, as it became known after 1920), and their properties could in principle be derived from the theory. Alas, this was in principle only, for the form of the world function that entered the equations was unknown. Had the values of the charges and masses of the two elementary particles been different, the situation would have been the same. On the other hand, neither could it be proved that there were no world functions compatible with the existence of electrons and protons, and so believers in Mie's programme could argue that future developments might eventually lead to success. At any rate, it was the spirit and aims of the theory, rather than its details, that appealed to some mathematical physicists. As Hermann Weyl expressed it in 1919: 'These [Mie's] laws of nature, then, enable us to calculate the mass and charge of the electrons, and the atomic weights and atomic charges of the individual elements whereas, hitherto, we have always accepted these ultimate constituents of nature as things given with their numerical properties. All this, of course, is merely a suggested plan of action as long as the world-function L is not known.'[41]

The aim of a fundamental and unified theory is to understand the richness and diversity of the world in terms of a single theoretical scheme. The mass and charge of the electron, for example, are usually considered to be contingent properties, that is, quantities that just happen to be what they are ('things given'). They do not follow uniquely from any law of physics and for this reason they could presumably be different from what they are. According to the view of the unificationists, the mass and charge of the electron, and generally the properties of all elementary particles and forces, must ultimately follow from theory—they must be turned from contingent into law-governed quantities. Not only that, but the number and kinds of elementary particles must follow from theory too; not only those particles that happen to be known at any given time, but also particles not yet discovered. This is a formidable task, especially because physical theories cannot avoid relying on what is known empirically and thus must reflect the state of art of the experimental and observational sciences. In 1913 the electron and the proton were known, and thus Mie and his contemporaries designed their unified theories in accordance with the existence of these particles and also with the known forces of electromagnetism and gravitation. But the impressive theories of the electromagnetic programme had no real predictive power.

For all its grandeur and impressive mathematical machinery, Mie's theory was a child of its age and totally unprepared for the avalanche of particle discoveries that occurred in the 1930s and later. Nor could it anticipate the existence of new forces of nature, the strong and weak interactions associated with the atomic nucleus. When Mie died in 1957, the world of particles and fields was radically different from what it had been in 1912. It was much more complex and much less inviting to the kind of grand unified theories that he and others had pioneered in the early part of the century.

3.5 PHYSICAL REALITY AND MATHEMATICAL HARMONY

Mie's theory was a failure, but not unimportant or without impact on other attempts to establish unified theories, which flourished in the wake of Einstein's general theory of relativity. It was valued by several leading theorists, including Hilbert, Pauli, Sommerfeld, Weyl, and Max Born. In his celebrated 1921 textbook on the theory of relativity, Max von Laue found it appropriate to include a detailed review of Mie's theory.[42] Weyl similarly dealt extensively with the theory in his important monograph *Raum-Zeit-Materie* (*Space-time-matter*), the first edition of which appeared in 1918. Although Weyl did not accept Mie's theory, he shared the ambitions and epistemic optimism of his colleague from Greifswald. 'I am bold enough to believe', Weyl wrote in 1918, 'that the whole of physical phenomena may be derived from one single universal world-law of greatest mathematical simplicity.'[43] The attitude was shared by the great Göttingen mathematician David Hilbert, who in 1915 admitted inspiration from 'the far reaching and original conceptions by means of which Mie constructs his electrodynamics'.[44]

As early as 1900, in a famous address in Paris, Hilbert had mentioned fundamental physics as one of the unsolved mathematical problems that needed the attention of mathematicians. About a decade later he started giving lecture courses on topics such as kinetic gas theory and radiation theory, areas of physics that he intended to provide with a logical and rigorous foundation. More generally he expressed his belief in a complete mathematization of physics, that is, that physics would eventually be turned into a branch of pure mathematics. As a result of the works of Mie, Einstein, and others his own engagement in theoretical physics greatly increased. Deeply immersed in foundational problems of physics, and especially in the theory of gravitation, Hilbert actually presented the basic equations of general relativity a few days before Einstein.[45] He ended his important address of 20 November 1915 on the foundations of physics by stating that he was convinced that

> ... through the basic equations presented here the most intimate, presently hidden processes in the interior of the atom will receive an explanation, and in particular that generally a reduction of all physical constants to mathematical constants must be possible—even as in the overall view thereby the possibility approaches that physics in principle becomes a science of the type of geometry: surely the highest glory of the axiomatic method, which as

we have seen takes the powerful instruments of analysis, namely variational calculus and theory of invariants, into its service.[46]

Whatever the role of Hilbert in the construction of the equations of general relativity, he was deeply involved in the process and the subsequent clarification of the theory. Shortly after his breakthrough to a covariant theory of gravitation, Einstein wrote in a letter that 'only *one* of my colleagues has understood it'.[47] That colleague was Hilbert.

I shall not deal with the complex story of unified field theories based on general relativity that occupied a good many physicists and mathematicians in the years after 1916. Suffice it to say that Einstein's theory established a whole new framework for theories that attempted to unify gravitation and electromagnetism and might possibly also include atomic and quantum phenomena.[48] Such unified theories based on the general theory of relativity were very popular in the 1920s, when they were examined in about 400 papers. Nearly one third of all these papers on relativity theory dealt with aspects of unified field theories. Hilbert was among the first and most serious of the relativist unificationists. In 1924 he published a major paper on the subject, characteristically in a mathematics and not a physics journal.

The character of Hilbert's ambitious and axiomatically structured theory may be illustrated by a series of lectures on the foundational problems of physics he gave in Hamburg the previous year. Here he suggested the possibility of an all-embracing physics based on what he called 'world equations'. These equations, he argued, would allow the deduction of all known (and as yet unknown) experimental facts without the necessity of stipulating contingent boundary or initial conditions. Among the examples he discussed was Bohr's recent atomic theory and its extension to cover the atomic structure of all the elements of the periodic system. Hilbert apparently misunderstood Bohr's theory of complex atoms as being fundamental and complete, not recognizing its semiclassical and eclectic nature. He thought it was possible 'to deduce the intimate properties of matter, even the characteristic particulars of the chemical elements, from the laws of field and motion alone, and to recognize them as necessary thought consequences [*denknotwendige Folgerungen*]'.[49]

Not only would the microworld be fully explained by the laws of mathematical physics, so, ultimately, would the macroworld: 'Also the existence and continual evolution of the stars seem to be, in the last resort, a consequence of the world equations.' Although he did not mention cosmology as a field of application, Hilbert seriously believed that his world equations amounted to a theory of everything. As he said to his audience in Hamburg:

> If these world equations, and with them the framework of concepts, is now complete, and we know that it fits in its totality with reality, then in fact one needs only thinking and conceptual deduction in order to acquire all physical knowledge; in this respect Hegel was right to claim that all natural events can be deduced from concepts.

This sounds much like *a priori* physics, but Hilbert maintained that the validity of the world equations and the logical structure of the theory had to depend on observed facts. He explicitly denied that the world equations were of an *a priori* nature.

The ultimate aim of the ambitious programme of Hilbert and some other German mathematicians and physicists was to base all of physics deductively on a fundamental law that had the character of an axiom. From this single law of nature all physical phenomena were supposed to flow, much like geometric theorems flow from the basic axioms and rules of inference. It was a programme that for a period appealed to scientists and non-scientists alike. In 1922 a German author described the ultimate theory that was under construction by his compatriots:

> Physics, as the recent attempts of Mie, Hilbert, Weyl and others seem to indicate, is on the way to establishing certain all embracing, mathematically invariant fundamental laws, a kind of world formula, from which, with logical consequence, the whole edifice of the world can be theoretically constructed—from the atom or even the nuclei up to the system of the Milky Way, in all their details. It goes without saying that this sounds like faraway music of spheres; but in principle the possibility exists that the world can be grasped theoretically, exactly as in geometry.[50]

These were grand words, but they did not correspond to reality any more closely than did the equally grand formulations of earlier scientists such as Hicks (1895) and Mie (1907). Yet, formulations of the same kind would soon be proposed by Eddington and later by several other theoretical physicists.

Advanced mathematics was very much part of the attempts in the early twentieth century to formulate a fundamental and unified framework for all of physics, as exemplified by the theories of Mie and Hilbert. German scientists were particularly receptive to the old doctrine of a 'pre-established harmony' between mathematics and physics, an idea traditionally traced back to Pythagoras and Plato but made explicit by Leibniz in particular.[51] According to the modern formulation of Leibniz's idea, mathematical forms and symmetries are the true guides to discovering nature's laws. It was a doctrine accepted by most of the leading German mathematicians and theoretical physicists, although they subscribed to it in somewhat different versions.

The Göttingen mathematician Minkowski, a colleague of Hilbert, was convinced that mathematics was the royal road to progress in physics. In a lecture course given in 1904, he asserted that 'through a peculiar, pre-established harmony, it has been shown that, by trying logically to elaborate the existing edifice of mathematics, one is directed on exactly the same path as by having responded to questions arising from the facts of physics and astronomy'.[52] Minkowski's encounter with Einstein's theory of relativity only confirmed his belief that what physicists do is to verify the actual existence of truths that mathematicians have already established. He spelled out this somewhat arrogant belief in a lecture of 1907:

> To some extent, the physicist needs to invent these concepts [of the theory of relativity] from scratch and must laboriously carve a path through a primeval forest of obscurity; at

the same time the mathematician travels nearby on an excellently designed road.... It will become apparent, to the glory of the mathematicians and to the boundless astonishment of the rest of mankind, that the mathematicians have created purely within their imagination a grand domain that should have arrived at a real and most perfect existence, and this without any such intention on their part.[53]

Although Minkowski thus argued that the creative power of physics resided in mathematics, he conceded that the final arbiter of physical theory, conceived as a theory of natural phenomena, must be experiment.

Einstein, who in his youth had been leaning towards a version of Machian empiricism, came to believe in a pre-established harmony between physics and mathematics that was close to the one espoused by Minkowski. He had not always held such a view, but his work with the general theory of relativity convinced him of the crucial role of mathematics in the construction of physical theories. As he wrote to Sommerfeld in 1912: 'Never before in my life have I toiled any where near as much, and I have gained enormous respect for mathematics, whose more subtle parts I considered until now, in my ignorance, as pure luxury.'[54] Yet, it spite of this enormous respect, Einstein was at the time fully aware that a theory's mathematical qualities do not automatically translate into physical validity. The same year he commented on a new four-dimensional theory of gravitation proposed by Abraham, saying that it was 'created out of thin air, i.e. out of nothing but considerations of mathematical beauty, and is completely untenable'.[55] Mathematical beauty was one thing, physical truth might be quite another thing.

In a lecture before the Prussian Academy of Sciences of 1921, Einstein reflected on the question of how 'mathematics, being after all a product of human thought, which is independent of experience, is so admirably appropriate to the objects of reality?' Einstein was impressed by the power and certainty of mathematics, but he did not conclude that nature was mathematical in any fundamental sense: 'As far as the propositions of mathematics refer to reality, they are not certain; and as far as they are certain, they do not refer to reality.'[56] At several later occasions he explained what by then had become his rationalist credo, namely that the basic structure of physical theory must be constructed from mathematical ideas without considerations of experienced nature. This was an important message of the Herbert Spencer Lecture that he delivered in Oxford in 1933. Given that some physical theories, such as classical mechanics, agree very well with experience, and yet do not get to 'the root of matter', can experience ever show the way to truth? Not according to Einstein, but there was another way, a 'right way':

> Our experience hitherto justifies us in believing that nature is the realization of the simplest conceivable mathematical ideas. I am convinced that we can discover by means of pure mathematical constructions the concepts and the laws connecting them with each other, which furnish the key to the understanding of natural phenomena. Experience may suggest the appropriate mathematical concepts, but they most certainly cannot be deduced from it. Experience remains, of course, the sole criterion of the physical utility of a mathematical

construction. But the creative principle resides in mathematics. In a certain sense, therefore I hold it true that pure thought can grasp reality, as the ancients dreamed.[57]

This and similar statements reflect a rationalist attitude to the methods of science and the construction of scientific theories. In a letter to the Hungarian–German physicist Cornelius Lanczos he said: 'The problem of gravitation converted me into a believing rationalist, that is, into someone who searches for the only reliable source of truth in mathematical simplicity.'[58] But Einstein was never a rationalist in the stronger sense that mathematical considerations should also decide the validity of physical theories. When it came to this crucial issue, he never wavered from his belief that 'Experience remains, of course, the sole criterion of the physical utility....'

In the early attempts to create a unified theory of fundamental physics, quantum theory was the weak point. It was generally agreed that, somehow, the quantum of action had to be incorporated together with relativity, gravitation and electromagnetism, but attempts to do so were few, half-hearted, and unsuccessful. Most unificationists preferred to ignore the troublesome quanta. Just at the time when Einstein presented his covariant theory of general relativity in Berlin, Sommerfeld in Munich completed a remarkable synthesis of Bohr's quantum theory of atoms and the *special* theory of relativity. By introducing the Lorentz–Einstein $m(v)$ expression into Bohr's atomic theory, he found after lengthy calculations an expression for the energy levels of the hydrogen atom that agreed beautifully with spectroscopic experiments performed by his colleague Friedrich Paschen at the University of Tübingen.[59] First and foremost, Sommerfeld's *Zauberformel* (magic formula) explained the fine structure of the hydrogen lines, a phenomenon that had eluded Bohr, as a relativistic effect. It also introduced a new dimensionless quantity in physics, the fine-structure constant $2\pi e^2/hc \cong 1/137$, which eventually would come to occupy a central position in fundamental physics and to which I shall return in some of the later chapters.

Einstein understood immediately the significance of Sommerfeld's work. 'Your investigations of the spectra belongs to my most beautiful experiences in physics', he wrote to his colleague in Munich. 'Only through it do Bohr's ideas become completely convincing. If I only knew what little bolts the Lord had used for it!'[60] However, neither Einstein nor other unificationists at the time found the fine structure to be relevant for what they were aiming at. (Mie did not refer to Sommerfeld's work, and nor did Hilbert.) To their minds, the synthesis of special relativity and the quantum atom was much too limited in scope.

Notes for Chapter 3

1. Rutherford 1914, p. 337.
2. For accounts of the system of energetics and its historical development, see Leegwater 1986, Deltete 2005, and Jungnickel and McCormmach 1986, pp. 217–27.

3. Planck 1896, p. 77.

4. Energetics and monism as attempts to create a new scientific world view are dealt with in Görs, Psarros and Ziche 2005.

5. Larmor 1927, p. 475. On the history of the electron, see e.g. Arabatzis 2006 and the contributions in Buchwald and Warwick 2001.

6. Larmor 1900, p. 78.

7. The electron theories of Larmor and Lorentz are analyzed in Darrigol 1994. Larmor noticed that the ratio between the 'orbital velocities of electrons' and the velocity of the radiation emitted was of the same order as the ratio between atomic dimensions and the wavelength of light: $v/c \sim a/\lambda \sim 10^{-3}$ (Larmor 1900, p. 233). See also Whyte 1960, who speculates that this 'corresponds' to the later fine-structure constant.

8. For the role of the Zeeman effect in the process that led to the discovery of the electron, see Robotti and Pastorino 1998.

9. Letter to Oliver Lodge, 8 May 1897, quoted in Arabatzis 2006, p. 93.

10. FitzGerald 1897, p. 104.

11. Letter to Oliver Lodge of 11 April 1904, quoted in Kragh 2001b, p. 207.

12. Dirac 1930, p. 605. See also Chapter 7.

13. Jeans 1901, p. 426.

14. Ibid., p. 454. During the first decade of the twentieth century a few physicists believed to have discovered positively charged electrons, mirror particles of the ordinary electron. The French physicist Jean Becquerel (a son of Henri Becquerel of radioactivity fame) claimed to have identified the particles in magneto-optical experiments and also in discharge tubes, but his discovery claim failed to win recognition among contemporary physicists. On this episode in the history of physics, see Kragh 1989b.

15. Heaviside 1970, p. 528.

16. Wiechert 1896, p. 3. The title of Dyson 2004 (first published 1988) is taken from Wiechert's paper. I have used the translation that appears on p. 36 in Dyson's book. On Wiechert and his early work on electron theory, see Mulligan 2001.

17. On the emergence and development of the electromagnetic world view, see McCormmach 1970, Jungnickel and McCormmach 1986, pp. 227–44, and Hirosige 1966. See also Kragh 1999a, pp. 105–19.

18. Quoted in Jungnickel and McCormmach 1986, p. 238.

19. Kaufmann 1901, p. 97.

20. For a detailed analysis of Abraham's theory, see Goldberg 1970.

21. Quoted in Sopka and Moyer 1986, p. 292 and p. 230.

22. Quoted in Pyenson 1982, p. 140. On the pre-established harmony, see below.

23. Illy 1981a, p. 182.

24. Letter to Friedrich Kuntze, quoted in Kragh 2001b, p. 211.

25. Thomson 1908, p. 550.

26. Thomson 1909, p. 15 and p. 20. Thomson's picture of the energy-rich ether filling all of space may remind a modern reader of the dark energy of twenty-first century cosmology.

27. LeBon 1905, p. 315.

28. Quoted from the online edition of *Materialism and Empirio-Criticism* available on http://www.marxists.org/archive/lenin/works/1908/mec/index.htm. Precisely what Lenin meant is unclear, but somehow the statement became famous and in some circles seen as proof of his deep

insight into the structure of the world. But it hardly went deeper than that of LeBon, who suggested that the electron is 'itself of a structure as complicated as that now attributed to the atom, and may...form a veritable planetary system. In the infinity of worlds, magnitude and minuteness have only a relative value' (LeBon 1905, p. 227). There is some similarity between Lenin's statement and the view of Wiechert quoted above, but I consider it unlikely that Lenin received inspiration from Wiechert's paper. On Lenin and the electromagnetic world view, see also Illy 1981b.

29. See the list of energy-mass relationships 1885–1914 in Fadner 1988. The mass–energy relations of the electromagnetic theories were much more restricted than the one proposed by Einstein. According to Einstein, *any* kind of energy change would result in an equivalent change in mass and vice versa.

30. For a detailed account of the origin and early development of the special theory of relativity, including the experiments of Kaufmann, Bucherer and others, see Miller 1981.

31. On the relationship between the electromagnetic world view and early quantum theory, see Seth 2004 and McCormmach 1970, pp. 485–88.

32. Lorentz 1952, p. 287. Originally published 1909, the book was based on a series of lectures given at Columbia University in 1906.

33. Letter to Sommerfeld of 15 June 1908, quoted in Seth 2004, p. 81.

34. On Mie's theory, see Vizgin 1994, pp. 26–38, Corry 1999, and Smeenk and Martin 2007. Excerpts of Mie's papers in the *Annalen* are translated into English in Renn 2007, vol. 4, pp. 633–97.

35. Mie 2007, p. 634.

36. Mie 1907, translations from Vizgin 1994, p. 18 and p. 27.

37. Mie 1912, p. 512.

38. Mie 2007, p. 696.

39. In a letter of 1906 Planck had pointed out that h has nearly the same order of magnitude as e^2/c, and Einstein did the same in a paper of 1909. In retrospect, they referred to the value of the dimensionless fine-structure constant. See Kragh 2003, pp. 397–98.

40. Einstein to Weyl, 6 June 1922, as quoted in Vizgin 1994, p. 37.

41. Weyl 1922, p. 214.

42. Laue 1921, pp. 239–46.

43. Weyl 1918, p. 385.

44. Hilbert's paper on 'The Foundations of Physics' is translated in Renn 2007, vol. 4, pp. 1003–17. On Hilbert's works on the foundations of physics, and his inspiration from Mie, see Corry 1999 and Corry 2004.

45. This does not mean that the priority of general relativity belongs to Hilbert rather than Einstein. The question of priority is complicated, but has been clarified by historians of science. For an accessible account of the complex relationship between the works of Hilbert and Einstein, see Rowe 2001.

46. Renn 2007, vol. 4, p. 1015.

47. Einstein to Heinrich Zangger, 26 November 1915, in Einstein 1998, p. 205.

48. Much has been written about this class of theories. A helpful overview is presented in Goenner 2004.

49. From Hilbert's manuscript of his 1923 lectures, as reproduced in Sauer and Majer 2009, on p. 414. The two following quotations are from the same source, p. 415 and p. 423.

50. Paul Gruner, *Das moderne physikalische Weltbild und der christliche Glaube* (Berlin: Furche, 1922), slightly modified from quotation in Goenner 2001, p. 5.

51. On the importance of the idea to German mathematicians and physicists in the early part of the twentieth century, see Pyenson 1982.

52. Quoted in Corry 2004, p. 186.

53. Minkowski 1915, pp. 927–28. Minkowski died in 1909. The article, based on the lecture notes he left at his death, was published by Sommerfeld.

54. Letter of 29 October 1912, in Einstein 1993, p. 505.

55. Letter to Michele Besso of 26 March 1912, in Einstein 1993, p. 436.

56. Einstein 1982, p. 233. On Einstein's changing attitudes to the epistemic power of mathematics, see Norton 2000.

57. Einstein 1982, p. 274.

58. Letter of 24 January 1938, quoted in Jammer 1999, p. 40.

59. The authentic history of the relationship between theory and experiment was more complex. For this history and references to the case of the fine structure of the spectra of hydrogen and helium, see Kragh 1985. Sommerfeld's celebrated fine-structure formula was pre-quantum mechanical and did not make use of the electron's spin, a concept only introduced several years later. Yet it was identical to the later formula derived from Dirac's relativistic spin quantum mechanics. An attempt to solve this 'Sommerfeld puzzle' is given in Biedenharn 1983.

60. Einstein to Sommerfeld, 3 August 1916, in Einstein 1998, p. 326.

4

Rationalist Cosmologies

He thought he saw electrons swift
Their charge and mass combine.
He looked again and saw it was
The cosmic sounding line.
The population then, said he,
Must be 10^{79}.

Herbert Dingle, in Eddington, *The Expanding Universe*, 1933, p. 113.

The unified theories of physics considered in the two previous chapters were primarily theories of matter and fields as known from terrestrial experiments. Although they were also assumed to be valid beyond the Earth, they did not include a proper cosmological perspective. In the 1930s this situation changed with the emergence of a new class of physical ideas that comprised cosmology, indeed that were principally of a cosmological nature. These ideas were mostly discussed in the United Kingdom, just as the earlier ideas of a universe governed by the laws of electromagnetism were mostly discussed in Germany. One can identify in British theoretical physics and cosmology in the period from about 1930 to the late 1940s a strong rationalistic trend opposed to the empirical attitude that is traditionally associated with British scientists and philosophers. Remarkably, only a few traces of this trend can be found in other countries. Equally remarkably, a somewhat similar trend can be identified in the United Kingdom in the 1950s in connection with the steady-state system and again in the 1970s with the introduction of the anthropic principle and ideas related to it.

The two leading system builders in the interwar period were Arthur Eddington and Edward Milne, whose theories will be reviewed below. They were to some extent followed by other scientists who adopted parts of their research programmes or at least were sympathetic to and inspired by them. This group includes Paul Dirac, Arthur G. Walker, George J. Temple, and Gerald J. Whitrow. The physical systems raised by Milne and Eddington were different, but also shared many features with regard to aims and methodology.[1] Thus, their theories were mainly about metrical science, meaning those parts of science that are intimately connected with measurements. Both argued for deductivism, a sort of synoptic thinking based on *a priori* principles from which the laws of nature were deducible by pure reason. It followed that experimental tests were assigned a much more modest role than in ordinary physics, essentially being subordinated mathematical and logical arguments.

What I call *cosmophysics* can to some extent be seen as a continuation of an anti-empirical and speculative tradition in British intellectual life, represented by philosophers such as John McTaggart, Robin G. Collingwood, and Alfred North Whitehead. Although these philosophers did not propose cosmologies in the scientific sense, they did construct cosmological world views and their ideas had a general affinity with the ambitions of the British cosmophysicists. To characterize the systems of Milne and Eddington as physical theories is really to underrate their scope and breadth. They were more like *Weltanschauungen*: grand and enormously ambitious projects aimed at a full reconstruction of physics modelled on abstract mathematics. Still they were not total cosmologies in the even broader sense of Whitehead, whose main work, *Process and Reality*, offered a metaphysical world system, a unified picture of all of reality, rather than a scientific theory of the observed or physical universe. Generally speaking, when cosmologists speak of the universe they study as comprising everything, they mean everything of a physical nature, which from a philosophical point of view may only be a small slice of reality.

4.1 PHYSICS MEETS COSMOLOGY

If a fundamental theory of physics is to be truly universal, it must not only cover terrestrial matter and forces but also what goes on in the heavens; it must not only deal with what is *in* the universe, but also with the structure and evolution of the universe itself. Einstein's general theory of relativity ushered in a new and fruitful era in unified field physics and in addition it served as the foundation of a whole new understanding of the universe at large. For the first time in the history of human thought it became possible to deal with the universe in precise and well-defined scientific terms. It is no exaggeration to claim that the theory of relativity gave a new meaning to the concept of the universe. Of course, it was not the first time that scientists and philosophers had tried to understand the cosmos by applying the laws of physics to it. Newton's law of gravitation was supposed to have unlimited validity and hence to be applicable to the universe at large. The same was the case with the two principal laws of thermodynamics: energy conservation and the law of entropy increase. In the nineteenth century both gravitation theory and thermodynamics were applied in a cosmological context, but without yielding results that were physically convincing or could be compared with observations in any direct way.[2]

Back in 1785, the great astronomer William Herschel had spoken of the enigmatic nebular clusters as 'the laboratories of the universe', and with the rise of astrophysics in the nineteenth century it became common to look towards the stars for an extension of the empirical base of physics. After all, can we be sure that the matter and forces known from Earth are the only ones in the universe? There was no compelling reason for accepting the assumption of uniformity on a cosmological scale. Speculative astrophysics and astrochemistry led to hypotheses concerning the evolution of matter and the existence of non-terrestrial chemical elements, in the stars, the atmosphere of the Sun, the distant nebulae, the aurora borealis, or in the earlier history of the universe. For example, in the Silliman Lectures of 1903, J. J. Thomson speculated that atoms much lighter than hydrogen had existed in the cosmic past, but that they had now formed aggregations, the smallest of which happened to be hydrogen. Although a speculation, it was not a particularly fanciful one before the periodic system of the chemical elements was understood in terms of atomic theory. Two years later he indulged in speculations of what later would be called physical eschatology. On the basis of his electron model of the atom, he suggested that 'the general trend of the universe would be towards simplification of the atom ... The final stage would be that in which all atoms contained only one corpuscle.'[3]

Inspired by spectroscopic data, some years later Thomson's compatriot John William Nicholson, at the time working at the Cavendish Laboratory, suggested that 'proto-elements' still existed in stellar atmospheres and that astrophysics provided evidence for the complex structure of atoms. The planetary atomic model he proposed in 1911 (and which made use of Planck's constant) was directly inspired by astrospectroscopy and also relied on elements of the electromagnetic world view in so far as he assumed that the mass of the elementary particles had an electromagnetic origin. Foreshadowing many later developments in elementary particle physics, he suggested that the fundamental physics of the future would rely on and benefit from astrophysics:

> Astronomy, in the wider interpretation of its scope which is now general, owes much to Physics.... A point appears to have been reached in its [astronomy's] development at which it becomes capable of repaying some of this debt, and of placing Physical Science in its turn under an obligation. For it would seem that the most satisfactory test of the newer physical theories is to be derived from a discussion of the accumulated results of astrophysical observations.... The reason for the possibility for this position of astrophysics—as an arbiter of the destinies of ultimate physical theories—is ... [that] the terrestrial atoms are apparently in every case too complex to be dealt with by first principles ... [but] when an astrophysicist discovers hydrogen in a spectrum, he is dealing with hydrogen in a simpler and more primordial form than any known to a terrestrial observer.[4]

Nicholson concluded that further progress in physical theory could only be obtained 'by the continued help of astrophysics'. The general idea that, as astronomy could learn from the physico-chemical sciences, so physics and chemistry could learn from astronomy, continued to occupy the minds of a few scientists. Gilbert N. Lewis, a distinguished American physical chemist, expressed the theme as follows:

> While the laboratory affords means of investigating only a minute range of conditions under which chemical reactions occur, experiments of enormous significance are being carried out in the great laboratories of the stars. It is true, the chemist can synthesize the

particular substances which he wishes to investigate and can expose them at will to the various agencies which are at his command; we cannot plan the processes occurring in the stars, but their variety is so great and our methods of investigation have become so refined that we are furnished an almost unbounded field of investigation.[5]

However, the stars are not the same as the universe, and astrophysics not the same as cosmology. Under the impact of quantum theory, astrophysics flourished in the 1920s, but with almost no connection to the new field of relativistic cosmology.

The cosmological model that Einstein proposed in 1917—the prototype of all later models—pictured a finite and spatially closed universe homogeneously filled with an unspecified dilute matter. Moreover, the 'spherical' Einstein universe was static, in the sense that its radius of curvature did not vary with time. It was described by the field equations of general relativity, but in a modified form that included a term given by the so-called cosmological constant (Λ) or what Einstein originally called a 'for the time being unknown universal factor'. This constant of nature, which appeared here for the first time, expressed the mean density of matter in the universe ρ and also the radius of curvature R, both assumed to be constant quantities. The relations given by Einstein were

$$\Lambda = \frac{4\pi G}{c^2}\rho = \frac{1}{R^2}$$

The Λ constant had to be extremely small and have the dimension of an inverse area, but apart from this almost nothing was known about the quantity.

Apart from its basis in the fundamental theory of general relativity, the Einstein model had little connection to physics and none to the physics of matter. The mathematical and abstract character of early relativistic cosmology was even more pronounced in the alternative relativistic model proposed by the Dutch astronomer Willem de Sitter, also in 1917. De Sitter's model was certainly unphysical in the sense that it contained no matter at all. In spite of being empty, like Einstein's model it was spatially closed, static, and included a cosmological constant.[6] Although theoretical cosmology in the 1920s was predominantly of a mathematical nature, there were a few attempts to introduce a more physical perspective in the study of the universe.[7] For example, the closed Einstein model was studied from a thermodynamical point of view by the German physicist Wilhelm Lenz in 1926 and later, in a much improved form, by the American physicist and cosmologist Richard Tolman.

Speculations about a possible connection between quantum theory and general relativity were in a few cases carried over into cosmology, but the early attempts to establish links between the two very different theories were few and attracted very little interest. Inspired by Eddington's extension of the general theory of relativity to cover electromagnetism, in 1925 the British physicist James Rice at Liverpool University suggested a possible connection between the spectroscopic fine structure constant (e^2/hc) and cosmic units such as the density and curvature radius of the Einstein radius.[8]

In the same year, the Hungarian physicist Cornelius Lanczos examined wave propagation in the Einstein model by relating the quantum phenomena of the microcosmos to the structure of the macrocosmos. He was led to a cosmological interpretation of the quantum of action: 'We arrive at the assumption that the quantum behaviour of natural processes is not to be traced back to the microcosmos, but... that the solution to the quantum riddles is hidden in the spatial and temporal closedness of the universe.'[9] Lanczos' early attempt to establish a bridge between quantum theory and cosmology attracted no interest at all among contemporary physicists.

The discovery of the expanding universe about 1930 offered new possibilities for combining cosmological models with physical processes. The most popular expanding model in the 1930s was the closed Lemaître–Eddington model, according to which the universe evolved asymptotically from a static Einstein state. The cause of the original expansion was studied by several physicists in the 1930s, resulting in hypotheses of gravitational perturbations and instabilities caused by matter annihilation. The latter idea, in the form of hypothetical proton–electron annihilation processes, was discussed by Tolman, who also examined in detail the thermodynamical behaviour of various world models, including a class of cyclic models (see Chapter 8). However slowly and hesitatingly, cosmology was on its way to becoming a physical science.

In retrospect, the most important of the early expanding models was the 'primeval atom' model suggested by the Belgian pioneer cosmologist (and Catholic priest) Georges Lemaître in 1931, the first example of a physical big-bang model.[10] In this daring picture of the absolute origin of the cosmic expansion, the initial universe was likened to a huge atomic nucleus which, being subject to the laws of quantum mechanics, exploded radioactively. Quantum mechanics and nuclear physics were thus parts of Lemaître's scenario, but not in a way that could be worked out quantitatively. The original scenario was to a large extent a piece of cosmopoetry rather than cosmophysics. In reality, the exploding universe of the 1930s was as separate from quantum physics as other cosmological models of the period. Besides, Lemaître's idea of a 'fireworks universe' was scarcely taken seriously by the majority of scientists involved in cosmological research. Not only was the notion conceptually strange and hard to swallow, it also lacked convincing empirical evidence. From the point of view of physicists and astronomers in the 1930s, it had little to offer.

The cosmological constant introduced by Einstein in 1917 served in a few cases as a link between cosmology and the physics of matter and fields. As early as 1919, in an attempt to connect gravitation with electromagnetism, Einstein suggested that the enigmatic constant might play a role in atomic theory. In a later paper, of 1927, he considered a classical model of electrically charged particles with a negative pressure in the interior. This pressure he related to the cosmological constant. At that time, shortly after the introduction of quantum mechanics, a few physicists vaguely conceived that there might be some connection between the cosmological constant and the new quantum theory of atomic structure. In a letter to Einstein, Weyl said that 'all the properties I had so far attributed to matter by means of Λ [the cosmological constant] are now to be taken over by quantum mechanics'.[11] He did not elaborate.

After 1931, when Einstein publicly abandoned the cosmological constant as a mistake, its role in cosmology diminished. Lemaître and Eddington were among the few experts who disagreed with Einstein and, with different motivations, continued to champion the cosmological constant. In a paper of 1934 Lemaître argued that the constant could be understood, not only as a repulsive cosmic force, but also as a vacuum energy density. He suggested that the cosmological constant corresponded to the vacuum having a negative pressure given by:

$$p = -\rho c^2 \text{ with } \rho = \frac{\Lambda c}{4\pi G}$$

where ρ is the energy density. The idea eventually became of great importance in cosmological models (see Chapter 8), but in the 1930s it was ignored.

The early ideas of Lanczos, Lemaître and a few others were not followed up until much later, and during the 1930s theoretical cosmology developed largely as a science concerned with the solutions of Einstein's field equations. Its connection to observations was limited to astronomical measurements of the redshifts of galaxies. The one grand attempt to build up a comprehensive theory encompassing both quantum mechanics and cosmology, due to Eddington, was heterodox and controversial.

4.2 QUANTUM MECHANICS AND THE UNIVERSE

In a book published in 1939, the crystallographer, Marxist, and pioneering sociologist of science, John D. Bernal, expressed his worries about what he saw as a tendency in British intellectual life towards the 'mysticism and abandonment of rational thought...[that] penetrates far into the structure of science itself'. As an example, he mentioned that 'those metaphysical and mystical theories which touch on the universe at large or the nature of life, which had been laughed out of court in the 18th and 19th centuries, are attempting to win their way back into scientific acceptance'.[12] Among those who had so betrayed the true nature of science, apart from the usual crowd of idealistic philosophers and theologians, he singled out Eddington, one of Britain's most distinguished and popular scientists.

Arthur Stanley Eddington did very important work in astronomy and, not least, theoretical astrophysics. In 1924 he predicted the existence of white dwarf stars of exceptionally high density, at the same time arguing that the prediction verified the gravitational redshift associated with general relativity. Two years later he published a classical work on astrophysics, *The Internal Constitution of the Stars*, in which he suggested that the Sun and other stars are powered by nuclear reactions. An early convert to and expert of the general theory of relativity, Eddington was one of the leaders of the famous eclipse expedition, which in 1919 verified Einstein's prediction of the bending of light and thereby catapulted general relativity into the limelight. Moreover, he was among the small group of theorists who attempted to generalize Einstein's theory into a unified theory of gravitational and electromagnetic fields, which he did in works from the early 1920s. As a leading cosmologist of the relativistic school, he advocated and developed the expanding universe in both scientific and popular works. However, he never came to terms with Lemaître's idea of a finite-age universe starting in an explosive event.

What is of interest in the present context is not Eddington's important contributions to astrophysics and relativity, but his grand attempt to construct a new framework for physics with the aim of unifying the quantum world and cosmology.[13] This project, which he pursued between 1929 and his premature death in 1944, resulted in a long series of papers, mostly published in the *Proceedings of the Royal Society*, and in the monograph *Relativity Theory of Protons and Electrons*, which appeared in 1936. He prepared a systematic account of his theory and its mathematical foundations, but *Fundamental Theory* only appeared after his death, edited by the mathematician Edmund Whittaker.[14] In addition to his scientific works, he made his theory known to the public in several books of a more popular and philosophical nature. The most important of these was *The Philosophy of Physical Science*, based on his Tarner Lectures of 1938. Although there were considerable differences in Eddington's methods and results as his research developed over the years, I shall disregard these and focus only on some of the characteristic features and main results.

Eddington was not the first to consider possible relations between fundamental atomic and cosmic quantities. In his generalization of Einstein's general theory of relativity, Weyl discussed the electromagnetic radius of the electron ($r_e \approx 10^{-15}$ m) in relation to its so-called gravitational radius ($r_g \approx 10^{-56}$ m), noting that the ratio r_e/r_g was of the order of magnitude 10^{40}. The huge size of this dimensionless number had been pointed out earlier, but Weyl went further by suggesting that 'the ratio of the electron radius to the ratio of the universe may be of similar proportions'.[15] That is, he suggested the relationship:

$$\frac{r_e}{r_g} = \frac{e^2/mc^2}{Gm/c^2} = \frac{e^2}{Gm^2} \approx \frac{R}{r_e}$$

Although Eddington was quick to take up Weyl's speculation, his more ambitious research programme only had its beginnings in the late 1920s, directly inspired by progress in the new quantum mechanics. In his earlier work Eddington had dealt extensively with general relativity theory and proposed a unification of this theory and the theory of electromagnetism. Like other unificationists, he realized that the early attempts at formulating a unified theory, including his own, were incomplete because they left out atomic and quantum physics. As he expressed it in his important and influential work of 1923, *Mathematical Theory of Relativity*: 'We offer no explanation of the occurrence of electrons or of quanta; but in other respects the theory appears to cover fairly adequately the phenomena of physics. The excluded domain forms a large part of modern physics, but it is one in which all explanation has apparently been baffled hitherto.'[16] Five years later he saw in Paul Dirac's new wave equation of the electron a possibility to establish the required link from general relativity to quantum theory. Dirac's theory took him by surprise and caused him to reconsider his ideas concerning the foundations of physics.

Whereas the original quantum mechanics of Heisenberg and Schrödinger was non-relativistic, in early 1928 Dirac succeeded in finding a fundamental equation that agreed with the special theory of relativity and also provided a natural explanation of the electron's spin. Contrary to Schrödinger's wave equation, Dirac's was of the first order in both the time and the space derivatives of the wave function ψ. Dirac's theory was a great step forward in unifying physics, but according to Eddington it did not go far enough. In his view, it made no sense to apply the equation to a single electron because: 'An electron has no individuality and is not separable from all the other electrons in the universe in the way that the classical picture supposed. I take the view that the mass of an electron is an interchange energy with all the other charges in the universe suitably averaged.'[17] Having thus introduced the universe, he deduced that the dimensionless fine-structure constant α, defined as $2\pi e^2/hc$, was related in a simple way to both atomic and cosmological constants. The equation he arrived at in 1929 was:

$$\alpha = \frac{mcR}{\hbar\sqrt{N}}$$

where R is the radius of the Einstein universe and N the number of electrons (or electrons and protons) in the universe, what he called the cosmical number and hestimated to be about 10^{79}. The symbol m stands for the mass of the electron and ($\equiv h/2\pi$, a symbol introduced by Dirac in 1930) for Planck's constant. The important point is the connection between microphysical and cosmological quantities.

Like no physicist before him, and probably no one after him, Eddington was obsessed by the constants of nature and their theoretical significance. Indeed, he was instrumental in giving these constants the elevated significance they have in modern physics and to which we shall return in Chapter 7. 'One may,' he wrote, 'look on the universe as a symphony played on the seven notes of a scale.'[18] As the seven fundamental constants he took e, m, M (mass of the proton), h, c (speed of light), G (Newton's constant of gravitation), and Λ, setting himself the task of deducing the values of as many of them as possible. Since he needed three constants to fix the units of length, mass, and time, he was left with four quantities. He paid particular attention to the dimensionless ratios between the constants. Apart from M/m and the inverse fine-structure constant, he considered the very large constants

$$\frac{e^2}{GmM} \quad \text{and} \quad \frac{c}{\hbar}\sqrt{\frac{mM}{\Lambda}}$$

Rationalist Cosmologies

'All four constants are obtained by purely theoretical calculation', he proudly announced.[19] Eddington not only calculated the ratios of natural constants, but also some of the constants themselves. Among the theoretically determined constants appearing in *Fundamental Theory* were Planck's constant, the mass of the electron, the constant of gravitation, and the range constant of nuclear forces. He even claimed to have determined unambiguously the number of dimensions of space-time (3 + 1), which he sometimes regarded as a 'fifth natural constant'.

Eddington introduced yet another pure number, which he considered to be of great importance. As mentioned, the cosmical number N was the number of electrons and protons in the universe. However, N was not merely the number of particles in the universe, for it had a more general significance as a fundamental constant that entered many of the physical formulae. Eddington's aim was to show, 'not that there are N particles in the universe, but that anyone who accepts certain elementary principles of measurement must, if he is consistent, think there are'.[20] He insisted that N was truly a constant of nature, of the same character as Avogadro's number but much more fundamental. Like other fundamental constants its value was determined by the very fabric of the cosmos. Whereas the traditional view was and still is that the number of particles in the universe is accidental, according to Eddington the number could not have been different from what it is.

The idea of connecting pure numbers relating to microphysics with the number of particles in the world had earlier appealed to him. In his important book on the theory of relativity from 1923, Eddington had briefly dealt with the theme, pointing to the 'attractiveness in the idea that the total number of the particles may play a part in determining the constants of the laws of nature'.[21] As his work progressed, he claimed to be able to deduce the number rigorously from theory, with the result that

$$N = 2 \times 136 \times 2^{256} \cong 3.15 \times 10^{79},$$

or, as he expressed it in one of his books: 'I believe there are 15,747,724,136,275,002,577, 605,653,961,181,555, 468,044, 717,914,527,116,709, 366,231,425,076,185,631,031,296 protons in the universe, and the same number of electrons.'[22] No more and no less.

The numbers themselves were important, and the relationships between them even more so because they tied together the microcosmos and the macrocosmos. The precise formulations of the relations changed during the course of his research programme, but not drastically. To give the flavour, consider two characteristic examples:

$$\frac{e^2}{GmM} = \frac{136}{137\pi}\sqrt{\frac{5N}{3}} \quad \text{and} \quad \sqrt{\frac{N}{30}} = \frac{2\pi c}{h}\sqrt{\frac{mM}{\Lambda}}$$

Personally, Eddington was a modest man; scientifically, he was not. His fundamental theory was meant to be a framework for a final theory about nature in its entirety. At the end of his main work, *Relativity Theory of Protons and Electrons*, he expressed his ambitious goal as follows:

> Unless the structure of the [atomic] nucleus has a surprise in store for us, the conclusion seems plain—there is nothing in the whole system of laws of physics that cannot be deduced unambiguously from epistemological considerations. An intelligence, unacquainted with our universe, but acquainted with the system of thought by which the human mind interprets to itself the content of its sensory experience, should be able to attain all the knowledge of physics that we have attained by experiment. He would not deduce the particular events and objects of our experience, but he would deduce the generalisations we have based on them. For example, he would infer the existence and properties of radium, but not the dimensions of the earth.[23]

Here we have one more version of Laplace's intelligence or demon, but with the difference that the demon must be provided with or at least have complete knowledge of *our* mental faculties. In effect, the demon is human.

Contrary to most other theories of everything, Eddington's was far from empirically sterile. It resulted in a number of predictions, many of them of a precise and testable nature. Before mentioning some of them, it is worth noting how he looked at empirical testing. Although Eddington was particularly pleased to quote those of his predictions that agreed with experiments, he emphasized that the theory did not rest on either these tests or other empirical data. It was, he said on many occasions, epistemological considerations and not empirical facts that provided the key to unlocking the secrets of nature. As proved by his own theory, it was possible to obtain knowledge of the fundamental laws of nature by pure deductions from the peculiarities of the human mind. Consequently these laws could not refer to an objective world and therefore not be objective. 'All the laws of nature that are usually classed as fundamental can be foreseen wholly from epistemological considerations. They correspond to *a priori* knowledge, and are therefore wholly subjective.'[24]

Eddington thought that he was able to explain numerical coincidences between the constants of nature by an epistemic analysis of the nature of observation. In doing so he appealed to selection arguments somewhat similar to those that would later be associated with the anthropic principle. He thought that the cosmic number and most other constants were determined by mental and therefore human factors, namely 'the influence of the sensory equipment with which we observe, and the intellectual equipment with which we formulate the results of observation of knowledge'. 'This influence,' he said, 'is so far-reaching that by itself it *decides* the number of particles into which the matter of the universe appears to be divided.'[25]

If the laws of nature are not the contributions of an external world, but essentially the subjective constructions of physicists, little or no room is left for empirical–inductive methods in fundamental physics. The new theory of protons and electrons, Eddington wrote, is 'even more purely epistemological than macroscopic relativity theory; and I think it contains no physical hypotheses—certainly no new hypotheses—to be tested'. He continued:

> It should be possible to judge whether the mathematical treatment and solutions are correct, without turning up the answer in the book of nature. My task is to show that our theoretical resources are sufficient and our methods powerful enough to calculate the constants exactly—so that the observational test will be the same kind of perfunctory verification that we apply sometimes to theorems in geometry.[26]

This was of course a view in provocative contrast to what had been the standard conception of physical science since the scientific revolution. Understandably, Whittaker compared Eddington's unusual position with the one of Descartes, the rationalist natural philosopher, and his theory of vortices.[27]

Among the theoretical predictions that flowed in a steady stream from Eddington's fundamental theory were values of the cosmological constant, the Hubble expansion parameter, the proton-to-electron mass ratio, and the fine-structure constant. From certain analogies between cosmological theory and Dirac's wave equation of the electron, he found in 1931 that the 'cosmical constant' could be written as

$$\Lambda = \left(\frac{2GM}{\pi}\right)^2 \left(\frac{mc}{e^2}\right)^4 = 9.79 \times 10^{-55} cm^{-2}$$

This was not an unreasonable value, but it could not be compared with observational determinations of the constant since there were no such determinations. Instead he used it to derive a limiting speed of the recession of nebulae, that is, their speed disregarding gravitating matter. Various versions of his theory resulted in values of 528 (in 1931), 432 (in 1936), 572 (in 1944), and 585 (in 1946), all in the unit km s^{-1} Mpc^{-1}, that is, kilometres per second per megaparsec.[28] Given that the generally accepted value of the Hubble constant was about 500 km s^{-1} Mpc^{-1}, as determined by Edwin Hubble and Milton Humason, these theoretical values could be considered reasonably successful. But the agreement was not all that important to Eddington, who was unwilling to let a conflict between a beautiful theory and empirical data ruin the theory. This attitude may be illustrated by a 'definitive' revision of his 1931 result, which he published in 1935. He now obtained 865 km s^{-1} Mpc^{-1}, which definitely disagreed with observations. Rather than putting the blame on the theory, he suggested that 'the present scale of distances of the remote nebulæ is rather too great and should be reduced by 20–30 per cent'.[29] This was not an attitude that helped create confidence in Eddington's project. One wonders how he would have responded to the drastic reduction of the Hubble constant that occurred in the 1950s and eventually led to its present value of 74 ± 4 km s^{-1} Mpc^{-1}.

Among the dimensionless constants that appealed to Eddington was the mass ratio between the proton and electron M/m, still considered to be an important and enigmatic number. By an involved and somewhat obscure chain of reasoning he deduced in 1933 that the two masses should be roots of the equation

$$10x^2 - 136\,\mu x + \mu^2 = 0$$

where the quantity μ was expressed by the radius of curvature of the universe and the number of particles in it, $\mu = \sqrt{N}/R$. From the equation he found a theoretical value of $M/m = 1847.6$, which agreed nicely with the experimental value known at the time, 1847.0. When later and more precise experiments resulted in the lower value 1834.1 (the present value is 1836.2), Eddington argued that the measurements were based on a definition of mass somewhat different from the one used theoretically and that it was 'therefore not necessarily in conflict with our determination of the mass-ratio'.[30] The same kind of explaining away appeared in Eddington's notoriously famous analysis of the fine-structure constant.[31] In the table below, a comparison is presented between the observed values of natural constants and those calculated by Eddington at the end of his life.

According to Eddington, the inverse fine-structure constant (which I shall denote β) expressed a certain number of algebraic degrees of freedom and therefore had to be an integer. It did not specifically relate to electrons but had a much wider significance. In his early paper of 1929 he found $\beta = 16 + \tfrac{1}{2} \times 16 \times (16-1) = 136$, a value which did not agree with the then known experimental value of 137.1. The discrepancy did not worry Eddington: 'Although the discrepancy is about three times the probable error attributed to the experimental value, I cannot persuade myself that the fault lies with the theory', he wrote.[33] Yet, although his theory was admittedly of an *a priori* nature, of course he could not ignore how its predictions compared with experiments. In a subsequent paper he came up with a revised argument that resulted in $\beta = 137$, still a whole number. He argued that the extra number was a result of the exclusion principle, arising from the indistinguishability of two particles. For the rest of his life, he stuck to the integer 137, which he claimed to have 'obtained by pure deduction, employing only hypotheses already accepted as fundamental in wave mechanics'.[34]

As late as 1944, in his very last publication, he quoted the experimental value $\beta = 137.009$ and concluded that the small discrepancy was a problem for the experiment rather than the theory. His saving device was a liberal use of the so-called Bond factor of 136/137, a quantity that supposedly had been missed by the experimentalists. Although the Bond correction resulted in a better agreement between theory and experiment, to the majority of physicists it appeared coincidental and Eddington's use of it conspicuously *post hoc*. In fact, in his theory the factor might occur raised to any power, and his arguments that it entered precisely as 136/137 remained obscure and unpersuasive.

Eddington had started his grand project in 1929, at a time when it was universally believed that there were only two material elementary particles, the proton and the electron. (The massless photon might enter as a third particle.) A decade later the world of particles and fields looked very different and much less inviting to unificationists of

Calculated and observed values of constants of nature[32]

Constant	Calculated	Observed
Mass of electron	9.10924×10^{-28} g	9.1066
Mass of proton	1.67277×10^{-24} g	1.67248
Proton–electron mass ratio	1836.34	1836.27
Elementary charge	4.80333×10^{-10} esu	4.8025
Planck's constant	6.62504×10^{-27} erg s	6.6242
Inverse fine-structure constant	137	137.009
Gravitational constant	6.6665×10^{-8} cm^3 g^{-1} s^2	6.670 ± 0.005
Hubble constant	572.36 km s^{-1} Mpc^{-1}	560
Magnetic moment of H atom	2.7899 nuclear magnetons	2.7896 ± 0.0008
Magnetic moment of neutron	1.9371 nuclear magnetons	1.935 ± 0.02

Eddington's brand. The positron and neutron had entered the scene in 1932–33 and Pauli's neutrino, although undetected, was generally accepted as a real particle. Much to the surprise of the theorists, studies of the cosmic rays revealed the existence of 'heavy electrons', at the time often known as mesotrons and today called muons. The mass of the new particle was estimated to be about 130 electron masses, but a proper mass determination had to wait until about 1950 (when the mass of the muon was found to be about 215 electron masses).

Eddington was as unprepared for the new particles as he was unprepared for other discoveries in nuclear and atomic physics, fields in which he was not at home. As to the neutrino, he more or less ignored it, suggesting that it probably did not exist.[35] Nor did he accept Dirac's idea of antiparticles, and he chose to identify the positron as 'minus one electron'. In *Fundamental Theory* he made an attempt to cope with the new development, but in a somewhat amateurish way. He included sections on deuterium, the helium atom, metastable states of hydrogen, the radii of atomic nuclei, and the magnetic moments of the proton and neutron, attempting to show that his theory could be used in these more mundane areas of physics. He even calculated the mass of the new mesotron, getting 173.98 electron masses, and saw no reason 'why there should not also exist *heavy mesotrons* which decay into protons and negatrons [electrons]'.[36] For the hypothetical heavy meson he calculated the mass 2.38 proton masses or 4360 electron masses.

Although Eddington did not apply his theory to calculate the maximum number of chemical elements, a few other scientists did.[37] In 1932, the Indian mathematician Vishnu Narlikar (the father of the cosmologist Jayant Narlikar) used a modified version of Eddington's theory to argue that the highest possible number of electrons in an atom is 92, corresponding to uranium. In 1947, presumably influenced by the discovery of the first transuranium metals, a Spanish physicist, D. M. Masriera, deduced from the same theory that the maximum atomic number was 96. For the record, element number 97, berkelium, was manufactured in 1949.

Eddington's physical theory cannot be sharply separated from his general philosophy of science, such as he presented it in *The Philosophy of Physical Science* and elsewhere.[38] In spite of the rationalistic elements in his conception of physics, Eddington's rationalism was tightly connected with basic experimental operations. In principle, he said, all a physicist needs is a scale, a clock, and an eye; and then, of course, a mind to produce the mathematical structure for a correlation of the empirical information.

An important element in his philosophy was what he called selective subjectivism, by which he meant that the mind largely determines the nature and extent of the knowledge of what we think of as an external world. We force the phenomena to fit into forms that reflect the observer's intellectual equipment and the instruments he uses. For this reason, discovering a phenomenon is as much a matter of manufacturing it. 'In the end what we comprehend about the universe is precisely that which we put into the universe to make it comprehensible.' Eddington insisted not only that *a priori* knowledge was justified in physics, indeed necessary, but also that the subject (the physicist) could not be separated from the object (the physical phenomenon). 'The physicist', he wrote, 'might be likened to a scientific Procrustes, whose anthropological studies of the stature of travellers reveal the dimensions of the bed in which he has compelled them to sleep.'[39]

The emphasis of the human mind as an active part in the acquisition of knowledge in the physical sciences was not a new theme in Eddington's thinking. He had said as much in an essay of 1920, in which he stated that the physicists' laws of nature are indirectly imposed by the mind. There were, he thought, true natural laws inherent in the external world, but these could neither be expressed rationally nor be discovered empirically.[40] This is also what he said in his popular account of relativity from the same year, *Space, Time and Gravitation*, which concluded as follows:

> We have found that where science has progressed the farthest, the mind has but regained from nature that which the mind has put into nature. We have found a strange foot-print on the shores of the unknown. We have devised profound theories, one after another, to account for its origin. At last, we have succeeded in reconstructing the creature that made the foot-print. And Lo! It is our own.[41]

Eddington's frequent references to fundamental physics being metrical was no trivial point, but a way of distinguishing physics from the much larger (and to him more important) non-metrical or spiritual world. As Stewart and Tait had argued in the Victorian era, so the deeply religious Eddington was convinced that physics implies a vast domain beyond its frontiers. This domain he discussed in his *Science and the Unseen World* of 1929, a title which may deliberately have been an allusion to *The Unseen Universe*. However, Eddington insisted that the unseen spiritual world was not ruled by the laws of physics or otherwise amenable to scientific investigation. He always insisted on keeping religion and science separate.[42]

4.3 WORLD PHYSICS

To the extent that Edward Arthur Milne has a name in modern physics it is because of his important contributions to astrophysics and not for his serious attempt to establish a new physics on the basis of general cosmological principles.[43] Yet in the 1930s and 1940s he was considered a leading reformer of cosmological thought and his research programme attracted much scientific attention—in fact much more than Eddington's. A comparison between the two systems of cosmophysics is inevitable but not a straightforward task. There were certainly significant similarities, especially as concerned the methods and aspirations of the two physicists. They both aimed at reconstructing the very foundation of physics and doing so by allowing *a priori* reasoning a legitimate and important role. If Eddington's approach to fundamental physics was a blend of rationalism and idealism, Milne's can be characterized as a blend of rationalism and positivism. Their systems had shared elements of rationalism and deductivism, with pure mathematics serving as a model for theoretical physics. As the philosopher Charlie Broad wrote: 'For Descartes the laws of motion were deducible from the perfection of God, whilst for Eddington they are deducible from the peculiarities of the human mind.... For both philosophers the experiments are rather a concession to our muddle-headedness and lack of insight.'[44] This description also covers Milne, to a large extent.

In spite of the similarities, the world systems developed by Milne and Eddington in the 1930s were very different. As we have seen, Eddington aimed at extending the general theory of relativity and constructing a unified system that comprised quantum mechanics as well as relativistic cosmology. Milne, on the other hand, did not accept general relativity and wanted to do with the special theory only. Moreover, his world physics was largely restricted to cosmology and for this reason was not really a unified theory. By and large, atomic and quantum physics were foreign to his research programme and he made only feeble attempts to incorporate these fields in it. He opposed reductionist attempts to understand the laws of the universe in terms of atomic theory and found it a waste of time to try to unify microphysics and cosmic physics. The scattered attempts he nonetheless made in this direction were singularly unsuccessful and not taken seriously by experts in quantum physics. As to the structure of the universe, Eddington's view was orthodox in the sense that he favoured a relativistic expanding model of finite size and without a sudden origin in time. Milne's view was very different: he argued that the universe behaved in accordance with simple kinematic laws, that space is Euclidean and contains an infinity of particles, and that the expansion of the universe, indeed time itself, started a finite time ago.

As far as sociology is concerned, the theoretical systems of Milne and Eddington were both grand failures, but Eddington's failure was even greater than Milne's. They were also individual projects, neither of the physicists having started anything that amounted to a school of thought or having many students. Their ideas were unconventional and much criticized even in their own days—misunderstood, they believed. Eddington was perplexed that his great work went unappreciated. 'In the case of Einstein and Dirac people have thought it worth while to penetrate the obscurity', he complained in a letter shortly before his death in 1944. 'I believe they will understand me right when they realise they got to do so'.[45] (Fig. 4.1). Milne was no less frustrated over the lack of recognition of what he thought was a promising new world view. As he confided to his friend, the eminent astrophysicist Subrahmanyan Chandrasekhar, his theory was 'utterly and surprisingly different from usual mathematical physics, and... when it comes to be recognised, it will be regarded as revolutionary'.[46] But this never happened.

Milne's first scientific work was in atmospheric and solar–terrestrial physics, soon to be followed by important investigations on the theory of stellar atmospheres and the radiation equilibrium of stars. It was primarily for this work, in part done in collaboration with Ralph Fowler of Cambridge University, that he became recognized as one of the period's most outstanding astrophysicists. In 1926, only 30 years old, he was elected a fellow of the Royal Society. Appointed professor of applied mathematics at the University of Oxford in 1928, a few years later he took up an interest in cosmology. The first major result of this interest came in 1933, a 95-page article in which he discussed the basic features of his new system of cosmology. He gave a full and systematic account of it in a book of 1935, entitled *Relativity, Gravitation and World-Structure*. In several later articles and books he extended and refined his theory in various ways. In *Kinematic Relativity*, which appeared in 1948, only two years before his untimely death, he applied the theory to gravitation, electromagnetism, and some phenomena of microphysics.

Although Milne only developed his cosmophysics in the 1930s, some of the characteristic methodological features of his research programme had roots farther back in time. This was the case, not least, with his fascination for the epistemic power of pure mathematics. In an early address of 1922, he praised the mathematician for being the only one who 'has the whole of space and time for his habitation' and 'knows anything about nature and the universe at all'. Contrariwise, he had only scorn for the experimentalist's 'gluttony of accumulation'.[47] This theme—the relative insignificance of observation and experiment and the predominating role of mathematical reasoning—remained an important and persistent part of his cosmophysics. On several occasions he likened the proper method of theoretical physics to that of the mathematician, and the laws of physics to geometric theorems. His cosmophysics was, from the beginning, designed as an axiomatic and deductive system, to be 'a complete reconstruction of physics from the bottom upward, on an axiomatic basis'.[48]

Fig. 4.1. Einstein in conversation with Eddington in 1930. Both physicists followed a programme of unification, but in different ways. Einstein had no confidence in Eddington's attempt to merge microphysics and cosmology into a unified theory. Photograph by Winifred Eddington.
Source: Allie V. Douglas, *The Life of Arthur Stanley Eddington*. London: Thomas Nelson, 1956, Plate 11.

Often accused of being a Cartesian rationalist, he claimed that the laws of physics could ultimately be established from pure reasoning and processes of inference. When his system was completed the contingency of the laws of physics would have vanished and it would turn out that they were no more arbitrary than the theorems of geometry. The similarity between his cosmophysical system and pure mathematics even extended to the concept of truth. In the ordinary understanding of physics, truth relates to a theory's empirical power, its ability to explain and predict natural phenomena, but not so with Milne's physics: 'Just as the mathematician never needs to ask whether a constructed geometry is true, so there is no need to ask whether our kinematical and dynamical theorems are true. It is sufficient that the structure is self-consistent and free from contradiction; these are in fact the only criteria applied to a modern algebra or geometry'.[49] As far as mathematical rationalism goes, Milne was as radical as Eddington and both were more radical than Einstein.

Although well versed in both the mathematical and conceptual aspects of general relativity, Milne was unwilling to admit Einstein's theory as a basis for cosmology. Space was for him not an object of observation, not something physical, but just an abstract system of reference. As such it could have no structure, nor could it expand or contract. He deemed the concept of space curvature to be irrelevant as well as illegimate. What was important and relevant was the existence of elementary observations, in order to base cosmology on possible measurements of length and time intervals. Milne's own system, as he presented it most fully in his book of 1935, supposed a flat, infinite space and simple kinematic considerations. From such considerations, boiled down to a few general principles, he argued that all the basic laws of cosmic physics could be deduced. Milne obtained these principles by analysing the concepts used to order temporal experiences, and to communicate them by means of light signals. He considered time measurement to ultimately be the only form of measurement, arguing that the measurement of length derived from that of time. The physics arising from such reasoning would naturally be restricted to relations of distance and time; they could involve no initial appeal to dynamical or gravitational assumptions.

As an important result, Milne was able to show that a system of particles, unaffected by collisions or gravitational forces, would evolve in a way quite similar to the galaxies in the picture of the expanding universe. He found an outward velocity in agreement with Hubble's law and thus explained the expansion of his model universe without Einstein's gravitation theory—indeed without any gravitation at all. According to Milne, there was no more to the expansion of the universe than the recession of galaxies. Expanding space had no place in his theory. As he pointed out, not only was his kinematic explanation much simpler than the one based on general relativity, it also presented the expansion as an inevitable phenomenon (which it is not in relativistic cosmology). In Milne's expanding universe the distance between any two particles moving with relative velocity v, would increase in time as $r = vt$. Identifying the Hubble parameter with the inverse of t, it corresponded to the Hubble law $v = Hr$. The age of the universe should thus be the same as the Hubble time $1/H$, which in the 1930s was believed to be a little less than two billion years.

The 'particles' that occupied a central position in Milne's system were strange and abstract creatures. In most cases they could be understood simply as galaxies, but they had human-like properties in so far that they had to be able to reckon the flow of time. According to his critics, by associating particles with a 'perceiving ego' he introduced an anthropomorphic element in his system. But Milne found it to be justified, for without these egos there would be nothing to perceive. 'Till then, space does not exist. It is a map I invent for the location in it of objects I perceive. If there are no objects to perceive, there is no space.'[50]

One might believe that Eddington recognized in Milne a brother of spirit, a scientific thinker who pursued a vision of a rationalistic physics not unlike his own. But this was not the case at all. Eddington seems to have nourished a deep dislike of Milne's alternative world physics, which he saw as entirely different from his own system, both in spirit and in its technical details. He mostly ignored his rival from Oxford, and the few times he commented on kinematic cosmology it was to criticize it. Eddington found Milne's theory to be contrived and 'perverted from the start' because it built on what he saw as mistaken definitions of length and time. He also argued that it was fundamentally wrong to disregard atomic and quantum physics in a cosmological context, such as Milne did. As to the uniform expansion of the universe assumed by Milne, Eddington dismissed it as unfounded and inferior to 'the acceleration of expansion which relativity theory predicts'.[51] (In fact, the acceleration is an effect of the cosmological constant, not of general relativity theory as such.)

Milne's system of the world built on two fundamental principles or postulates. The one, taken over from the special theory of relativity and therefore uncontroversial, was the principle of the constancy of the speed of light. The other postulate, soon known as the *cosmological principle*, had a particular status in his system. In 1937 he described his programme as follows: 'I have endeavoured to develop the consequences [of the cosmological principle] without any empirical appeal save to the existence of a temporal experience, an awareness of a before-and-after relation, for each individual observer'.[52] According to Milne, the world would, as a matter of principle, appear to be centred around each observer and would look the same to each of them. All observers would see the same events occurring, and agree upon the basic laws of nature. These laws themselves depended on the cosmological principle: 'What had previously been regarded as empirical laws of nature are deducible as consequences of the existence of a temporal experience for each individual observer, together with an [homogeneity] assumption about the distribution of such observers'.[53]

His version of the cosmological principle corresponded to the commonly held assumption of large-scale homogeneity and isotropy, but Milne believed that his version differed from the one adopted by relativist cosmologists. And indeed it did, for in standard cosmology it was an extrapolation of observations, a hypothesis which needed to be abandoned if future observations spoke against it. As the American cosmologist and specialist in relativity theory Howard Percy Robertson pointed out in a critical review of Milne's theory, in this theory it appeared as an *a priori* principle. Milne sometimes spoke of it in Kantian terms, as a precondition for having knowledge of the universe. For example: 'If the universe is not homogeneous (in our extended usage of the word), then... there will be no world-wide dynamics possible'.[54] He admitted the possibility, if only as a thought experiment, that the universe might not satisfy the cosmological principle, but argued that in this case the laws of phenomena would vary from place to place, hence not be proper laws. In other words, physics was only possible because of the cosmological principle.

Among the new results communicated by Milne in his monograph of 1935 was that the value of the constant of gravitation increases linearly with the cosmic epoch: $G = kt$. For the relative change he found that it was extremely small, about $\Delta G/G = 5 \times 10^{-10}$ per year. However, according to Milne the varying-G law did not apply to local gravitation, such as the solar system. He believed that it was not meaningful to ask if G *really* increases with time or not, and also that it was not meaningful to ask if the universe *really* expands or not. To a large extent he adopted a conventionalist perspective that implied that many questions of physics were merely conventions and therefore did not have an objective meaning independent of the chosen conventions. Thus, he operated with two time scales, which he called kinematic time (t) and dynamic time (T). These two time scales were connected by a simple logarithmic transformation,

$$T = t \ log \left(\frac{t}{t_0}\right) + t_0$$

where t_0 signifies the present epoch. Some phenomena are best described in t-time, others in T-time, but it is largely a matter of convenience which scale to use. Milne sometimes spoke of the two time scales as two distinct but translatable languages. The proportionality between G and t does not hold for observers using T-time, where G reduces to a constant. Whereas on the t-scale, the universe is expanding from an origin at a point, on the T-scale it is static. Understandably, the conventionalist framework and the two time scales made it difficult to confront Milne's physics directly with observational results.

Milne's somewhat eclectic philosophy of physics is not easily characterized, except that it included strong doses of rationalism and deductivism, and positivism (or perhaps better operationalism) as well. He was not unconcerned with observations, but thought they were mainly of value in testing independently proposed theories. In a discussion of 1937 he said:

> The role of observation would again be to attempt to verify in Nature the existence of entities corresponding to those mentioned in the axioms or any afterwards constructed, by ascertaining whether the resulting theorems hold good in the external world. The relevance of the theorems to Nature would require to be established by observation, but these theorems or laws of Nature would hold good in their own right, as the inevitable consequences of our having said exactly what we are talking about.[55]

Milne thought that his own theory not only offered a simpler and more comprehensible picture of the world, but also that it fared better when it came to tests and explanatory power. For example, he argued that it provided unique explanations in areas such as cosmic rays and the spiral structure of galaxies, and he was keen to compare these explanations with observations. 'Theories differ simply and solely when their predictions as to phenomena differ', he said in 1934, urging rival cosmologists within the relativist programme to come up with explanations that uniquely depended on their notion of curved space. 'The non-Euclidean geometers must state what they expect to observe.'[56] As Milne saw it, in the ultimate comparison of cosmological theories geometry was irrelevant. What mattered was observable phenomena only. In this respect, both rhetorically and substantially, there was a great deal of similarity between Milne's cosmology and the one of the later steady-state theory that will be reviewed in Chapter 5.

Although Milne often spoke of verification and falsification, and stated that propositions about nature necessarily had to be verifiable, this was to some extent limited to the rhetorical level. At any rate, his system of world physics was not a theory in the ordinary sense. It could not be proved wrong in any simple manner, among other reasons because it referred to non-observational concepts such as 'universe', 'observer–particle', 'equivalence' and 'substratum'. These were abstract kinematic concepts whose properties were deducible from their definitions only and had nothing to do with what might or might not exist in nature. Reason and logic, rather than observation and experiment, were the key elements, both when it came to discovery and justification in world physics. In *Relativity, Gravitation and World Structure* he expressed this view as follows:

> Whether the universe contains an infinite number of systems can never be answered by observation in the affirmative, for we can at most count a finite number. I have already given general considerations which lead us to conclude that that the universe must include an infinite number of particles, but bearing in mind the historic vulnerability of general considerations I only point out here that whilst observation could conceivably verify the existence of a finite number of objects in the universe it could never conceivably verify the existence of an infinite number. The philosopher may take comfort from the fact that, in spite of the much vaunted sway and dominance of pure observation and experiment in ordinary physics, world-physics propounds questions of an objective, non-metaphysical character which cannot be answered by observation but must be answered, if at all, by pure reason; natural philosophy is something bigger than the totality of conceivable observations.[57]

The ultimate aim of Milne's research programme was, much like Eddington's, to provide a rational explanation of the world, leaving nothing out and allowing no irreducible laws or brute facts. In other words, to reduce the contingency maximally. 'The universe is rational', Milne stated, explaining what he meant by this credo, which was certainly not derived from experience:

> Given the mere statement of *what is*, the laws obeyed can be deduced by a process of inference. There would then not be two creations [of matter and law] but one, and we should be left only with the supreme irrationality of creation, in Whitehead's phrase. We can only test this belief by the act of renunciation, by exploring the possibility of deducing from some assumed description of just *what is* the laws which *what is* obeys, avoiding so far as possible all appeal to empirically ascertained laws. Laws of nature would then be no more arbitrary than geometrical theorems. God's creation would be subject to laws not at God's further disposal.[58]

Milne's credo has obvious similarities to Descartes' philosophy of nature, which is hardly a coincidence. Nor is it a coincidence that Milne (like Descartes) referred to God's will and possibilities, for God was an active agent in Milne's cosmological system. This he spelled out most fully in the posthumously published *Modern Cosmology and the Christian Idea of God*, surely one of the most remarkable works of modern physics and cosmology, but not one that had any influence on the course of science or, for that matter, on the course of theology.[59]

The rationalistic or constructive elements that appeared so visibly in the theories of Eddington and Milne were not their inventions. Apart from their long philosophical heritage, they were integrated parts of the grand project of understanding the universe that Einstein had pioneered two decades earlier. The question was not so much the legitimacy of theoretical principles as how to obtain the principles and how much weight these should be given relative to knowledge obtained inductively from observations. What is the proper balance? In an address of 1932 to the Philosophical Union, a club at the University of California at Los Angeles, the leading cosmologist Richard Tolman distinguished two ways of constructing theories of the universe. The first way—where the theoretical principles are suggested as 'immediate generalizations of experimental findings'—corresponded to an empiricist–inductivist methodology. Although Tolman had great respect for this traditional method, he used the occasion to emphasize that principles might also be suggested by 'desiderata for the inner harmony and simplicity of the theoretical structure he [the] physicist is attempting to build'.[60] This method or second way he saw illustrated in Einstein's discovery of the general theory of relativity, a theory which could not possibly have been discovered by the first way alone. Foreshadowing the later cosmophysical debate centering on the theories of Milne, Dirac, and Eddington, he said:

> We must admire Galileo for insisting on observational fact as the ultimate arbiter, ... [and] for his power and skill in obtaining physical principles from the immediate generalization of experimental facts. But we must not let this just admiration blind us to the power and skill of those other theoretical physicists who obtain the suggestion for physical principles from the inner workings of the mind and then present their conclusions to the arbitrament of experimental test.[61]

Tolman did not advocate an extreme deductivist or rationalist approach to science, but he did allow mathematical constructions an important and in some cases indispensable role. When it came to the context of justification, he (much like Einstein) kept to 'observational fact as the ultimate arbiter'. The questions addressed by Tolman not only continued to be discussed throughout the 1930s, they are also relevant to the modern arena of cosmology and fundamental physics.

4.4 THE RECEPTION OF COSMOPHYSICS

Seen over a longer time span, the theories of Eddington and Milne were just failures, two more examples of attempted revolutions in physics that ended up in the graveyard of theories of everything. By 1950 the two theories of cosmophysics were largely forgotten and current interest in them is essentially restricted to historians and philosophers of science. While Hermann Bondi found some of the elements in Milne's theory valuable, he dismissed Eddington's way of thinking: 'A deductive theory that aims at showing that an observationally known fact is a necessary consequence of the process of human thought is immediately suspected of fitting the means to the end'.[62] This was undoubtedly a common attitude among physicists, not only after Eddington's death but also during his lifetime. The involved nature and unconventional mathematical basis of his arguments did little to lessen the suspicion. To the limited extent that Eddington's ideas attract attention among modern physicists it is as a system of philosophy of science and not because they are believed to be scientifically fruitful.

In the period from about 1930 to 1948 the situation was different. Although controversial, the ideas of Eddington and Milne attracted a great deal of attention, both scientifically and philosophically. The interest was by far the greatest in Britain, but physicists and astronomers elsewhere also found it worthwhile to look into the remarkable thought constructions of Eddington and Milne, if for no other reason than to criticize them. Indeed, criticism was more common than support.

Milne's theory had a considerable influence on theoretical cosmology in the 1930s. Even an observational cosmologist like Edwin Hubble could, for a time, express a positive interest in Milne's kinematic model of the universe, which he found to 'possess unusually significant features'.[63] Among the scientists who were inspired by his theory and for some time worked within the problem area defined by it, were William McCrea, Martin Johnson, Gerald Whitrow, Arthur G. Walker, and John B. S. Haldane. Also Bondi, in his approach to the steady-state theory of the universe, was to some extent inspired by Milne's way of thinking. Walker's connection to Milne is worth recalling because it was a comparison of kinematic relativity and general–relativistic cosmology that in 1935 led him to the formulation of the Robertson–Walker metric, the most general form of metric for a space-time satisfying the cosmological principle. Walker's paper of 1935 carried the title 'On Milne's theory of world structure'.

Eddington's unified theory made only very little impact on mainstream physics and was generally dismissed as 'romantic poetry', as Pauli once called it.[64] Much of the opposition was methodological, a critique of the rationalistic elements in Eddington's fundamental theory, but it was also rooted in Eddington's unorthodox understanding of quantum mechanics. At a conference held in Warsaw in 1938, his discussion of the cosmological applications of quantum mechanics met united and stiff opposition from eminent quantum physicists such as Niels Bohr, Hendrik A. Kramers, Eugene Wigner, Oskar Klein, and Charles G. Darwin. None of them could recognize in Eddington's presentation what they knew as quantum mechanics. Only a few of the leading physicists looked upon Eddington's programme with milder eyes. Pascual Jordan believed that it was a valuable contribution to a renewal of fundamental physics, and Erwin Schrödinger was 'convinced that, for a long time to come, the most important research in physical theory will follow closely the lines of thought inaugurated by Sir Arthur Eddington'.[65] In works from the late 1930s, the father of wave mechanics eagerly developed some of the consequences of Eddington's theory, but his love affair with it was short. Schrödinger soon reached the conclusion that it was not possible to express the theory in language accessible to everyone—or, perhaps, anyone. In a letter to Eddington he wrote of his troubles in understanding the true meaning of the theory: 'My suspicion is, that there exist a few very important points, which you explain orderly in the right place, but for some reason or other we misinterpret your words just as if they were Chinese'.[66]

The only aspect of Eddington's theory that attracted widespread interest was his numerological arguments that the constants of nature are intimately related. Numerology of a more or less Eddingtonian flavour became popular in the 1930s when many physicists, often outside the mainstream tradition, engaged in the study of the constants of nature. Among the most industrious workers in this area were the Austrian Arthur E. Haas and the Germans Hans Ertel, and Reinhold Fürth. Whereas Eddington ignored Lemaître's model of a big-bang universe, Ertel considered it within the perspective of cosmonumerology.[67] None of the many Eddington epigones endorsed his more general theory or sought to develop it into a coherent and more understandable system of physics.

It would be unfair to characterize Dirac as an Eddington epigone, but the cosmological theory he developed in 1937–38 was admittedly inspired by Eddington's theory and also included elements of Milne's world physics. Dirac based his new cosmology on what he called the large number hypothesis, which is the postulate that whenever two very large numbers turn up in nature, or can be constructed from natural constants, they must be related in a simple way. It was known that the ratio of the electric to the gravitational force is a large dimensionless number, about 10^{39}. Dirac pointed out that if time is measured in units of the time it takes light to pass a classical electron, $\Delta t = e^2/mc^3$, then the age of the universe, as roughly given by the present Hubble time T_0, will be approximately the same number:

$$\frac{e^2}{GmM} \simeq \frac{T_0}{\Delta t} \simeq 10^{39}$$

That is, from this apparent coincidence, and by assuming e, m, M, and c to be true constants, Dirac concluded that the gravitational constant decreases slowly in time: $G \sim 1/t$. From similar arguments he made the radical proposal that the number of particles in the universe increases with the time.[68] The general idea that constants of nature may vary in time is still discussed in modern cosmology, if in a way that differs from Dirac's original arguments (see Chapter 7). Another influence on later physics is by way of the anthropic principle, which in its origin owed much to a critique of Dirac's cosmology (Chapter 9).

Dirac was to some extent influenced by Eddington, but he never felt tempted to build up a unified system of quantum physics and cosmology or otherwise to construct an all-encompassing theory of the physical world. This was not his style of physics. Instead he advocated a step-by-step approach in which a mathematically appealing theory was gradually improved to cover more and more areas. In an address of 1981, a few years before his death, he dissociated himself from the idea of a theory of everything: 'A good many people are trying to find an ultimate theory which will explain all the difficulties, maybe a grand unified theory. I believe that it is hopeless. It is quite beyond human ingenuity to think of such a theory'.[69] Incidentally, this was a view in harmony with that of his great predecessor as Lucasian professor, Isaac Newton, as quoted at the beginning of Chapter 1.

The cosmophysical theories of Eddington and Milne were the works of distinguished scientists and therefore supposedly of a scientific nature. But not all scientists thought it was good science, and some questioned if it were science at all. In England the emergence of cosmophysics gave rise to an interesting and heated debate in which scientists and philosophers were forced to discuss foundational issues, including the very definition of science. The theoretical astronomer George McVittie, a leading exponent of standard relativistic cosmology and at the time working at King's College in London, thought that Milne's elaborate system was closer to pseudoscience than to science. It was a theoretical construction that was intellectually impressive but had little to do with nature as observed. In a critical review of Milne's and Walker's expositions of kinematic relativity, he complained about 'the absence of any attempt to compare the results of their theory with observation'. Then: 'It is eventually borne in on the puzzled reader that Milne and Walker are not trying to understand Nature but rather are telling Nature what she ought to be. If Nature is recalcitrant and refuses to fall in with their pattern so much the worse for her.'[70]

In a more elaborate and even sharper form, Milne's physics was criticized by Herbert Dingle, an astrophysicist and philosopher of science at Imperial College, University of London. In a book of 1937, *Through Science to Philosophy*, Dingle had argued that science was organized common sense, and philosophy an attempt to extract critically and systematically the results of the special sciences. This positivistic view of science was flatly contradicted by the physical theories of Milne and other cosmophysicists. The target of Dingle's vitriolic attack was not only Milne but also Eddington and Dirac, whom he saw as the chief representatives of what he called 'modern Aristotelianism'. Basically a positivist and supporter of the inductive–empirical method, Dingle insisted that he had no problems with speculation and imagination so long they entered legitimately in the discovery process. But he did have problems with the kind of unrestrained speculative physics offered by his three distinguished opponents:

> The question presented to us now is whether the *foundations* of science shall be observation or invention. Newton did not lack imagination, but he chose to examine pebbles rather than follow the Gadarene swine, even when the ocean before him was truth. Milne and Dirac, on the other hand, plunge headlong into an ocean of 'principles' of their own making, and either ignore the pebbles or regard them as encumbrances. Instead of the induction of principles from phenomena we are given a pseudo-science of invertebrate cosmythology, and invited to commit suicide to avoid the need of dying.[71]

In a later address, delivered after the deaths of Eddington and Mine, Dingle repeated his complaint that the science of cosmology was in danger of becoming a non-science. As he saw it, it was governed by considerations based on personal taste and pure reason, not on inductions from observational data. Dingle pointed out that the laws of nature cannot possibly be derived by reason alone, without recourse to experience. His argument was simply that laws of *nature* are by definition laws about what we experience. We can derive laws of 'nature' without worrying about experiential facts, but we cannot know that they are laws of nature—that 'nature' = nature—without recourse to such facts. So, he concluded, the claim of a purely rational physics is self-contradictory.[72] Although simple, the argument is not without force.

In the debate that followed Dingle's attack of 1937, the distinguished Edinburgh physicist Charles Galton Darwin felt obliged to defend Dirac's speculative cosmology, not because he believed in it but because he considered it a legitimate scientific hypothesis. 'It is surely hard enough to make discoveries in science without having to obey arbitrary rules in doing so', he said; 'in discovering the laws of Nature, foul means are perfectly fair.'[73] But he restricted his advocacy of foul means to the creative phase of science, the so-called context of discovery. Eddington used the occasion to repeat his belief in *a priori* science, maintaining that he had calculated 'the mass-ratio of the proton and electron by an *a priori* method which does not involve any observational measurements'. From this he concluded that knowledge of the mass ratio did not belong to the objective universe. A true law, according to him, was not merely an empirical regularity with no known exceptions, it had a compulsory character: violation of a law of nature is strictly impossible. 'The general laws of Nature (embodied in the fundamental equations of mathematical physics), including the universal constants associated with them, do not express knowledge of an objective universe.' This interpretation was very much Eddington's own and not, I think, shared by other physicists either then or later.

The debate in *Nature* resulted in some clarification but not in a consensus opinion as to the proper methods and standards of theoretical physics. A few of the participants, notably the mathematician and geophysicist Harold Jeffreys, tended to agree with Dingle. As Jeffreys argued, although the much-criticized inductive method was no magic wand neither could it be totally dispensed with, not even in fundamental physics. 'Without using induction, Milne and Eddington could not order their lives for a day', he quipped. Jeffreys suggested a short list of criteria that a physical theory or system must satisfy in order to be recognized as scientific. Among the criteria were that the system must not treat any hypothesis as *a priori* certain, that it must be possible to choose between hypotheses on an observational basis, and that it must be possible that the consequences of the system were wrong.[74] Clearly, the theories of Milne and Eddington did not comply with these standards. They agreed somewhat better with the view of William McCrea, an astrophysicist and cosmologist who had worked with both Milne and Eddington. A system of mathematical physics was to him simply the working out of the mathematical consequences of certain hypotheses. 'The worth of the theory is judged ... by the closeness of the agreement of its predictions with the results of observation, and also the number of phenomena which it can so predict from the one set of hypotheses.' He added that 'The scientific attitude is, not to cavil at the attempt, but to see if it is successful'.[75]

The debate initiated in *Nature* in 1937 continued four years later in the same journal. This time Eddington's rationalism was in the centre, being criticized by Jeans and Dingle, among others. Whereas Jeans held that the finiteness of the speed of light was an empirical fact, hence contingent, Eddington maintained that it was true by necessity. It was knowledge about nature that we have prior to the actual observation of it:

> We know *a priori* that the velocity of light is not infinite, just as we know *a priori* that the velocity of light is not blue or hexagonal or totalitarian; it is not the sort of thing to which these terms apply. ... When an *observer* sets out to determine the velocity of light, an infinite result is not among the possible alternatives; and if he announces that he has found the velocity to be, not merely exceedingly large, but actually infinite, we know *a priori* that the announcement is untrue.[76]

Dingle sided with Jeans, arguing the standard view that physics is in essence an empirical science whose laws can conceivably be violated by experiments. Eddington would have nothing of it: 'If the fundamental ("inviolable") laws are not assertions about experience, they cannot in any circumstances be violated by experience.'[77]

It is noteworthy that there were no similarly broad discussions of cosmophysics outside Britain. The research tradition, if it can be so called, was largely confined to the United Kingdom, where it may have been cultivated actively by at most a dozen physicists, astronomers, and mathematicians. The only leading physicist outside Britain who contributed to the cosmophysical tradition was Pascual Jordan in Germany, and he considered it in an entirely different light to Milne and Eddington. A positivist and empiricist, Jordan emphasized that there was nothing speculative or *a priori* about numerological reasoning based on the constants of nature. Far from being cosmythology, as Dingle would have it, he was convinced that the Eddington–Dirac approach was a promising road toward a truly empirically based understanding of the universe.[78]

Notes for Chapter 4

1. For a clear and accessible comparison of Eddington's and Milne's theories, see Burniston Brown 1949.
2. On these attempts, see the survey in Kragh 2007, pp. 100–10.
3. Thomson 1905, p. 7.
4. Nicholson 1913, pp. 103–105. On Nicholson's atomic theory and its relation to the models of Rutherford and Bohr, see McCormmach 1966.
5. Lewis 1922, p. 309.

6. Many of the classical papers of cosmology, including the works of 1917 by Einstein and de Sitter, are translated in Bernstein and Feinberg 1986. From a modern point of view, the de Sitter model is exponentially expanding, but this is not how the model was seen in the 1920s.

7. Kragh 1996, pp. 42–44, 81–101.

8. Rice 1925.

9. Lanczos 1925, p. 80.

10. On Lemaître's primeval-atom model see Kragh and Lambert 2007, which includes references to the primary sources.

11. Quoted in Kerzberg 1989, p. 334. The history of the cosmological constant is reviewed in Earman 2001.

12. Bernal 1939, pp. 3–5.

13. Eddington's philosophy of physics has attracted more interest than his fundamental theory of physics. A good guide to the latter is Kilmister 1994. See also Douglas 1956, Singh 1970, pp. 239–53, and Durham 2003.

14. Eddington 1946. Some of Eddington's earlier drafts not included in *Fundamental Theory* were published and discussed in Slater 1958.

15. Weyl 1919, p. 129. Eddington described Weyl's suggestion the following year (Eddington 1920b, p. 178).

16. Eddington 1923, p. 237.

17. Eddington 1931a, p. 606.

18. Eddington 1935a, p. 227. The significance of the constants of nature will be further dealt with in Chapter 7.

19. Eddington 1936, p. 3.

20. Eddington 1946, p. 265.

21. Eddington 1923, p. 167.

22. Eddington 1939a, p. 170.

23. Eddington 1936, p. 327.

24. Eddington 1939a, p. 57.

25. Ibid., p. 60. Emphasis added. Eddington can reasonably be counted as one of the precursors of the anthropic principle that will be examined in Chapter 9.

26. Eddington 1936, pp. 3–4.

27. In an extensive review of Eddington 1936 appearing in *The Observatory* **60** (1937), 14–23.

28. The sources are Eddington 1931a, Eddington 1936 (p. 279), Eddington 1944, and Eddington 1946 (p. 10). Due to the gravitational effects of matter, the real values would be somewhat smaller. In 1954 Noel Slater found an error in Eddington's calculations that would reduce his value of 1946 by a factor of 4/9.

29. Eddington 1935b, p. 636.

30. Eddington 1936, p. 304. For the mass ratio of 1847.6, see Eddington 1933, p. 117. As late as 1970, an American physicist, suggesting that Eddington's fundamental theory 'is likely to be partly right', used Eddingtonian arguments to deduce the value $M/m = 1836.12$ (Good 1970). For a comparison of Eddington's values for the constants of nature with those obtained by later scientists using somewhat similar numerological methods, see Eagles 1976.

31. For a history of the fine-structure constant, see Kragh 2003. Eddington's view of the constant is also considered in Bekenstein 1986.

32. The table is adapted from Burniston Brown 1949.

33. Eddington 1929, p. 358.
34. Eddington 1932, p. 41.
35. Eddington 1939a, p. 112: 'In an ordinary way I might say that I do not believe in neutrinos.' In a paper of 1936, Dirac rejected the neutrino hypothesis, arguing that the neutrino was unobservable and introduced only to save energy conservation in beta decay (see Kragh 1990, pp. 170–72). However, this was a minority view. By the late 1930s almost all physicists accepted the neutrino as a real if undetected elementary particle.
36. Eddington 1946, p. 214. The first evidence of a heavy meson (a kaon) was reported in 1944, undoubtedly unknown to Eddington. The class of heavy mesons (kaons and hyperons) was discovered 1948–52, with masses up to 1.3 proton mass. A. H. Klotz, a mathematician at the University of Sydney, later calculated from Eddington's theory the mass and lifetime of the neutral pion π^0 to be 282 times the electron mass and 2.4×10^{-16} s, respectively (Klotz 1969). The presently known values are 267.3 and 8.4×10^{-17} s.
37. See Kragh and Carazza 1995 for comments and references.
38. For a modern analysis, see French 2003.
39. Eddington 1936, pp. 328–29.
40. Eddington 1920a.
41. Eddington 1920b, pp. 200–201. Whereas Eddington suggested that there were no genuine and cognizable laws in the external world, he was willing to consider the 'law of atomicity' to have its seat in external nature: 'The world may be so constituted that the laws of atomicity must necessarily hold; but, as far as the mind is concerned, there seems no reason why it should have been constituted in that way. We can conceive a world constituted otherwise.' Ibid., p. 199.
42. Eddington was a devoted Quaker and convinced that his science was complementary to his religious faith. For a scholarly study of Eddington's mysticism and religious faith, see Stanley 2007. A briefer account is provided in Batten 1994.
43. Analyses and descriptions of Milne's cosmophysics include Harder 1974, Lepeltier 2006, Urani and Gale 1994, and Kragh 2004, pp. 200–29. For an account of his cosmological theory, see also North 1990, pp. 149–85 and Singh 1970, pp. 168–91.
44. Broad 1940, p. 312.
45. Letter to Herbert Dingle, quoted in Douglas 1956, p. 178.
46. Letter of 6 July 1943, quoted in Chandrasekhar 1987, p. 85.
47. Rebsdorf and Kragh 2002.
48. Milne 1940, p. 133.
49. Milne 1948, p. 10.
50. Milne 1943, as reproduced in Munitz 1957a, on p. 369.
51. Eddington 1939b, p. 231. Recall that the Lemaître–Eddington universe includes an accelerating phase after its slow beginning in an Einstein state.
52. Milne 1937, p. 999.
53. Milne 1939, p. 211.
54. Milne 1940, p. 151.
55. Milne 1937, p. 998.
56. Milne 1934, p. 27.
57. Milne 1935, p. 266.
58. Milne 1937, p. 999.

59. On the connection between Milne's Christian belief and his cosmophysics, see Milne 1952a and Kragh 2004, pp. 212–19.

60. Tolman 1932, p. 373.

61. Ibid. Historians of science have long ago dismissed the picture of Galileo as an empiricist who obtained his results 'from the immediate generalization of experimental facts'. Galileo followed the second way no less than he followed the first way.

62. Bondi 1952, p. 158. Eddington's legacy is briefly considered in Wesson 2000.

63. Hubble 1936, p. 199.

64. Letter from Pauli to Oskar Klein, 18 February 1929, quoted in Pauli 1979, p. 491. Milne, more charitably, called Eddington's fundamental theory 'a notable work of art' (Douglas 1956, p. 176).

65. Schrödinger 1937, p. 744. For Schrödinger's inspiration from Eddington, see Rüger 1988.

66. Letter of 23 October 1937, quoted in Kragh 1982, p. 105.

67. See Schröder and Treder 1996.

68. Dirac 1937. On Dirac's cosmology and his use of the large number hypothesis, see Kragh 1990, pp. 223–46 and also Chapter 7. The history of large numbers and their interrelations is also covered in Barrow 1990.

69. Quoted in Kragh 1990, p. 361.

70. McVittie 1940, p. 281. McVittie's views on the methodology and nature of cosmology is analysed in Sánchez-Ron 2005.

71. Dingle 1937a, p. 785. The debate is analysed in Kragh 1982, Gale 2002 and Gale and Urani 1999.

72. Dingle 1953, pp. 397–98.

73. From supplement to *Nature* on 'Physical science and philosophy', *Nature* **139** (1937), 1000–12. The following quotations are from the same source. Charles G. Darwin was a grandson of the 'real Darwin' of evolutionary fame.

74. Jeffreys gave a full exposition of his view of science in Jeffreys 1931, which appeared in a second edition in 1957.

75. McCrea 1937, p. 1002.

76. *Nature* **148** (1941), p. 256. In the form of letters to the editor, the discussion took place under the general title 'The Philosophy of Physical Science' from August to October 1941. Eddington not only denied that the speed of light could possibly be infinite, he also denied that it could vary in time. Such speculations, he said, were 'nonsensical because a change of the velocity of light is self-contradictory.' Eddington 1946, p. 8. On the question of a possible variation of the speed of light, see also Chapter 7.

77. *Nature* **148** (1941), p. 505.

78. On Jordan as a cosmophysicist, see Kragh 2004, pp. 175–85.

5

Cosmology and Controversy

> *In cosmology two impostors have usurped the throne of science, worn her crown and taken her name. Whereas the source of final court of appeal in science is experience, that of one impostor is personal taste, and that of the other, pure reason.... Let our younger cosmologists forget cosmology for a space of three years—the universe is patient, it can wait—and instead read the history of science.*
>
> Herbert Dingle, *Science and modern cosmology*, 1953, p. 398 and p. 407.

By its very nature, cosmology invites speculation. As a science that aims to describe and explain the universe in terms of physical law, it depends on astronomical observations, both as input for constructing theories and as data for evaluating them. However, few observations are cosmologically relevant and it is to a large extent the underlying theory that determines the role and relevance of observations. If modern cosmology was born in the interwar period, it went through a difficult period of adolescence in the first two decades after World War II. In this process of maturation the steady-state theory of the universe played an important role. Although proved wrong by observations in the 1960s, the theory of an eternally expanding yet stationary universe is methodologically instructive and of great historical interest.[1] There are even remnants of it in some of the current ideas about the universe.

The development of the steady-state theory illustrates in a pure form several of the features that can be found in later theories of the universe, such as the delicate interplay between theory and observation and the role played by general principles of a regulative nature. What was at stake during the cosmological controversy that raged in the 1950s was not merely an astronomical world picture, it was also the very standards on which scientific cosmology should be judged. Foundational issues were part and parcel of the controversy. In such a situation it can be difficult to distinguish between scientific and metascientific or philosophical arguments—the line of distinction was part of the controversy. Indeed, the role of philosophical reasoning, whether coming from scientists or philosophers, is one of the most interesting features of the debate and one we can still learn from today.

Like the earlier theory of Milne, the original steady-state theory was designed as a theory of the one and only universe, not as a theory including or significantly relying upon microphysics. It was not meant to be a theory of everything. Nonetheless, the dream of connecting the universe at large with the local physics of matter and fields was not foreign to steady-state physicists. Fred Hoyle, in particular, sought to develop his version of steady-state cosmology in directions that included both gravitation and quantum physics. Another of the steady-state pioneers, Hermann Bondi, also saw cosmology as linked to microphysics, but not in the traditional sense of microphysics leading to cosmophysics. According to his point of view, probably inspired by Milne, 'cosmology is the most fundamental of the physical sciences, the proper starting-point of all scientific considerations.... Cosmology [is] the truly all-embracing science, the long-sought link between philosophy and the physical world.'[2]

5.1 THE STEADY-STATE THEORY OF THE UNIVERSE

In 1948, when the new steady-state cosmology was introduced, there was no consensus about the large-scale structure and evolution of the universe except that a majority of astronomers agreed that the universe expands in conformity with the field equations of general relativity. Most believed that the expansion took place without the cosmological constant, but the presence of $\Lambda \neq 0$ remained a possibility. It is to be emphasized that a continually expanding and therefore increasingly empty universe does not necessarily imply a big-bang universe of the type that George Gamow and his collaborators proposed in the period 1946–53. This is exemplified by the Lemaître–Eddington model of the universe, which is continually expanding but cannot be assigned a definite age. The hypothesis of an exploding universe with an origin in time, whether in the version of Lemaître or Gamow, was well known but far from generally accepted. Many astronomers and physicists considered it to be speculative and conceptually weird. First and foremost, they lacked convincing empirical evidence that the universe had once been in a ultracompact state entirely different from the present one. As late as 1956, three distinguished British physicists concluded of Gamow's theory that it 'cannot be regarded as more than a bold hypothesis'.[3]

In England, the three young physicists, Fred Hoyle, Hermann Bondi, and Thomas Gold, suggested a very different conception of the universe, namely that it had existed for an eternity of time in about the same form that we observe it today and that it would continue to do so in the future. In a general and qualitative sense, similar ideas had been put forward before World War II, mostly by scientists who did not accept the expansion of the universe, such as the Chicago astronomer William MacMillan and the distinguished physicist Robert Millikan, a Nobel laureate of 1923.[4] The theory of Hoyle, Bondi, and Gold was very different from these older ideas, among other reasons because it took the expanding universe for granted. The steady-state theory, as it was soon called, was developed in two rather different versions, one by Hoyle and the other jointly by Bondi and Gold. Both of the founding papers appeared in the summer of 1948 in the *Monthly Notices* of the Royal Astronomical Society, the Bondi–Gold paper slightly earlier than Hoyle's paper.

Bondi had recently published a review of cosmological theories in which he distinguished between what he called the 'extrapolatory' and 'deductive' approaches to the study of the universe.[5] In the first approach, which he said was characteristic of relativistic cosmology, the physicist extrapolates ordinary terrestrial physics to form a comprehensive theory, *in casu* the general theory of relativity, which is then assumed to be valid for the entire universe. The consequences of this assumption are then compared with measurements. The alternative deductive approach, which Bondi saw realized in Milne's kinematic relativity, is to start from a small number of cosmological axioms or postulates and from these to deduce the corresponding physical theories. The Bondi–Gold theory clearly belonged to the deductive type and in a general way shared some of the methodological characteristics of Milne's cosmology, although the similarity was not deep.[6] While Milne based his theory on the ordinary cosmological principle, relating to locations in space, Bondi and Gold argued that only in a universe which is stationary on large scales can it be safely assumed that the laws of physics are constant in time. Such constancy, both spatial and temporal, was at the bottom of what Bondi called the 'repeatability principle' and therefore a *sine qua non* for obtaining scientific knowledge: 'Unless we postulate that position in space and time is irrelevant,... the repetition of an experiment becomes quite impossible since the condition of space and time cannot be repeated.'[7]

Bondi and Gold consequently proposed extending the ordinary cosmological principle into what they called the *perfect cosmological principle*. This principle or assumption remained the defining foundation of the steady-state theory, at least as conceived by Bondi and Gold and their few followers. It states that the large-scale features of the universe do not vary with either space or time. According to Bondi and Gold it was a postulate or fundamental hypothesis. One could argue for it philosophically or in terms of the consequences it implied, but not derive it from other physical laws. Although it could be supported by observations, it could not be proved observationally. On the other hand, it could be disproved by observations, namely if the consequences of the principle were unequivocally contradicted by observations.

The original aim of Bondi and Gold was to base the science of the universe on cosmological considerations rather than on established physical theories, and if these disagreed the two men were not much concerned. This claim of an independent cosmological science, which as mentioned was probably inspired by Milne, was radical and provided ammunition for critics of the steady-state theory. As one of the critics charged: 'By severing the continuity with physics, the "New Cosmology" abandons the tradition of scientific cosmology and takes the road of all crackpot theories, which are characteristically isolated from the bulk of science. Nothing can justify the rejection of physics in the name of cosmological considerations.'[8]

Whereas Hoyle did not think very highly of the perfect cosmological principle, Bondi and Gold did: 'We regard the principle as of such fundamental importance that we shall be willing if necessary to reject theoretical extrapolations from experimental results if they conflict with the perfect cosmological principle even if the theories concerned are generally accepted.'[9] They added, in a more cautionary mode: 'Of course we shall never disregard any direct observational or experimental evidence and we shall see that we can easily satisfy all such requirements.' What might look to be an *a priori* principle, and was often accused of being one, was not really of an *a priori* nature. On the other hand, Bondi and Gold assigned it such importance that it came close to being so:

> We do not claim that this principle must be true, but we say that if it does not hold, one's choice of the variability of the physical laws becomes so wide that cosmology is no longer a science. One can then no longer use laboratory physics without relying on some arbitrary principle for their extrapolation. But if the perfect cosmological principle is satisfied in our universe then we can base ourselves confidently on the permanent validity of all our experiments and observations and explore the consequences of the principle. Unless and until any disagreement appears we therefore accept the principle, since this is the only assumption on the basis of which progress is possible without further hypothesis.[10]

Now it is obvious that the perfect cosmological principle alone is insufficient to construct a new cosmological model; witness that Einstein's original static model of 1917 also satisfies the principle. But, as shown by the continual increase of entropy in the universe (the thermodynamic arrow of time) and independently by Hubble's redshift data, the universe is not static. It is expanding. Apparently the expansion of the universe contradicts the perfect cosmological principle because the expansion implies that the average density of matter decreases with time. Rather than admitting a contradiction, the steady-state theorists drew the conclusion that matter is continually created throughout the universe at such a rate that it precisely compensates the thinning out caused by the expansion. Bondi and Gold easily showed that the creation rate must be given by $3\rho H \approx 10^{-43}$ g s^{-1} cm^{-3}, where ρ is the average density of matter and H the Hubble constant. The small creation rate made it impossible to detect the creation of new matter by direct experiment, yet the hypothesis did have detectable consequences. The new matter was supposed to be created in the form of hydrogen atoms, or perhaps neutrons or electrons and protons separately, but this was nothing more than a reasonable assumption.

It hardly needs to be emphasized that *spontaneous* matter creation is a most radical claim, violating as it apparently does one of the most sacred principles of physics, the law of energy conservation. Whereas many physicists and astronomers considered the creation of matter to be a grave violation of the norms of physics, something of the nature of a criminal offence, the steady-state advocates saw no great problem. After all, do we know empirically to *any* degree of accuracy that matter (or matter–energy) is conserved? Of course not: we only know that it is conserved to a very high degree of accuracy. This is what experiments tell us, and no more. In addition, if matter creation were a problem in the steady-state theory, it could be argued that it was no less a problem in the big-bang versions of relativistic cosmology where all matter was created suddenly and mysteriously at the beginning of time. Gold considered the tiny violation in steady-state theory to be 'a matter of complete indifference' and justified by the perfect cosmological principle being a simpler and more powerful explanatory principle than the law of energy–mass conservation.[11] According to Hoyle, matter creation did not really violate the law of energy conservation, but he agreed with Gold that the hypothesis was scientific and preferable on methodological grounds. In *The Nature of the Universe*, his best-selling popular book of 1950, he said of the continuous creation of matter: 'This may seem a very strange idea and I agree that it is, but in science it does not matter how strange an idea may be so long as its works—that is to say, so long as the idea can be expressed in a precise form and so long as its consequences are found to be in agreement with observation.'[12]

Hoyle's version of the steady-state theory differed in important respects from the one of Bondi and Gold, but led to the same results and predictions. Contrary to his two friends and colleagues, he did not introduce the perfect cosmological principle axiomatically. He argued that it followed from his theory, which also led to matter creation and a constant density of the cosmos. The theory he proposed in 1948 and developed over the following years was based on field equations that formally had a great deal of similarity to the relativistic equations but included a 'creation tensor' added for the purpose. Hoyle's creation tensor replaced the cosmological constant in Einstein's equation, but in such a way that it allowed for a violation of energy conservation. It followed from Hoyle's equations not only that the mass density must remain constant, but also that it had the definite value of

$$\rho = \frac{3H^2}{8\pi G} = \frac{3}{8\pi G}\frac{1}{T^2}$$

The value happened to be equal to the so-called critical density characteristic of the 1932 Einstein–de Sitter model. While in the relativistic model this density implies a slowing down of the expansion, in the steady-state model it corresponds to an exponentially growing expansion, $R(t) \sim \exp(Ht)$. The density predicted by Hoyle, given by the value of H known at the time, was $\rho \approx 5 \times 10^{-28}$ g cm^{-3}, much larger than the average density of matter estimated from observations of stars and galaxies. According to Hoyle, this only showed that most of the matter in the universe existed in other, non-luminating forms.

In relativistic cosmology the Hubble time, $T = 1/H$, is a measure of the age of the universe, and therefore by its very nature is increasing in time. In steady-state cosmology, which presupposes an eternal universe, T is a true constant characteristic for the universe but has nothing to do with cosmic age. It further follows from the perfect cosmological principle that space is flat, meaning that there are infinitely many galaxies (and stars, atoms, electrons...) in the steady-state universe. The appearance of an actual infinity of objects is in many ways philosophically troubling, but this feature went unnoticed in the early phase of the theory. Finally, Hoyle, as well as Bondi and Gold, assumed that the cosmological constant must be zero. The results mentioned above were common to both versions and can be considered the basic features of the classical steady-state theory.

It is quite remarkable that the simple steady-state theory of the universe led to several definite and testable predictions; and, in addition to these, to several consequences of a less definite nature: qualitative expectations of what new observations would be like. The theory predicted unambiguously that we live in a flat, exponentially expanding universe. From this it follows that the so-called deceleration parameter, a measurable quantity that expresses the rate of slowing down of the expansion, must have the value $q = -1$. (The parameter is defined as $q = -R''/RH^2$, with $R' = dR/dt$.) It also follows from the steady-state theory, as first shown by Hoyle, that the angular diameter of a distant object will vary with the redshift of the object in a definite way. This relationship is testable too, if not easily. As to galaxies, the theory predicted that there is no systematic difference between galaxies nearby and far away. The ages would be distributed according to a certain statistical law, giving an average age for galaxies, or more generally matter in a large volume, of $T/3$. Of course, the time scale difficulty of the universe being younger than some of its constituent parts did not exist in the steady-state theory. What matters here is not so much the relationship between these and other predictions and actual observations, but that the predictions existed. In short, the steady-state theory was eminently testable.

The original steady-state theory, as it appeared in 1948, was developed by a small group of physicists and astronomers over most of the next two decades. Most of these contributions related to Hoyle's field-theoretic version, whereas it proved difficult to develop the more closed and deductive version of Bondi and Gold. Almost all the attempts to extend and develop the theory took place in England. Much like the earlier cosmophysical theories of Milne and Eddington, steady-state cosmology was very much a British theory and it attracted almost no positive response outside the United Kingdom. The critique, too, was mainly confined to the British scene. I have counted 40 scientists actively interested in the steady-state theory from 1948 to about 1970, of whom about 30 can be classified as supporters or sympathizers of the theory. Almost all of them were Britons.

Among the earliest and most important contributions was a theory proposed in 1951 by William McCrea, the Irish-born theoretical physicist who had earlier worked with both Eddington and Milne. An early advocate of the steady-state theory, McCrea felt that the status ascribed to the perfect cosmological principle by Bondi and Gold was exaggerated and unfruitful. As he saw it, the principle was not really a fundamental principle of nature because it was not applicable to small-scale physics. In his paper of 1951 McCrea suggested a way to explain the mysterious creation of matter that was an alternative to Hoyle's formal explanation by means of the creation tensor. By introducing a uniform negative pressure given by $p = -\rho c^2$ he explained matter creation as a kind of transmutation process rather than a genuine creation *ex nihilo*.[13] In this way he could argue that the process, often criticized for being *ad hoc* and plainly unscientific, was compatible with the notion of energy conservation in the general theory of relativity.

Another interesting innovation was the 'electrical cosmology' proposed by Bondi and the Cambridge astronomer Raymond Lyttleton in 1959.[14] This model was based on the hypothesis of a universal charge excess, meaning that the numerical charge of the electron differs slightly from that of the proton. In that case a hydrogen atom would not be neutral, but carry a tiny charge ϵe, where ϵ was found to be at least 10^{-19}. Bondi, Lyttleton, and Hoyle argued that the hypothesis supplied the source for the negative pressure out of which, according to McCrea, matter was created. For a year or so the electrical steady-state theory was thought to be an important advance in cosmology. Alas, it was mercilessly shot down when high-precision experiments proved that the proton–electron charge excess is smaller than the 10^{-19} required by the steady-state theorists. The short-lived electrical universe is an interesting case of an ambitious cosmological theory that was proved wrong by precision experiments in the laboratory.

The spontaneous creation of matter out of nothing continued to plague the Bondi–Gold theory. Consequently, physicists sympathetic to the steady-state universe came up with various suggestions of how to account for the necessary creation of matter without violating the law of energy conservation. McCrea's theory of 1951 was one of them, and in 1959 V. A. Bailey from the University of Sydney proposed a different way in which steady-state matter creation could be harmonized with the old dictum *ex nihilo nihil fit*. Bailey suggested that the apparent continual supply of matter came from a five-dimensional hyperspace, in which the ordinary four-dimensional space–time was embedded. Associating the idea with the Kaluza–Klein theory of the 1920s, he imagined that the law of energy conservation would hold in the five-dimensional world but not manifestly in our world of four space–time dimensions. Bailey thought the hypothesis of a fifth dimension was fruitful 'since it serves not only to account for the steady-state universe of Bondi and Gold but also to unify three other great branches of physics', namely electromagnetism, gravitation, and quantum mechanics.[15] Bailey's hypothesis was ignored by contemporary cosmologists, but many years later Kaluza–Klein theories and similar ideas of extra dimensions became popular in theoretical cosmology.

Although a theory of the cosmos, and not one aiming at unifying cosmology and microphysics, the steady-state theory naturally related also to atomic and quantum physics. Bondi's widely used textbook *Cosmology* included a brief chapter on 'Microphysics and Cosmology' in which he discussed the Eddington–Dirac numerical coincidences. 'It is difficult to resist the conclusion', he said, 'that they represent the expression of a deep relation between the cosmos and microphysics, a relation the nature of which is not understood.... In any case it is clear that the atomic structure of matter is a most important and significant characteristic of the physical world which any comprehensive theory of cosmology must ultimately explain.'[16] At the end of the book he returned to the hope that when such an understanding had been obtained it would benefit both cosmic physics and local physics: 'It would be particularly interesting if cosmology, as a large-scale testing ground of physics, were to acquire an influence on the development of local physical theories.'[17] This is indeed what happened over the next few decades, but then by way of the big bang theory of the universe and not the steady-state cosmology that Bondi defended. Bondi's discussion of the numerical relations between physical parameters would later serve as an inspiration for Brandon Carter in his formulation of the anthropic principle (see Chapter 9).

The creation of particles of matter was often assumed to provide a link between the two realms of nature, microphysics and cosmic physics, although it was hard to say exactly how. In his original paper of 1948, Hoyle noted that if the quantity $cT/3$ (one third of the Hubble length) is divided with the range constant of nuclear forces k, one obtains a dimensionless constant of magnitude 10^{39}. Since this is a large number in Dirac's sense, it suggested an invariant relationship between gravitational, electromagnetic, and nuclear constants:

$$\frac{cT}{3k} \cong \frac{e^2}{GmM}$$

Hoyle suggested that the relationship might be of 'deep significance' and referred to earlier considerations of the same kind by Weyl and Eddington.[18]

The interest in combining cosmology and subatomic physics was also expressed by McCrea in his development of the steady-state theory. In a general sense he was convinced that 'the parameters of the whole observable universe, its size, its mass, and Hubble's constant for the recession of the galaxies, must be related to the fundamental constants of atomic physics'. He thought that the steady-state approach was particularly promising in this respect because of the key role played by the creation rate of new matter. What the new cosmological theory promised was nothing less than an answer to 'why the universe is what it is observed to be—all in terms of an extended quantum mechanics'.[19] This was a big promise, indeed much too big. Similar hopes were expressed in connection with the Bondi–Lyttleton theory of an electric steady-state universe, which suggested that the ratio of the observable universe might be determined by the charge difference between electrons and protons. According to Lyttleton, the theory 'seems to promise just those links between cosmology, quantum theory, and relativity that have for so long been dimly perceived peeping over the horizon'.[20] However, the promises remained promises. In spite of scattered attempts, a bridge between cosmology and microphysics was never established on the basis of the classical steady-state theory.

5.2 IS COSMOLOGY SCIENTIFIC?

The steady-state theory of the universe was controversial from its very beginning, criticized by both scientists and philosophers. A substantial part of the debate that occurred in the 1950s related specifically to the steady-state theory, and especially to its claim of continual creation of matter, but it was also concerned with cosmology in general. At the time it was far from obvious that cosmology was a legitimate scientific project on a par with accepted disciplines such as astronomy, physics, and geology. The scarcity of observational data of cosmological relevance furnished a fertile soil for discussions that involved philosophical and other extrascientific arguments. In the absence of data that unambiguously spoke for or against one of the rival theories, much interest was directed toward their logical and conceptual structure. Although it was generally recognized that such considerations had only limited value in evaluating a theory, they were not unimportant in the controversy between the two conceptions of the world. If it could be convincingly argued that one of the theories was 'philosophically more appealing' or 'more scientific' than its rival, it would have an effect on the balance of power between the two cosmological theories.[21]

Arguments that appeal to metascientific principles such as unity, simplicity, and coherence are by their very nature ambiguous. What some scientists saw as methodological virtues of the steady-state theory, others saw as deficiencies and reasons to distrust it. According to the British physicist Martin Johnson, who was generally sympathetic to the steady-state programme, the choice between cosmological models was 'an aesthetic or imaginative choice' rather than a rational one. In his opinion cosmology had 'more in common with the poetic or artistic attitude towards experience than with the solely logical', and he rated the steady-state theory highly by these standards.[22] Evaluations of this kind were common at the time. The esteemed Estonian–Irish astronomer Ernst Öpik agreed that the choice between cosmological models would remain 'a matter for esthetic judgment' and that the steady-state theory 'can at present be addressed only from the standpoint of esthetic value'.[23] Öpik had a preference for cyclic models and he found the theory of Hoyle, Bondi, and Gold to be both wrong and ugly. Again, the eminent Swedish theoretical physicist Oskar Klein stated in 1953 that cosmology was a field where 'personal taste will greatly influence the choice of basic hypotheses'.[24] Klein's taste was for neither big-bang nor steady-state models. He objected to the latter theory by criticizing from a methodological perspective its foundation in the perfect cosmological principle.

Although the response of most astronomers and physicists to the steady-state theory focused on its predictions and their relation to observations, philosophically oriented objections were very much part of the response as well. McVittie was among those who positively disliked the theory, which he accused of being 'unscientific' because he thought it violated a unified description of the universe. A supporter of mainstream relativistic cosmology, he complained about the steady-state scientists' use of *a priori* assumptions and what he took to be rationalistic dogmas. From a long career in cosmology, he knew that 'the temptation to substitute logic for observation is peculiarly hard to resist in astronomy and especially so in cosmology'. Given the scarcity of relevant and reliable data, the researcher might be tempted to 'supplement what is directly observed by additional items of information based on the *absence* of detectable phenomena'. This approach, which he associated with the steady-state theory, he found to be totally unacceptable. Instead, the proper approach would be to emphasize positive and inductively obtained knowledge, to 'discover how much can be found out about the universe through measurements that yield non-null results rather than by the consideration of logical possibilities which might conceivably be the case'.[25]

Many years later, looking back on the then defunct steady-state theory, McVittie commented that its element of matter creation was a violation of the basic rules of scientific reasoning. 'It's like breaking the rules when you are playing a game', he said. 'If you allow yourself in the game of American football to take knives on board with you and stab your opponent, now and again, of course the results will be very remarkable, particularly if one side only has the knives and the other is merely the recipient.'[26] McVittie was not the only scientist who felt that the hypothetical continual creation of matter (which was thought to imply a small-scale violation of energy conservation) was reason enough to dismiss the steady-state theory.

Another critic of the theory of the eternally stationary universe was the British astronomer and mathematician Gerald Whitrow, a lecturer at Imperial College and a former collaborator of Milne. In 1954 he discussed with Bondi the scientific nature of physical cosmology in the pages of the newly founded *British Journal for the Philosophy of Science*. The subject of this interesting dialogue was not limited to Bondi's favoured theory, but concerned a broader question: can modern physical cosmology be considered a science or not? By and large, Whitrow argued that cosmology was not truly scientific and, moreover, it was unlikely ever to become so. It was and would remain semi-philosophical, among other reasons because of the unique nature of its domain, the universe. Bondi, on the other hand, suggested that physical cosmology was quickly on its way to becoming a proper scientific endeavour and was indeed already a science. Whereas Whitrow argued that philosophical and aesthetic values would always remain an essential part of the study of the universe, according to Bondi such values had now largely been replaced with comparison with observations. A firm basis for demarcating science from non-science could only be found in the possibility of empirical disproof, and Bondi thought that cosmology lived up to this crucial criterion. As he expressed it in his debate with Whitrow:

> Although the adherents of some theories are particularly concerned with pointing out the logical beauty and simplicity of their arguments, every advocate of any theory will always be found to stress especially the supposedly excellent agreement between the forecasts of this theory and the sparse observational results. The acceptance of the possibility of experimental and observational disproof of any theory is as universal and undisputed in cosmology as in any other science, and, though the possibility of logical disproof is not denied in cosmology, it is not denied in any other science either. By this test, the cardinal test of any science, modern cosmology must be regarded a science.[27]

Several of the critics of the steady-state theory sought to undermine its credibility by means of thought experiments or by arguing that it led to highly bizarre and perhaps even contradictory consequences. Non-empirical 'testing' of this kind played an important role in the controversy, just as it had done in some earlier cases and would continue to do in later cases. For example, Whitrow felt that the theory behaved epistemically so weirdly that it just could not be the correct description of the universe.

It can be shown that the steady-state model, describing an exponentially expanding space, will at all times contain regions of galaxies that cannot be observed, not even in principle. The universe will thus contain infinitely many subuniverses, which are causally unconnected. As Whitrow remarked, this is indeed a puzzling situation since the model then postulates the existence of galaxies that are, always have been, and always will be unknowable by empirical means.[28] Not only are there galaxies that are unobservable, there are also galaxies of a gigantic mass, and in both cases an infinite number of them. Since extremely old galaxies continue to grow in mass by accreting the material continuously formed throughout space, there will always be a large number of enormously massive galaxies, possibly an infinity of them. That such objects are not observed may be explained by the fact that they have long ago passed beyond the optical horizon, but even so there will be problems. As McVittie pointed out, an observer placed near such a supermassive galaxy would experience a very different world picture than observers in our observable universe. The observer would not see, even at a very large scale, a homogeneous universe, and this contradicts the cosmological principle on which the steady-state theory builds. According to Whitrow and McVittie, this kind of *reductio ad absurdum* argument was a serious conceptual difficulty, which the steady-state theory could only escape by modifying the original assumptions by means of non-testable hypotheses.[29]

These conceptual objections were attempts to show that the steady-state theory led to consequences that were contradictory, or at least highly bizarre. An objection of a similar kind was raised by the American physicist Richard Schlegel in 1962, namely that the infinite number of created atoms in the steady-state universe leads to contradictions. Basing his arguments on the mathematical theory of transfinite cardinal numbers, he purportedly showed that in the infinite past time of the steady-state universe a number of atoms greater than denumerably infinite would be produced. This, he argued, was impossible both from a mathematical and physical point of view, and he consequently suggested that the theory was contradictory.[30] Schlegel's claim was ignored by the steady-state physicists, but resulted in a minor dispute among philosophers and mathematicians.

Related logical objections have been raised with respect to the infinite past timescale of the universe. Generally, actual infinities, whether of a spatial, numerical or temporal kind, always create conceptual problems. Although they can be handled mathematically, they make no sense in physics. The past temporal infinity of the steady-state universe was not much discussed in the 1950s, but after the demise of the theory some philosophers and scientists argued that actual infinities cannot possibly occur in nature and that this rules out the steady-state theory.[31] The conceptual and logical problems caused by the infinities of steady-state models have continued to attract interest. For example, it has been argued that the perfect cosmological principle is self-contradictory because it leads to the paradox of non-evolving intelligent beings.[32] Philosophically interesting as this category of arguments is, they were mostly discussed *post hoc* and are therefore of less importance from a historical point of view.

Bondi's emphasis on the possibility of experimental disproof as a hallmark of science reflected the philosophical views of the famous philosopher Karl Popper (who, like Bondi and Gold, was an Austrian emigrant). The basic message of Popper, spelled out in his main work *The Logic of Scientific Discovery*, can be condensed into two statements: (i) it must be possible for a scientific theory to be refuted by empirical means; (ii) a theory with more possibilities of refutation—more 'potential falsificators'—is preferable over a rival theory with less possibilities of refutation. It is understandable that steady-state protagonists welcomed Popper's programme. Since the very beginning of the steady-state theory, they had complained that the relativistic evolution theories did not allow precise predictions of a kind which, if contradicted by observations, would lead to their refutation. Contrariwise, steady-state cosmology led directly to a number of such crucial predictions, such as a definite space–time metric, the average density of matter in the universe, and the age variation of galaxies. Although Bondi and his allies could not use Popper in defeating the evolutionary models, they could use him in arguing that, from a philosophical point of view, the steady-state theory was superior to its rivals, meaning that it was more scientific.

Bondi, more so than Gold and Hoyle, deeply admired Popper's philosophy, which he considered most relevant to working scientists. In a glowing review of *The Logic of Scientific Discovery*, he and the mathematical physicist Clive W. Kilmister pointed out that 'it rings true' and was directly applicable to the contemporary situation in cosmology. 'For here the correct argument has always been that the steady-state model was the one that could be disproved most easily by observation. Therefore, it should take predecence over other less disprovable ones until it has been disproved.' Bondi and Kilmister generally agreed with Popper that 'the notion of falsifiability in particular is a concept of the most direct significance to science'.[33] On several later occasions, Bondi described himself as a loyal follower of Popper and praised his theory of science in similar words. His admiration for the Austrian-British philosopher was boundless. In 1992, on the occasion of Popper's ninetieth birthday, he summarized his view in a single sentence: 'There is no more to science than its method, and there is no more to its method than Popper has said.'[34]

Popperian standards of science, or something close to them, were generally accepted during the cosmological controversy, but not everyone agreed with Bondi's rigid interpretation of them. Among those who crossed methodological swords with Bondi was William Bonnor, a chemistry-trained professor of mathematics at the University of London. Bonnor was a specialist in general relativity and held a critical attitude to both steady-state and big bang models. He was by no means against falsificationism, indeed he stressed that 'every scientific theory must be capable of disproof; otherwise it says nothing'. However, turning Popper's philosophy against the steady-state theory, he noticed that, contrary to the relativistic models of cosmology, the steady-state theory was from its very beginning constructed specifically as a theory of the universe. 'Now both Milne's theory and the steady-state theory were designed especially to solve the cosmological problem, so it is not remarkable that they yield unique models.'[35] The rival class of cosmologies built on Einstein's theory of general relativity, which was originally constructed as a theory of gravitation and not as a cosmological theory. Neither had McVittie much respect for Bondi's appeal to Popperian falsificationism, which he parodied by saying that, if it were accepted in its strict form, then 'we should be justified in inventing a theory of gravitation which would prove that the orbit of every planet was necessarily a circle. The theory would be most vulnerable to observation and could, indeed, be immediately shot down.'[36]

Although physical cosmology was not seen as nearly as philosophically interesting as, say, quantum mechanics, several philosophers and philosophically minded scientists commented on the controversy between the two world pictures. Most of them found the steady-state theory to be unacceptable, without therefore accepting the rival theory of an explosive finite-age universe. As Dingle had earlier attacked the cosmophysical theories of Eddington, Milne, and Dirac, he now went on the warpath against the 'unscientific romanticism' of Bondi, Hoyle, and Gold. In an unusually polemical presidential address of 1953 delivered before the Royal Astronomical Society he made it plain that, in his view, the steady-state theory was no less *a priori* than Eddington's discarded cosmophysics, hence no more scientific. Like it, the new cosmological model was *cosmythology*, a mathematical dream that had no credible connection with physical reality:

> It is hard for those unacquainted with the mathematics of the subject, and trained in the scientific tradition, to credit that the elementary principles of science are being so openly outraged as they are here. One naturally inclines to think that the idea of the continual creation of matter has somehow emerged from mathematical discussion based on scientific observation, and that right or wrong, it is a legitimate inference from what we know. It is nothing of the kind, and it is necessary that that should be clearly understood. It has no other basis than the fancy of a few mathematicians who think how nice it would be if the world were made that way.[37]

Given that Dingle had earlier dismissed the ordinary cosmological principle as pseudoscientific, it comes as no surprise that he judged the perfect cosmological principle totally unacceptable, to be *ad hoc* as well as *a priori*. According to Dingle, the principle invented by Gold and Bondi had precisely the same dubious nature as the perfectly circular orbits and immutable heavens of Aristotelian cosmology. These were more than mere hypotheses, they were central elements in the paradigm that ruled ancient and medieval cosmology, and as such they were inviolable within the framework of the paradigm. Dingle claimed that the perfect cosmological principle had a similar status. He further suggested that the steady-state theory somehow reflected a more general zeitgeist, in this case an ideological and exaggerated version of the Copernican principle:

> In every age there is a certain climate of opinion that predisposes thinkers towards a certain type of view...Now it is the exceptional that is out of favour. By a sort of cosmic democracy we are predisposed to deny any unique characteristic to anything, and whatever we happen to see now, all the universe must see at all times.[38]

A somewhat similar critique of the Copernican uniformity principle would later lie behind the introduction of the anthropic cosmological principle (see Chapter 9).

Not to be outdone by Dingle's unrestrained rhetoric, Mario Bunge, a respected Argentine philosopher and physicist, suggested that the steady-state theory should be grouped together with 'Eddingtonian neo-Pythagoreanism, ESP, psychoanalysis, and philosophical psychology'.[39] To his mind, it was just one more example of science-fiction cosmology. According to Bunge, the magical character of steady-state cosmology was most clearly seen in its reliance on the continual creation of matter in the form of *creatio ex nihilo*. 'The concept of emergence out of nothing is characteristically theological or magical—even if clothed in mathematical form', he barked.[40] As mentioned, Hoyle and McCrea denied that steady-state creation of matter was *ex nihilo*.

As a third example of philosophical critique of the steady-state theory, consider the American philosopher Milton Munitz, at the New York University, who in the 1950s provided a careful analysis of the theory's philosophical basis and implications. Like most other critics, he focused on the continual creation of matter. The basis of his objections was what he claimed to be the ultimate and irreducible nature of steady-state creation. According to Munitz, this made the theory dogmatic and unscientific, the reason being that no ordinary scientific explanation could be given for the creation of matter, not even in principle. Creation of matter could be verified (if not in practice, then in principle), but Munitz nonetheless found it to be scientifically meaningless because it was unexplainable: 'To say that matter is found in the universe leaves open the possibility of explaining its appearance, whereas to say that it is created not only denies such a possibility but also employs a term without any significant content.'[41]

The philosophical critique of steady-state theory (whether coming from professional philosophers or physicists), and also the few philosophical voices in support of it, are interesting in their own right and as illuminating the historical situation at the time. However, the views of critics such as Schlegel, Dingle, Bunge, and Munitz had very little impact on the scientific controversy between the steady-state cosmologists and those advocating an evolutionary universe governed by general relativity. The physicists and astronomers involved in the controversy preferred to follow the standard scientific procedure, that is, to keep to comparisons of observational data against theoretical predictions. To the extent that philosophical arguments played a role, these were arguments suggested by the participating scientists themselves. Then as now, scientists are not in the habit of paying much attention to what philosophers say. The only philosopher who made an impact during the controversy in the 1950s was Popper, and his role was indirect, mostly mediated through Bondi. Popper never commented on the cosmological controversy.[42]

5.3 BEYOND CLASSICAL STEADY-STATE THEORY

The steady-state theory of the universe had a high degree of falsifiability, and it was this methodological virtue which killed it. It turned out that the theory disagreed with data from radioastronomical observations produced by the Cambridge astronomer Martin Ryle and others, it was unable to account for the observed amount of helium in the universe, it was contradicted by the discovery in 1965 of a microwave background radiation, and it failed to explain the distribution of the newly discovered quasars. However, all this did not imply that the general idea of the universe being in some kind of eternal steady state had to be given up. By 1966 the large majority of cosmologists thought that this idea belonged to the past, but not all shared the majority view.

Whereas the Bondi–Gold theory of the steady-state universe was quietly abandoned in the mid-1960s, Hoyle continued to develop his much more flexible field theory into a number of versions that could accommodate the microwave background and other conflicting observations. Most of this work, which continued until Hoyle's death in 2001, was done in collaboration with his former student, the Indian-born Jayant Narlikar. Their series of theories not only led to a picture of the universe that differed drastically from the classical steady-state picture, it also included connections between cosmology and microphysics in a stronger sense than earlier. In conformity with the perfect cosmological principle, the Hubble parameter had originally been regarded a true constant of nature, but in a modified theory Hoyle presented in 1960 it was allowed to vary in time.[43] The creation or C-field that appeared in the now covariant field equations was assumed to arise from unspecified microscopic processes of fundamental physics. Hoyle's C-field came into effect only at the times when particles were created. However, the mechanism that Hoyle proposed had too much the character of a *deus ex machina* to convince critics that it provided a proper explanation of particle creation.

In a further development of the theory, Hoyle and 24-year-old Narlikar associated the C-field with a negative energy density, somewhat along the lines earlier proposed by McCrea.[44] They suggested a mechanism in which matter creation fully complied with conservation of energy, essentially by having the negative energy of the C-field compensate for the positive energy of the created particles: Whenever a particle with positive mass–energy was created, a C-field of equal but negative energy was created, thus securing energy conservation. As in McCrea's theory, the negative energy density acted as a repulsive effect, or an internal pressure. With the new theory, the perfect cosmological principle was definitively abandoned and the universe was pictured as having developed into an exponentially expanding steady state from some initial but non-singular state. In this asymptotic situation, creation, and expansion would be in exact balance, given by the same creation rate as in the original steady-state theory, namely $3\rho H$. The balance was not assumed, but proved. Since the Hubble parameter could be expressed in terms of the coupling constant of the C-field, important empirical quantities such as the rate of expansion of the universe and the mean density of matter were expressed by the elementary creation process. Unfortunately, since the value of the coupling constant could not be determined independently, this result, like the *C*-field itself, was a postulate of no real empirical significance.

Whereas Hoyle and Narlikar had hitherto followed the standard assumption of a smoothed-out, homogeneous universe, in 1966 they abandoned this assumption, too, and with it the idea of matter being created uniformly through space. As an alternative, they explored the possibility of matter pouring into space from areas with already existing supermassive bodies. When confronted with the question of whether the simplicity and philosophical appeal of the original steady-state theory was not lost in the new version, the two Cambridge physicists merely shrugged their shoulders. Rather than arguing that the new theory lived up to the criterion of simplicity, if only properly interpreted, Hoyle answered: 'To the somewhat pallid objection that the theory thereby loses much of its attractive simplicity, we would answer that nothing known to us in physics or astronomy is simple and that we see no reason at all why phenomena on a large scale should be simple.'[45] Physicists may have different ideas of what it means that nature is simple, but it is not often that they denounce the principle of simplicity in such an explicit way.

In another paper of 1966, Hoyle and Narlikar suggested a model of the universe as consisting of separate 'bubble universes' in which the creation process was temporarily cut off, and which would therefore expand much more rapidly than their surroundings. The bubbles would eventually be filled with matter and *C*-fields from the denser outside, but other evacuated regions would be formed at the same time and develop as new bubble universes. Hoyle and Narlikar found that if the bubbles developed synchronously, the universe would follow a series of expansions and contractions; not as in oscillating models, but developing around an exponential steady-state expansion. The two cosmologists expressed great confidence in the pulsating universe built on *C*-field theory, which they considered a strong alternative to the victorious big-bang theory. Among other advantages they stressed that the cycles would contain no singularities. Cosmic and other singularities were much discussed in about 1965, when the so-called singularity theorems in general relativity were proved by Roger Penrose and Stephen Hawking, and were generally interpreted making the conclusion that the big bang started in a singular state unavoidable. This was a notion that Hoyle would have nothing of. Rather than admitting a cosmic singularity, he preferred to change the laws of physics:

> [A universal singularity] seems as objectionable to me as if phenomena should be discovered in the laboratory which not only defied present physical laws but which also defied all possible physical laws. On the other hand, I see no objection to supposing that present laws are incomplete, for they are almost surely incomplete. The issue therefore presents itself as to how the physical laws must be modified in order to prevent a universal singularity, in other words how to prevent a collapse of physics.[46]

Hoyle thought his and Narlikar's theory had prevented not only the universe from collapsing, but also physics from collapsing.

One reason for their confidence (which was not shared by most other cosmologists) was that they could relate their theory to a new theory of gravitation. The theory they proposed was equivalent to that of Einstein for all observable phenomena, but it had certain theoretically satisfactory features that went beyond standard gravitation theory. For example, it required a positive gravitational constant, whereas $G > 0$ is introduced empirically in the ordinary theory to account for the fact that bodies gravitate rather than repel. Moreover, in the new theory there was no room for the cosmological constant, which in relativity theory might be there or not. While it makes perfect sense to speak of an empty world within the framework of the theory of general relativity—recall that de Sitter's universe of 1917 was completely empty—according to Hoyle and Narlikar the concept was meaningless: 'In the present theory emptiness demands no world at all. Nor can there be a world containing a single particle, the least number of particles is two.' They further illustrated the Machian basis of the new theory of gravitation by considering 'what would happen to the solar system if the rest of the universe were removed'. Whereas almost nothing would happen in Newtonian and Einsteinian theory, the effect would be nearly inconceivable in the new theory. Local gravitation would become much stronger, with the result that the Sun's emission of energy would increase drastically. 'Take away half of the distant parts of the universe and the Earth would be fried to crisp', they said.[47]

Another reason for the confidence of Hoyle and Narlikar rested on complex theoretical arguments that in order to understand the electrodynamic arrow of time—that we only observe the retarded solutions of the wave equation and not the advanced solutions (arriving from the future)—matter has to be created continually in accordance with steady-state theories. Although these arguments were purely theoretical, Hoyle and Narlikar believed that they provided strong support for a modified steady-state universe and made relativistic evolution cosmologies unlikely. According to Narlikar, the understanding of the arrow of time had great advantages as a cosmological test over the usual observational tests. As he expressed it, the arrow of time was 'a powerful test of cosmological theories' because 'it has the merit of being clear cut and free from... observational uncertainties'.[48] But he and Hoyle were largely alone in thinking so. To almost all other astronomers and physicists observational tests were far more important than conceptual and theoretical 'tests' such as the one related to the unidirectionality of time.

Nor were cosmologists of the 1960s impressed by the appeal to combinations of constants of nature that was part of the Hoyle–Narlikar programme and which expressed their belief in a necessary connection between the cosmos and the subatomic world. To give but one example of this belief, evidently with roots in the ideas of Weyl, Eddington and Dirac, in 1968 Hoyle suggested a relationship between the Hubble radius cT and other constants of nature,

$$\frac{c}{H} = \frac{e^2}{GmM}\frac{4\pi e^2}{mc^2} \cong 10^{10} \text{ light years}$$

that was held to be satisfied exactly. Mainstream cosmologists, meaning protagonists of the standard hot big-bang model, did not doubt the importance of the microphysics–cosmophysics connection, but they found the Hoyle–Narlikar approach to be uninteresting and unproductive. Consequently they tended to ignore it.

Together with the distinguished British–American astrophysicist Geoffrey Burbidge, in the 1990s Hoyle and Narlikar developed the C-field approach to cosmology into a comprehensive model of the universe, what they called the quasi-steady-state cosmology or QSSC.[49] This theory is still being defended by Narlikar and Burbidge and a few other physicists as an alternative to the standard big-bang theory, but it has not succeeded in attracting much interest in the community of cosmologists. Methodologically, the new theory of the universe had much in common with Hoyle's original theory proposed 45 years earlier. Hoyle, Burbidge, and Narlikar were uniformitarianists who found it unscientific to base a theory on 'some mystical state, wholly unattested either by observation or experiment'. The willingness to do so, they said, was characteristic of the 'breed of new theologians, who like the old theologians claim to be privy to revelations'.[50]

The universe of QSSC is eternal and creative. As in the earlier versions, the C-field provides a negative pressure, which secures the creation of matter in conformity with the law of energy conservation. The creation of matter has the consequence that the universe expands. Although on a very long timescale the size of the observable universe increases exponentially as in classical steady-state theory, oscillations on a smaller timescale are superimposed on this expansion (Fig. 5.1). A typical cycle has a period of about 40 billion years, of the same order of magnitude as the age of the big-bang universe.

The energy density of the C-field varies with the size of the visible universe, as given by the scale factor, being greatest at the minima and smallest at the maxima. Matter creation is supposed to occur only in epochs when the universe is relatively dense, near the oscillatory minima. Once matter is created in a 'minibang', it is followed by a compensatory negative-energy creation field that causes the universe to expand at a high rate. As the scale factor increases, creation diminishes and the expansion rate slows down. The created matter is assumed to appear as hypothetical, very massive, and unstable particles that decay into quarks. These elementary particles, sometimes known as Planck particles, have a mass (5×10^{-5} g) corresponding to the Planck unification energy of about 10^{19} GeV and a decay time of about 10^{-24} seconds. Although radiation is produced together with the matter, the universe is never radiation-dominated, which is one more difference from the big-bang model.

The QSSC theory is claimed to succeed where the classical steady-state theory failed and in general to account for all observational data, including the cosmic microwave background. According to QSSC this radiation is not a fossil from the cosmic past, but is

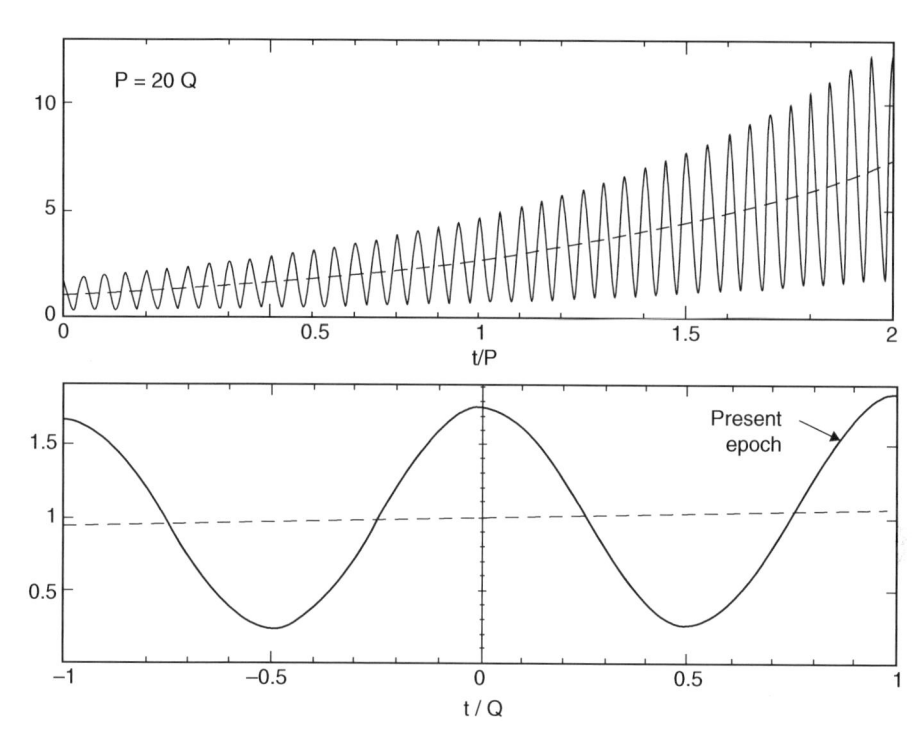

Fig. 5.1. The QSSC model of Fred Hoyle, Geoffrey Burbidge, and Jayant Narlikar. The variation of the scale factor R with time is shown for two different time scales, $Q = 40$ billion years and $P = 20Q$. The universe evolves cyclically, with no singularities and no big bang.

Source: Fred Hoyle, Geoffrey Burbidge and Jayant Narlikar, 'The basic theory underlying the quasi-steady-state cosmology', Proceedings of the Royal Society A **448** (1995), 191–212. Reproduced by permission of the Royal Society.

due to starlight that is thermalized by tiny grains of carbon and iron floating around in space. Whereas the observed temperature fluctuations of the cosmic microwave background is normally explained in terms of the inflationary big-bang theory, Hoyle, Narlikar, and Burbidge saw no need to appeal to inflation. The accelerated expansion of the universe, as demonstrated by supernovae observations in the late 1990s, is generally seen as evidence for a repulsive cosmic force (or dark energy) due to a positive cosmological constant. Things look quite different from a QSSC perspective, where the acceleration is explained as an effect of the negative-energy C-field. The cosmological constant has a place in the QSSC framework, but it is different from the one in standard cosmology. Reflecting the role of Mach's principle in QSSC, both the cosmological constant and the gravitational constant are determined by the large-scale distribution of matter in the universe. This is done in such a way that $G > 0$ and $\varLambda < 0$. The negative cosmological constant acts as an attractive force and is responsible for the shifts from expansion to contraction that occur when the scale factor is at its maximum value.

The quasi-steady-state model is testable insofar that it yields predictions that may turn out to be wrong. For example, the form of the cosmic microwave spectrum should deviate from a Planck distribution at long wavelengths. Furthermore, contrary to the standard cosmology the model does not operate with non-baryonic dark matter, which presumably means that if some of the candidates for such matter were found, the theory would have to be abandoned. On the other hand, compared with the original steady-state theory QSSC is not very vulnerable to new discoveries and lacks clear predictions of the Popperian type. The methodological weaknesses of QSSC are admitted by its few protagonists. According to Narlikar and Thanu Padmanabhan, the theory 'cannot continue to survive and hope to make an impact based purely on the shortcomings of SC [standard cosmology]. It needs to be developed to such a level that it can make clear predictions rather than provide post facto explanations for already observed phenomena.'[51] This level has not been reached and it is, I assume, unlikely that it will ever happen.

It will come as no surprise that there is a sociological dimension to the continuing efforts of the few remaining steady-state cosmologists and their feeling of being unfairly ignored by mainstream cosmology. This kind of complaint is not new, having been a feature of steady-state theory (and some other unorthodox theories of the universe) ever since the establishing of the hot big-bang paradigm in the late 1960s. As late as 2008 Narlikar and Burbidge wrote, not without some bitterness, about this sociological dimension and the possible tests of the QSSC:

> Many of the tests are being attempted, but ... most of the observers who make them know that they will get more favourable treatment from their colleagues, editors, funding agencies and others who assign telescope time if they concentrate on tests confirming the big bang. For example, no time has ever been assigned on the Hubble Space Telescope to observers who are thought to favour the QSSC, and everyone who designs, plans, builds, and observes for the microwave background believes from the beginning that it originated in the big bang. So much for unbiased observers.[52]

The new oscillatory steady-state theory of the universe has not won much support within the communities of physicists and astronomers. According to Kumar Theckedath, a Marxist and mathematics teacher at the University of Mumbai, the theory confirms the basic principles of dialectical materialism and is therefore superior to the standard big-bang model.[53] I imagine that this is a kind of support that Narlikar and other QSSC advocates would rather be without.

Although the elaborate attempts of Hoyle, Narlikar and a few others to create a new basis for cosmology did not succeed, in some respects their efforts were not quite in vain. As it only turned out much later, the Hoyle–Narlikar theory of the 1960s had many features in common with the inflationary models developed from the early 1980s onwards. From a technical point of view, the revised steady-state theory may been described as inflation *avant le lettre*, an inflation universe in which inflation always occurs. Thus, in 1963 Hoyle and Narlikar explained from their theory why the universe is so isotropic and homogeneous, an achievement repeated with much bigger impact by the inflationary scenario of Alan Guth and others in the 1981–82. 'Provided the continuous creation of matter is allowed', Hoyle and Narlikar wrote, 'the creation acts in such a way as to smooth out an initial anisotropy or inhomogeneity over any specified proper volume... In other words, any finite portion of the universe gradually loses its 'memory' of an initially imposed anisotropy or inhomogeneity.... It seems that the universe attains the observed regularity irrespective of initial boundary conditions.'[54]

Understandably, after inflationary cosmology had obtained a nearly paradigmatic status, Hoyle and Narlikar often pointed out the similarity to their old and discarded theory of a steady-state universe. At least some of the pioneers of the inflationary scenario have admitted the connection and their debts to the now-defunct theory of the steady-state universe. However, although there are substantial scientific similarities between the Hoyle–Narlikar steady-state theory and the later inflation theory, the historical trajectories of the two research traditions were quite separate. Inflation theory did not grow out of the steady-state theory.

Notes for Chapter 5

1. A detailed account of the steady-state theory and its relationship to the early big-bang theory is given in Kragh 1996.
2. Bondi 1952, p. 5.
3. Jones, Rotblat, and Whitrow 1956, p. 228.
4. On this tradition, see Kragh 1995.
5. Bondi 1948 and similarly in Bondi 1952, pp. 3–5. Bondi's distinction between the two schools of cosmological thinking was not particularly original. Using different names, a similar one was earlier made by Tolman and Milne, and later Gamow and McVittie would suggest about the same distinction.
6. Gale and Urani 1999 argue that Bondi's view of cosmology was strongly indebted to Milne's theory and that the two physicists shared the same methodology of science. In my opinion, they considerably overstate their case. Bondi was not another Milne.
7. Bondi 1952, p. 11.
8. Bunge 1962, p. 139.
9. Bondi and Gold 1948, p. 255. The other paper that founded the steady-state theory was Hoyle 1948.
10. Bondi and Gold 1948, p. 255.
11. Gold 1949.
12. Hoyle 1950, pp. 123–24.
13. McCrea 1951. On McCrea's theory and other attempts to reconcile steady-state cosmology with general relativity, see Kragh 1999b.
14. For analysis, references and details, see Kragh 1997.

15. Bailey 1959. On the Kaluza–Klein theory, see Chapter 11. A Japanese physicist, Toshima Araki, made a similar suggestion in 1953. The idea of energy or matter flowing from some hidden dimension to the world we experience is vaguely reminiscent to the speculations of Stewart, Tait, and Pearson in the Victorian era, such as mentioned in Chapter 2.

16. Bondi 1952, pp. 61–62.
17. Ibid., p. 169.
18. Hoyle 1948, p. 381.
19. McCrea 1950, p. 886.
20. Bondi et al. 1960, p. 33.
21. The subjects dealt with in this section are covered in greater detail in Kragh 1996, pp. 202–50. On the role of philosophical considerations in the steady-state theory, see also Balashov 1994.
22. Johnson 1951.
23. Öpik 1954, p. 91 and p. 106.
24. Klein 1954, p. 43. As late as 2000, the cosmologist Edward R. Harrison suggested that 'when a cosmologist presents an argument for a particular type of universe, perhaps we should not read too much into the science but wonder about that person's religion, philosophy, and even psychology' (Harrison 2000, p. 297). Apparently, 'taste' remains a factor to be considered in at least some parts of cosmology.
25. McVittie 1965, pp. 8–9.
26. McVittie, interview by David DeVorkin, American Institute of Physics, 21 March 1978.
27. Whitrow and Bondi 1954, p. 279.
28. Whitrow 1959, pp. 138–41. See also Kragh 1996, pp. 233–36. Whitrow was essentially pointing out that the steady-state universe was what later would be called a multiverse (see Chapter 10).
29. The argument was apparently first suggested in Jones, Rotblat, and Whitrow 1956, p. 222.
30. Schlegel 1962 and Schlegel 1964.
31. Examples are Huby 1971, Whitrow 1978, and Schlegel 1964. Arguments against an infinite past are old and were in earlier periods often forwarded for apologetic reasons, to prove that God had created the world a finite time ago. Arguments of this type were suggested by the Christian philosopher John Philoponus in the sixth century against the Aristotelian view of an eternal universe, and they have since been repeated in a variety of versions by many philosophers, scientists, and theologians. As will be apparent from later chapters, this classical discussion lives on to this very day, if, of course, in forms different from those of the past.
32. Barrow and Tipler 1986, pp. 601–608. See also Chapter 9.
33. Bondi and Kilmister 1959, pp. 56–57.
34. Bondi 1992.
35. Bonnor 1964, p. 158.
36. McVittie 1961, p. 1231.
37. Dingle 1953, p. 403.
38. Ibid., p. 405.
39. Bunge 1962, p. 122.
40. Bunge 1959, p. 24.
41. Munitz 1957b, p. 162.
42. Popper did have an interest in cosmology, such as shown by a little-known paper of 1940 in which he discussed various non-Doppler interpretations of the galactic redshifts, including

Milne's hypothesis of different time scales and the hypothesis that the speed of light decreases with time. See Popper 1940. Elsewhere in his philosophical writings, Popper often referred to 'cosmology', but essentially in the sense of natural philosophy and not in the sense of the scientific study of the universe.

43. Hoyle 1960.
44. Hoyle and Narlikar 1962.
45. Burbidge, Burbidge, and Hoyle 1963, p. 878.
46. Hoyle 1965a, p. 113.
47. Hoyle and Narlikar 1964, p. 191 and p. 204.
48. Narlikar 1973, p. 83.
49. Hoyle, Burbidge, and Narlikar 2000. For a historical perspective, see Kragh 2010a.
50. Hoyle, Burbidge, and Narlikar 2000, p. 227.
51. Narlikar and Padmanabhan 2001, p. 246.
52. Narlikar and Burbidge 2008, p. 249.
53. Theckedath 2003, according to whom the hot big-bang model 'fits very well with the claims of theologians in general, and with the claims of the Catholic Church in particular' (p. 61). The charge of an unholy alliance between Christianity and the big-bang theory was an old theme in Hoyle's works.
54. Hoyle and Narlikar 1963, p. 11.

6
The Rise and Fall of the Bootstrap Programme

Consistency is the last refuge of the unimaginative.

Oscar Wilde

The traditional domain for unified theories of physics has been matter and forces rather than the universe at large. According to an old tradition in natural philosophy, matter ultimately consists of fundamental particles ('atoms') which cannot themselves be decomposed or conceived as composite bodies. Perhaps there is only one type of truly elementary particle, a *protyle*, which is the ultimate building block of all matter. This dream of the natural philosophers became part of experimental science in the twentieth century when it was understood that matter consists of internally structured particles held together by attractive forces. With the emergence of quantum mechanics it became possible to explain in quantitative details the behaviour and interactions of elementary particles such as electrons and protons. For example, a hydrogen atom is nothing but a bound state between a proton and an electron. From the properties of the constituent particles and the electrical force that keeps them together we can infer all the properties of hydrogen.

This general picture of matter was challenged in the 1960s by a radically new way of understanding nuclear and other strongly interacting particles, known as hadrons. Contrary to the 'aristocratic' conception of particles, where some are more fundamental and hence more important than others, the new theory understood all particles as equally elementary or equally composite. According to the principle of 'nuclear democracy', strongly interacting particles were dynamical structures that owed their existence to the same force through which they interacted, and no structure was fundamental in an absolute sense. Another important feature of the bootstrap hypothesis, as it came to be known, was that it denied the legitimacy of a field description of particles—indeed, it denied the legitimacy of the space–time continuum on which the field concept rests. The formal core of the theory was a mathematical object, the S-matrix, from which all knowledge of hadronic experiments was supposed to flow. Launched as an alternative to the 'fundamentalism' of standard quantum field theory, the S-matrix or bootstrap programme was itself a fundamental theory based on principles from which all observable facts of strong interactions could be derived. It pictured a world of particles that was in a sense necessary for the existence of the world.

The bootstrap programme was ambitious and radical, but it was neither a unified theory nor one building on alternative epistemic standards. Although it did present a unified framework for strong interactions, the unity did not extend to the other fundamental interactions of which the theory had nothing or only very little to say. Methodologically, it built on a positivistic criterion of observability, not on *a priori* principles. On the other hand, the research programme also invited speculations of a 'complete bootstrap' that would in principle explain the world as necessary, but these speculations were recognised as philosophical rather than scientific. The bootstrap alternative flourished for about a decade, after which it declined, to be taken over by more conventional field theories of the aristocratic type. In 1970 the leading bootstrap theorist Geoffrey Chew said: 'I would find it a crushing disappointment if in 1980 all of hadron physics could be explained in terms of new arbitrary entities. We should then be in essentially the same posture as in 1930, when it seemed that neutrons and protons were the basic building blocks of nuclear matter.'[1] The development must have disappointed him, for it led to a revival of the traditional hierarchic picture of matter, now with quarks and gluons as the basic constituents of nuclear matter.

6.1 QUANTUM THEORY OF PARTICLES AND PROCESSES

During the formative years of quantum mechanics, the period from 1925 to about 1928, all matter was supposed to consist of just two elementary particles, the proton and the electron.[2] It was often assumed that alpha particles, made up of four protons and two electrons, were additional nuclear constituents, but no one suggested including genuinely new particles. (The hypothetical 'neutron' of the 1920s was considered a proton–electron composite.) The attractive and short-range nuclear force had been known to exist since Rutherford's discovery of the atomic nucleus in 1911, but almost nothing was known about it. Gravitation, the oldest of the forces of nature, was considered to be external to atomic and quantum physics. (However, a few physicists disagreed.[3]) As to the weak force, associated with the beta decay of radioactive nuclei, it was still not recognized as a force of its own. Consequently, physicists in the late 1920s focused more on the electron and electromagnetic processes. What was needed, they felt, was to establish a quantum electrodynamics (QED) as part of a more general theory of quantum fields (QFT, quantum field theory).

Among the pioneers in this early development was Paul Dirac, whose works may be used to illustrate two of the themes that figured prominently in the period, reductionism and unity. In 1929 Dirac expressed the period's general epistemic optimism, including the somewhat arrogant view that chemistry had now been nearly fully explained by quantum physics. The physicists were ready to go on and conquer the domain of biology. Dirac famously wrote:

> The general theory of quantum mechanics is now almost complete, the imperfections that still remain being in connection with the exact fitting in of the theory with relativity ideas.... The underlying physical laws necessary for the mathematical theory of a large part of physics and the whole of chemistry are thus completely known, and the difficulty is only that the exact application of these laws leads to equations much too complicated to be soluble.[4]

The following year Dirac suggested on the basis of his new relativistic wave equation of the electron that two of the solutions to this equation represented protons (in two spin states), which he at this stage conceived as antiparticles of the electrons. Dirac's short-lived proposal was in part motivated by a philosophical desire to understand all matter in terms of a single particle described by a single equation. The identification of protons with antielectrons (in Dirac's picture: vacant negative-energy states) was highly attractive because it promised a reduction of the then known elementary particles to just one fundamental entity, the electron. As he stated in an address of 1930 to the British Association for the Advancement of Science:

> It has always been the dream of philosophers to have all matter built up from one fundamental kind of particle, so that it is not altogether satisfactory to have two in our theory, the electron and the proton. There are, however, reasons for believing that the electron and proton are really not independent, but are just two manifestations of one elementary kind of particle. This connexion between the electron and proton is, in fact, rather forced upon us by general considerations about the symmetry between positive and negative electric charge, which symmetry prevents us from building up a theory of the negatively charged electrons without bringing in also the positively charged protons.[5]

Alas, Dirac's dreams remained dreams. While his claim that chemistry had been effectively reduced to quantum physics was at least a realistic programme for the future—and one which to a large extent would be realized with the advent of computational quantum chemistry in the 1950s—his idea of an ultimate elementary particle was a capital mistake. What happened over the next few years was not a reduction of particles to a single primary object, but on the contrary a confusing proliferation of elementary particles.

The generally held belief in the two-particle paradigm broke down in the 1930s when, much to the physicists' surprise, several new elementary particles were discovered and others were hypothesized. The positron, the true antiparticle of the electron, showed up in 1932, the same year as the neutron was discovered to be an independent constituent of the atomic nucleus. Five years later the 'heavy electron' (mesotron, μ-meson, or muon) was detected in cosmic rays, leaving the theorists utterly confused. Isidor Rabi, later a Nobel laureate of physics, is to have exclaimed, 'Who ordered that?'

In addition to these detected particles, the nearly massless neutrino, predicted by Pauli in 1929, was widely believed to exist and to play a crucial role in beta decay. The recognition of the neutrino was an important element in Enrico Fermi's theory of beta radioactivity from 1933, which effectively established weak interactions as a new and fundamental kind of force.

The ambitious attempts to establish a relativistic theory of quantum electrodynamics, starting in 1929 with an important work by Heisenberg and Pauli, soon ran into difficulties. The Heisenberg–Pauli theory and other theories that built on it were in many ways theoretically satisfactory, but they were not of much use when it came to accounting for new experimental facts. In addition, the quantum electrodynamics of the early 1930s faced grave problems of a logical and conceptual nature. First and foremost, it led to an infinite value of the electron's self-energy, that is, the energy of an electron in its own electromagnetic field. Infinities or 'divergencies' are anathema in physics insofar that they refer to measurable phenomena and are not purely abstract quantities. The infinite self-energy of the electron was followed by other divergent quantities, which caused many physicists to doubt if existing quantum theory was the correct framework for the more general quantum physics of the future. The conceptual problems were serious indeed, and so were experimental anomalies arising from nuclear physics and the study of cosmic rays. Analysis of the high-energy part of cosmic rays apparently disagreed violently with the quantum field theory of charged particles, which some physicists took to be a sign of failure of quantum electrodynamics at high energies.

Responses to what was widely perceived as a crisis in quantum theory varied.[6] Some of the leading physicists, including Bohr, Dirac, and Heisenberg, believed that the troubles necessitated a drastic revision of existing physical theory, perhaps as drastic as the one that had occurred when classical mechanics was replaced by quantum mechanics. Other physicists adopted a more pragmatic attitude, arguing that the problems might be solved without leaving the framework of existing quantum theory. Whatever the attitude, in the mid-1930s it was generally thought that fundamental quantum physics was in a state of crisis. This feeling was given voice by Robert Oppenheimer, who in a letter of 1934 wrote: 'As you undoubtedly know, theoretical physics—what with the haunting ghosts of neutrinos, the Copenhagen conviction, against all evidence, that cosmic rays are protons, Born's absolutely unquantizable field theory, the divergence difficulties with the positron, and the utter impossibility of making a rigorous calculation of anything at all—is in a hell of a way.'[7]

The problems of quantum electrodynamics led to several proposals of new theoretical ideas, some of them quite unorthodox and reflecting the sense of despair in the community of theoretical physicists. One of them was due to the ever innovative Heisenberg, who as early as 1930 contemplated that space might consist of cubic cells with a length of h/Mc, where M is the mass of a proton. Although he soon abandoned the idea, in the late 1930s he would return to it.[8] In 1938 he suggested solving the divergence problems by introducing a new fundamental constant, a 'smallest length'.

For the characteristic length constant he took the value $h/m_\pi c \cong 10^{-15}$ m, where m_π is the mass of the meson (now pion) proposed by the Japanese theorist Hideki Yukawa as the carrier particle of the nuclear force. Heisenberg hoped on this basis to construct a new quantum theory that was relativistically invariant and would contain the old one as a limiting case. The new theory was supposed to differ from the old one only where the fundamental length could not be considered an infinitesimal quantity.

Another proposal, no less radical, was made by Dirac, who in 1941 introduced negative energies and probabilities into the formalism of quantum mechanics. In ordinary physics, energy is a positive quantity and a probability is limited to being a number in the closed interval between zero and one. Dirac's negative probabilities did not appear as the initial or final state of a physical process, but he argued that they had to be included among the intermediate states in the processes. He did not introduce the counterintuitive concept of negative probabilities as just a trick to avoid infinities, but also because he found it to be mathematically well defined. Generally he was inclined to think that if a quantity was mathematically consistent it would somehow appear in nature's scheme, if not necessarily in the form of a detectable quantity. As he said in his Bakerian Lecture of 1941: 'Negative energies and probabilities should not be considered as nonsense. They are well-defined concepts mathematically ... [and] should be considered simply as things which do not appear in experimental results.'[9]

A breakthrough in the formulation of a divergence-free quantum electrodynamics only appeared in the late 1940s with the so-called renormalization theory proposed and developed in different versions by Sin-Itiro Tomonaga, Julian Schwinger, Richard Feynman, and Freeman Dyson. The basic result of this work was a consistent and coherent quantum-mechanical theory of electrons and photons where infinities no longer occurred in the measurable quantities. The 'renormalization' occurred in a sense by subtracting infinities in such a way that only finite quantities remained. These calculated finite quantities, such as the electron's magnetic moment and the energy levels of the hydrogen atom, agreed convincingly with precision experiments. The new theory was instrumentally satisfactory insofar as it provided a complete and definite method for the calculation of physical processes and properties. Although some physicists, including Dirac (and later also Feynman), concluded that the theory was objectionable from a logical and mathematical point of view, most were persuaded by the impressive agreement between calculated and experimental results. In an article on field theory in *Scientific American* of 1953, Dyson expressed the pragmatic and positivistic credo of the majority of quantum electrodynamicists as follows:

> Quantum electrodynamics occupies a unique position in contemporary physics. It is the only field in which we can choose a hypothetical experiment and predict the result to five places of decimals confident that the theory takes into account all the factors that are involved. Quantum electrodynamics gives us a complete description of what an electron does; therefore in a certain sense it gives us an understanding of what an electron is. It is

only in quantum electrodynamics that our knowledge is so exact that we can feel we have some grasp of the nature of an elementary particle.[10]

By the 1950s, at the time when quantum electrodynamics was celebrating its victories, new elementary particles, some of them detected in the cosmic rays and others in high-energy accelerators, attracted increasing attention. Mesons, hyperons, and resonance particles challenged the theorists. Based on Yukawa's meson theory, many physicists sought to understand the nuclear force and the interactions between the new particles, the natural framework of explanation being quantum field theory. However, it soon turned out that this was extremely difficult and might perhaps even be impossible.

In strong interactions the carrier particle is not the massless photon, as in electromagnetic processes, but massive mesons. Whereas the coupling constant (a measure of the strength of the interaction) of electromagnetic processes is small, $\alpha_{EM} \approx 0.007$, this is not the case with the corresponding coupling constant of strong interactions, which is $\alpha_s \approx 1$. This means that the perturbation calculations of quantum electrodynamics cannot be duplicated. This was the basic reason why attempts to extend quantum field theory into the realm of strong interactions failed. To make things worse, it turned out that the infinities of the field theory of weak interactions could not be removed by the technique of renormalization either. To the mind of some physicists, the problems were so serious that they justified a break with quantum field theory, at least in the domain of strong interactions.

The famous Russian physicist Lev Landau was an early and prominent exponent of the view that quantum fields are scientifically illegitimate because they contain quantities that are unobservable in principle. By the summer of 1959 he was ready to bury quantum field theory, arguing that it should be replaced by other methods and ideas that did not rely on the concept of the continuous field. Among these other methods he mentioned the S-matrix. Several years later, in a famous textbook series coauthored by Evgeny Lifshiftz, he expressed the S-matrix programme as follows:

> The concepts of 'elementary' and 'composite' particles lose their earlier significance; the problem of 'what consists of what' cannot be formulated without considering the process of interaction between particles, and if this is not done the whole problem becomes meaningless. All particles which occur as initial or final particles in any physical collision phenomenon must appear in the theory on an equal footing. In this sense the difference between those particles usually said to be 'composite' and those said to be 'elementary' is only a quantitative one, and amounts to the value of the mass defect with respect to decay into specified 'component parts'. For example, the statement that the deuteron is composite... differs only quantitatively from the statement that the neutron 'consists of' a proton and a pion.[11]

The concept of a scattering or S-matrix can be traced back to the American theorist John Wheeler, who in 1937 introduced the idea in the context of nuclear physics. In a series of papers published between 1943 and 1946 the idea was independently developed by Heisenberg in an ambitious attempt to formulate a relativistic quantum electrodynamics without infinite quantities.[12] As he saw it, many of the divergence problems of field quantum electrodynamics could be traced back to its inability to incorporate into its framework the notion of a fundamental length. This notion, taken over from his earlier work, was part of the new S-matrix theory. As he wrote in his first paper on the new theory:

> The known divergence problems in the theory of elementary particles indicate that the future theory will contain in its foundation a universal constant of the dimension of a length, which in the existing form of the theory cannot be built in any obvious manner without contradiction.... In a manner of speaking, a new universal constant of the dimension of a length plays a decisive role, which has not been considered in the existing theory.[13]

Heisenberg's programme of 1943 was consciously modelled on his quantum mechanics of 1925 and, like this pioneering theory, based wholly in terms of observable quantities. As such quantities he chose the scattering or S-matrix representing the transition of a physical system from an initial state Ψ to a final state Φ. The formal connection is $\Phi = S\Psi$, where the square of the elements in the scattering matrix S or S_{ab} gives the transition probabilities. (The 'S' in S-matrix stands for 'Streuung', the German word for 'scattering'.)

The rationale behind Heisenberg's early S-matrix programme was to ignore what causes the interaction between particles and also what takes place inside the region of interaction given by the fundamental length. Instead, he suggested focusing only on the observable states before and after the interaction, far away from the region of interaction. Here the particles are free, with constant and well defined momenta, which are the only variables that enter the theory; due to the uncertainty principle, the space coordinates are unknown. In this picture, time does not appear as a significant variable except that it distinguishes between the initial and final states. Heisenberg conceived the S-matrix as the fundamental object of study in particle physics. As he argued, all observable quantities—such as the scattering cross-sections, the energies of bound states, and decay lifetimes—could in principle be calculated from it. Among the results established by Heisenberg was that the S-matrix or operator was unitary, meaning that $S^\dagger S = 1$ ($S^\dagger$ is the adjoint operator of S). The physical interpretation of this property is that, given some initial state, the probabilities of the transitions to all possible final states add up to unity.

Although Heisenberg's S-matrix theory was taken up and developed by a several physicists, including Walter Heitler and Ernst Stueckelberg, by 1950 it had largely come to a halt. One physicist who never liked the theory was Pauli, who found it to be complicated and unnecessary. Pauli was no less interested in establishing a fundamental theory of elementary particles than Heisenberg, but he had no confidence in the S-matrix approach. In late 1946, shortly after having received the Nobel Prize in Stockholm, Pauli to wrote to his old friend and collaborator: 'I believe that no real progress is possible any more without having a theory that determines the quantity $e^2/\hbar c$. In your point of view of the universal length, I miss, however, the connection with the $e^2/\hbar c$ problem; in addition I do not know whether perhaps different lengths are associated with different particles.'[14]

With the advent of renormalization quantum electrodynamics, the mathematically complex theory of the S-matrix no longer seemed either necessary or particularly attractive. However, when the S-matrix programme was incorporated into a new theory of strong interactions its fate changed rapidly. By that time, Heisenberg had abandoned his earlier research programme and changed to other areas of work. He eventually turned to non-linear field theory, hoping to find a single fundamental matter field that would underlie all of particle physics. In Heisenberg's later research programme the S-matrix was not fundamental, but something that had to be derived from the field equations. Heisenberg's unified theory was based on a 'world formula'—the term was coined by journalists, not by Heisenberg—which would purportedly describe all elementary particles occurring in nature and all fundamental laws of physics. However, in agreement with the bootstrap idea that was developed at about the same time, he argued that the concept of 'elementary particle' could not be distinguished from that of a compound system. Heisenberg suggested that the question of whether the proton is elementary or composite had no answer.[15]

Together with Hans-Peter Dürr and other collaborators in Munich, Heisenberg developed the theory through the 1960s in the vain hope that it would provide a unified framework for particles and fields. Like Eddington many years earlier, Heisenberg considered the old and still enigmatic problem of calculating the fine-structure constant from fundamental theory. He arrived at a value in fair agreement with the one known experimentally:[16]

$$\alpha \cong 0.386\left(\frac{m_\pi}{M}\right) \cong \frac{1}{120}$$

However, in spite of this and a few other results, the ambitious theory did not live up to its promise. Although the so-called world formula made headlines in German newspapers, it did not impress the majority of Heisenberg's colleagues in theoretical physics. To put it briefly, it was a failure.

Heisenberg's theory was ambitious indeed and a kind of theory of everything, but it did not share the reductionism of many other proposals of a similar kind. 'The fundamental equation does not determine completely the laws in all other parts of physics', he said in 1966, and he further stated that the equation provided no answers to problems at the borderline between physics and biology. 'Therefore the apparent universality of the natural law underlying elementary particle phenomena should not lead to a misinterpretation of its relevance in other parts of natural science. The possibility and necessity of a unified field theory of matter should be judged primarily on its value for understanding the complicated phenomena of elementary particle physics.'[17]

6.2 THE BOOTSTRAP REVOLUTION

In spite of the decline of interest in Heisenberg's S-matrix programme, the S-matrix and related techniques lived on and were in the 1950s revived and further developed by a new generation of physicists, most of them Americans. Among them were Stanley Mandelstam, Murray Gell-Mann, Steven Frautschi, and Francis Low. The strongest and most articulate proponent of the new 'analytic S-matrix theory', and also the most radical one, was, however, Geoffrey Chew, a former student of Enrico Fermi, who in 1955, at the age of 31, became full professor at the University of Illinois, Urbana. Two years later he moved to the University of California, Berkeley. Unlike many other S-matrix physicists, Chew was convinced that the theory was in direct opposition to quantum field theory and not just complementary to it. Throughout the 1960s he developed the idea of an autonomous S-matrix, which he presented as a superior alternative to field theories. He referred to the new way of understanding high-energy physics as the 'bootstrap hypothesis', a name he first used in 1961. The general idea behind the bootstrap approach was to consider strongly interacting particles as dynamical structures which owed their existence to the same force through which they interacted. Each particle helped to generate other particles, which in turn generated it.[18]

The name 'bootstrap' alludes to the adventurous and fabulous eighteenth-century Baron von Münchhausen, who once, having fallen to the bottom of a deep lake, picked himself up by his own bootstraps. It should be noted that the term also appears in scientific contexts outside hadron physics. For example, it may refer to a method of statistical inference that is widely used in medicine and other fields. It may even appear in cosmology, in relation to models relying on Mach's principle that the inertia of a body is determined by the distribution of all matter in the universe. The cosmologist Edward Harrison considers Mach's principle to be an instance of the general bootstrap idea that 'all things are immanent within one another, and the nature of any one thing is determined by the universe as a whole'.[19]

Chew, who was not originally aware of Heisenberg's earlier theory, was convinced that the conventional association of fields with hadronic particles was a mistake. Indeed, he crusaded against quantum field theory insofar as it related to hadrons and their interactions. At a conference in 1961 he asserted that field theory 'is sterile with respect to strong interactions and that, like an old soldier, it is destined not to die but just to fade away'.[20] On other occasions he referred condescendingly to quantum field theory as 'an old, but rather friendly mistress who I would even be willing to recognize on the street if I should encounter her again'. Using the same metaphor, in 1963 he stressed the revolutionary nature of the bootstrap programme:

> As you can see, the new mistress [S-matrix or bootstrap theory] is full of mystery but correspondingly full of promise. The old mistress is clawing and scratching to maintain her status, but her day is past. Twentieth-century physics already has undergone two

breath-taking revolutions—in relativity and in quantum mechanics. We are standing on the threshold of a third.[21]

The Berkeley research programme shared with Heisenberg's earlier theory, and also with the views of Landau, that it was influenced by a positivistic conception of science. As Steven Weinberg has noted, Chew's objections to quantum field theory represented 'a positivist sense of guilt', namely that 'in speaking of the values of the electric and magnetic fields at a point in space occupied by an electron they were committing the sin of introducing elements into physics that in principle cannot be observed'.[22] A similar 'sense of guilt' lay behind Heisenberg's introduction of the S-matrix in the 1940s.

To Chew and a few other physicists, the bootstrap or autonomous S-matrix programme was more than just another theory of high-energy physics; it was envisaged as no less than the beginning of a new chapter in the history of physical thought. The programme appealed greatly to him, not only rationally but also emotionally. In a talk in 1962, he started as follows: 'My belief is that a major breakthrough has occurred and that within a relatively short period we are going to achieve a depth of understanding of strong interactions that I, at least, did not expect to see within my lifetime... I am bursting with excitement, as are a number of other theorists in this game.'[23]

From a methodological point of view the bootstrap programme had similarities to the observability criterion that served as a background for Heisenberg's introduction of quantum mechanics in 1925. According to young Heisenberg, the new quantum mechanics should be founded exclusively on relationships between quantities that were in principle observable.[24] Just like Heisenberg, Chew emphasized that only observable elements should enter into a theory's structure, which he took to imply that a theory of strong interactions could not be based on the space–time continuum. This was *a priori* an unobservable concept, he argued, and therefore of no more use in microphysics than the ether of the nineteenth-century physicists had been in macroscopic physics. In accordance with this view, and again in agreement with Heisenberg's philosophy in the 1920s, he advocated an instrumentalistic conception of physical science, namely that the validity of a physical theory is to be judged solely by its capacity for experimental predictions.

The bootstrap hypothesis was sometimes referred to as the self-consistent particle model of strong interactions. Even though it did not operate with *elementary* particles, it was a particle model of sorts, if of strange sorts. Although Chew continued to use the terminology of particles, he thought of nucleons, pions, kaons, etc. as dynamical entities, manifestations of the underlying strong force. All the hadronic particles, he wrote, 'are dynamical structures in the sense that they represent a delicate balance of forces; indeed they owe their existence to the same forces through which they mutually interact'.[25] For example, the proton was not conceived as a fundamental particle, as the traditional view would have it, but merely as a low-lying state of the strong force or 'strongly interacting matter'. In a sense, Chew's hadrons were non-corpuscular dynamical entities of a nature somewhat analogous to the picture of atoms envisaged a long time ago by natural philosophers such as Leibniz, Boscovich, and Ørsted.[26]

As to the crucial element of self-consistency, it meant that no aspect of the hadronic world was arbitrary. The characteristics of the particles found in nature, such as their masses, electrical charges, and spin quantum numbers, should be uniquely determined by self-consistency. The S-matrix should thus describe the only possible set of hadrons, the same set as found in nature. Not only the particles and their properties, but also the laws and symmetry groups governing the strong interactions were assumed to be settled by the requirement of self-consistency in the unitary and analytic S-matrix: 'To know the S-matrix is to know all that can be known about the microscopic world; a theory that allows the S-matrix to be calculated is complete since the result of any experiment can be predicted.'[27] This did not imply Laplacean determinism, for as a quantum-mechanical theory the S-matrix could only give information about the probabilities of the outcomes of future experiments. Moreover, if there was a demon associated with the S-matrix theory, it would be a seriously handicapped one with access only to the hadronic part of the world.

The idea that the existence of a hadronic system is given by self-consistency leads inevitably to a holistic picture of nature. Particles do not bootstrap themselves individually, but only in concert with the family of particles of which they are composed, and these other particles are themselves composites. As Chew expressed it: 'The self-consistency hypothesis makes it possible to imagine that for *any* hadrons at all to exist, the entire family must exist in a mutually supporting framework.'[28] Not only must the family of particles participate in the bootstrap, all of these particles must themselves be bootstrapped into existence: it follows that the whole hadronic part of the universe is being generated by the self-consistency of the S-matrix. This adds a cosmological dimension to the bootstrap programme, as it seems to imply that the universe of strong particles must have come into existence all at once. Moreover, being grounded in self-consistency, the hadronic universe cannot be even slightly different to what it is. Such a universe, with no evolutionary history, is a strange one and of course it does not qualify as a model of the universe in the ordinary sense of cosmology. During the 1960s cosmological models of the universe were much discussed, and so were conceptions of the strong interactions, but apparently there were no connections between the two. As far as I know, no-one pointed out the inconsistency between the bootstrap picture and the big-bang scenario of the early universe, where hadronic particles constantly appear and disappear.

According to the traditional view of particle physics, some particles are elementary and the building blocks of other less elementary, composite particles. For example, the deuteron (the nucleus of the heavy hydrogen isotope) is a composite particle made up of two elementary particles, the proton and the neutron. Chew denied this 'fundamentalist' picture and indeed the very notion of elementary particles. According to the bootstrap programme, no hadron was more elementary than any other. Each hadron was conceived as a composite of all the others, an idea which he, following a suggestion by Gell-Mann, called 'nuclear democracy' and contrasted with the fundamentalist or 'aristocratic' picture characteristic of quantum electrodynamics. In 1964 Chew formulated his democratic programme as 'every nuclear particle should receive equal treatment under the law'![29] Since all particles were thought to be generated through the exchange of other particles, nuclear democracy implied that there are no truly elementary strongly interacting particles. The idea involved the apparently paradoxical notion of two particles that are constituents of each other, or a composite particle that is one of its own constituents. Paradoxical as these notions may appear, they made sense within the framework of the autonomous S-matrix. For example, bootstrappers viewed mesons as states of a baryon–antibaryon system, and states of baryons as combinations of other baryon states and meson states.

To illustrate the difference between the 'democratic' bootstrap picture and the more traditional hierarchic or fundamentalist picture, consider a particle A, which collides with another particle X, with the result that two new particles, B and C, are created:

$$A + X \rightarrow B + C + X$$

Then consider the same process with A and B interchanged:

$$B + X \rightarrow A + C + X$$

The traditional understanding is that, in the first process, A is a composite particle that is decomposed into the more elementary B and C. But if B is a constituent of A, A cannot be a constituent of B, and then the second process cannot occur. According to the bootstrap picture, there is no inconsistency between the two reactions. In the first one, B and C are simply created out of the available energy of the colliding system, and in the second one A and C are created out of the corresponding energy. Therefore one may say that B is a constituent of A *and* A is a constituent of B.

Although there never was a proof of Chew's conjecture that there exists a unique solution to the equations of the S-matrix, the approximate bootstrap approach led to a few results that agreed promisingly with experiments and created further interest in the approach. For example, in 1962 two young American physicists working with Chew, Fredrik Zachariasen and Charles Zemach, used the bootstrap method to calculate the properties of the ρ meson, a resonance state appearing in pion scattering that had recently been discovered in accelerator experiments (Fig. 6.1). They found the mass of the particle to be about 650 MeV, in reasonable agreement with experimental evidence. Assuming the ρ meson to be a 'self-generating particle' they described the bootstrap view as 'the system of particles produces itself, in that the various particles give rise to forces among themselves making bound states which are the particles'.[30]

The bootstrap conjecture and its associated idea of nuclear democracy challenged the traditional notion of high-energy physics as identical to the physics of elementary particles. It even promised a new kind of science, the contours of which could only be dimly perceived but was thought to hold great promise. Chew explained the vision as follows:

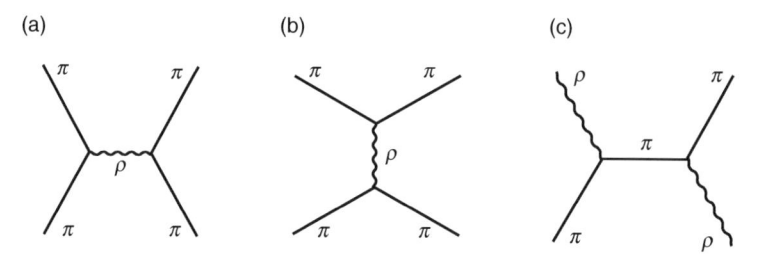

Fig. 6.1. Feynman diagrams played an important role in S-matrix theory, used and interpreted in accordance with the 'democratic' bootstrap idea. The figure shows examples from Zachariasen and Zemach 1962. In (a) a ρ meson is exchanged between two incoming pions, giving rise to a strong force between them. In (b) the ρ meson does not appear as a force, but as a composite particle formed by two pions and which quickly decays into a new pair of pions: $\pi\pi \to \rho \to \pi\pi$. In (c) a pion and a ρ meson, conceived as an elementary particle, scatter, with a pion appearing as the force-carrying particle. Zachariasen and Zemach considered several other Feynman graphs, some of them with the ρ as elementary and others where it was conceived as composite. According to the bootstrap philosophy there was no real difference between the two notions. The use of Feynman diagrams in S-matrix physics is detailed in Kaiser 2005.

Reprinted figure with permission from F. Zachariasen and C. Zemach, 'Pion resonances', *Physical Review* **128** (1962), 849–858. Copyright 2010 by the American Physical Society.

> Ultimately, however, science must answer questions of 'Why', as well as 'How?', and it is difficult to imagine answers that do not involve self-consistency. Why is space three-dimensional? Why are physical laws relativistically invariant? Why do physical constants have values that make it possible to understand sub-classes of natural phenomena without understanding all phenomena at once? In other words, why is science possible in the first place? The bootstrap idea for hadrons does not directly touch on these profound questions but there is a similarity in spirit in concern for the notion that the laws of nature may be the only possible laws. . . . Perhaps the hadron dilemma is the precursor of a new science, so radically different in spirit from what we have known as to be indescribable with existing language.[31]

Yet, ambitious and bold as the bootstrap hypothesis was, in the 1960s it was put forward as a restricted scientific research programme and not as a new world picture. Chew distinguished between two versions, the complete- and the partial-bootstrap hypotheses, where the first one asserted that 'nature is as it is because this is the only possible nature consistent with itself'.[32] This broad and essentially metaphysical version, with its vague association to Leibnizian philosophy, he found to be fascinating and to have 'enormous esthetic appeal'. If the bootstrap could somehow be enlarged to cover also electromagnetic and weak interactions it might lead to a bootstrap understanding of even macroscopic space–time itself. Moreover, it might cast light on the role of consciousness in quantum-mechanical measurements, a problem that Eugene Wigner, among others, had recently called attention to.[33] The complete bootstrap clearly appealed to Chew's imagination:

> Such a future step would be immensely more profound than anything comprising the hadron bootstrap; we would be obliged to confront the elusive concept of observation and, possibly, even that of consciousness. Our current struggle with the hadron bootstrap may

thus be only a foretaste of a completely new form of human intellectual endeavor, one that will not only lie outside physics but will not even be describable as 'scientific'.[34]

However, in his more sober moments Chew realized that it was a dream of the future and that it was a philosophical, not a scientific dream. For one thing, he subscribed to the Popperian formula that 'to be scientific, a hypothesis must be susceptible of experimental disproof',[35] a criterion that the complete bootstrap could not possibly live up to. For another thing, he did not believe that attempts to explain *all* concepts in terms of self-consistency of the whole could be of a scientific nature. Even though the contingent features in nature might be reduced drastically, some unquestioned and, in a sense, arbitrary framework would always be needed.

On several occasions Chew pointed out that the all-or-nothing philosophy was counterproductive and basically unscientific. Scientific progress—indeed the very meaning of scientific investigation as traditionally understood—necessitated an approximate or partial approach to the study of nature. Remarkably, although ultimately all phenomena were presumably interconnected, the history of science had proved that the partial approach was not only possible, but that it seemed to be the only possible way to get scientific insight in nature. For these and other reasons, he, as a physicist (and not a philosopher), preferred to focus on the limited hadronic bootstrap hypothesis. This was a testable scientific hypothesis, although it could not be tested either directly or easily. Chew was careful to point out that certain future discoveries in high-energy physics *might* demolish the hadronic bootstrap, thus securing its scientific nature. However, he was confident that such falsifying discoveries would not, in fact, turn up.

6.3 DECLINE OF A RESEARCH PROGRAMME

S-matrix theory was very influential in strong interaction physics in the 1960s, when it appealed to many physicists and scored several scientific successes. Although very few physicists followed Chew all the way in his crusade against field theory and for the bootstrap hypothesis, the general idea of S-matrix theory enjoyed wide popularity.

Gell-Mann, who in 1969 received the Nobel Prize for his work in elementary particle physics, was among those who were inspired by, and significantly contributed to, S-matrix theory. It would seem that the associated ideas of the bootstrap and nuclear democracy were in obvious conflict with Gell-Mann's 'aristocratic' quark model of hadrons, but this was not how Gell-Mann saw the situation. In fact, when he developed the quark concept he was quite enthusiastic about the bootstrap approach, including the idea of nuclear democracy, which helped him conceptualize the new concept of unobservable and fractionally charged quarks. 'The quark model', he wrote in 1967, 'may perfectly well be compatible with the bootstrap hypothesis, that hadrons are made up of one another.'[36] Twenty years later he recalled: 'In 1963, when I developed the quark proposal, with confined quarks, I realized that the bootstrap idea for hadrons and the quark idea with confined quarks can be compatible with each other, and that both proposals result in "nuclear democracy", with no observable hadron being any more fundamental than any other.'[37] From Gell-Mann's perspective in the mid-1960s there was no irreconcilable conflict between quantum field theory and some versions of S-matrix theory. Of course, Chew disagreed.

As the bootstrap hypothesis attracted much positive attention, so it attracted much criticism. The critical responses were to some extent the result of Chew's enthusiastic proselytizing of the idea. Not all physicists found it appropriate to market a theory as a new revolution in scientific thinking. According to John Polkinghorne, an English physicist active in S-matrix theory, Chew was a great 'salesman of ideas' who advocated his cause with a fervour of an 'impassioned evangelist'.[38] Another physicist described him as 'the Billy Graham of physics', a reference to the influential and charismatic American evangelist.[39] One of the early critics, the American physicist Alexander Stern, pointed to two serious deficiencies in the bootstrap programme. First, it was effectively limited to strong interactions and had nothing to say about the weak and electromagnetic forces. For this reason the theory, which was claimed to be fundamental, went against the ideal of the unity of science. Second, because it was inconsistent with the electromagnetic field, and this field is (according to the standard view) crucially involved in all measurements and observations, 'the measurement process has no logical place in the S-matrix theory'.[40] Yet another reason for doubting the bootstrap theory was that it could be used only for scattering amplitudes and not for the internal structure of particles.[41]

Feynman did not like the bootstrap approach to physics, which he tended to see as a fashion that one-sidedly focused on one particular theoretical framework. In an interview of 1988, shortly before his death, he mentioned S-matrix theory, Wheeler's quantum geometrodynamics, and string theory as examples of fashions he had experienced in his long career in theoretical physics. 'Once Geoffrey Chew believed that everything was going to come out of the S-matrix; then he gathered a coterie around him, who believed that they would find the answer by following that route.' As Feynman saw it, the Berkeley S-matrix school of the 1960s represented an unhealthy trend in theoretical physics: 'Somebody notices a good idea and starts working on it, and all the other guys rush in believing that this is *the* thing. They expect to be at the forefront of the thing; they follow the leaders who say that by working on this thing they'll be on the cutting edge.'[42] As we shall see in Chapter 11, this characterization may also include modern string theory, as indeed Feynman thought it did.

In hierarchic theories operating with elementary particles one may always ask why and if the elementary particles are truly elementary. Hydrogen, once believed to be a primitive substance, consists of protons and electrons, and protons consist of quarks and gluons, and quarks may consist of superstrings. Who knows? If so, is there a layer below the strings? Probably not, but in terms of what can strings then be explained? Questions of this kind, concerning the ultimate explanation of things, are old and they continue to be asked. In the fourth edition of his famous textbook in quantum mechanics, *Principles of Quantum Mechanics*, Dirac reflected as follows:

> In a classical explanation of the constitution of matter, one would assume it to be made up of a large number of small constituent parts and one would postulate laws for the behaviour of these parts, from which the laws of the matter in bulk could be deduced. This would not complete the explanation, however, since the question of the structure and stability of the constituent parts is left untouched. To go into this question, it becomes necessary to postulate that each constituent part is itself made up of smaller parts, in terms of which its behaviour is to be explained. There is clearly no end to this procedure, so that one can never arrive at the ultimate structure of matter on these lines.[43]

The bootstrap hypothesis was not faced with this problem as it did not operate with a hierarchical structure of hadronic particles but explained particles by requirements of self-consistency rather than in terms of more elementary particles. Dirac was generally in favour of the philosophy behind the *S*-matrix programme, including its emphasis on the principle that only observable quantities should appear in a fundamental theory. In fact, much of the essence of *S*-matrix philosophy can be found in a paper on relativistic quantum mechanics he wrote in 1932.[44] However, although Dirac found the standard theory of particles and fields to be unsatisfactory, he did not find the bootstrap alternative to be attractive. His principal objection was that it destroyed the unity of physics, a value he (like most other physicists) held in high regard. Dirac explained his disbelief in *S*-matrix theory as follows:

> High energy physics forms only a very small part of the whole of physics—solid state physics, spectroscopy, the theory of bound states, atoms and molecules interacting with each other, and chemistry. All these subjects form a domain vastly greater than high energy physics, and all of them are based on equations of motion. We have reason to think of physics as a whole and we need to have the same underlying basis for the whole of physical theory.[45]

From about 1970 Chew's autonomous *S*-matrix programme ran into trouble and was soon relegated to a small corner of physics. By the end of the decade the number of physics papers dealing with the bootstrap had dwindled to less than a handful per year. There were several reasons for the decline, some scientific and some of a more sociological nature. Remarkably, the theory was never directly refuted by experiments. What happened was rather that the ambitious theory proved unable to deliver what it promised. When calculations of actual high-energy collision processes were attempted on the basis of the first principles of the theory, such as unitarity and analyticity, they turned out to be hopelessly complex. The holism of bootstrap philosophy implied that, in principle, one could not understand anything without understanding everything, which obviously creates problems. This is what Chew referred to in a paper of 1971: 'Since all hadrons are mutually dependent in a bootstrap, an attempt to completely understand any individual strongly interacting particle requires an understanding of all.

This "all or nothing" character of the hypothesis makes its experimental predictive content extraordinarily elusive.'[46] The remedy was to introduce approximations and special assumptions, but these refinements, made to bring it into closer contact with experiments, spoiled the original simplicity of the theory. More than one physicist in the early 1970s compared it with the late-medieval versions of Ptolemaic astronomy, with their many added epicycles, shortly before the Copernican revolution.

In short, the autonomous S-matrix theory collapsed under the weight of its own complexity, a feature known also from other physical theories of an ambitious and foundational nature. This does not mean that it vanished, for in various modifications (such as 'topological S-matrix theory') it has continued to this very day. A minority of physicists, including Chew, continued to investigate the bootstrap S-matrix and extend it to new areas, arguing that it might be applicable also to weak and electromagnetic interactions.[47] In the early 1980s Chew believed that electromagnetism was inevitable in a consistent S-matrix and that it would lead to an understanding of the numerical value of the elementary electric charge, that is, the fine-structure constant $\alpha \cong 1/137$. 'Bootstrappers expect eventually to understand our familiar space–time world of "real" objects as one aspect of a self-sustaining universal "mosaic".'[48] He thought that the bootstrap mosaic would eventually also encompass gravitation.

Other physicists have argued that the confrontation between field theory and S-matrix theory is largely artificial and that the two approaches are better seen as complementary than contradictory. Some of them have developed bootstrap theories that purportedly explain quarks. But neither this nor other lines of work have succeeded in bringing the S-matrix back into the centre of physics. It is ironic that Chew's 1961 comparison of quantum field theory with an old soldier who 'is destined not to die but just to fade away' should become valid, not with respect to quantum field theory but with respect to S-matrix theory itself. On the other hand, the importance of the theory was not restricted to an episode of hadronic physics in the 1960s. It is worth noting that modern string theory has its origin in theoretical models of hadrons that were closely related to ideas of the S-matrix. Indeed, it has been suggested that had it not been for the S-matrix theory, string theory would not have emerged at the time it did (see Chapter 11).

It was not only internal problems in the S-matrix programme that caused its decline. Even more important was the emergence of powerful new field theories in the 1970s that made the bootstrap message appear much less convincing. One very important innovation was the gauge theory of quantum fields, which scored its first significant success with the unified theory of electroweak interactions developed in the 1960s by Sheldon Glashow, Steven Weinberg, and Abdus Salam. (In 1979 they shared the Nobel Prize for their contributions to the theory.) The Weinberg–Salam electroweak theory ignored the non-renormalizable infinities but succeeded in the more ambitious goal of unifying the electromagnetic and weak interactions. Although it took a few years until the electroweak theory won general recognition, by the early 1970s the 'dark age' of quantum field theory was over. Gerardus t' Hooft, at the time a graduate student at the University of Utrecht, The Netherlands, proved in 1971 that the electroweak theory was no more plagued by infinities than quantum electrodynamics. When it further turned out that the theory agreed with high-energy experiments, it quickly became recognized as a pioneer contribution to a new unification programme in fundamental physics.

The successes of gauge field theory were extended by the development in the late 1970s of the so-called standard model, which included electroweak as well as strong interactions (quantum chromodynamics). Quarks and gluons were the aristocratic constituents of the new class of grand unified theories, so different from the democratic S-matrix theory. As seen from the perspective of mainstream particle physics, there was no need to pay attention to an alternative theory of strong interactions in which quantum fields had no role.

6.4 PHILOSOPHY AND IDEOLOGY OF THE BOOTSTRAP

From the late 1960s, when the bootstrap hypothesis began to lose its scientific attraction, its supporters increasingly emphasized its more philosophical virtues, which Chew had hinted at for several years. 'The bootstrap idea has, to date, made few inroads into the present predominantly fundamentalist thinking of the high energy physics community', one physicist noted with regret in 1973. 'Yet,' he went on, 'its possible importance not only to particle physics but to the whole Western cultural tradition of scientific thought cannot be overstated.'[49]

Given the philosophical flavour of the bootstrap idea one might expect that Chew originally came to the idea inspired by philosophers or philosophical reasoning. But this was not the case. Initially Chew had no interest in questions of a philosophical nature, which played no role in his early conception of the bootstrap. The idea had its origin in technical problems and Chew only found it to be philosophically interesting at a later stage.[50] On various occasions he reflected on the larger implications of the theory, such as the possibility that it might form the basis of a new physics (or natural philosophy). The complete bootstrap, he conjectured, would imply that the existence of consciousness is necessary for self-consistency in the universe. However, Chew never developed a systematic bootstrap philosophy.

A few other physicists took an interest in the theory from a philosophical point of view, often in connection with the eternally discussed interpretation of quantum mechanics. One of them was Henry Stapp, an S-matrix physicist and a colleague of Chew in Berkeley. Stapp advocated a 'web philosophy' according to which interrelated events and processes, rather than localized objects, were the fundamental constituents of the world. As he noted, this general view was in harmony with the process philosophy of Alfred North Whitehead. According to Stapp, the view that 'the world cannot be understood as a construction built from a set of unanalyzable basic entities or qualities' was just the view of the bootstrap theory. 'The web or bootstrap philosophy represents the final rejection of the mechanistic ideal', he opined.[51]

What was perhaps the most extensive popular exposition of the bootstrap philosophy—or what may be more appropriately be called bootstrap ideology—was also one of the most controversial. Fritjof Capra, an American physicist and former student of Chew, published in 1975 *The Tao of Physics*, a book that became immensely popular because it resonated so well with the New Age spirit of the time. Capra's basic message was that the insights of quantum physics, if only properly interpreted, were the very same as those reached by Eastern mystics through spiritual meditation and intuition. His insights included the suggestion that physics was a reflection of the human mind rather than describing an objective nature made up of particles and fields. Capra found the democratic *S*-matrix theory to be the far most convincing parallel with Eastern mysticism, especially as represented by Mahayana Buddhism, a complex and other-worldly version of Buddhist thought. He consequently went to some lengths to explain the meaning and broader implications of the bootstrap programme and its associated world picture, which he described as follows:

> In the new world-view, the universe is seen as a dynamic web of interrelated events. None of the properties of any part of this web is fundamental.... The bootstrap hypothesis not only denies the existence of fundamental constituents of matter, but accepts no fundamental entities whatsoever—no fundamental laws, equations, or principles—and thus abandons another idea which has been an essential part of natural science for hundreds of years.... To discover the ultimate fundamental laws of nature remained the aim of natural scientists for the three centuries following Newton. In modern physics, a very different attitude has now developed. Physicists have come to see that all their theories of natural phenomena, including the 'laws' they describe, are creations of the human mind; properties of our conceptual map of reality, rather than of reality itself.[52]

It was certainly true that the bootstrap programme employed concepts and methodology that differed from conventional physics, but it was a wild exaggeration to claim that 'modern physics' had adopted the new way of thinking. On the contrary, the unconventional nature of bootstrap physics made it difficult for it to be accepted as a respectable scientific theory. Chew felt that the clash between two cultures of science was in part responsible for the decline of the bootstrap programme: 'Because of the long tradition of Western science the bootstrap approach has not become reputable yet among scientists', he said in 1983. 'It is not recognized as science precisely because of its lack of a firm foundation.'[53] All the same, in the later editions of *The Tao of Physics*, Capra continued to insist that *S*-matrix physics was a better and more successful theory than the outdated quantum field theory. He further suggested that Chew's bootstrap theory was a revolutionary step in the history of physics of the same magnitude as the theory of relativity and quantum mechanics, only surpassing them in radicality. Indeed, in the third edition of 1991 he claimed that it 'represents a radical break with the entire Western approach to fundamental science'.[54]

Postmodernists and social constructivists have on some occasions expressed interest in the bootstrap programme of the past, which from their perspective has as much or more right to recognition as the undemocratic and fundamentalist dogma of quantum chromodynamics. In a paper with the telling title 'The irrelevance of reality', Robert Markley, a professor of English at the University of Illinois at Urbana-Champaign, suggested that the reasons for the decline of the bootstrap programme were ideological rather than scientific:

> It is not surprising... that bootstrap theory has fallen into relative disfavor among physicists seeking a GUT (Grand Unified Theory) or TOE (Theory of Everything) to explain the structure of the universe. Comprehensive theories that explain 'everything' are products of the privileging of coherence and order in western science. The choice between bootstrap theory and theories of everything that confronts physicists does not have to do primarily with the truth-value offered by these accounts of available data but with the narrative structures—indeterminate or deterministic—into which these data are placed and by which they are interpreted.[55]

As there were philosophical elements involved in the bootstrap story, so there appeared to be political and ideological elements. This is indicated in particular by the vocabulary associated with the new theory, such as the 'nuclear democracy' that Chew contrasted with the traditional 'fundamentalist' theories of matter and fields that gave 'special status' to a select 'aristocracy' of particles. He used these and similar terms consistently, and not only to characterize two competing research programmes.

Chew also thought that the S-matrix theory was, in a sense, more democratic because it was open for all physicists and did not exclude those who had not mastered the complex perturbation calculations of quantum field theory. The dominant mathematical framework of S-matrix physics was the theory of analytic functions, a branch of the theory of complex variables that went back to the French mathematician Augustin Cauchy in the early part of the nineteenth century and with which most physicists were familiar. In a book of 1961 on S-matrix theory, Chew emphasized that readers did not have to be 'conversant with the subtleties of field theory, and a certain innocence in this respect is perhaps even desirable'. He added that experts in field theory 'seem to find current trends in S-matrix research more baffling than do nonexperts'.[56] What appealed to Chew did not appeal to other physicists with a more mathematical mindset. Dyson acknowledged that S-matrix theory was impressive in its ability to interpret experiments and offer guidance to experimenters. But he contrasted unfavourably the mathematical methods of S-matrix theory with the deeper and more sophisticated methods that governed quantum field theory. 'I find S-matrix theory too simple, too lacking in mathematical depth', he said. 'If the S-matrix theory turned out to explain everything, then I would feel disappointed that the Creator had after all been rather unsophisticated.'[57] It should be added that S-matrix theory was not really elementary, in the ordinary sense of the word, or easily accessible from a mathematical point of view.

It was definitely not a theory suited for the 'science for the people' movement that was popular in radical circles in the 1970s.

Chew was politically active during the 1950s, mostly in relation to the atomic scientists' movement and the Federation of American Scientists, but he did not subscribe to any particular political ideology. While there is no reason to assume that his ideas about hadronic physics reflected or were caused by his political views in some direct way, he did draw on intellectual resources with roots in those views, as they reflected the political situation in the United States at the time.[58]

Notes for Chapter 6

1. Chew 1970, p. 25.

2. For a brief survey of the two-particle paradigm and the new particles that followed in the 1930s and later, see Mukherji and Roy 1982.

3. In the 1920s a few physicists argued that the motion of electrons in the Bohr–Sommerfeld model should be treated in analogy with the motion of planets, hence that gravitation should be taken into account. The idea of using Einsteinian gravitation in the study of atomic structure was followed in particular by Manuel Vallarta, a Mexican theorist at the Massachusetts Institute of Technology. However, nothing useful came out of either his or related efforts. For a brief survey of this episode in the history of atomic physics, see Kragh 1985, pp. 100–102 and Kragh 1986.

4. Dirac 1929, p. 714. For an analysis of the claim and the chemists' response to it, see Simões 2002.

5. Dirac 1930, p. 605. As pointed out in Chapter 3, at the turn of the century J. J. Thomson entertained a similar hope of reducing all matter to electrons.

6. Rueger 1992. On the history of quantum field theories, see also Weinberg 1977 and the detailed and technical account in Schweber 1994.

7. Letter to Frank Oppenheimer, 4 June 1934, as quoted in Kragh 1999, p. 198.

8. On Heisenberg's and other physicists' early ideas of a smallest length, sometimes called a 'hodon', see Kragh 1995a. Other ideas of a fundamental length will be considered in later chapters.

9. Dirac 1942, p. 8.

10. Dyson 1953, as quoted in Schweber 1994, p. 568.

11. Landau and Lifschitz 1974, pp. 262–63. The book was based on a Russian original which appeared in 1972.

12. On Heisenberg's S-matrix physics, see Rechenberg 1989. See also Cushing 1986 and Cushing 1990, pp. 30–42.

13. Heisenberg 1943, p. 513.

14. Pauli to Heisenberg, 25 December 1946, quoted in Rechenberg 1989, p. 566.

15. Heisenberg 1957.

16. Heisenberg 1966, p. 118. For the 'average mass of leptons' (electrons and muons), Heisenberg calculated the value 40 MeV. In Heisenberg 1957, he reported $a \cong 1/267$.

17. Heisenberg 1966, pp. 134–35.

18. The philosophical aspects of the bootstrap hypothesis are dealt with in Freundlich 1980, Cushing 1985, and Redhead 2005. On Chew and his career, see Kaiser 2002 and Kaiser 2005.

19. Harrison 2000, p. 5.

20. Quoted in Cushing 1990, p. 143.

21. Cushing 1990, p. 169 and Chew 1963, p. 539.

22. Weinberg 1992, p. 181.

23. Chew 1962, p. 394.

24. As Heisenberg expressed it in his seminal paper, his aim was to find 'a basis for theoretical quantum mechanics founded exclusively upon relationships between quantities which in principle are observable'. Heisenberg 1925, p. 879.

25. Chew, Gell-Mann and Rosenfeld 1964, p. 79.

26. See Gale 1974, who argues that there is a close analogy between Chew's hadronic particles and Leibniz's system of monads. On dynamical conceptions of atoms, see also Chapter 1.

27. Chew 1963, p. 534.

28. Quoted in Gale 1974, p. 346.

29. Quoted in Kaiser 2005, p. 314. Gell-Mann later suggested that 'nuclear egalitarianism' might be a better term than 'nuclear democracy'. Gell-Mann 1987, p. 482.

30. Zachariasen and Zemach 1962, p. 849. See also Kaiser 2005, pp. 307–12.

31. Unpublished paper of 1966, quoted in Cushing 1990, p. 180.

32. Chew 1968, p. 762.

33. Wigner 1983 (first published 1961).

34. Chew 1968, p. 765.

35. Ibid., p. 763. Chew did not mention Popper by name.

36. Quoted in Cushing 1990, p. 156.

37. Gell-Mann 1987, p. 482.

38. Polkinghorne 1985, p. 24. In 1979 John Polkinghorne, a Cambridge professor of mathematical physics and a fellow of the Royal Society, gave up his chair to become an Anglican priest. He has written extensively on the interactions between science and theology.

39. Letter from Marvin Goldberger to Murray Gell-Mann, 27 January 1962, quoted in Kaiser 2002, p. 263.

40. Stern 1964, 43.

41. Using high-energy electrons rather than alpha particles, Rutherford-like experiments showed in the 1950s that nucleons have a particular structure and charge distribution. These experiments, which in 1961 resulted in a Nobel Prize to the Stanford physicist Robert Hofstadter, were of great importance but outside the reach of the S-matrix theory.

42. Interview conducted by Jagdish Mehra in January 1988, quoted in Mehra 1994, p. 591.

43. Dirac 1958, p. 3.

44. See Schweber 1994, p. 48.

45. Dirac 1969, p. 4.

46. Chew 1971, p. 2334.

47. For example Chew 1981 and Chew 1983. For topological S-matrix theory, see Cushing 1990, pp. 200–203.

48. Chew 1983, p. 219.

49. Swetman 1973.

50. Interview conducted by Fritjof Capra in July 1983, in DeTar *et al.*, 1985, pp. 247–386.

51. Stapp 1971, p. 1319. A somewhat similar view, again with explicit reference to Chew's theory, was at the same time espoused by Gerald Feinberg at Columbia University (Feinberg 1972).

52. Capra 1976, pp. 276–77. The book ran through 43 editions and 23 translations. For an in-depth analysis of Capra's and others' attempts to establish significant parallels between Eastern mysticism and modern physics, see Restivo 1985. In later books and articles, Capra has continued to employ the framework of bootstrap theory and to suggest that life, consciousness, and human values become in this way parts of a new holistic physics.

53. DeTar et al.,1985, p. 250.

54. On p. 257.

55. Markley 1992, p. 269.

56. Chew 1961, p. vii.

57. Dyson 1964, p. 136.

58. See the careful investigation in Kaiser 2002 and Kaiser 2005. There are other cases in the physical sciences where social and political metaphors appear in the scientific vocabulary. For an example from quantum physics in the Soviet Union (including such well known terms as 'collective phenomena' and 'collective excitations'), see Kojevnikov 1999 and Kojevnikov 2004, pp. 47–72, 248–75.

PART II
The Present Scene

7

Varying Constants of Nature

At the beginning of time the laws of Nature were probably very different from what they are now. Thus we should consider the laws of Nature as continually changing with the epoch, instead of as holding uniformly throughout space–time.

Paul Dirac, *The relation between mathematics and physics*, 1939, p. 913.

At the bottom of the physical sciences stand the fundamental laws of nature and their associated constants, parameters that, as far as we know, cannot be deduced from theory but need to be determined by experiment. Since the constants moved to the forefront of physics in the late nineteenth century, their nature and significance have been much discussed. To claim that a quantity is constant usually means that it does not depend on either position or time, which is clearly a non-verifiable claim. If it is merely a generalization from local experience it is conceivable that some of the constants, however fundamental, may vary in time (it is also conceivable that they have different values at different places in the universe). This idea popped up in the 1930s, first in connection with Newton's constant of gravitation, the speed of light, and the dimensionless fine-structure constant. On the basis of his hypothesis of a varying gravitational constant Dirac developed a new model of the expanding universe which stimulated further speculations of varying constants. Such speculations or hypotheses have traditionally been connected to cosmological theories and they continue to play an important role in cosmologies of the less conventional kind.

Although known and discussed by a few physicists before World War II, the idea of changing laws of physics was heterodox and not well received by the majority of physicists. For a long time it was dismissed as a speculation with no foundation in either theory or experiment. For example, Richard Feynman considered the 'historical question' in his series of physics lectures in the early 1960s, but found it to be highly unlikely. Yet, should it nonetheless turn out that the laws were not fixed in time, then 'the historical question of physics will be wrapped up with the rest of the history of the universe'.[1] Physicists would then be talking the same language as astronomers, geologists, and biologists.

Hypotheses of varying laws or natural constants only began to be taken seriously in the 1970s, and during the last two decades they have proliferated and even gained a status of semi-respectability.[2] However, the relative popularity of the idea in theoretical cosmology is not based on any solid experimental evidence that some of the constants do in fact vary. Such evidence is still missing and thus the question remains unsettled. In this chapter, I examine three classes of hypothesis in their historical context. The case of a varying gravitational constant was much discussed in the earlier period, but at the turn of the century, physicists' interest focused increasingly on two other cases. One was the possibility of a varying fine-structure constant, such as was supported by some astronomical observations, and the other was the possibility that the speed of light had changed in the past. These ideas are still being discussed and it is too early to say whether they are viable or not. Generally speaking, the climate in theoretical physics and cosmology is today much more favourable to ideas of varying constants than it was just a few decades ago. But the verdict lies with experimental confirmation, which is still missing.

7.1 THE CONCEPT OF NATURAL CONSTANTS

Modern physics recognizes that the fundamental constants of nature have a special place in the fabric of the universe, that they are the very building blocks of fundamental physical theory and therefore of central importance in the endeavour to understand nature on the most basic level.[3] The constants of nature—or the constants of physics (the two terms are generally used synonymously)—typically appear in the fundamental laws of physics. Being *constants*, they are assumed to be independent of space and time, but they are not so by definition or necessity.

There are different kinds of constants, some regarded as more fundamental than others. Some of them are characteristic of particular objects or physical phenomena, some refer to a whole class of phenomena, and others again are constants of a strictly universal significance. The acceleration of gravity at the surface of the Earth ($g = 9.8$ cm s^{-2}) is a useful and non-trivial constant, but it is not nearly as fundamental as Newton's constant of gravitation G, by which quantity it can be expressed ($g = GM/R^2$, where M denotes the mass of the Earth and R its radius). It is with the exclusive class of universal constants that physicists and cosmologists are primarily occupied. The status of a constant is not given *per se*, but varies with the state of knowledge and therefore changes in unpredictable ways through history.

For example, consider the speed of light, which turned up in physics in 1676 when the Danish astronomer Ole Rømer discovered from an analysis of Jupiter's innermost moon that it was a finite quantity. After James Bradley's discovery of the aberration of the fixed stars in 1728 it was realized that light propagates with the same velocity throughout the universe, meaning that it is a universal constant of nature. This recognition was further established by Maxwell's electromagnetic theory of light, which expressed the velocity in terms of new electric and magnetic constants characteristic of the ether or vacuum, $c^2 = (\epsilon_0 \mu_0)^{-1}$. Before Maxwell, the speed of light referred to just a particular phenomenon of nature, while after him it got a much wider meaning. Finally, Einstein realized in his special theory of relativity that the speed of light is independent of the motion of the light source and also of the reference frame of the observer. After 1905 c became truly fundamental and universal.

A somewhat similar story can be told about the elementary charge, which first turned up in Faraday's electrolytic experiments of the 1830s and later became an essential part of electrodynamics and atomic physics. On the other hand, there are many physical constants which have been 'degraded' in the same way as was the acceleration due to gravity. The Rydberg constant of spectroscopy ($\approx 10^7$ m^{-1}), so named after the Swedish physicist Johannes Rydberg who introduced it in 1888, was for a time considered a constant of deep significance. It appeared as a constant of proportionality in spectral series formulae and was known only empirically. With Bohr's atomic theory of 1913 the quantity was realized to be just a combination of other, more fundamental constants, its value being given by $2\pi^2 me^4/h^3c$.

Physicists are particularly interested in the *fundamental* constants of nature, their numerical values and their interrelations. A truly fundamental constant is contingent in the sense that it does not follow from any known physical theory and cannot be explained by other constants in the manner that g can be explained from G. It is irreducible, a so-called 'free parameter'. In this sense it has the same character as the basic objects of matter, such as leptons and quarks. We may say that a fundamental constant, say the elementary charge, is defined by our ignorance of it—not of its value but of its very existence. Why does it have the value it has? Why are all electrical charges multiples of a smallest charge? Similarly, why do only certain quarks and leptons exist? Obviously, 'ignorance' is a relative term, depending on the state of knowledge at any given time.

Then the question arises of which are the fundamental constants of nature? Although there is no consensus among physicists, it is generally agreed that the following constants are fundamental in a deep sense:[4]

speed of light $c = 2.9989 \times 10^8$ m s^{-1}
Planck's constant $h = 6.567 \times 10^{-34}$ J s
gravitational constant $G = 6.673 \times 10^{-11}$ m^3 kg^{-1} s^{-2}
mass of the electron $m = 9.109 \times 10^{-31}$ kg
mass of the proton $M = 1.673 \times 10^{-27}$ kg
elementary charge $e = 1.602 \times 10^{-29}$ C

In addition to these classical constants, the cosmological constant Λ has more recently come to be seen as fundamental. Arguments based on relativity theory combined with astronomical measurements indicate a value of $\Lambda \approx 10^{-35}$ s^{-2}, but this is a contested issue in present-day physics to which we shall return. A few scientists, Eddington among them, have suggested that the dimensionality N of space or space–time (that is, 3 or 4) should also count as a constant of nature. For example, in 1984 the Russian physicist Iosif Leonidovich Rozental proposed including $N = 3$ among his list of fundamental dimensionless constants, the other six numbers being

a_s (strong coupling constant) $\sim 10^{-11}$

a_e (fine-structure constant) $\sim 10^{-2}$

a_w (weak coupling constant) $\sim 10^{-11}$

a_g (gravitational coupling constant) $\sim 10^{-38}$

m/M (electron-proton mass ratio) $\sim 10^{-3}$

$(M_n - M)/M$ (neutron–proton mass difference ratio) $\sim 10^{-3}$

In the spirit of Eddington, Rozental wrote: 'The structure of the Universe is determined by the above numbers. In a certain sense, this statement signifies a return to the concept of the ancient natural philosophers (Plato, Pythagoras), assigning to numbers a crucial role in the structure of the universe.'[5] He was neither the first nor the last to suggest that the essence of physics is to be found in pure numbers (see Chapter 10).

Most lists of physical constants typically include more quantities than those mentioned above, such as Boltzmann's constant ($k = 1.381 \times 10^{-23}$ J K^{-1}) and Avogadro's number ($N_A = 6.022 \times 10^{23}$ mol^{-1}), but these are not fundamental in the same elevated sense. Boltzmann's constant is often regarded as merely a conversion factor between temperature and energy (by $E = kT$) and Avogadro's number has a more limited significance. Moreover, it is related to Boltzmann's constant by $N_A = k/R$, where R ($= 8.314$ J kg^{-1} mol^{-1}) is the so-called gas constant appearing in the equation of ideal gases, $pV = nRT$. Fundamental physics endeavours to reduce the contingency of nature maximally, whether this contingency is related to material objects or to the laws of nature and their associated constants. There is no agreement among physicists of how far this reduction can go. Some physicists argue that there are three absolute and irreducible constants, typically c, G, and h, while others believe that the number can be reduced to two. There are even those who hold that all natural constants are to be understood as conversion factors, in which case the number of fundamental constants will be zero.[6]

The recognition of the importance of the constants of nature is of relatively recent origin, as it can be traced back only to the last decades of the nineteenth century. Although some of the constants had been recognized and measured earlier (such as G, c, and e), their significance was first highlighted by George Johnstone Stoney in a prescient paper of 1881 read to the British Association for the Advancement of Science. Based on Faraday's laws of electrolysis, Stoney recognized the elementary nature of electricity and suggested that the charge of the 'electrine' (10 years later he called it an electron) should be considered a constant of nature. He argued that this constant or unit was of the same fundamental significance as the constant of gravitation and the speed of light and that the values of these fundamental constants were the same all over the universe. They were the units chosen by nature, not merely units convenient from a human point of view.

If we take these as our fundamental units instead of choosing them arbitrarily, we shall bring our quantitative expressions into a more convenient, and doubtless into a more intimate, relation with Nature as it actually exists.... Hence we have very good reason to suppose that in V_1, G_1, and E_1 we have three in a series of systematic units that in an eminent sense are the units of nature, and stand in an intimate relation with the work which goes on in her mighty laboratory.[7]

Moreover, Stoney showed that the three quantities could be combined in 'a truly Natural Series of Physical Units', namely in such a way that they resulted in new universal units for mass, length, and time. For example, the Stoney time unit was the inconceivably small number

$$t_S = \frac{e}{c^3}\sqrt{G} = 3 \times 10^{-46} \text{s}$$

Stoney's attempt to construct natural and universal units attracted little attention at a time when the electron, conceived as an elementary particle and not a unit charge, was still unknown. It was considered an interesting speculation, but nothing more than that. However, 18 years later a similar idea was independently suggested by Max Planck, the father of quantum theory.

In his quest to understand the nature of universal blackbody radiation, Planck derived in 1899 a spectral formula that included two constants a and b, the values of which he was able to calculate. His improved theory of the following year, now based on the revolutionary hypothesis of energy quantization, also included two constants: Boltzmann's constant k and a new enigmatic energy constant h of the dimension of an action (the product of energy and time). The relations between the two sets of constants were simply that $a = h/k$ and $b = h$. The agreement between theory and experiment was of course important to Planck, but he found the constants appearing in his theory to be no less important. In his 1899 paper he wrote of the constants a and b that they 'provide the possibility of identifying the new units of length, mass, time and temperature, which, independently of specific bodies or particular circumstances, necessarily maintain their meaning at all times and for all cultures, even for extra-terrestrial or non-human life forms, and can therefore be considered as "natural units of measurements"'.[8] As the new de-anthropomorphic units, he proposed what later became known as the Planck units. Written in terms of k and h, these are:

$$l_{Pl} = \sqrt{\frac{Gh}{c^3}} = 4.13 \times 10^{-35} \text{m}$$

$$m_{Pl} = \sqrt{\frac{ch}{G}} = 5.56 \times 10^{-8} \text{kg}$$

$$t_{Pl} = \sqrt{\frac{Gh}{c^5}} = 1.38 \times 10^{-43}\,\text{s}$$

$$T_{Pl} = \frac{1}{k}\sqrt{\frac{c^5 h}{G}} = 3.5 \times 10^{32}\,\text{K}$$

These quantities, Planck wrote at the end of his 1899 paper, 'will retain their natural meaning for as long as the laws of gravity, the propagation of light in vacuum, and the two principles of the theory of heat hold, and, even if measured by different intelligences and using different methods, must always remain the same'. Planck's conception of the nature of the fundamental constants was shared by Paul Drude who, in a textbook of 1900, suggested a modified 'truly absolute system of units...[which] does not depend on any particular properties of any body'.[9] In Drude's system the constants G, c, h, and k all had the value 1.

Even more than Stoney, of whose work he seems to have been unaware, Planck stressed the objective and absolute character of the natural constants. This was not only crucial from a scientific point of view, but also held a moral value for him. In his autobiography of 1949 he drew a parallel between the values of the absolute constants of nature and 'absolute values in ethics'.[10] An outspoken critic of positivism and relativism (not to be confused with relativity!), it was important for Planck to present fundamental physics as completely detached from the human mind. In an address of 1908 he spoke of the constants of physics as signifying an 'emancipation from anthropomorphic elements'. Because they were universal in character and independent of particular substances, they could be used to 'determine units of length, time, mass, and temperature, and these units must necessarily retain their meaning for all time and for all extra-terrestrial and superhuman *kulturs*...[They] are such that the inhabitants of Mars, and indeed all intelligent beings in our universe, must encounter at some time—if they have not already done so'.[11] It is ironic that the anthropic principle introduced in physics and cosmology in the 1970s built upon the combinations of and relations between natural constants. While Planck proposed new units in order to make physics as objective and non-anthropomorphic as possible, in the 1970s the constants of nature came to serve the opposite purpose. My guess is that Planck would have completely dismissed the anthropic principle in any of its several meanings.

Although Planck's quantum of action entered physical theory in 1900, for several years the nature and meaning of the constant remained unclear. Most physicists did not recognize its fundamental nature and consequently sought to explain it by reducing it to other constants. The quantity h was seen as derivative rather than fundamental. For example, in connection with an atomic model he proposed in 1910, the Austrian physicist Arthur E. Haas suggested writing the constant as $h = 4\pi e(ma)^{1/2}$, where m is the mass of the electron and a the radius of the hydrogen atom. On the other hand, in his more advanced atomic theory of 1913 Bohr stressed that Planck's constant was an irreducible quantity, mysterious and inexplicable. For the hydrogen atom in its ground state he happened to get the same result as Haas, but significantly he expressed the radius in terms of h (and m and e) and not the other way around. There is a whole lot of difference between the expressions $h^2 = 4\pi^2 e^2 ma$ and $a = h^2/4\pi^2 e^2 m$. The latter formula is known as the Bohr radius and has the value 5.292×10^{-11} m. It is still recognized as an important constant of physics.

Planck's fundamental system of units did not receive much attention, but it was advocated—or possibly reinvented—by Eddington in his *Report on the Relativity Theory of Gravitation* of 1918 and at some later occasions.[12] In Russia, the three young Leningrad physicists George Gamow, Dmitrii Ivanenko, and Lev Landau were aware of Planck's system and deeply fascinated by the relationships between the constants of nature. In 1927, a few years before Eddington embarked on the research programme that eventually led to his fundamental theory, the three physicists speculated that the basic constants h, c, and G were connected with the mass of the electron.[13] They suspected a relation of the form:

$$m = q\sqrt{\frac{hc}{G}}$$

where q is a numerical factor. Their compatriot Matvei Bronstein reported similar kinds of connections in several works from the early 1930s. In one of his papers, carrying the impressive title 'On the question of the possible theory of the world as a whole', he argued that 'If a theory can explain dimensionless constants its task would be more or less complete, since the value of these constants are responsible for the specific picture of the world'. He ambitiously dreamt of formulating a framework of the unified '*cGh* physics' of the future, a theory in which quantum mechanics would merge with electrodynamics and relativistic gravitation theory: 'After relativistic quantum theory is formulated, the next task would be to realize the third step, ... namely, to merge quantum theory (h constant) with special relativity (c constant) and the theory of gravitation (G constant) into a single whole.'[14]

Although Bronstein did not succeed in creating a unified *cGh* theory, his dream was not forgotten. The Planck values and *cGh* physics reappeared in the 1950s and are, directly or indirectly, parts of modern theories of quantum gravity. In some cases the Planck length, or something close to it, was discussed as an absolute minimum length, for example in the sense of the smallest possible uncertainty in the position of a particle.[15] According to the usual interpretation of Heisenberg's uncertainty relations (such as $\Delta q \Delta p \geq h$), the position variable can be precisely determined ($\Delta q = 0$) if only the conjugate momentum variable is completely undetermined ($\Delta p = \infty$). In the alternative view, independent of one of the variables the other cannot possibly be smaller than a certain value. Such ideas have been discussed by several physicists.[16]

Interestingly, Bronstein also considered the cosmological constant Λ, which he proposed was not really constant. In a paper of 1933 he suggested incorporating into Einstein's cosmological field equations a time-varying cosmological constant that would act as an arrow of time.[17] The suggestion of $\Lambda = \Lambda(t)$ implied a violation of the law of energy conservation, a drastic consequence which Bronstein nonetheless found justified in view of Bohr's contemporary doubts about the universal validity of strict energy conservation. Moreover, by introducing a decreasing cosmological constant he could provide an explanation of why the universe expands rather than contracts.

Gamow also retained an interest in the basic constants of nature after his emigration from Soviet Russia to the United States. In a popular article of 1949 he reflected on the possibility of an end of physics. Advocating a system of four constants, he somewhat surprisingly included Boltzmann's constant and excluded the constant of gravitation in the quadrumvirate. In addition to c, h, and k he argued in favour of a new universal constant in the form of an elementary length. Admitting inspiration from 'Heisenberg and other physicists', Gamow proposed the minimum length to be of nuclear dimensions, approximately 10^{-15} m. This length would be 'the smallest numerical value of the distance that can be used in any physical sense to characterize the space separation of two elementary particles'. Of the constants of nature and the end of physics he said:

> If and when all the laws governing physical phenomena are finally discovered, and all the empirical constants occurring in these laws are finally expressed through the four independent basic constants, we will be able to say that physical science has reached its end, that no excitement is left in further explorations, and that all that remains to a physicist is either tedious work on minor details or the self-educational study and adoration of the magnificence of the completed system.[18]

But Gamow did not think that physics was approaching its end and he certainly did not hope so, for then 'physical science will enter from the epoch of Colombus and Magellan into the epoch of the National Geographic Magazine!'

Einstein, too, maintained an interest in the constants of nature, as demonstrated by a passage in his autobiographical notes, written in 1946 and instigated by a correspondence with the philosopher Ilse Rosenthal-Schneider. Having stressed the particular importance of dimensionless constants in the basic equations of physics, Einstein expressed his belief in the simplicity of nature as follows: 'There are no *arbitrary* constants of this kind; that is to say, nature is so constituted that it is possible logically to lay down such strongly determined laws that within these laws only rationally, completely determined constants occur (not constants, therefore, whose numerical value could be changed without destroying the theory).'[19] This was not only Einstein's view, but one shared by several other physicists at the time. According to Gamow, there was 'a deep-seated wishful thinking among theoretical physicists to the effect that in nature there are no incidental numerical constants, i.e. pure numbers which cannot be derived by purely mathematical reasoning'.[20] Later physicists in the tradition of string theory are most definitely in agreement with Gamow's credo.

7.2 DOES GRAVITY CHANGE IN TIME?

Until now it has been tacitly assumed that the constants of nature are indeed constant, meaning that they do not vary in time. The possibility of such variation was first discussed in a scientific context in the 1930s, but more philosophical ideas about varying laws of nature can be found further back in time. For example, in an address of 1872 read before the British Association for the Advancement of Science, the mathematician William K. Clifford insisted that we have no right to assume that the laws of nature are theoretically exact. We do not know for sure that the laws are exactly true, nor can we ever know. According to Clifford, the assumption of the absolute immutability of the laws of nature was just that—an assumption or postulate from which no universal conclusions about past or future could justifiably be drawn. 'A law would be theoretically universal if it were true of all cases whatever; and this is what we do not know of any law at all.'[21]

The subject was brought up by Henri Poincaré in an address of 1911 in which the French mathematician concluded that the idea was inconsistent with the very meaning of natural law. According to Poincaré, law-based knowledge about the past, as in geology and astronomy, necessitated immutable laws that can link the past with the present: 'If...the immutability of the laws plays a part in the premises of all our reasoning process, it is bound to occur in our conclusions.'[22] Thus, he held that invariable natural laws are a necessary condition for any scientific knowledge. It is of interest to note that a somewhat similar point was made by Bondi and Gold in their original version of steady-state cosmology from 1948. This theory was explicitly motivated by the requirement that the laws of physics must be constant. For, as Bondi and Gold argued, this was necessary for experiments to be repeatable, a precondition for all science (see further in Chapter 5).

Neither Poincaré nor Bondi and Gold referred specifically to the constants associated with the laws. In principle, a law of nature may vary in time because its mathematical form changes, while the associated constant remains unaltered, or because the constant changes, while the form remains constant. Of the two possibilities it is the second that is almost always thought of in connection with varying laws. From this point of view, there is no difference between speaking of varying laws and varying constants of nature.

As mentioned in Chapter 4, in 1937 Dirac suggested that the gravitational constant slowly decreases in time, and the following year he developed the hypothesis into a cosmological model. It followed from the model and its foundation in the so-called large number hypothesis that the cosmological constant must be zero, that space is flat, and that the age of the universe is one-third of the present Hubble time. In a prize lecture of 1939, Dirac extended his hypothesis of a connection between natural laws and the very large numbers occurring in nature. On this occasion he not only considered 'the laws of Nature as continually changing with the epoch', he also speculated that 'we should expect them also to depend on position in space, in order to preserve the beautiful idea of relativity that there is fundamental similarity between space and time'.[23]

Dirac's address deserves further attention. Its main theme was the classical question of the relationship between mathematics and physics, a question that his brother-in-law Eugene Wigner would later highlight. Impressed by the power of mathematical reasoning, Dirac dreamt of an ultimate unification, 'every branch of pure mathematics then having its physical application, its importance in physics being proportional to its interest in mathematics'. According to Dirac, the theoretical physicist should endeavour to express *all* of nature in mathematical terms. From this perspective he rejected as philosophically unsatisfactory the ideal of mechanical determinism. His argument was this: classical mechanics operates with two types of parameters, a complete system of equations of motion and a complete set of initial conditions. With these provided, the development of any dynamical system is completely determined. However, while the equations of motion are amenable to mathematical treatment, the initial conditions are not. They are contingent quantities, determinable only by observation. But then an asymmetry arises in the description of the universe, which methodologically is divided into two spheres. This Dirac found to be intolerable because it ran contrary to the expectation of unity in nature. As he saw it, all the initial conditions—including the elementary particles, their masses and numbers, and the constants of nature—must be subject to mathematical theory.

Although Dirac realized that quantum mechanics might seem to be incompatible with a complete mathematical description of the world, he foresaw a mathematical physics of the future in which 'the whole of the description of the universe has its mathematical counterpart'. Laplace's supreme intelligence had to have recourse to the initial conditions in order to predict the development of the universe. In Dirac's philosophical vision, Laplace's demon reappeared in an even more powerful version, as one being able to deduce *everything* in the universe by pure mathematical reasoning:

> We must suppose that a person with a complete knowledge of mathematics could deduce, not only astronomical data, but also all the historical events that take place in the world, even the most trivial ones. Of course, it must be beyond human power actually to make these deductions, . . . but the methods of making them would have to be well defined. The scheme could not be subject to the principle of simplicity since it would have to be extremely complicated, but it may well be subject to the principle of mathematical beauty.[24]

After this digression, I return to the varying constants of nature. Although Dirac only returned to cosmology in the 1970s, his unorthodox ideas were well known, if far from accepted. In fact, they were generally rejected as unfounded and speculative. Pascual

Jordan was about the only physicist who took the varying-G hypothesis seriously and tried to develop it into a more comprehensive theory in accordance with the framework of general relativity. In works published in the 1950s he and his collaborators applied the hypothesis to a number of astro- and geophysical problems, including Alfred Wegener's still controversial hypothesis of drifting continents.[25] Jordan also applied his Dirac-inspired theory to the formation of stars, arguing that they all started as supernovae formed by spontaneously created nuclear matter. This led him to a rate of supernova formation of about one supernova per galaxy per year, a value much greater than the observed rate. In general the response from physicists, astronomers, and geologists to the ideas of Dirac and Jordan remained cool. In what follows I shall focus on the classical Dirac cosmology and its connections to empirical testing. This is far from an exemplary case of the use of scientific methods, but it is an instructive case.

The basis of Dirac's cosmological theory was the large number hypothesis, a general principle that he valued highly and continued to believe in until his death in 1984. Although it does not follow unequivocally from the principle that the gravitational constant decreases with the cosmic era, Dirac argued that this was the only reasonable interpretation. As he saw it, the atomic constants such as the mass and charge of the electron and proton were true constants. In reality, the proposal of a decreasing gravitational constant was an independent hypothesis with no previous observational or theoretical support and no motivation except its basis in the large number hypothesis. It had to be evaluated by its consequences, both theoretical and experimental. Although it did not fare well on either of the counts, Dirac was so convinced of its truth that he was unwilling to take the objections against it very seriously. The story of Dirac's attitude to the mounting evidence against the hypothesis is a case study of how far a scientist may go in order to keep alive an idea in which he believes.[26] Undoubtedly, Dirac went too far.

A time-varying constant of gravitation is not allowed by classical general relativity, which meant that Dirac's cosmology was in conflict with the generally accepted fundamental theory of gravitation. It also led to an absurdly low value for the age of the universe: with the accepted value of the Hubble constant in the late 1930s of about 500 km s^{-1} Mpc^{-1}, the age would be only about 700 million years, much less than the age of the Earth as reliably estimated by radiometric methods. This timescale difficulty was a serious problem, but it was a problem that Dirac's model shared with other cosmological models, although it appeared more seriously in Dirac's theory. Thus, in the generally recognized Einstein–de Sitter model the age was two thirds of the Hubble time, a result that did not escape the difficulty.[27] For this reason the problem was not considered a crucial argument against the varying-G hypothesis in particular. A direct test was preferable and possible in principle, but at the time it was not possible in practice. According to Dirac's theory the decrease in G was given by $\Delta G / \Delta t = -3HG$, which, with the accepted value of the recession constant, gives about 10^{-11} per year. This predicted variation was far too small to be measured with the technology available at the time.

Although direct testing seemed impossible, one could test the hypothesis indirectly, namely by focusing on its consequences. This is what the Hungarian–American nuclear physicist Edward Teller did in 1948, when he examined its consequences with respect to early life forms on Earth.[28] The structure of Teller's argument, a kind of *Gedankenexperiment*, is typical for a large part of science that cannot be subjected to direct experimentation because it relates to an inaccessible past. Historical geology, astronomy, cosmology, and palaeontology often argue retrodictively by suggesting scenarios of what the world would have looked like in the past if certain conditions were altered. This hypothetical past, or rather the present consequences of this past, is then compared with our actual world. The hypothesis is judged deductively on the agreement between present empirical knowledge and these consequences. In Teller's case, he argued from general astrophysical relations that the surface temperature of the Earth varied as $G^{9/4}M^{7/4}$, where M is the mass of the Sun. Assuming the solar mass to be constant, on the $G(t)$ hypothesis it implies a terrestrial temperature in the past given by

$$T = T_0 \left(\frac{t_0}{t}\right)^{9/4}$$

Here T_0 denotes the present absolute temperature, t_0 the present age, and t the age in the past. For the Cambrian era, some 300 million years ago, this yielded a temperature about 110°C, which conflicted with palaeontological evidence of life—such as trilobites—on Earth at that time. Consequently Teller concluded that Dirac's hypothesis was most likely wrong. However, he realized that his argument 'cannot disprove completely the suggestion of Dirac, [which]... is, because of the nature of the subject matter, vague and difficult to disprove'.

In spite of Teller's caution, his work was generally regarded a mortal blow against cosmologies of the Dirac type. Yet it rested on a number of extrapolations and simplicity and, *ceteris paribus*, assumptions that either turned out to be unfounded or could be questioned. Thus, the extended cosmic timescale that was established in the 1950s—with a nearly ten-fold increase in the Hubble time as a result—not only eased the timescale difficulty, it also weakened Teller's objection. Because, with an age of the universe of 12 billion years the temperature of the Earth in the Cambrian era would, according to Teller's argument, no longer be 110°C but only 40°C and thus allow life on Earth.

Gamow shared Dirac's fascination of the large number hypothesis, but not his conviction that it implied a varying gravitational constant. From the perspective of Gamow and most other physicists, the only way to evaluate the hypothesis was to test it experimentally; if evidence spoke clearly against it, it would have to be abandoned. Rather than referring to early life on Earth, which involves many questionable assumptions and extrapolations, in the 1960s he and a few other astrophysicists examined the consequences of Dirac's hypothesis on stellar evolution. The conclusion of this work was that with the $G(t)$ hypothesis the Sun's luminosity in the past would have been too great to account for its present state as a main sequence star. The Sun would have evolved so quickly that it would now be a red giant, which is obviously not the case. Gamow's verdict of 1967 was crystal clear, namely that 'the possibility of the change of gravity forces in inverse proportion to the age of the universe has been completely ruled out'.[29]

In spite of the strong empirical evidence against the G(t) hypothesis, Dirac remained untouched. His response to the objections of Teller, Gamow, and others was to suggest new special assumptions or, in some cases, simply to disregard the objections. For example, at one point he proposed getting over the difficulties by suggesting that the Sun increased its mass by accretion of interstellar dust. As a means of theory-saving this auxiliary hypothesis had the advantage of not being very definite: the accretion rate was not well known and could thus be assigned a value fitting with the wanted result. On the other hand, this advantage was methodologically a blemish since it made the hypothesis completely *ad hoc*, a feature Gamow immediately pointed to in his correspondence with Dirac. It is obvious from their correspondence that although in principle they shared the same epistemic standards, in practice they did not.[30] What Gamow found contrived and unacceptable, Dirac did not. Gamow indicated as much in a letter to Dirac of 15 October 1967:

> I suppose that it is possible to devise such a rate of accretion that it would do simultaneously two quite different jobs: (1) prevent the sun from cooling because of the decreasing g [G], (2) prevent the earth from getting farther away from the sun, and to do it in such a way that the combined effect will just keep the surface temperature of the earth between 0 and 100°C.... I am afraid that, doing all that, the sun will have an adverse effect on the observable properties of planetary system. It must be an angry Creator who made such complicated arrangements with the single purpose to conceal from the poor scientists of the earth that gravity decreases with time![31]

When Dirac returned to cosmology in the 1970s he retained the large number hypothesis and its consequence of a decreasing gravitational constant. But now he argued (returning to the view he had originally expressed in 1937) that the large number hypothesis also implied continuous creation of matter, namely that the number of particles in the universe would increase with the square of the cosmic age. He argued that the hypothesis would solve some of the problems of his model, including Teller's objection. However, the scene of cosmology had changed drastically with the recognition of the hot big bang, a concept that was not known in the late 1930s and that Dirac did not appreciate. By the 1970s an alternative cosmological model would have to explain the main evidence in favour of the hot big bang, in particular the large amount of helium in the universe and the existence of a Planck-distributed microwave background. Dirac ignored the first problem and admitted that he was unable to explain the second. Since in Dirac's theory photons were continuously created according to $N_\gamma \sim t^2$, he was forced to assume that the blackbody radiation background was just an accidental result of the present age, perhaps resulting from intergalactic ionized hydrogen. Obviously this was not a satisfactory argument or one that made the theory convincing.

Since the varying gravitational constant was at the heart of Dirac's theory, it was important to establish by means of real experiments, and not just indirect reasoning, whether or not Newton's constant decreases at the predicted rate. Such experiments were made in the 1970s by Irving Shapiro, Thomas Van Vlandern, R. W. Hellings, and others, by studies of the motion of the Moon, by means of radar echo methods in the solar system, or by observations from the Viking landers on Mars. Although the results were not very clear, they did not support Dirac's prediction. What they and other experiments showed was that if the gravitational constant varies at all, it does so at a considerably smaller rate than predicted by Dirac's theory. In other words, the theory was falsified by the experiments, a conclusion substantiated by later experiments.[32] Remarkably, and a bit sadly too, even these results failed to convince Dirac, who continued to believe in the basic correctness of his hypothesis.

The failure of Dirac's $G(t)$ hypothesis did not mean that the question of a possible variation of the gravitational constant was thereby solved. A minority of physicists, closer to mainstream cosmology than Dirac, continued to investigate the hypothesis and construct more advanced theories without the weaknesses of the original Dirac theory. For example, in 1977 Vittorio Canuto and co-workers constructed a cosmological model based on Dirac's large number hypothesis, including a varying gravitational constant. They argued that the model could account for both the microwave background and primordial helium production. Another physicist inspired by Dirac's thinking commented on the large number hypothesis as follows:

> It is a very unconventional type of hypothesis for it does not at all appear to be a fundamental principle. One's immediate inclination is to explain it in terms of some other more fundamental principles, as we have attempted to do. This is only because standard physical laws have been local, differential laws. The LNH [large number hypothesis] may be a pioneering example of global laws of nature: By considering the universe as a whole, we may be led to a deeper understanding of the 'fundamental' interactions.[33]

Apart from Dirac cosmologies, the Hoyle–Narlikar version of steady-state cosmology also led to a slow decrease of the gravitational constant, at roughly the same rate as in Dirac's theory but for entirely different reasons.[34] Most of the more recent works on varying-G cosmologies have focused on theories based on generalizations of Einstein's theory of general relativity, so-called scalar-tensor theories. Although Dirac's version of varying gravitation is no longer tenable, ideas of a gravitational constant changing in time are still being considered by a minority of physicists.

7.3 THE FINE-STRUCTURE CONSTANT REVISITED

As mentioned in Chapter 4, Eddington emphasized the importance of the dimensionless fine-structure constant ($2\pi e^2/hc$) and claimed to be able to derive its value purely by means of theoretical arguments. The name of this fundamental constant relates to its origin in spectroscopy, where it appears as a measure of the close doublet lines in the hydrogen spectrum. An explanation of the doublet structure on the basis of quantum theory was first achieved by Sommerfeld in 1916 by combining Bohr's atomic theory with the mass variation of the electron described by the special theory of relativity. Although introduced as a spectroscopic quantity, Sommerfeld realized early on that it had a wider significance: it symbolized some deep and as yet mysterious connection between electromagnetism (e), relativity (c), and quantum theory (h), and therefore pointed towards a future theory of quantum electrodynamics.

About seventy years later, Feynman said of the value of the fine-structure constant: 'It's one of the *greatest* damn mysteries of physics: a *magic number* that comes to us with no understanding by man.... We know what kind of a dance to do experimentally to measure this number very accurately, but we don't know what kind of a dance to do on a computer to make this magic number come out—without putting it in secretly!'[35] The status of the fine-structure constant has changed during the last few decades, so that it is no longer considered quite as fundamental as it used to be. Today it is seen as evolving in a predictable way with the energy scale and derivable from the coupling constants of electroweak interactions or grand unified theory. At the grand unification energy governing the very early universe (about 10^{15} GeV) it is supposed to have been equal in value to the coupling constants of the weak and strong interactions. From a modern point of view, the magical number 137 is not all that magical. 'It is absurd to try to derive $a_{em}(0)$ from some purported number of degrees of freedom of the electron, as suggested by Eddington', notes a modern physicist.[36] In spite of its changed status, the fine-structure constant remains a very important quantity in fundamental physics.

The idea of a possible variation of physical constants was only widely noticed after the appearance of Dirac's paper of 1937, but it had turned up in the physics literature a few years earlier. The first time (apart from Bronstein's suggestion of 1933 relating to the cosmological constant) may have been in 1935, in connection with an attempt to explain Hubble's redshift-distance data without interpreting the redshifts as due to a Doppler effect. In the early 1930s several scientists resisted the strange idea of an expanding universe, which they not only found conceptually weird—how can space expand? Expand into what?—but also to be in conflict with the age of the stars. They consequently suggested alternative ways in which the Hubble relation could be understood on the basis of a static universe. There were several such ways, one of them involving the inclusion of a variation of the elementary constants of nature. According to J. A. Chalmers and B. Chalmers, two British physicists inspired by Eddington, a solution might be to assume that Planck's constant varied exponentially in time:

> It has been shown experimentally that h is a constant here now, but there is nothing except the Hubble effect that can possibly give information on the value of h in the remote past. We are thus interpreting the Hubble effect as an alteration of h as between emission in the nebula and emission here now.[37]

Since they assumed the charge of the electron and the speed of light to remain fixed, the proposal implied a varying fine-structure constant. The two physicists suggested a doubling time of Planck's constant of 1.4 billion years and also considered the possibility that the total number of particles in the universe might vary with the cosmic epoch. A few other physicists in the 1930s, most of them working under the spell of Eddington's numerological ideas, entertained similar (non-Eddingtonian!) speculations about changing constants of nature. However, they were ignored by most mainstream physicists and astronomers.

An exception was Pascual Jordan, who had no problem with either the expanding or the exploding universe and was open to the possibility of changing constants of nature. In a paper of 1939 he argued that the question of a varying fine-structure constant could be solved empirically, namely, by examining the spectra of distant galaxies. If the fine-structure constant depended on time, it would result in a complex spectral shift that had not been observed, and for this reason he concluded that the constant had not changed over cosmological periods. Although Jordan did not believe in a varying fine structure, he thought that the coupling constant of weak interactions governing beta radioactivity might vary with the epoch. After World War II he developed the idea, which amounted to replacing the ordinary decay law $N = N_0\exp(-kt)$ with the expression $N = N_0\exp(-2k\sqrt{t})$. The hypothesis had consequences for geophysics and radioactive dating methods, but failed to win recognition. Later investigations have not changed the general belief in the constancy of weak and strong interactions.

At this point it should be noted that there is a non-controversial sense in which the dimensionless coupling constants vary in time. As mentioned, these constants are now recognized to be 'running' constants, which depend on the energy or temperature of the universe. According to grand unified theories, such as had been known since the early 1980s, all three fundamental interactions (electromagnetic, weak, strong) were of the same strength, at an energy of about 10^{15} GeV (or temperature 10^{27} K), just 10^{-35} s after the big bang. As the universe expanded and cooled, the fine-structure constant increased and at a temperature of about 10^{13} K it was approximately 1/128. It has its present value only because we live in a cold universe. Since the energy or temperature of the universe is a measure of its age, there is a sense in which the coupling constants depend on cosmic time. However, it is not this kind of change, which concerns only the very early universe, that is of interest in the present context.

The possibility of a fine-structure constant varying in time was seriously considered only in the late 1960s, when it was discussed by Gamow, Freeman Dyson, and a few others. Although Gamow rejected the varying-G hypothesis on observational grounds, he found Dirac's large number argument, which related e^2/GmM to the cosmic time, to be so 'elegant and attractive' that he was loath to abandon it. He consequently suggested reviving it by assuming e^2 or α to vary linearly with cosmic time.[38] As he showed, the charge-variation hypothesis would have no serious geophysical and astrophysical consequences, as Dirac's varying-G hypothesis had. As to spectroscopy, Gamow noted that the hypothesis implied a redshift in all galactic spectra because of its influence on the Rydberg constant, $R \sim t^2$. While such a uniform shift cannot be distinguished from the Doppler redshift, the fine-structure splitting should be able to serve as a test. 'It may be that the large red shifts now observed in quasars (or rather, qualiaxies) are partially due to the change e^{2}'. However, the varying-e hypothesis was short-lived as it was immediately contradicted by measurements of the fine-structure splitting in the recently discovered quasars. Maarten Schmidt and his collaborators from Caltech reported measurements of the fine-structure splitting in the spectra of quasars that showed that the fine structure did not change even for very large redshifts, that is, for very old objects. For example, for photons that had travelled for about 2 billion years they found that $a/a_{lab} = 1.001 \pm 0.0002$, whereas Gamow's hypothesis required $a/a_{lab} = 0.8$.

Gamow quickly accepted that his hypothesis was untenable. Rather than trying to rescue it by means of auxiliary hypotheses, or by questioning the observational data, he acted as a good Popperian: when faced with contradicting experimental data of high quality he accepted that the hypothesis was wrong. Short-lived as the varying-e (or varying-a) hypothesis was, it stimulated much work on the subject and on varying constants in general. Since that time hundreds of papers have been written on the possible variation of constants of nature, some of them theoretical and others experimental.

As far as the fine-structure constant is concerned, for about 30 years there was no observational evidence at all that indicated a change in time. Using a variety of methods principally based on geological evidence, astrophysical observations, and laboratory experiments, the question was carefully examined.[39] No change in the fine-structure constant was found. It therefore came as a big surprise when the Australian astrophysicist John Webb and his collaborators, on the basis of a series of quasar observations from 1999–2001, reported a smaller value for the constant in the past. Using what is known as the many multiplet method, Webb found that for quasar redshifts in the interval $0.5 < z < 3$, the change in the fine-structure constant was about $\Delta a/a = -6 \times 10^{-6}$. Moreover, the decrease seemed to be limited to a distant period in the cosmic past, from about 10 to 12 billion years ago, and not to continue to this day (Fig. 7.1). The announcement caused great excitement, and theoretical cosmologists immediately developed cosmological models that accommodated a varying fine-structure constant. They also explored another of Eddington's magical numbers, the proton-to-electron mass ratio, but without finding any variation over a cosmic era of 10 billion years.

The announcement of the quasar results was considered important in the physics community and in general inspired further interest in the question of varying constants of nature.[40] However, enthusiasm cooled when the much-publicized observations were contradicted by new, still more precise measurements made in 2004 by two different international teams. According to a press release from the European Southern Observatory of 31 March 2004, 'New quasar studies keep fundamental constant constant'. It seems to be too early to write off a varying a-constant but also too early to conclude that it is truly constant. Of course, experiments can never tell if a quantity is absolutely constant, it can only establish an upper limit for its variation in time.

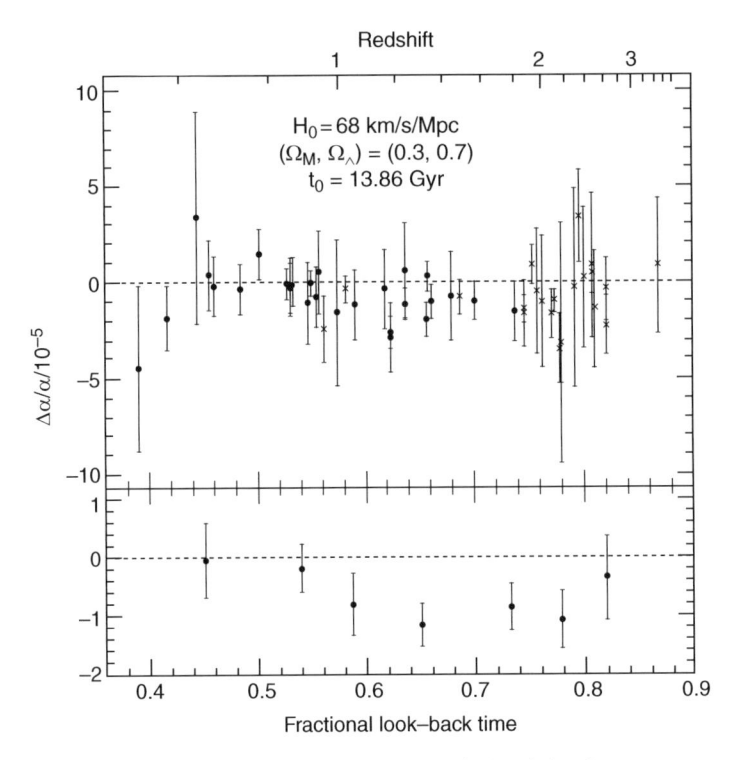

Fig. 7.1. Data from 2001 showing the relative change $\Delta\alpha/\alpha$ of the fine-structure constant as inferred from the light received from distant quasars. The upper graph shows all raw data with error bars, while in the lower one the same data are collected in groups. They are shown as a function of redshift and look-back time, that is, the time it takes before the light reaches us. The data indicates a negative shift in the range $1 < z < 3$.

Source: M. T. Murphy et al., 'Possible evidence for a variable fine-structure constant for QSO absorption lines: Motivations, analysis and results,' *Monthly Notices of the Royal Astronomical Society* **327** (2001), 1208–1222. Reproduced by permission of John Wiley & Sons Ltd.

It is one thing to propose that one of the constants of nature varies in time, quite another to suggest that constants and their associated laws vary in general. In 1970 William McCrea concluded that the notion of changing laws was 'useless'.[41] This was because, he argued, the only way to know of such a change would be by virtue of a law that described how the laws change. But such a super-law would have to be unchanging, thus leading to a contradiction. McCrea's argument is not a valid one against Dirac's hypothesis of a changing gravitational constant and similar proposals, but a reminder that ideas that *all* laws of nature vary in time are nonsensical. Ideas of this kind have nonetheless been discussed by a few philosophers, including Charles S. Peirce and Alfred N. Whitehead. Peirce believed that the laws of nature are the results of a cosmic evolutionary process that is governed by some principle that must itself be an evolving law or 'generalizing tendency'. He suggested that by studying the operations of the mind, one can discover the laws of the world. In his major work, *Process and Reality* of 1929, Whitehead argued for a worldview where processes were of higher ontological status than objects and where everything was in a state of flux, including the laws of nature. The expansion of the universe only confirmed him in this general idea. As he wrote in 1933:

> Since the laws of nature depend on the individual characters of the things constituting nature, as the things change, then correspondingly the laws will change.... The modern evolutionary view of the physical Universe should conceive of the laws of nature as evolving concurrently with the things constituting the environment. Thus the conception of the Universe, as evolving subject to fixed eternal laws regulating all behaviour should be abandoned.[42]

However, Whitehead's philosophy of nature was not a contribution to physics in its ordinary meaning and it made almost no impact at all on physicists and cosmologists.[43]

7.4 VARYING SPEED OF LIGHT COSMOLOGIES

As briefly mentioned in Chapter 4, Eddington was convinced that the constants of nature do not vary in time. In his *Fundamental Theory* he stressed this necessary feature with respect to the speed of light:

> The ratio of the wave-length to the period of $H\alpha$ light is the velocity of $H\alpha$ light. Thus it follows from the definition of the ultimate standards of length and time that the velocity of light is constant everywhere and everywhen.... The speculation of various writers that the velocity of light has changed slowly in the long periods of cosmological time... has seriously distracted the sane development of cosmological theory. The speculation is nonsensical because a change of the velocity of light is self-contradictory.[44]

Whether a varying speed of light is considered 'nonsensical' or not, the quantity itself is certainly a fundamental constant of nature. With Einstein's special theory of relativity, its constancy became a postulate of relativity physics rather than just an empirical fact. To accept a time-varying speed of light might therefore seem to jeopardize one of the cornerstones of modern physics, the principle of relativity. Nonetheless, in the 1930s a few scientists did propose a change in the speed of light. There is no doubt that they were the 'various writers' Eddington referred to. However, Eddington exaggerated their role in the development of cosmology, which was scarcely noticed by the majority of astronomers and physicists.

The first proposals of a slowly decreasing speed of light were made 1930–31, in particular by Tokio Takeuchi, a Japanese physicist who was also a pioneer of cyclic models of the universe. Four years later P. Wold, a physicist at Union College, New York, suggested that if the speed of light decreased with cosmic time it might explain the Hubble relation without the hypothesis of an expanding universe. That is, his motivation was of the same kind as the one that caused some scientists to propose a varying electrical charge or fine-structure constant. Whereas neither Takeuchi nor Wold offered independent evidence in favour of their hypotheses, the British amateur scientist M. Gheury de Bray argued that measurements since the 1850s provided such evidence. The evidence he claimed was historical, not astronomical. By collecting experimental data during an 80-year period he convinced himself that the speed of light decreased slowly and linearly with time. Therefore physicists needed to reconsider the theories, such as Einstein's, that built on the presumed constancy.[45] In a paper of 1939 he suggested from his collected historical data the relation $c(t) = 299,774 - 173t$, where c is in km s^{-1} and t is measured in millions of years. Like Wold he saw his hypothesis (which he insisted on calling a 'fact') as an alternative to the 'fantastic' idea of an expanding universe.

The possibility of a varying speed of light was ignored or forgotten for nearly four decades. Only in the late 1970s did it reappear in the scientific literature, but without attracting much attention. When the idea finally caught on, in the late 1990s, it was not as an alternative to the expanding universe but as an alternative to the popular inflationary theory of the early universe. According to this theory or scenario, as proposed by Alan Guth in 1981, the universe started in a state of 'false vacuum' that expanded at a phenomenal rate during a very short period of time (about 10^{-30} s) and then decayed to a normal vacuum filled with hot radiation energy. In spite of the briefness of the inflation phase, the early universe inflated by the amazing factor of about 10^{40}. What made Guth's inflation theory appealing was primarily that it was able to explain two problems that the conventional big-bang theory did not address. One was the horizon problem, which is the problem that in the very early universe distant regions could not have been in causal contact, that is, they could not communicate by means of light signals. So why is the universe so uniform? The other problem, known as the flatness problem, concerns the initial density of the universe, which must have been extremely close to the critical value in order to result in the present universe. Within a few years the inflationary scenario became very popular and broadly accepted as an integral part of the consensus model of the universe.[46] However, in spite of its popularity not all cosmologists found inflation to be an acceptable hypothesis.

The idea of introducing a drastically increased value for the speed of light in the very early universe was first suggested by the Canadian physicist John Moffat in a paper of 1993.[47] Moffat showed that on this hypothesis all points in the early universe would have been in communication with one another, thus offering a solution to the horizon problem without the need of inflation. His contribution to early-universe cosmology was little noticed until six years later when a similar approach was independently adopted by Andreas Albrecht and João Magueijo at Imperial College, London.[48] Contrary to Moffat's ill-fated paper, the work of Magueijo and Albrecht caused a minor sensation and resulted in much theoretical work on what is called the varying speed of light (VSL) theory of cosmology. The many papers that have been written on VSL cosmology have in common that they assume a time-varying speed of light in the past but in other ways they differ a great deal, as present VSL theory is a collection of theories rather than one unified theory. Initially seen as highly unorthodox, even a provocation, VSL theory soon became recognized as a respectable, if also unconventional and somewhat speculative cosmological theory. In 2003 Magueijo presented the new theory to a wider audience in a popular book entitled *Faster than the Speed of Light* and subtitled *The Story of a Scientific Speculation*.

Magueijo and Albrecht's VSL theory was deliberately constructed as an alternative to the dominant inflation theory and designed to solve, as a minimum, the same problems that inflation took care of, such as the horizon and flatness problems. Both physicists had serious reservations with regard to the inflation paradigm, which is interesting because Albrecht was himself one of the founders of the highly successful, so-called new inflation theory of 1982. Although he had a vested interest in the inflation theory he decided to leave it and search for an alternative. And he was not the only one, for at about the same time another of the architects of the new inflation theory, Paul Steinhardt, looked for an alternative in the form of cyclic cosmology (see Chapter 8).

Rather than assuming an early inflation phase, VSL keeps to an early radiation-dominated universe with expansion and geometry as in the standard big-bang models of the Friedmann type. The difference is that the local speed of light is assumed to decrease drastically at a time close to the Planck epoch, from an exceedingly large value of perhaps 10^{38} km s^{-1} to its current value of just 300,000 km s^{-1}. Other VSL models assume a much slower and gentler decrease in the speed of light. The varying speed of light means that the second of the postulates on which Einstein founded special relativity is violated. It further follows from the theory that energy conservation is violated in the very early universe, so that matter can be created and destroyed without energy compensation.

The observational claim of a possible variation of the fine-structure constant was announced shortly after the VSL theory and inspired further interest in it. Indeed, VSL theorists were quick to devise phenomenological models that explained the reported variations in a on the assumption of a suitably chosen $c(t)$ variation. If a varies in time, is it due to a variation in c or in e? In order to decide between the two alternatives, observations restricted to electromagnetism will not do, but according to Paul Davies and his collaborators the question might be settled by comparing the consequences of the two possibilities with the thermodynamics of black holes. 'Black-hole thermodynamics may provide a stringent criterion against which contending theories for varying 'constants' should be tested', they argued.[49] Considering that the thermodynamics of black holes is completely theoretical, this is clearly a notion of testing that differs from the ordinary one based on experiment and observation.

The first versions of VSL theory were exclusively associated with the early universe, seen as an alternative to the inflationary scenario. As the theory was explored by a growing number of physicists its nature and perspective changed. In its more modern versions it has turned into a much broader and general type of theory, with potential applications also outside cosmology. Among other things, the VSL approach has resulted in new theories of quantum gravity and black-hole physics. On the empirical front, always the Achilles' heel of this kind of exotic theory, VSL has led to several predictions which can, at least in principle, be tested. As usual, it is not very clear if they can also be tested in practice. As Magueijo admits, all the predicted effects 'are invariably either well beyond the reach of current technology or at best on the threshold'.[50] One difficulty for VSL theory compared with inflation big-bang cosmology was that it did not provide a theory of structure formation. This difficulty is avoided in recent versions of the theory, which include a mechanism for generating density fluctuations in the early universe.

Although far from empirically empty, VSL covers so many different theories and approaches that it is difficult to think of observations that falsify the theory as a whole. On the other hand, it is easy to think of observations that confirm the theory, which according to Magueijo is really 'an umbrella for many different theories'.[51] Among these many theories are some that predict a variation in time of the speed of light and others that predict a variation with colour or frequency.

VSL cosmology is not merely an abstract physical theory of the early universe, it also involves problems of a conceptual and philosophical nature. Does it make sense at all to operate with a varying speed of light? In agreement with Eddington's view, the cosmologist George Ellis has objected that 'It is ... not possible for the speed of light to vary, because it is the very basis of measuring distance'.[52] It should be recalled that since 1983, the speed of light has been *defined* to be $c = 299,792.458$ km s^{-1}, thus leading to a value of the metre from the value of the second. The value, then, is not really a measured property, as indicated by the lack of experimental uncertainty. It is only for reasons of convenience that the value has been chosen so as to agree with earlier measurements—in principle, one could have chosen $c = 1$ m s^{-1}. The speed of light is the only fundamental constant of this type.

Another critic, the American physicist Michael Duff at the University of Michigan, argues that it is confusing to speak of a time variation of the speed of light as if it were a meaningful and measurable property: 'Asking whether c has varied over cosmic history ... is like asking whether the number of liters to the gallon has varied.'[53] It is often held that only dimensionless combinations of fundamental constants have physical significance and that experimental results can only be sensitive to such combinations. Magueijo and Albrecht were aware of this kind of objection in their 1999 paper, where they argued that physics necessarily involves dimensional quantities and that a time-variation of these can be determined on grounds of conventionalism.

Magueijo has continued to stress the unavoidable element of conventionalism in cosmological theories, arguing that the choice between constant and varying quantities can only be determined by considerations of simplicity. According to this point of view, by a change of units a VSL theory can, for example, be turned into a theory in which c is constant but e varies. Still the choice is not arbitrary, for only units that imply a varying speed of light will result in a reasonably simple theory. In a recent paper Magueijo repeats the appeal to conventionalism:

It is often stated that units can always be defined so that the speed of light is a constant. This is true in the same way that units can always be defined so that the Hubble 'constant' is indeed a constant, the acceleration of gravity is the same everywhere as it is on Earth, the Universe is not expanding, etc. The argument is a perfect twin and likewise the reason it makes sense to talk about varying c is identical to why it makes sense to talk about the expansion of the Universe, Newton's laws, a varying G, etc.: they are dimensional statements justified by the simplicity of the overall picture. For example, by choosing units where the Universe is not expanding all the equations of physics would become ridiculously complicated.[54]

During its brief lifetime VSL theory has become an interesting alternative to inflation theory, but it has not succeeded in its ambitious goal of replacing this theory. Citation data indicate that between its origin in 1999 and 2003, VSL cosmology was seen as an attractive alternative, but that interest in the theory then declined.[55]

Criticism of the theory of a varying speed of light has in part been of a methodological nature and a response to the rather aggressive way in which it was advertised. According to two critics, Ellis and Jean-Philippe Uzan at the Institute of Astrophysics in Paris, VSL theory 'cannot be taken seriously... as a proposal for the speed of light to vary'.[56] Certainly, Magueijo and his colleagues in the VSL cottage industry are not modest in their claims for what the new theory offers. Not only does it explain the origin of matter, it also explains the origin of the universe, something even the inflation theory cannot do—'How did the universe come into being? VSL provided the very answer'.[57]

Notes for Chapter 7

1. Feynman, Leighton, and Sands 1963, p. 9.
2. According to Philip Mirowski, a philosopher of science and economic thought, the recent interest in varying constants reflects cultural trends and is an 'indication of the postmodernist mind-set' (Mirowski 1992, p. 184).
3. Comprehensive treatments of the constants of nature in their historical contexts can be found in Barrow 2002 and Uzan and Lehoucq 2005. See also the popular account in Uzan and Leclercq 2008. Uzan and Lehoucq 2005 includes excerpts of primary sources.
4. The values are known with greater precision than given here. For precise values for the constants of nature, see http://physics.nist.gov/cuu/Constants/.
5. Rozental 1988, p. 88 (translation of Russian original of 1984).
6. See the discussion in Duff, Okun, and Veneziano 2002.
7. Stoney 1881, pp. 384–85. See also Barrow 2002, pp. 16–22. The three symbols V_1, G_1, and E_1 correspond to c, G and e, respectively.
8. A detailed examination of Planck's ideas of fundamental constants and their place in his theory of blackbody radiation can be found in Robotti and Badino 2001. Quotation from Planck on p. 146.
9. Drude 1902, p. 527.

10. Planck 1949, p. 77.

11. Planck 1960, p. 18. The reference to Martians may reflect an issue much discussed at the time, the possible existence of intelligent life on Mars. This was the period in which the American astronomer Percival Lowell campaigned vigorously for his idea of canals on Mars constructed by an advanced civilization. On natural constants and Planck's attempt to purge physics from anthropomorhic elements, see Mirowski 1992.

12. Eddington 1918, p. 91, who referred to the Planck length but neither to Planck nor the other Planck units.

13. Gamow, Ivanenko, and Landau 2002, a translation of the Russian original from 1928.

14. Quoted in Gorelik and Frenkel 1994, p. 90.

15. See, for example, Mead 1964.

16. The idea of absolute uncertainties goes back to the late 1920s, shortly after Heisenberg's introduction of the uncertainty principle, when it was discussed by Arthur Ruark, Henry Flint, Gleb Wataghin, and others. The favoured absolute position uncertainty was $\Delta q = h/mc$, with m denoting the mass of either an electron or a proton. See Kragh and Carazza 1994.

17. Bronstein 1933.

18. Gamow 1949, pp. 18–19.

19. Einstein 1949, p. 63. Einstein elaborated in a letter of 13 October 1945, which is reproduced in Rosenthal-Schneider 1980, pp. 36–38.

20. Gamow 1949, p. 18.

21. Clifford 1947, p. 12.

22. Poincaré 1913, p. 51. For a philosophical perspective on Poincaré's position and the idea of evolving laws, see Balashov 1992.

23. Dirac 1939, p. 128. The following quotations are from the same source. Dirac returned to these ideas over the years. For example, in an address delivered in Trieste in 1968 on the methods of theoretical physics he advocated to investigate cosmological models in which the laws and constants of nature vary with cosmic time.

24. Dirac 1939, p. 129.

25. Jordan 1952. Jordan was not the only cosmologist who considered the geophysical consequences of cosmological theories. Just at the time when the new and revolutionary theory of plate tectonics began to be developed, Robert Dicke wrote a paper on the subject in which he examined, among other things, the terrestrial effects of a decreasing gravitational constant (Dicke 1962).

26. For the historical development of Dirac's cosmological views, see Kragh 1990, pp. 223–46.

27. On the time scale difficulty and physicists' and astronomers' attitude to it, see Kragh 1996, pp. 73–79. The problem only diminished in the 1950s, when it was realized that the Hubble time was substantially greater than the 'authoritative' value of about 2 billion years. See also Brush 2001.

28. Teller 1948. Dicke 1957a reconsidered and criticized Teller's argument. For example, he pointed out that Teller had ignored the effect of an increased solar constant on the heat balance of the Earth's atmosphere.

29. Gamow 1967, p. 761.

30. The correspondence between Gamow and Dirac in the 1960s concerned not only a varying gravitational constant, but also the possibility that the electron's charge varied in cosmic time, such as Gamow advocated for a brief period of time. See Kragh 1991.

31. Quoted in Kragh 1991, p. 120.

32. See the list of the measured constraints on the time variation of the gravitational constant 1948–1998 in Uzan 2003, p. 55.

33. Hsieh 1982, p. 682.

34. See the survey in Narlikar 1983.

35. Feynman 1985, p. 129.

36. Cahn 1996, p. 954. Of course, this is irrelevant from a historical point of view. On Eddington and the fine-structure constant, see Chapter 4.

37. Chalmers and Chalmers 1935.

38. Gamow 1967.

39. Uzan 2003, pp. 8–25. Barrow 2002, pp. 251–68.

40. For example Olive and Qian 2004.

41. McCrea 1970, p. 23.

42. Whitehead 1933, p. 143.

43. On Whitehead's cosmology and its possible relevance for modern ideas of many worlds, see Barrow and Tipler 1986, pp. 191–94. A few physicists hold views that are as extreme as and have some affinity to Whitehead's. Notably, Wheeler suggested that 'the only law is that there is no law' and that the physical world is inherently mutable on *all* levels. See Wheeler 1973 and Wheeler and Zurek 1983, pp. 182–215. For critical comments, see Deutsch 1986.

44. Eddington 1946, p. 8. 'H_α light' refers to the red line in the hydrogen spectrum of wavelength 656.3 nm. Commenting on the statement, Magueijo found it to be 'smelling of religion' and an expression of Eddington's 'religious fervor' (Magueijo 2003, p. 202). I find it difficult to see what it has to do with religion.

45. Gheury de Bray 1936.

46. Guth 1997 tells the story of the inflationary universe. For an excellent review of the theory, see Guth and Kaiser 2005.

47. Moffat 1993. On cosmologies with varying speed of light, see also Kragh 2006, which includes references to the literature and forms the basis of this section.

48. Albrecht and Magueijo 1999.

49. Davies, Davis, and Lineweaver 2002, p. 603.

50. Magueijo 2003, p. 2055.

51. Magueijo 2004, p. 254.

52. Ellis 2003.

53. Duff 2004. See also Rich 2003, p. 1046, who concludes that, 'If an experiment shows that a dimensionless combination varies with time, any question about which constant is varying is only a matter of taste.' For an early plea to base physics on dimensionless quantities, see Whyte 1954.

54. Magueijo 2009, p. 7.

55. ISI Web of Science. Whereas the number of citations to the 1999 Albrecht-Magueijo paper peaked in 2003 with 30 citations, in the period 2004–07 the average number of citations per year was only 5.

56. Ellis and Uzan 2005, p. 11.

57. Magueijo 2004, p. 159.

8
New Cyclic Models of the Universe

To bring the dead to life
Is no great magic.
Few are wholly dead:
Blow on a dead man's embers
And a live flame will start.

Robert Graves, *To Bring the Dead to Life*, 1936.

One of the more remarkable trends in recent cosmology is the reconsideration of cyclic and other eternal models of the universe. With the establishment of the hot big-bang model in the 1960s, the picture of an explosive and finite-age universe achieved a paradigmatic status. Models of the steady-state type were no longer taken seriously, and astronomers and physicists spoke with increasing confidence of the big bang as if it were the absolute beginning of everything. Within the now-dominant tradition of mainstream cosmology, questions of the universe before the big bang were not asked, indeed could not be asked. If the question was nonetheless raised (and it did happen), it was deemed meaningless or scientifically irrelevant. This was because, if the big bang was the beginning of everything, it was also the beginning of time. And how can there possibly be something 'before time'?

During the last decade or two this situation has changed quite considerably. Although the basic features of the big-bang picture have remained robust and are rarely questioned, the status of the bang some 14 billion years ago has changed: An increasing number of cosmologists, although undoubtedly a minority, now believe that it might have been the outcome of a previous state and that our present universe may well give birth to a successor universe. The cyclic—or pulsating, oscillating, bouncing, periodic— universe has returned, and in a confusing number of versions. The quasi-steady-state model mentioned in Chapter 5 belonged to the cyclic class and shared some features with other of the models proposed in the same period.

This chapter starts with a brief account of the history of cyclic cosmological models, focusing on the twentieth century. Although classical relativistic models of a closed cyclic universe were plagued by serious problems of a theoretical, conceptual, and observational nature, they continued to attract interest. The main part of the chapter considers very different versions of the cyclic or bouncing universe model that only emerged in the years around the turn of the millennium. These models differ widely, but on a qualitative level they share some characteristics, for example presenting cosmic time as stretching from minus infinity to plus infinity. Moreover, relying in various degrees on theories of quantum gravity, they challenge the paradigmatic inflation model of the very early universe, although they are not necessarily irreconcilable with inflation.

Theories of cosmic cycles have often been accused of being speculative, proposed for philosophical reasons and not because they solve scientific problems. Speculative or hypothetical elements are indeed parts of the new cyclic models, but such elements are to be found in other cosmological models as well. Speculative or not, the cyclic models of the twenty-first century are proposed as scientific hypotheses, not philosophical world views. As such they result in predictions that can be tested—if not easily and sometimes perhaps only in principle.

8.1 FROM PHILOSOPHICAL SPECULATIONS TO RELATIVISTIC MODELS

Ideas that the universe is cyclic, that over immense periods of time it alternates between creative and destructive phases without ever coming to an end, can be traced back to the earliest mythological world views.[1] Many of the ancient cosmologies imagined a future cosmic catastrophe represented by a huge battle between the good and evil forces of nature. The catastrophe did not necessarily imply an absolute end of the world, for out of the ashes of the old world a new one might rise. In some cultures, notably in India, the process was thought to go on endlessly. This is the archetypical conception of the cyclic or oscillatory universe, an idea which has fascinated humans throughout history and can be found in mythical as well as scientific cosmologies right up to the present. For example, the Stoic philosophers in ancient Greece saw the world as a gigantic sphere oscillating through cycles of expansion and contraction in the void surrounding it.

According to tradition, the church father Augustine of Hippo (354–430) was once asked what God did before he created the universe. He is supposed to have answered, 'He was preparing Hell, just for those who pry too deeply.' In his main work, *Confessions* (Book XI, chapter 12), Augustine did discuss the question of God's activity before the creation of the world and he also included the quoted answer, but only to distance himself from it. Admitting that he did not know, he preferred to think that before God made the world, he made nothing at all.

At any rate, he was preceded by another Christian thinker, Origen, who in the early third century discussed the same question, thought to be of great theological significance. In accordance with Aristotelian doctrines Origen argued that infinity was incomprehensible, but he added that the existence of something without a beginning was equally incomprehensible. The dilemma caused him to speculate about what God was doing before he created the cosmos. How did he avoid idleness? It is in this context that Origen suggested that God, far from being idle, busied himself with the creation of new worlds: 'God did not begin to work for the first time when he made this visible world, but that just as after the dissolution of this world there will be another one, so also we believe that there were others before this one existed.'[2] Origen was well aware of the theological problems of assuming a series of similar worlds, for which reason he stressed that the earlier worlds were different from the present one. 'It is', he said, 'impossible for a world to be restored for the second time, with the same order and with the same amounts of births, and deaths, and actions.' In this way he thought that the freedom of the will could be maintained and also that an endless series of crucifixions could be avoided. As for Augustine, he rejected the notion of a cyclic universe altogether and it was his view that won the day. Origen's bold speculation of an eternal cyclic universe was condemned by the church at the second Council of Constantinople in 553.

Speculations of roughly the same kind as those discussed by the ancient thinkers were considered by many later natural philosophers, including giants such as Newton and Kant. With the emergence in the nineteenth century of the second law of thermodynamics it seemed to follow that eternally cyclic conceptions of the universe were ruled out. Indeed, as early as 1868 Rudolf Clausius, the inventor of the concept and coiner of the name of entropy, concluded that cyclic world models were contradicted by the fundamental law of entropy increase. He thought he had disproved the view that 'the same conditions constantly recur, and in the long run the state of the world remains unchanged'.[3] However, his conclusion in no way deterred philosophers and amateur scientists from advocating such models which were, in fact, very popular in the period from about 1850 to 1910.

In this period it was common, especially in atheistic and materialistic circles, to conceive the universe as a *perpetuum mobile*, a clock that was never wound and would continue to run for an eternity. According to Rudolf Falb, an Austrian writer and amateur scientist, the universe consisted of a huge but limited number of stars and nebulae and was therefore bound to collapse gravitationally into a giant solar body. The mega-sun, comprising all matter in the universe, would, however, immediately evaporate and form an expanding gas, a nebulous system of the same kind that had originally given rise to the celestial bodies. Contending that the process would continue indefinitely, Falb spoke of the life of the universe as 'a recurrence of expansion and contraction, like the breaths of a monstrous colossus'.[4] More than 30 years later the English pioneer in radiochemistry Frederick Soddy, a Nobel laureate of 1921, also subscribed to the cyclic world view, which he thought received support from the newly discovered subatomic energy. 'The most attractive and consistent explanation of the universe', he wrote, 'is perhaps that matter is breaking down and its energy being evolved and degraded in one part of a cycle of evolution, and in another part still unknown to us, the matter is being built up with the utilisation of waste energy.' As a consequence, 'in spite of the incessant changes, an equilibrium condition would result, and continue indefinitely'.[5] Views of the kind expressed by Falb and Soddy were shared by dozens of thinkers in the period before World War I, but they were qualitative and philosophically based views rather than cosmological models in a scientific sense.

The nineteenth-century view of the cyclic or oscillatory universe differed in essential aspects from the later one based on relativistic cosmology. According to the first view, even though parts of the universe might be running down and, for some period, end in a state of equilibrium, there would always be other parts in which constructive entropy-consuming processes dominated the cosmic scene. The recurring universe might well be infinite, as indeed it was usually conceived to be (Falb's view was exceptional). The classical view presupposed a Euclidean space. It did not operate with cosmic space varying in time, a notion which was foreign to the Victorian mind and would enter cosmology only in the 1920s.

A cyclic universe in the latter sense was introduced by Alexander Friedmann in his seminal paper of 1922 in which he analysed the dynamical solutions of Einstein's cosmological field equations. One class of these solutions, involving a closed space with a cosmological constant that was either zero or negative, was periodic, that is, with a radius of curvature that varied between zero and a maximum value. However, the Friedmann periodic model did not describe the universe as running through a series of cycles but only referred to a single cycle from a big bang to a big crunch. This was also the case with a similar model that Einstein discussed in 1931 and in which he assumed the cosmological constant to be zero.[6] Although the Einstein model described a single cycle only, it was often seen as a multicycle model and inspired new interest in the idea of endless cosmic oscillations both among scientists and lay persons. Einstein never expressed any interest in this kind of eternal universe.

A detailed investigation of cyclic models, going much beyond the simple Friedmann–Einstein model, was first undertaken by the American physicist Richard Tolman at the California Institute of Technology in a series of works from 1931 to 1934.[7] According to Tolman's analysis a continual series of cycles was not in conflict with the second law of thermodynamics in its generalized relativistic formulation. The entropy of the closed universe would increase from one cycle to the next, but without ever approaching a maximum limit. Each new cycle would be greater than the previous one, that is, have a longer period and a higher maximum value of the curvature radius. Moreover, he argued (but did not prove) that it was possible for a contracting universe to bounce into an expanding one and to do so an infinite number of times. This could not happen through the impenetrable singularities, but he suggested that these were mathematical artifacts that would disappear in physically more realistic models.

In spite of his great interest in cyclic models, Tolman was not emotionally committed to a cosmology based on periodic expansions and contractions. Well aware of the danger of confusing a cosmological model with the real universe, he chose to interpret the ever-oscillating universe as merely an interesting possibility that might or might not receive observational support. Tolman's general view concerning cosmological models, of no less relevance today than it was in the 1930s, is worth quoting:

> In studying the problem of cosmology we are immediately aware that the future of man is involved in the issue, and we must hence be particularly careful to keep our judgments uninfected by the demands of religion, and unswerved by human hopes and fears. Thus, for example, what appears now to be the mathematical possibility for a highly idealized conceptual model, which would expand and contract without ever coming to a final state of rest, must not be mistaken for an assertion as to the properties of the actual universe, concerning which we know all too little. To conclude then: Although I believe it is appropriate to approach the problems of cosmology with feelings of awe for their vastness and of exultation for their temerity of the human spirit in attempting their solution, they must also be approached at the same time with the keen, balanced, critical and skeptical objectivity of the scientist.[8]

Philosophical, religious (or rather anti-religious), and other extrascientific considerations were of paramount importance for the popularity of cyclic cosmologies in the late nineteenth century. On the other hand, they played almost no role with the scientists who first investigated such models within the framework of general relativity. None of the pioneers of this class of models—Friedmann, Einstein, and Tolman—assigned great philosophical significance to it or were emotionally attached to it. No physicist or astronomer in the early period expressed strong commitment to the idea of a cyclic universe, whereas a few reacted emotionally against it. Eddington disliked the eternally cyclic universe as much as he disliked the explosive finite-age universe of the kind suggested by Lemaître. His antipathy was not scientifically based but rooted in his religious and moral sentiments, as he made clear in 1935. 'From a *moral standpoint* the conception of a cyclic universe, continually running down and continually rejuvenating itself, seems to me wholly retrograde', he wrote. 'Must Sisyphus for ever roll his stone up the hill only for it to roll down again every time it approaches the top?'[9]

Although a possible model of the universe, the cyclic model had but few advocates and was not highly regarded by the majority of scientists. It suffered from several difficulties, among which were that it required a mean density of matter that seemed to be greater than observations could provide. Einstein found in 1931 that his model needed a density of about 10^{-26} g cm^{-3}, which was at least a factor of 100 more than observations indicated. It also resulted in an embarrassingly low age for the present universe, less than 1.2 billion years. As seen from a methodological point of view, it was most uneconomical to assume a multitude of previous universes to explain the one we live in. In addition, there was no physical justification for the postulated bounces at or near the periodically occurring singularities. In short, the cyclic universe was not seen as a convincing solution to the cosmological problem.

All the same, these and other difficulties were not enough to kill the idea of an endlessly regenerating universe, an idea which a minority of cosmologists found too attractive to be wrong. To mention but one example, in the 1950s the British physicist

William Bonnor reconsidered the model, in part because he found it to be more philosophically satisfactory than models of the big-bang type. He suggested that a contracting universe might change into an expanding one without passing through a singular state, namely if a negative pressure arising from radiation-to-matter conversion was built up near the minimum. Bonnor's favoured candidate was thus a cyclic model in which the universe oscillated smoothly, bouncing between states of high but not infinitely high density.[10] What mattered to him was not so much the cycles as the possibility of a relativistic, singularity-free model with an unlimited past and future. As he admitted, this was an epistemic desideratum, for he felt it was contrary to the spirit of science to restrict the scope of its hypotheses. If it were accepted that the universe was limited in time, as in the single-cycle model, questions about its state before $t = 0$ and after $t = t_{max}$ would be forbidden or meaningless, whereas this restriction was not imposed by the model of infinite oscillations. Incidentally, this kind of objection was the very same as Fred Hoyle used against the big-bang idea and for the steady-state model. However, as mentioned in Chapter 5, Bonnor disliked the steady-state theory of the universe.

The philosophical attraction of the cyclic universe can be illustrated even more clearly in the case of Herman Zanstra, a reputed Dutch astronomer who in 1957 published a detailed investigation of this kind of cosmological model.[11] Whereas Zanstra confirmed Tolman's result that new cycles would grow increasingly bigger, he also found from thermodynamic arguments that there could have been only a finite number of cycles before the present one (Fig. 8.1). Although he admitted that a pulsating universe was not supported by observations, he thought it was just observationally possible and at any rate to be preferred from a philosophical and aesthetic point of view. Contrary to most other scientists, Zanstra openly admitted the importance of extrascientific considerations, including what he called 'philosophical desires'. These

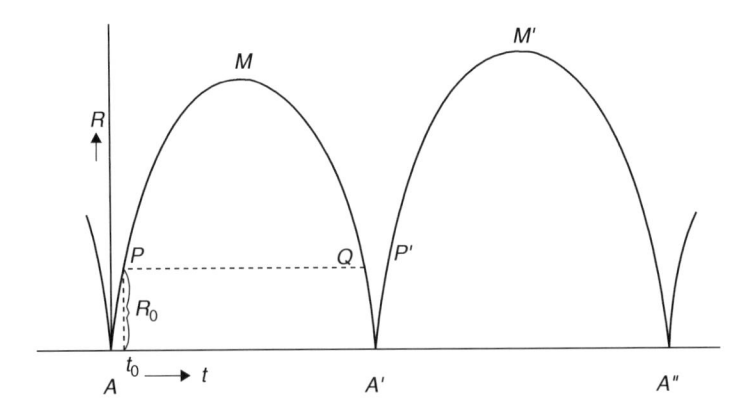

Fig. 8.1. The pulsating universe as depicted by Herman Zanstra in 1957. The cycle starts in A, where $t = 0$ arbitrarily, and bounces at maximum density (A', A''). The state of the present universe is represented by P, at radius R_0 and age t_0. Q marks the later contracting state. The second cycle $A'M''A''$ is greater than the first cycle AMA'.

Source: Zanstra 1957, p. 292.

were both unavoidable and legitimate in cosmological research, he thought. Of course Zanstra did not claim that scientific truth was dictated by philosophical desires, but he thought they should result in working hypotheses to be further investigated. Observations would then decide whether or not the hypothesis could be maintained. In the parlance of philosophers of science, philosophical desires should enter the context of discovery but have no role to play in the context of justification.

Zanstra's philosophical desire was an eternally recycling universe, which he felt was the only one to give 'a permanence and constancy of properties to the universe denied to human beings'.[12] Now he had found that a cyclic universe governed by the laws of thermodynamics and general relativity could have existed only for a finite time, and for this reason he did not consider the model to be entirely satisfactory. On the other hand, he thought it was much more appealing than the ever-expanding models of the big-bang type.

8.2 AN IDEA THAT WOULDN'T DIE

One of the problems of the cyclic universe was the bounce, which formally took place through a singular state and therefore could not be assigned physical reality. A solution might be to argue that at the extreme densities near the singularity a 'fluid' with negative pressure would be built up, with the result that the universe would never contract to zero size. Such a state, characterized by a pressure $p = -\rho c^2$ (or $w = -1$) where ρ is the energy density, had first been suggested by Lemaître in 1934 and it reappeared in the steady-state theory proposed by McCrea in the early 1950s (see Chapter 5). The hypothetical negative pressure was well suited to avoid the troublesome singularities in cyclic models and turn them into gently oscillating models without singular bangs or crunches. Jaroslav Pachner, a Polish physicist, may have been the first to apply the idea in the context of the cyclic universe. This he did in a work of 1965 in which he assumed that the density at the bounce was at least of the order of an atomic nucleus, that is, 10^{16} g cm^{-3} or more. During the next couple of decades several other negative-pressure cyclic models were proposed. Although the cosmic singularities disappeared in this way, it was at the expense of postulating a 'fluid' for which there was no independent evidence.

Even with the assumption of a negative pressure as a way to avoid the singularity at the bounce, oscillating models were not highly regarded by most cosmologists during the last third of the twentieth century. At a time when the standard hot big-bang theory, with or without inflation, dominated cosmology there seemed little reason to reconsider cyclic models. After all, models with a finite past time seemed to agree with observations, and the hypothetical series of endless cycles could be and often was dismissed as speculative. According to an article in the journal *Sky and Telescope*, 'The idea of an oscillating universe gained currency merely because it avoided the issue of creation—not because there was the slightest evidence for it.'[13] In 1997 Alan Guth noted, undoubtedly correctly, that the idea of an oscillating universe 'currently attracts very little interest among cosmologists'. He found this to be justified because 'since there is no reliable theory that describes how a universe might bounce, the basis of the oscillating universe relies solely on speculation'.[14]

The entropy problem first explicitly pointed out by Zanstra received support from later cosmologists, including Martin Rees, Igor Novikov, Robert Dicke, and Jim Peebles, who all confirmed that in each cycle the entropy increases by a finite amount. By the 1970s it was generally accepted that, as a result of the entropy increase, the universe cannot have lived through an infinite number of cycles. In itself this is not a strong argument against the cyclic universe, but it does conflict with the idea of eternal cycles. An oscillating universe that can be traced back to a first cycle loses much of its philosophical appeal, since the question of an absolute origin then reappears. The question of entropy in a cyclic or bouncing universe is old, going back to the days of Boltzmann, and it continues to be discussed by modern cosmologists. 'How does the chaotic high-entropy state of the big crunch get recycled into the low-entropy, nearly uniform, state of the next big bang? If it does not, then after an infinite number of cycles, why are we not in a universe with chaotic initial conditions?'[15]

Among those who found oscillating world models to be philosophically appealing was the Russian physicist and political dissident Andrei Sakharov, who investigated what he called 'multisheet models', another name for cyclic models. When he discovered that an oscillating universe can contain only a limited number of cycles, he was disappointed because it implied a first cycle beyond the limits of scientific enquiry. Still he continued to value the hypothesis of an infinite series of cycles, if perhaps more as part of a philosophical world view than a scientifically based picture of the universe. He expressed such sentiments at several occasions, one of them being the Nobel Peace Prize lecture of 1975, delivered *in absentia*. In a work from 1982 Sakharov found that his preferred multisheet model would lead to an over-production of black holes and therefore might be ruled out as a candidate for the real universe. But, clearly unwilling to abandon the model, he came up with various hypotheses to take care of the problem. Because, as he said, eternally cyclic models were associated with 'the hope that maybe nature realizes the picture of the Universe with infinite repetition in the past and future of cosmological cycles of expansion and contraction, a picture which is intrinsically more attractive for many people'.[16] Sakharov obviously belonged to this group of people.

Rozental was another Russian physicist who in the 1980s considered a cyclic model, in his case assuming that each of the cycles had its own set of physical constants and vacuum energy density. It was thus an example of a multiverse. Realizing that such a model could only have a finite number of cycles, and would therefore lead to the unpalatable concept of an absolute beginning, he did not commit himself to it.[17] Apart from theoretical and conceptual problems, the hypothesis of a perpetually cyclic universe faced observational problems such as those related to the mean density of matter in the universe. The classical cyclic models assumed space to be closed, and according to careful observations in the 1970s this assumption was contradicted by observational evidence. Data as well as theoretical considerations seemed to speak in favour of an open, ever-expanding universe. As one physicist concluded from an examination of the thermodynamics of closed-universe models: 'If the universe is closed and began with low entropy, it... lives only once, and can start with a whimper, but goes out with a bang.'[18]

Cyclic models were apparently popular among Russian scientists during the closing years of the Soviet Union. According to A. A. Logunov, a theorist at the Serpukhov Institute for High Energy Physics, Einstein's general theory of relativity was unsatisfactory and he consequently suggested an alternative theory of gravitation, which he somewhat confusingly called the 'relativistic theory of gravitation'.[19] When applied to the Friedmann universe, he and his collaborators found that its space was necessarily flat (contrary to ordinary cosmology, where the curvature of space is determined by the mass density). Although flat, the scale factor of Lugonov's universe oscillated eternally between a maximum value and a minimum value different from zero. To make his model agree with astronomical data, he was led to predict the existence of a large amount of hidden mass in the universe. However, cosmological models not agreeing with general relativity were not well received in the last decades of the twentieth century. Logunov's model was no exception.

The many problems that faced cyclic models were not enough to eradicate this conception of the universe, which continued to be explored by a minority of physicists and astronomers. As an example of a late model within the classical Friedmann–Tolman tradition, consider the theory developed by the two Israeli physicists Nathan Rosen and Mark Israelit about 1990. Rosen is better known as a collaborator of Einstein and one of the authors of the famous Einstein–Podolsky–Rosen (EPR) paper in which the Copenhagen interpretation of quantum mechanics was challenged by means of an ingenious thought experiment. He also worked with Einstein on gravitational waves and what became known as the 'Einstein–Rosen bridge', a precursor of so-called wormholes in general relativity. But this was back in 1935–37, more than half a century before his adventure in cyclic cosmology.[20]

The model of Rosen and Israelit was in part motivated by the two authors' dissatisfaction with the inflationary big-bang theory, which they found to be 'complicated' and 'highly speculative'.[21] As an alternative they offered a model that took care of the flatness and horizon problems without invoking a hypothetical inflationary phase. That is, their model was motivated by considerations similar to those which some years later led Moffat, Magueijo, and Albrecht to propose their varying-speed-of-light cosmology (see Chapter 7). According to Rosen and Israelit, at the arbitrary time $t = 0$ the universe was a hyperdense 'cosmic egg' with an incredibly high density given by the Planck value:

$$\rho_{\text{Pl}} = \frac{m_{\text{Pl}}}{l_{\text{Pl}}^3} = \frac{c^5}{hG^2} \cong 10^{93}\,\text{g cm}^{-3}$$

This extraordinary 'prematter', as they called it, was assumed to exhibit an enormous negative pressure (following $p = -\rho c^2$). Because of the negative pressure, which is equivalent to a tension, it would result in an explosion that in some ways corresponded to the inflationary phase in the standard theory. Whereas the initial radius of the cosmic egg was about 10^{-36} m, after a time of just 10^{-42} s it increased to a value of about 10^{-7} m. Then the prematter era ended with a conversion of prematter into radiation energy and ordinary elementary particles. At the same time the universe decelerated, following a course like the one in conventional big-bang theory. Rosen and Israelit calculated that it would continue to grow for some 600 billion years and then start to contract until the prematter state had been recreated, which builds up a new tension, etc., *ad infinitum*.

The Rosen–Israelit model was not seen as particularly important, but for a time it was well known and developed by a few other researchers who found it attractive as an alternative to the standard inflationary theory. In a study from around the same time, a Russian astrophysicist at Moscow University, Nicolai Kardashev, developed another non-singular oscillating model which, like the Rosen–Israelit model, appealed to a negative vacuum pressure. Among the merits of this model, he not only discussed its possible experimental tests, but also noted that it 'is most optimistic because it does not lead to the extermination of life as a result of the unlimited expansion of the Universe and of a density decrease or collapse to a singularity'.[22] Comments of this kind appear quite frequently in the cosmological literature, but they are rarely meant very seriously or as a justification of the theory.

Whereas belief in a spatially closed universe could be retained up to the end of the century, observations from 1998 of the redshifts of supernovae resulted in a consensus that space was flat or very nearly so. Moreover, the universe was now recognized to be in a phase of acceleration, driven by a repulsive cosmic force usually considered to be a manifestation of the cosmological constant. The new paradigm of a flat and accelerating universe seemed to herald the death of all realistic models of a cyclic universe, including the Rosen–Israelit model. However, this was not the end of the story of cosmic cycles. It was merely the beginning of a new chapter.

8.3 A NEW PHOENIX UNIVERSE

Twenty years after having co-fathered the revised inflation theory of 1982, Paul Steinhardt suggested, together with the British theorist Neil Turok, of Cambridge University, a radically different cosmological theory without inflation. The Steinhardt–Turok 'new cyclic theory' pictured the universe as eternally pulsating, with no original big bang, no inflationary era, and no singular points of infinite density and temperature. The model had its origin in many-dimensional string theory and the notion of brane-worlds, ideas that grew out of the fundamental M-theory found by Petr Hořava and Edward Witten in 1995 (see Chapter 11). According to this theory, as developed by string theorists in the late 1990s, our universe lies on a three-space dimensional 'brane-world' separated by a tiny gap in extra dimensions from another three-dimensional brane-world that is hidden from us. The hidden brane-world does not move in ordinary 3 + 1 space–time, but along an extra fifth dimension that can be conceived as a line segment connecting the two brane-worlds.

The connection to big-bang cosmology was established by studying what happens when two branes collide. In 2001 Steinhardt and Turok, investigating this problem in collaboration with Burt Ovrut and Justin Khoury, found that if there exists an attractive force between the branes, growing rapidly in strength as they approach, the collision will produce what, as seen from 'our brane', very much looks like a big bang.[23] When the brane-worlds collide, part of their kinetic energy will be converted into elementary particles and hot radiation in a flat universe. Although very hot, the temperature of the radiation stays well below the Planck temperature of 10^{32} K. As the branes are drawn together, the quantum-produced ripples of the brane colliding with the visible brane-world impresses a spectrum of density and temperature fluctuations on our universe. After lengthy calculations Steinhardt and his collaborators were able to show that these small variations agreed with the observed inhomogeneities of the microwave background radiation.

The model proposed in 2001 as an alternative to the inflationary theory was called the *ekpyrotic* model, a name derived from the Greek name for conflagration or 'out of fire'. Had the name not already been in use, it might also have been called the phoenix model—indeed, in later publications Steinhardt has stressed its qualitative similarity to earlier ideas of a phoenix universe.[24] The ekpyrotic model offered a mechanism for the big bang entirely different from the inflation mechanism, and yet it solved the same problems such as the horizon and flatness problems and also the temperature fluctuations in the microwave background. Steinhardt and his collaborators emphasized the advantage of basing the ekpyrotic scenario on concepts of string theory, which automatically ensured that the scenario comprised all known particles and symmetries. On the other hand, they found it a disadvantage of the inflationary scenario that it was not naturally connected to string theory or other fundamental theories of quantum gravity. The ekpyrotic model was basically an alternative theory of the big bang, but it also included a picture of the universe before the bang, namely that the universe began in the simplest possible state, cold and empty. In principle it could be asked how and why the universe started out in this way. In qualitative terms the ekpyrotic model was somewhat similar to another string-based cosmological scenario, the pre-big bang model suggested by Gabriele Veneziano and Maurizio Gasperini (see below).

The announcement of the ekpyrotic model caused an exchange of opinions between the authors of the new model and theorists committed to the inflationary scenario. Together with Renata Kallosh and Lev Kofman, Andrei Linde immediately responded to the new theory with an 'epicrisis of ekpyrosis', explaining that 'epicrisis' is a Greek word for critical evaluation. For one thing, the critics argued that the ekpyrotic model did not really follow from string theory and that the only way to modify it to do so would require additional fine tuning; for another, they argued that instead of resolving the homogeneity problem, the ekpyrotic scenario made its much worse. 'We believe', they stated, 'that at present inflation remains the only robust mechanism that produces density perturbations with a flat spectrum and simultaneously solves all major cosmological problems.'[25] Responding to the criticism, Steinhardt and his collaborators denied the characterization of inflation as 'robust' if taken as having the meaning that the observed universe is nearly inevitably described by inflation theory. Although inflation cosmology was indeed 'very appealing philosophically', they found it important to maintain a 'fair and open-minded attitude' that allowed for alternatives such as the ekpyrotic model.[26]

The cyclic model of the universe that Steinhardt and Turok presented in *Science* in 2002 and developed over the following years can be seen as an extension of the ekpyrotic model of the universe. Like this model, it was strongly motivated by string and M-theory, but Steinhardt and Turok stressed that their new theory did not depend crucially on this theory, which after all lacked empirical proof. The new cyclic theory could be largely formulated in the language of conventional four-dimensional field physics. By 2007 the connection to string theory was further loosened, as physicists working on the theory made calculations that reproduced the observed temperature fluctuations in the cosmic microwave background without the string-based assumptions of branes and extra dimensions.

Whatever the language of its formulation, the Steinhardt–Turok theory was an ambitious attempt to establish a new picture of cosmic evolution. When they announced it to the physics community, they did not speak modestly:

> The appeal of a cyclic model is that it provides a description of the history of the universe which applies arbitrarily far back into our past. The model presented here suggests novel answers to some of the most challenging issues in cosmology: How old is the universe—finite or infinite? What was the physical cause of its homogeneity, isotropy and flatness? What was the origin of the energy density inhomogeneities that seeded cosmic structure formation and are visible on the cosmic microwave sky? What is the resolution of the cosmic singularity puzzle? Was there time, and an arrow of time, before the big bang?[27]

Neither Steinhardt nor Turok had any previous predilection for the cyclic universe, no philosophical desires of the kind expressed by Bonnor and Zanstra in the 1950s. Turok, who used to be sceptical of such models, found their new model not only to be mathematically consistent but also aesthetically pleasing. 'Time is infinite, space is infinite, and they have always been there', he said, emphasizing the feature of eternity rather than cyclicity. He realized that many of the concepts and features of the new theory were closely similar to those of the old Hoyle–Narlikar steady-state theory.

Indeed, he and Steinhardt admitted that their cyclic theory can be considered 'a remarkable reincarnation of Fred Hoyle's steady-state model of the universe'.[28]

The new picture of a cyclic universe shows some similarity to the classical models of the Friedmann–Tolman type, but also differs from them in important respects. First of all, the Steinhardt–Turok model leads to an eternal or nearly eternal sequence of expansions and contractions without relying on a spatially closed universe.[29] On the contrary, the universe remains infinite and flat, in agreement with observations. Another noteworthy feature is that the cycles are not completely independent, as events that occurred before the last bang shape the large-scale structure of the universe and will continue to do so. On the other hand, matter, both dark and in the ordinary form, is produced afresh at each new bang. The infinitely old universe has one coherent history rather than an infinity of disconnected histories. Also contrary to the Tolman model and most later models in this tradition, the cycles are identical in the sense that a full cycle ago the physical conditions were the same as they are today, or very nearly so. Whereas in the Tolman model the lengths of the cycles increase continuously, according to the model of the twenty-first century all the cycles are of the same, extremely long duration ('trillions of years'). Again in contrast to the Tolman models, although the entropy increases from cycle to cycle, the entropy density remains constant.

As in the earlier ekpyrotic theory, in the cyclic model, density fluctuations of the same kind as those of the inflation theory are generated in the contracting and not in the expanding phase of the universe. As to the observed dark energy and cosmic acceleration, Steinhardt and Turok argued that they followed naturally from their theory, while they had to be added to the inflation theory. In an address to the American Philosophical Society of 2004, Steinhardt presented the new model of the universe in non-technical language. He summarized the essential features of the model as follows:

> The new cyclic model ... turns the conventional picture topsy-turvy. Space and time exist forever. The big bang is not the beginning of time. Rather, it is a bridge to a pre-existing contracting era. The universe undergoes an endless sequence of cycles in which it contracts in a big crunch and re-emerges in an expanding big bang, with trillions of years of evolution in between. The temperature and density of the universe do not become infinite at any point in the cycle. Indeed, they never exceed a finite bound (about a trillion trillion degrees). No high-energy inflation has taken place since the big bang. The current homogeneity and flatness were created by events that occurred before the most recent big bang.[30]

Steinhardt and Turok were eager to present their theory not only as an alternative to inflation cosmology, but as the *only* alternative. Ambitiously announcing their cyclic theory as a new paradigm, they suggested that cosmologists were now faced with a choice between two (and only two) widely different pictures of the universe. 'We now have two disparate possibilities: a universe with a definite beginning and a universe that is made and remade forever.'[31] Much the same rhetoric was heard in the 1950s, when the choice was between big bang and steady state. The same message was a prominent theme in *Endless Universe*, a popular book in which Steinhardt and Turok gave a broader introduction to their theory and also contrasted it to the new inflation-based theory of the multiverse. The picture of modern theoretical cosmology as a competition between just two rival theories, inflation and the cyclic model, is, however, flawed. Inflation is undoubtedly the predominant theory, but there are several alternatives among which the cyclic model is only one.[32]

Let us now briefly consider the merits *vis-à-vis* the inflation theory of the new cyclic cosmology, as perceived by Steinhardt and Turok. For one thing they argued that the cyclic theory was simpler and more parsimonious than the rival inflation theory. For example, the cyclic theory operates with only one period of accelerated expansion, the one which is observed. By contrast, in inflationary theory the observed acceleration is entirely different from the hyperexpansion assumed in the very early universe and therefore in need of a separate explanation. This feature was, however, questioned by the leading inflation theorist Andrei Linde, who expressed his dissatisfaction with Steinhardt and Turok having announced their anti-inflation theory in the 'popular press' (*Science!*).[33] According to Linde, resuming his earlier critique of the ekpyrotic model, the theory was not really an alternative to inflation, for a mechanism of the inflation type was also needed in the ekpyrotic-cyclic model. As he saw it, there was no need to introduce a speculative model with an infinity of inflationary stages separated by an infinite number of nearly singular states. Why not keep to the simpler version, the real inflationary scenario?[34]

The cyclic model also offers a complete history of the universe, dealing with both past and future, which inflation does not. Being a scenario of only the very early universe, the latter theory has nothing to say about the development at later stages. Steinhardt and Turok further called attention to what they thought was the superior explanatory and predictive power of their theory. It explains (so they argue) naturally the repulsive dark energy, a concept which was not foreseen by inflation, and in addition it explains the cosmic singularity as a transition from a contracting to an expanding phase. Inflation does not address the cosmic singularity problem in a similarly direct way but merely presupposes an initial creative event. In a more recent paper Steinhardt and Turok have offered an explanation of the cosmological constant problem, that is, why the constant is so tiny in comparison to the value expected from quantum mechanics.[35] They argue that in this respect, too, the cyclic model is superior to the inflation-based multiverse picture. According to Steinhardt and Turok, the cosmological constant decreases slowly over huge spans of time, meaning over many cycles. It was once much greater and has only attained its present value, small enough to sustain life, because it has decreased through many trillions of years. This kind of solution of the cosmological constant problem they consider one more reason to prefer the cyclic model over its inflationary rival.

The arguments mentioned above are largely of a methodological nature, important but not sufficient to choose between alternative theories. As Steinhardt and Turok are keenly aware, such a choice needs to be based on the standard procedure of comparing predictions with observational data. Generally speaking, inflation theory and cyclic cosmology are both stronger in explaining things than predicting things. But they do lead to predictions that differ, although the differences are so small that they escape detection by today's instruments. Both models predict the same spectrum of temperature fluctuations, but inflation also predicts a spectrum of gravitational waves that will produce a characteristic signature in the polarization of the cosmic microwave background. According to the cyclic (and also the original ekpyrotic) scenario, this gravitational signature should not exist. Thus, the new cyclic model is falsifiable as it predicts certain details in the microwave background that can be measured in very sensitive experiments and which differ from those predicted by the rival inflationary theory. If the gravitational signature is not found, 'then this would support the inflationary picture and *definitively* rule out the cyclic model'.[36]

The allegiance to Popperian standards of science is common among cosmologists, sometimes stated explicitly and at other times merely indicated between the lines. Thus, in a talk of 2007 Turok said that in general 'it is a healthy situation for science to have rival theories, which are as different as possible. This helps us to clearly identify which critical tests—be they observational or mathematical/logical—will be the key to distinguishing the theories and proving some of them wrong. Competition between models is good.' Noting that they key battleground between inflation theory and the new cyclic model is currently theoretical, he went on to stress 'the attempt to test the models observationally, because science is nothing without observational test'. And by this standard the cyclic scenario is held to be scientific: 'Our model with the colliding branes predicts that the Planck satellite and other similar experiments will detect nothing. So we can be proved wrong by experiment.'[37]

The new cyclic model is a work in progress, not a final theory. Its latest version, as developed by Steinhardt and Jean-Luc Lehners of Princeton University, differs in some respects from the original Steinhardt–Turok version.[38] For example, only a tiny fraction of the universe is now thought to be able to survive from one cycle to the next, a fraction that will expand hugely after the bounce. During each cycle vast new regions of space are created, not in the form of new universes, as in eternal inflation, but in the same universe. Also the dark energy plays a different role. Gravitational waves are expected from the bounce, but very weak and of a kind that may never be detected. Observationally the situation remains the same: if primordial gravitational waves of the kind predicted by inflation are found, it will rule out the new phoenix model; if they are not found, it will make the phoenix universe more attractive but not prove it.

Did Steinhardt and his collaborators succeed in their challenge to the dominant inflation paradigm? The cyclic theory, proposed by two physicists of high reputation, was fairly well received and 'marketed' with determination. While some cosmologists were unimpressed and found the theory to be both contrived and unnecessary, others thought it was an interesting idea that deserved to be further developed. Of course, there are also cosmologists who consider the new cyclic models and similar theories to be nothing but scholastic exercises that are of no relevance to what scientific cosmology should really be about. 'I think it's silly to make much of a production about this stuff', said Joel Primack, a physicist at the University of California at Santa Cruz. 'I'd much rather spend my time working on the really important questions observational cosmology has been handing us about dark matter and dark energy. The ideas in these papers [on the cyclic universe] are essentially untestable.'[39]

After seven years of existence the new cyclic world view does not seem to have made much of an impact on cosmology or been recognized as an important alternative to inflation. According to the ISI Web of Science, the Steinhardt–Turok paper in *Physical Review* of 2002 has been cited 261 times up until February 2010; the average number of citations per year is 29, with little variation over the period. This indicates that the theory is well known, but not that it is widely accepted nor nearly as popular as the standard inflation cosmology. By comparison, Guth's paper of 1981, which started the inflation industry has received 3492 citations and was cited 219 times in 2008 alone; in none of the years after 1982 was it cited less than 90 times. It is a real classic of science.

8.4 SOME OTHER MODELS

Although the Steinhardt–Turok model is presently the most discussed cyclic alternative to the standard theory, it is not the only game in town of its kind. There are several other oscillating and bounce models on the cosmological market, some of them relying on string concepts and unusual equations of state while others are closer to the classical Tolman tradition of the 1930s.

As an example, consider a model proposed in 2007 by two American physicists, Lauris Baum and Paul Frampton from the University of North Carolina, whose objections to the standard model were 'more aesthetic than motivated directly by observations'.[40] Among these aesthetic objections they mentioned the nature of the initial singularity and also that the fate of the universe in the far future is an indefinite expansion. Referring to the classical cyclic models of the Friedmann–Tolman type, they suggested that their model 'gives renewed hope for the infinitely oscillatory universe'.

The model of Baum and Frampton is based on so-called phantom energy, a hypothetical representative of dark energy with an equation of state given by $p < -\rho c^2$ (or $w < -1$), which was introduced in the early years of the twenty-first century. Phantom energy has remarkable properties and remarkable effects on the fate of the universe.[41] For example, its energy density increases in time and, if $w < -1$ persists, it will lead to the end of all bound objects from galactic clusters to atomic nuclei, indeed to the end of time. The phantom energy will blow up the universe almost to the point where the catastrophic 'big rip' would occur, tearing all matter in the universe apart, which shortly before would undergo a dramatic deflation. The universe would be left very small, very empty, and nearly without entropy. Immediately after the bounce, an inflationary phase would set in, as in the ordinary inflation scenario. At the turnaround the universe would have fragmented into an astronomical number of causally separate subuniverses, each of which would deflate. Since the model universe has cycled an infinite number of times, the number of subuniverses is infinite. Thus, the Baum–Frampton model is a multiverse as well as an eternal-cyclic scenario with links to the inflation theory. But whereas the proliferation of universes in eternal inflation occurs when the universe is very small, in the Baum–Frampton cyclic model it occurs when the universe has reached its maximum size.

Loop quantum gravity, often abbreviated LQG, is an ambitious quantum theory of space–time, an attempt to establish a unified framework of quantum mechanics and general relativity. As such it is a competitor of string theory and probably the most developed and serious of the several contenders. Contrary to string theory, loop quantum gravity does not incorporate higher dimensions and it does not assume unobserved supersymmetric particles.[42] Loop gravity (as it is also called) took off in 1986 when Abhay Ashtekar proposed a reformulation of Einstein's field equations of general relativity, and since then it has attracted much attention and been further developed by Ashtekar, Carlo Rovelli, Lee Smolin, and many other theorists. The application to cosmology, in the approach known as loop quantum cosmology (LQC), was pioneered in 1999 by Martin Bojowald at Pennsylvania State University. Loop quantum gravity leads to a picture of space–time very different from the ordinary one associated with general relativity. Space is not seen as continuous, but discrete or atomistic at a very small scale, consisting of tiny space atoms with a size about the Planck length. The smallest volume that can be assigned a physical meaning would then be the unimaginably small Planck volume, about 10^{-104} m^3. The general idea of atomistic space and time is far from new, but in the earlier scientific (and not merely philosophical) versions it was primarily rooted in quantum theory, without taking general relativity into regard.

The loop approaches to quantum gravity and cosmology are highly mathematical and it is not possible to test the theories in any direct way. Yet, although a proper empirical testing of loop quantum gravity requires experimental access to Planck-scale physics, which remains an airy dream of the future, the theory does have experimental consequences. One consequence of the discreteness of space, if not a strict prediction, is that photons of very high energy should travel slightly faster than those of low energy. It is believed that this effect, if it exists, should turn up in data from high-energetic cosmic rays and gamma-ray bursts. Although physicists associated with loop quantum cosmology have faith in their theory, they admit that 'Because this is science, in the end experiment will decide', as Smolin phrases it.[43]

The scenario of loop quantum cosmology, as developed by Bojowald in particular, implies that the universe has neither a beginning nor an end in the absolute sense. Within the framework of the theory it is possible to deal directly with the classical singularities, the big bang and the big crunch, and to continue across them. As the cosmic movie is played backwards and the big bang is approached, density and curvature will increase but without leading to infinite values. Quantum effects arising from the theory will create a repulsive force that overpowers the gravitational attraction and causes a bounce at finite values of the density, approximately of the order of the Planck density. After the quantum bounce the density quickly decreases and the repulsive quantum force is replaced by attractive gravitation. The repulsive force associated with the quantum nature of space is not introduced *ad hoc*, but follows from basic theory. Apart from allowing a transition from contraction to expansion the theory also offers an explanation of the rapidly expanding space in the inflationary period. In loop quantum cosmology, inflation is a natural consequence of the discreteness of space–time. This explanation is seen as superior to the one of standard inflation and other theories: 'Loop quantum cosmology thus presents a viable alternative to usual inflaton models which is not in conflict with current observations but can be verified or ruled out with future experiments such as the Planck satellite. Its attractive feature is that it does not require the introduction of an inflaton field with its special properties, but provides a fundamental explanation of acceleration from quantum gravity.'[44]

Loop quantum cosmology offers a precise description of the entire evolution of the universe through an endless series of cycles, including the transitions at the bouncing points. Bojowald first thought that the theory allowed calculation of the state of the universe before the bounce, but came to realize that a complete picture of the pre-bang universe would not be possible, even in principle. Although some of the properties of the earlier universe can be calculated, knowledge of other properties will be irretrievably lost in the bounce. 'An eternal recurrence of the same is prevented by intrinsic forgetfulness', Bojowald says. The intrinsic loss of memory, ultimately to be traced back to the uncertainties of quantum mechanics, implies that 'assumptions (or prejudice) will remain necessary for knowing the precise state of the Universe, which cannot be fully justified within science itself'.[45] The good news is that loop quantum cosmology predicts which aspects of the pre-bang universe can be known and which cannot. There are unavoidable limitations to what can be observed of the properties of the universe before the big bang, but these can be derived from the theory and are not limitations to it. According to Ashtekar, the important message of loop quantum cosmology is that the universe was not created a finite time ago but exists eternally. Contrary to earlier times, when this was largely a matter of taste, today 'one does not need an aesthetic or a philosophical preference to select one possibility over another'.[46] The equations tell us.

The ekpyrotic model was not the first cosmological model based on string theory. In the early 1990s the two Italian string theorists, Gabriele Veneziano, a co-developer of the original string theory, and Maurizio Gasperini, introduced a string-based alternative to the standard model in the form of what they called pre-big-bang cosmology. Although this kind of cosmological scenario does not belong to the cyclic class, in principle it is possible to extend it to a series of cosmic cycles. As Veneziano and Gasperini emphasized, their picture of an era before the big bang was meant as a completion of the standard model, which cannot itself be extended beyond the Planck era, rather than a replacement of it.[47] Indeed, their main objection to the standard model, including inflation, was that it had nothing to say about the cosmic singularity and did not address the question of the origin of the inflationary phase. In their paper of 1993, as well as in many later papers, they argued from string theory that the big bang is not to be understood as a singular event but as the moment in cosmic time when a previous state of very high curvature transformed into a state of rapidly decreasing curvature. As one goes back in time, one will not meet an impenetrable wall at $t = 0$ where the density and curvature are infinite. The curvature will increase, but only up to a maximum, after which (in reversed time) it will begin to shrink and approach zero. String theory includes a fundamental length of about 10^{-34} m, or some ten times the Planck length, and consequently the Hubble radius c/H, which is a measure of the radius of curvature, cannot be smaller than this length.

In a popular article of 2004, announcing the pre-big-bang theory as 'the latest frontier of cosmology', Veneziano explained:

> In the standard theory, acceleration occurs after the big bang because of an *ad hoc* inflaton field. In the pre-big-bang scenario, it occurs before the bang as a natural outcome of the novel symmetries of string theory. According to the scenario, the pre-big-bang universe was almost a perfect mirror-image of the post-bang one. If the universe is eternal into the future, its contents thinning to a meager gruel, it is also eternal into the past. Infinitely long ago it was nearly empty, filled only with a tenuous, widely dispersed, chaotic gas of radiation and matter.[48]

That is, the universe is infinite in time, 'starting' in a cold, flat and nearly empty state, which is not itself explained or supposedly in need of explanation—one has to start somewhere and with something. The initial string vacuum is unstable and decays into a non-static configuration, a pre-big-bang universe governed by a so-called dilaton field. It evolves into separate regions of high density, eventually forming a number of disconnected black holes. Our universe emerged in a big bang from the interior of one of these black holes, namely when the density reached the maximum value allowed by string theory. In the initial state of the pre-bang universe all the coupling parameters of the forces of nature were close to zero. In particular, the Newtonian constant of gravitation was almost zero and only increased towards its present value when the universe entered the regime of high curvature.

Veneziano, Gasperini, and other advocates of the pre-big-bang scenario realize of course that it is not sufficiently justified by its foundation in the theory of superstrings, however authoritative this theory is considered to be. After all, string theory has not been empirically tested and it does not lead uniquely to a particular cosmological model (why pre-big-bang rather than, say, the ekpyrotic scenario?). Without observable physical consequences the string-based theory of Veneziano and Gasperini would merely be an interesting exercise in mathematics, and for this reason pre-big-bang physicists have since the origin of the model struggled to establish such consequences. In its modern versions the model can account for the measured variations in the cosmic microwave background, as can standard inflation and the ekpyrotic-cyclic model. But the pre-big-bang model also leads to a distinct background of gravitational waves that should be detectable. The relative energy density of the cosmic gravitational background has in this model a maximum theoretical value of $\Omega_G \cong 10^{-6}$ (one millionth of the critical density) and is concentrated in the high-frequency range. This is contrary to the inflationary background, where the intensity of gravitational waves in this region is very low.

As usual in these branches of esoteric physics, 'detection' mostly refers to in-principle detection and to experiments with instrument technologies of the future. On the other hand, direct experimental detection and investigation of weak gravitational waves may be just around the corner. Recent data from the advanced and very expensive American LIGO—an acronym for Laser Interferometer Gravitational-wave Observatory—project have already ruled out some string-based cosmological models, but not the pre-big-bang model.[49] The predictions of this model are not within the reach of LIGO, but they are expected to be tested in experiments in the near future, perhaps with the Advanced LIGO detector, which may be operational in 2014. The even more advanced LISA (Laser Interferometer Space Antenna) project, planned as a detector in orbit around the Sun, will most likely be sensitive enough to distinguish between some of today's rival cosmological models.

According to Gasperini, the situation in modern cosmology resembles the one of the 1960s, when observations supported the hot big-bang model and ruled out the rival steady-state theory. Referring to the possibility of discriminating between standard inflation and the pre-big-bang model on the basis of gravitational waves of high frequency, he says that 'once again we expect the choice between the different scenarios to be made on the basis of experiment'.[50] As far as testability as a defining feature of science is concerned, the rhetoric of pre-big-bang cosmologists is strikingly similar to that of the proponents of other exotic cosmologies, whether eternal inflation, varying speed of light, loop quantum cosmology, or the new cyclic model. It remains to be seen how much of it is just rhetoric.

The pre-big-bang universe is an example of a bouncing rather than a cyclic model. Models of this kind were considered a long time before quantum gravity became an issue, for example by Gamow in the years about 1950. Gamow seems have been fascinated by the idea that the universe evolved from infinite rarefaction, over 'the big squeeze' of a superdense state, towards a new expanding state that would eventually recreate the infinitely rarefied universe. From nothingness to nothingness. However, he realized that the idea, however fascinating, was a speculation that was unlikely to result in testable consequences. As a physicist, he did not take it quite seriously:

> Mathematically we may say that the observed expansion of the universe is nothing but the bouncing back which resulted from a collapse prior to the zero of time a few billion years ago. Physically, however, there is no sense in speaking about the 'prehistoric state' of the universe, since indeed during the stage of maximum compression everything was squeezed into the pulp, or rather into ylem, and no information could have been left from the earlier time if there ever was one.[51]

Somewhat similar ideas were suggested on a few later occasions, both within the framework of relativistic cosmology and in more unorthodox cosmologies, such as the plasma universe suggested by the Swedish physicist Hannes Alfvén.[52]

At about the same time that Israelit and Rosen proposed their pulsating 'cosmic egg' model, two German astrophysicists, Wolfgang Priester and Hans-Joachim Blome, examined the possibility that the expanding universe started in a non-singular 'big bounce' of an earlier contracting universe.[53] (They first used the term 'big bounce' in 1987.) According to the model of Priester and Blome, the original state of the universe was a quantum vacuum of density $\rho \cong 10^{76}$ g cm^{-3}. As a result of the decay of the vacuum, it would have transformed into the particles and matter of an expanding state. Like the Israelit–Rosen model, the bouncing universe of Priester and Blome presupposed a closed space and a large negative pressure at the bounce. The minimum curvature radius at the bounce, expressed by the energy density of the primordial quantum vacuum, was found to lie in the interval $\ell_{Pl} \leq R_{min} \leq 10^8 \ell_{Pl}$, where ℓ_{Pl} is the Planck length.

Finally, and without any intention of completeness, it should be noted that since about 2005 Roger Penrose has worked with cosmological ideas that have certain similarities to those mentioned above, such as the pre-big-bang model, the ekpyrotic-cyclic model of Steinhardt and Turok, and loop quantum cosmology. Penrose's so-called conformal cyclic cosmology does not make use of either inflation or of extra spatial dimensions, but it belongs to the class of cyclic cosmic scenarios. According to this model, 'the history of the universe consists of a (perhaps infinite) succession of *aeons*, where the indefinitely expanding remote future of each aeon can be joined smoothly as a conformal manifold to a big bang for the succeeding one'.[54]

Notes for Chapter 8

1. The idea of eternally recurring universes, whether in the context of myth, philosophy, or science, is well described in the literature. See, for example, Jaki 1974 and Eliade 1954. For the modern development, see Kragh 2009a and Kragh 2010b.

2. Jaki 1974, p. 171, who quotes a work by Origen known as *On First Principles*. This work can be found online as http://www.newadvent.org/fathers/04122.htm.

3. Clausius 1868, p. 419.

4. Falb 1875, p. 202. On nineteenth-century ideas of cyclic universes and their relations to the law of entropy increase, see Kragh 2008.

5. Soddy 1909, pp. 241–42.

6. There is no English translation of Einstein's paper of 1931. For an account of his and others' early cyclic models, see Kragh 2010b and Nussbaumer and Bieri 2009, pp. 147–49.

7. On the history of the early cyclic models of Tolman and others, see Kragh 2010b. See also the reconsideration of Tolman models in Heller and Szydlowski 1983.

8. Tolman 1932, p. 373. See also Tolman's detailed discussion of cyclic models in his monograph on thermodynamics, relativity and cosmology (Tolman 1934a).

9. Eddington 1935a, p. 59. Emphasis added.
10. Bonnor 1964.
11. Zanstra 1957.
12. Ibid., p. 302.
13. MacRobert 1983.
14. Guth 1997, p. 26.

15. Gott and Li 1998, p. 3, who suggested that in the contracting phase entropy would decrease with respect to time. The question of the relationship between the cosmological and thermodynamic arrows of time has been discussed by numerous physicists since the 1930s. The counterintuitive idea that the two arrows always point in the same direction was first argued by Thomas Gold in the late 1950s, but has not been generally accepted.

16. Sakharov 1982, p. 404.

17. Rozental 1988, p. 121, who seems to have been in favour of the infinite bubble universe which at the time was suggested by his compatriot Andrei Linde.

18. Bludman 1984, p. 319. This is an allusion to *The Hollow Men*, a major poem by T. S. Eliot from 1925, which includes the lines: 'This is the way the world ends/Not with a bang but a whimper.'

19. Logunov, Loskutov, and Mestrvirishvili 1988.
20. See the obituary in Israelit 1996.
21. Israelit and Rosen 1989.
22. Kardashev 1990, p. 256.
23. Khoury, Ovrut, Steinhardt and Turok 2001a.

24. For example Lehners and Steinhardt 2009. A 'phoenix universe' is basically the same as a bouncing universe. The name derives from the ancient myth of the sacred bird phoenix which, after having been burnt to ashes, would reappear from the fire to live again. The myth originated in Persia but was known in many other cultures, including Greece. In the literature of modern cosmology, the phoenix universe is often associated with Lemaître, who referred to it in a paper of 1933. However, Lemaître never endorsed the idea of a bouncing or cyclic universe.

25. Kallosh, Kofman, and Linde 2001, p. 2. The 'pyrotechnic universe' discussed by the three theorists as a string-based alternative to the ekpyrotic model included inflation. They insisted that it is 'very hard or even impossible' to avoid inflation and still solve all the major problems of cosmology.

26. Khoury, Ovrut, Steinhardt, and Turok 2001b.
27. Steinhardt and Turok 2002b, p. 2.
28. Turok, as quoted in Seife 2002. Steinhardt and Turok 2002b, p. 18.

29. It is not clear to me if the Steinhardt–Turok theory allows an infinite number of cycles or just a very large number of them. In some of their works (e.g. Steinhardt and Turok 2004) they seem to favour the latter option. Although of no observational relevance, from a philosophical perspective there is a great deal of difference between the two cases.

30. Steinhardt 2004, p. 466.
31. Steinhardt and Turok 2002a, p. 1439.

32. See Brandenberger 2008, which evaluates some of the alternatives. In addition to ekpyrotic-cyclic models, he mentions varying-speed-of-light scenarios, the pre-big-bang scenario, and string gas cosmology. See also Hollands and Wald 2002, who show that the density perturbation spectrum derived from inflationary models can be obtained also by means of a mechanism that

does not presuppose an era of inflation. Scoular 2007, p. 354 includes a number of quotes from physicists and science writers critical of the inflationary theory.

33. Linde 2003.
34. Ibid.
35. Steinhardt and Turok 2006, p. 1182. On the problem of the cosmological constant, see also Chapters 9 and 10.
36. Steinhardt 2004, p. 469. Emphasis added.
37. http://www.edge.org/3rd_culture/turok07_index.html.
38. Lehners 2008.
39. Quoted in Lemonick 2004.
40. Baum and Frampton 2007.
41. Caldwell, Kamionkowski, and Weinberg 2003. Phantom energy is an extreme form of 'quintessence,' a hypothetical kind of cosmic dark energy with density $w < 0$ that can vary in space and time. Quintessence or the Q-field was introduced as an alternative to the cosmological constant in Caldwell, Dave, and Steinhardt 1998. The name, of course, alludes to the fifth element (*quinta essentia*) of Aristotelian cosmology, a pure and frictionless substance which was restricted to the celestial realms.
42. A brief introduction to loop quantum gravity is given in Rovelli 2003. See also Smolin 2001 and further in Chapter 11.
43. Smolin 2004, p. 72.
44. Bojowald 2005, p. 407.
45. Bojowald 2007, p. 523. See also Bojowald 2008. Back in the 1960s, William McCrea suggested a kind of cosmic uncertainty principle, arguing that the universe as a whole was necessarily unknowable (see Kragh 1996, pp. 240–42).
46. Ashtekar 2010 (forthcoming).
47. Gasperini and Veneziano 1993.
48. Veneziano 2004, p. 62.
49. Abbott *et al.* 2009. See also Gasperini 2008, pp. 103–23.
50. Gasperini 2008, p. 197.
51. Gamow 1951, pp. 405–406. 'Ylem', a name introduced by Ralph Alpher in 1948, was Gamow's favourite term for the primordial substance out of which all matter and energy were formed. Gamow also described the bouncing universe in his popular book, *The Creation of the Universe*, in which he introduced the term 'big squeeze' for the maximum compression of the pre-big-bang universe (Gamow 1952, p. 36).
52. Alfvén, a Nobel laureate of 1970, argued that the primordial universe consisted of a very dilute mixture of particles and antiparticles which, due to gravitational attraction and annihilation processes, would contract and produce a radiation pressure that would halt the collapse and turn it into an expansion. For an account of the plasma universe, and an attempt to revive interest in it, see Lerner 1991.
53. Blome and Priester 1991. On Priester and his important contributions to cosmology, see Overduin, Blome, and Hoell 2007.
54. A recent account of conformal cyclic cosmology appears in Penrose 2009.

9

Anthropic Science

It's this simple paradox. The Universe is very old and very large. Humankind, by comparison, is only a tiny disturbance in one small corner of it—and a very recent one. Yet the universe is only very large and very old because we are here to say it is... And yet, of course, we all know perfectly well that it is what it is whether we are here or not.

Michael Frayn, *The Human Touch*, 2006.

In a general sense the anthropic principle is an attempt to deduce non-trivial consequences about nature from the consideration that what we observe must be compatible with our existence. The explicit reference to human observers, such as underlined by the term 'anthropic' (derived from the Greek word *anthropos* for 'man'), is what makes the principle unusual and controversial. Since it was explicitly formulated in the mid 1970s, the anthropic principle and modes of reasoning based on it have attracted a very great deal of attention, primarily in physics and cosmology, but far from only in the domain of science.[1] Although as controversial as ever, the anthropic principle has proved remarkably robust and resistant to criticism. (Critics may argue that its resistance to challenge is due to its lack of falsifiability.) Contrary to the expectation of many, it has not only survived but become increasingly popular as a tool for reasoning in some areas of science. When I reviewed John Barrow and Frank Tipler's encyclopedic work *The Anthropic Cosmological Principle* in 1987, I was convinced that within a decade physicists would lose interest in anthropic considerations and leave them to where they properly belonged, namely to philosophers and theologians.[2] I was seriously wrong.

Like most other general principles in science, the anthropic principle exists in several versions and with meanings that have changed over time. While in the early period it was closely associated with the numerical coincidences appearing in nature, today it is mostly seen as a selection principle operating in the context of the multiverse, a concept that will be dealt with more fully in Chapter 11. From a modern point of view, anthropic considerations and the multiverse hypothesis are often regarded as two aspects of the same issue. The close connection to the multiverse may have provided the anthropic principle with some measure of explanatory power, but it has done nothing to make it less controversial. Quite the contrary.

9.1 FROM DESIGN ARGUMENT TO SELECTION PRINCIPLE

The general idea that the world we experience is special and somehow conditioned by the presence of humans to observe it can be traced far back in time. With some goodwill it can be found in ancient Greece. But then, what can't? In his famous poem *De Rerum Natura*, composed about 50 BC, the poet and natural philosopher Titus Lucretius Carus advocated an atomistic cosmology inspired by the ideas of Democritus and Epicurus, presenting the universe as infinite in space but finite in time. Based on the shortness of human history, he argued that 'the whole of the world is of comparatively modern date, and recent in its origin; and had its beginning but a short time ago'.[3] Scholastic philosophers in the late middle ages discussed ideas of an anthropic nature in connection with God's free will and omnipotence—did he design the universe with the purpose that it should accommodate his chosen creatures?

A kind of anthropic-like argument was common in the past, but in the popular and at the time uncontroversial form of natural theology. An early example is provided by Bernard le Bovier de Fontenelle, permanent secretary of the Royal Academy of Sciences in Paris and the author of the immensely popular *Entretiens sur la Pluralité des Mondes*. In this work, a classic of pluralist literature published in 1686, Fontenelle explained the inclination of cometary orbits relative to the ecliptic by arguing that otherwise the comets would have destroyed life on Earth. Had the orbits been 'normal'—similar to those of the planets—we would not be here.[4] This may be interpreted as an early example of anthropic reasoning, but it may also and perhaps with better justification be read as just an example of the kind of old-fashioned teleology of which the period was so rich. Somewhat similar ideas were ventured by Edmund Halley, William Whiston, and a few other natural philosophers.

Much later, the naturalist and pioneer evolutionist Alfred Russell Wallace published a controversial work, *Man's Place in the Universe*, in which he proposed an anthropocentric cosmology. Holding that a universe without humans would be an absurdity, he argued that the purpose of the universe was man as a spiritual being. Wallace suggested that the position of the Earth, which he thought was in the centre of the Milky Way, could not be regarded a coincidence 'without any significance in relation to the culminating fact that the planet so situated has developed humanity'.[5] Another contemporary scientist who is sometimes mentioned as a precursor of the anthropic principle is Boltzmann, who in the 1890s suggested that the present low entropy, or high degree of cosmic organization, is the result of our world being in a statistically unlikely state. According to Boltzmann, the only reason that we witness this exceedingly unlikely situation of a deviation from high entropy is that our very existence depends upon it.

A contemporary of Wallace and Boltzmann, the British physicist Samuel Tolver Preston entertained ideas that had a similar anthropic flavour. In a paper of 1879 he observed that the region of the universe inhabited by man must be 'amply extensive enough to allow an amount of activity and variability of energy adapted to the conditions of life'. He furthermore suggested that 'we may happen to be in a part [of the universe] where the mean temperature of the component matter is exceptionally high, as, of course, from the fact of our being in existence, we must be in a part which is suited to the conditions of life'.[6]

Arguments somewhat resembling those of the anthropic principle were forwarded by James Jeans in the 1920s, when he gave several talks that included speculations about the role of life in the universe. For example, in a lecture delivered in 1926 he pointed out that 'the physical conditions under which life is possible form only a tiny fraction of the range of physical conditions which prevail in the universe as a whole'. As an example he mentioned that the liquid state (and hence ordinary water) required certain quite special conditions. Jeans elaborated:

> Primeval matter must go on transforming itself into radiation for millions and millions of years to produce an infinitesimal amount of the inert ash on which life can exist. Even then, this residue of ash must not be too hot or too cold, or life will be impossible. It is difficult to imagine life of any high order except on planets warmed by a sun, and even after a star has lived its life of millions of millions of years, the chance, so far as we can calculate it, is still about a hundred thousand to one against it being a sun surrounded by planets. In every respect—space, time, physical conditions—life is limited to an almost inconceivably small corner of the universe.[7]

Neither Jeans nor others at the time drew consequences with regard to the constants of nature or cosmic evolution from the fact that life does exist.

As pointed out in Chapter 4, Eddington's selective subjectivism, with its claim that our sensory equipment decides our observations of nature, may be seen as an anthropic selection principle. Again, a mild kind of anthropic reasoning can be found in the paper of 1948 in which Bondi and Gold introduced the steady-state theory of the universe. Referring to the classical problem of the heat death in an infinitely old static universe, they wrote: 'That our universe is not of this type is clear not only from astronomical observations but from local physics and indeed from *our very existence.*'[8] However, there is no reason to pay much attention to the statement or to believe that Bondi and Gold intended it to be a statement about the role played by human observers with cognitive faculties. Probably without knowing it, they were merely repeating what dozens of scientists and philosophers had said more than a century ago, in many cases with direct reference to the existence of human beings.

Although anthropic-like arguments can be found in much of the early history, they first appeared in the context of the evolving universe about 1960, when they were introduced in different ways by Grigory Idlis in the Soviet Union and Robert Dicke in the United States. Idlis forwarded his anthropic arguments in a speculative paper from the Kazakh Academy of Sciences in 1958, published in Russian only. Its title in English was 'Basic Features of the Observed Universe as Characteristic Properties of a Habitable

Cosmic System'. In a later historical review, Idlis referred to his early anthropic ideas, summarizing them as follows: 'Our entire infinite Metagalaxy... does not represent just one of the cosmologically possible, non-stationary Friedman worlds, but constitutes a precisely typical, habitable world at the stage of existence of intelligent life therein, since all corresponding properties of the Metagalaxy, directly observed by us, are, generally speaking, just the necessary and sufficient conditions for the natural origination and evolution of life to higher intelligence forms of matter, similar to man, finally aware of itself.'[9] Although Idlis later claimed to be the true discoverer of the anthropic principle, his paper of 1958 made almost no impact at all and remained unknown to scientists in the West.[10] It seems not to have aroused much attention in Russia either.

Another Russian, the physicist and cosmologist Abraham Zelmanov, has been mentioned as a possible father of the anthropic principle, many years before Carter. A specialist in relativistic cosmology, Zelmanov did important work on inhomogeneous cosmological models which he first published in a dissertation of 1944 from the Sternberg Astronomical Institute in Moscow. According to Dmitri Rabounski, by that time he had formulated a version of the anthropic principle, which he discussed with his colleagues but apparently without publishing his thoughts. He is to have said:

> Humanity exists at the present time and we observe world constants completely because the constants bear their specific numerical values at this time.... The Universe has the interior we observe, because we observe the Universe in this way. It is impossible to divorce the Universe from the observer. The observable Universe depends on the observer and the observer depends on the Universe.... If no observers exist then the observable Universe as well does not exist.[11]

This gives some of the flavour of the anthropic principle, but only in a very general way. Besides, the authenticity may be questioned. Russian predecessors apart, the path to the modern formulation of the anthropic principle went over Dirac's cosmological theory of 1937–38 based on the large number hypothesis. According to Dirac, the value of the Hubble time T (approximately the age of the universe) reflected the ratio of the electric and gravitational forces, from which he inferred that the constant of gravitation decreased in time (see Chapter 7).

The first physicist to consider Dirac's hypothesis in the light of the existence of human observers seems to have been Robert Dicke, a professor at Princeton University who was equally at home in quantum theory, general relativity, cosmology, and microwave instrument technology. Dismissing the possibility that 'nature is somewhat capricious', in papers of 1957 Dicke referred to Dirac's explanation of the large dimensionless numbers, which he found to be attractive. However, he also criticized the hypothesis because it lacked empirical evidence and disagreed with the equivalence principle on which Einstein had built his theory of gravitation.

From Dicke's point of view, the large numbers of the orders 10^{40} and 10^{80} could not have been much different. 'The age of the universe, "now", is not random but conditioned by biological factors', he said.[12] Humans could not have evolved had the age been much smaller, nor would they exist if the age was much greater. These comments were made in passing, in papers dealing with other subjects, and they did not attract any attention at the time. The following year Dicke repeated and amplified his argument that Dirac's reasoning contained a logical loophole since it assumed that the epoch of humans is random. From astrophysical estimates he argued that the epoch is limited by the lifetime of stars and the time it takes to produce carbon and distribute it in the surrounding space: 'The present epoch is conditioned by the fact that the biological conditions for the existence of man must be satisfied.'[13] According to Dicke, the present value of the Hubble time should be understood not as a result of the large number hypothesis but as a consequence of there being at least one habitable planet with human life. At the time he did not see his anthropic consideration as a clear alternative to the variability of physical constants.

In a later paper of 1961 Dicke considered the two large numbers,

$$N_1 = \frac{\hbar c}{GM^2} \approx 2 \times 10^{38} \text{ and } N_2 = \frac{T}{\hbar/Mc^2} \approx 10^{42}$$

where M denotes the mass of the proton. The first number is the inverse dimensionless gravitational coupling constant, which can also be written in terms of the Planck mass as $(m_{Pl}/M)^2$. The second number is the age of the universe T expressed in a dimensionless form. The numbers, including Planck's constant, are not quite the same as those Dirac had considered, but the difference is of no consequence. What matters is that they are both 'large' in Dirac's sense and that $N_1 \sim N_2$. Dicke now noted that 'carbon is required to make physicists', or that the present age is characterized by the existence of carbon and other elements heavier than helium. Without these elements, human observers would evidently not be around. The order of magnitude of the lifespan of a main sequence star can approximately be calculated as

$$T_{star} \sim \frac{\hbar}{Mc^2} \left(\frac{GM^2}{\hbar c}\right)^{-1}$$

Because $T \sim T_{star}$, this agrees with the result derived from Dirac's large number hypothesis, except that now it comes out as a consequence of our existence as observers. In this sense the relation is self-selected. More specifically, Dicke concluded that 'with the assumption of an evolutionary universe, T is not permitted to take one of an enormous range of values, but is somehow limited by the biological requirements to be met during the epoch of man'.[14] Notice that Dicke did not claim to have explained the age of the universe, only to have given an alternative explanation of the puzzling coincidence $N_1 \sim N_2$. Nor did his argument include any indication at all for some teleological design of the universe. Although Dicke did not explain the Hubble age, he did offer an argument that its value cannot be arbitrary. Far from being a deductive or predictive argument based on a fundamental hypothesis, in a sense he explained the past from the present, namely the undeniable existence of humans. This mode of reasoning, so different from the deductive mode traditionally used in the theoretical sciences, is characteristic for the methods associated with the anthropic principle.

As to the smallness of the gravitational coupling contant, Dicke appealed to Mach's principle, although realizing that 'this may not be a very satisfactory answer'. He concluded his brief paper by noting that Dirac's numerical coincidences could be explained by 'the existence of physicists now and the assumption of the validity of Mach's Principle'. The existence of *physicists* was presumably not essential to his argument. During the following years Dicke returned several times to Dirac's hypothesis, but without calling attention to the role of human observers. He suggested that the size of the gravitational coupling constant might be understood on the basis of Mach's principle and that this might imply a gravitational constant varying in time, but not following Dirac's expression $G \sim 1/t$.[15]

It was Brandon Carter, an Australian-born student of Sciama, who coined the name 'anthropic principle' and elevated it to such a status that cosmologists began to take it seriously. Carter had for some years been occupied with trying to understand the role of microphysical parameters in cosmology, and in 1967, while a 25-year-old PhD student at Cambridge University's Department of Applied Mathematics and Theoretical Physics (DAMTP), he wrote an extensive manuscript on the subject. Only the first part of the work, entitled 'The significance of numerical coincidences in nature', appeared in the form of a stencilled preprint that circulated among a small number of physicists without attracting much attention.[16] He also gave his first talks on the subject in Cambridge. It was Carter's intention to extend the notes, which mainly dealt with fine-tuning coincidences in nuclear physics and astrophysics, with a separate part dealing with cosmology, but this second part was only completed several years later. Carrying the title 'Large numbers in astrophysics and cosmogony', it only circulated as a regular DAMTP preprint after 1973.[17] The purpose of Carter's preprint of 1967 was 'to clarify the significance of the famous coincidence between the Hubble age of the universe and a certain combination of microphysical parameters',[18] that is, the relationship $N_1 \sim N_2$ considered by Dicke in 1961. However, he never got that far. The parameters he considered to be fundamental were the coupling constants of the electromagnetic, strong, and gravitational interactions, and the mass ratios of the proton, neutron, electron, and pion.

From the point of view of the later anthropic principle, what is interesting about the notes is not so much their content as what they do *not* contain. Surprisingly, Carter was at the time unacquainted with the works of Dicke, and consequently he did not follow up on Dicke's line of reasoning or mention Dirac's large number hypothesis. Nor did he formulate an anthropic principle or some general perspective like it, and he did not suggest any connection between the numerical coincidences and human observers. In short, there is little in the notes of 1967 that points toward the later anthropic principle.

Carter only came to know about Dicke's earlier works in 1968, when he spent some time as a postdoc in Princeton where Dicke was working.[19]

The first time Carter had an opportunity to present his more elaborated ideas was at a meeting at Princeton in 1970 commemorating the works of the British mathematician William Kingdon Clifford. This meeting was organized by Wheeler, with whom Carter had stayed as a postdoc in the spring of 1968. His interaction with Wheeler, both then and later, was important to the line of thought that eventually resulted in the anthropic princple. On 21 February, Carter presented a set of lectures notes on 'Large numbers in astrophysics and cosmology' in which he outlined some of the ideas that came to be known as the anthropic principle (Fig. 9.1). By that time he had become acquainted with Dicke's earlier work and also with Dirac's large number principle. While following an undergraduate course on stellar astrophysics in Cambridge he had learned about Dirac's hypothesis from Bondi's textbook *Cosmology*, which some years later motivated him to think of an alternative to it. Carter rejected Dirac's conclusion as 'an error of blatant wishful thinking', as he later expressed it.[20]

Although neither the notes of 1967 or of 1970 were published, their content was known to at least some cosmologists before 1974, when Carter, motivated by his reading of Dicke and Dirac and, not least, his discussions with Wheeler, finally

Fig. 9.1. Participants in the 1970 Clifford Memorial meeting in Princeton. In the front, from the left, John Wheeler, Robert Dicke, Eugene Wigner, Edwin Power with a picture of Clifford, Stephen Hawking, Brandon Carter and Cecile DeWitt. Standing behind Hawking and Carter are Charles Misner, Bryce DeWitt and Freeman Dyson.

Source: Courtesy B. Carter.

published his ideas. This occurred in the proceedings of a meeting of the International Astronomical Union held in Cracow 10–12 September 1973 and dedicated to the 500th anniversary of the birth of Copernicus. Wheeler, who chaired one of the sessions, suggested that Carter presented his ideas on the place of human observers in the universe, which he did. Introducing Carter's lecture, Wheeler said that 'The considerations of Hawking, Dicke, and Carter raise the question whether man is involved in the design of the Universe in a much more central way than one can previously imagine.'[21] It was an extended version of this lecture that appeared in print the following year. Much later Carter told of his reasons for making a public announcement of the anthropic principle:

> My motivation in bothering to formulate something that was (as I thought) so obvious as the anthropic principle in the form of an explicit precept, was partly provided by my later realisation that the source of such (patent) errors as that of Dirac was not limited to chance oversight or lack of information, but that it was also rooted in more deep seated emotional bias comparable with that responsible for early resistance to Darwinian ideas at the time of the 'apes or angels' debates in the last century.[22]

In his article on 'Large number coincidences and the anthropic principle in cosmology' Carter objected to an uncritical extension of the so-called Copernican principle, the doctrine that we do not occupy a privileged place in the universe.[23] Although there is indeed no such privileged place, there is a privileged time—contrary to the perfect cosmological principle of the steady-state theory—namely the epoch of life. Like Dicke, Carter was convinced that the large number coincidences were not evidence in favour of 'exotic theories' such as those proposed by Dirac and Jordan. 'I am now convinced', he said, that 'these coincidences should rather be considered as confirming "conventional" (General Relativistic Big Bang) physics and cosmology which could in principle have been used to predict them all in advance of their observation.'[24] He proposed that the coincidences could be understood by using 'what may be termed the *anthropic principle* to the effect that what we can expect to observe must be restricted by the conditions necessary for our presence as observers'.[25]

According to this 'weak' anthropic principle (WAP), human observers are a kind of measuring instrument; it is necessary to take into account our special properties when interpreting data, just as it is necessary for other measuring instruments. In addition to the weak selection principle, Carter also introduced a 'strong' version (SAP) stating that 'the Universe (and hence the fundamental parameters on which it depends) must be such as to admit the creation of observers within it at some stage'. He added, 'To paraphrase Descartes, "Cogito ergo mundus talist est."'[26]

The weak and strong anthropic principles introduced by Carter were the ones that proved to be of scientific interest, although the weak version is often seen as trivial and the strong version has generally been met with incredulity. The SAP is neither a principle nor a theory of physics, but merely a proposal put forward for consideration. The attitude of many physicists is that, whereas the weak principle is common sense, the strong principle is nonsense. Murray Gell-Mann considered the strong anthropic principle 'so ridiculous as to merit no further discussion', and Steven Weinberg has called it 'little more than mystical mumbo jumbo'.[27] Even Frank Tipler, an advocate of the strong proposal, found it necessary to 'warn the reader that any version of SAP is VERY speculative!'[28] Apart from these two versions there are many other formulations—according to one account there are no less than thirty![29]

Suffice it to say that in the mid-1970s Wheeler suggested a 'participatory anthropic principle' (PAP), which can be boiled down to the claim that observers are necessary to bring the universe into existence. Wheeler, who spoke of the universe as a 'self-excited circuit', conjectured that the requirement of observability will ultimately be sufficient to determine the laws of physics completely. His idea of participatory observers that give reality to the universe, which is obviously of a speculative nature, has not found much favour among scientists. Neither has the 'final anthropic principle' (FAP), stating that once life—or what anthropic scientists often refer to as 'intelligent information-processing'—has been brought into being, it will never die out. The final anthropic principle, also known as the eternal life postulate (ELP), assumes that the universe is endowed with a teleological imperative that ensures that intelligent life will continue for ever. According to Martin Gardner, a more appropriate acronym would be CRAP— the completely ridiculous anthropic principle.[30]

The word 'must' in Carter's formulation of the strong anthropic principle was problematic. Did he really mean that the universe *had* to be such as to make observers inevitable? If so it could at best be understood as a metaphysical claim. Carter did not offer any help for interpretation in his paper of 1974, but he did point out that by itself the strong anthropic principle was unable to explain things. As Barrow noted some years later, if restricted to a single universe it suggested design and was therefore 'religious in nature'.[31] Carter suggested that his weak anthropic principle would only have explanatory power if associated with the idea of a world ensemble, the assumption of many universes with all possible combinations of initial conditions and fundamental constants. These universes could not be mere possibilities, with our world as the only one actualized, for in that case it would be hard to avoid a teleological interpretation of the anthropic principle. Carter therefore assumed the other universes really existed. Although such a many-worlds hypothesis might seem 'philosophically undesirable', he said, it 'does not really go very much further than the Everett doctrine'.[32] Whereas Carter suggested that the fundamental parameters might vary from one universe to another, he did not admit the possibility that they might vary within our own universe.

Carter clearly believed that his new anthropic principle was a valuable scientific approach, but he also realized that it was extraordinary and potentially problematic. 'I would personally be happier', he said, 'with explanations of the values of the fundamental coupling constants etc. based on a deeper mathematical structure in which they would no longer be fundamental but would be derived.'[33] In later discussions, critics of anthropic explanations would accuse them of being cheap substitutes for explanations of the traditional, deductive-nomological kind. For example, if we want an explanation of why the Earth is covered with an ozone layer it may be argued that if such a layer, shielding the inhabitants of the Earth from lethal ultraviolet radiation, did not exist, advanced life would not have evolved. While this is an explanation of sorts, obviously it is not a satisfying one. If it is accepted as a valid explanation, why look for another explanation in terms of complex physical and chemical processes in the atmosphere? Carter was aware of the problem. 'An anthropic prediction', he wrote, 'will not be completely satisfying from a physicist's point of view since the possibility will remain of finding a deeper underlying theory explaining the relationships that have been predicted.'[34]

Among the physicists and cosmologists who were aware of and referred to Carter's anthropic ideas before the 1974 publication were Martin Rees, Freeman Dyson, John Wheeler, Barry Collins, and Stephen Hawking. Some of them (Dyson, Wheeler, and Hawking) had participated in the Clifford meeting at Princeton in 1970 at which Carter had talked about his new approach (Fig. 9.2). In a 1972 review of the possible time variation of the constants of nature, Dyson referred to Carter's 'principle of cognizability', namely 'the conclusion that the presence in the universe of conscious observers places limits on the absolute magnitudes of γ and δ and not only on their ratio'.[35] The two quantities mentioned by Dyson were the combinations of constants,

$$\gamma = \frac{GM^2}{\hbar c} \quad \text{and} \quad \delta = \frac{H\hbar}{Mc^2}$$

that is, the inverse of the quantities earlier considered by Dicke: $\gamma = 1/N_1$ and $\delta = 1/N_2$. The same year, while participating in a meeting in Trieste dedicated to the seventieth birthday of Dirac, Wheeler mentioned 'the explanation of Brandon Carter that many cycles of the universe are possible and the constants in this particular cycle are such as will permit life'.[36] Similar remarks appeared in a book of 1974, coauthored by Rees and Remo Ruffini, where Wheeler expressed his ideas of a universe selected by the presence of man. Referring to Dicke, he wrote:

> ...the right order of ideas may not be, here is the universe, so what must man be; but here is man, so what must the universe be.... So why on this view is the universe as big as it is? Because only so can man be here! In brief, the considerations of Carter and Dicke would seem to raise the idea of the 'biological selection of physical constants'. However, to 'select' is impossible unless there are options to select between. Exactly such options would seem for the first time to be held out by the only over-all picture of the gravitational collapse of the universe that one sees how to put forward today, the pregeometry black box model of the reprocessing of the universe.[37]

In fact, in his published article of 1974, Carter did not refer to the idea of a cyclic or oscillating universe but only to an ensemble of universes. Yet Carter was aware of the cyclic possibility, which he contemplated in discussions with Wheeler in the years 1968–72.[38]

Anthropic Science

PRINCETON UNIVERSITY
JOSEPH HENRY LABORATORIES

CLIFFORD DAY CELEBRATION

21 February, 1970

Theme: Where do we stand today, and what developments can we look forward to, on Clifford's vision of particles as made of curved empty space?

All sessions will take place in the Professors' Room, 218 Jadwin Hall

9:00 am William Kingdon Clifford   Edwin Power
                                      University College,
                                      University of London
         The Vision of Clifford and Einstein   John A. Wheeler,
                                      Joseph Henry Laboratories
                                      Princeton University
         The Large Numbers   Brandon Carter,
                                      Institute for Theoretical Astrophysic
                                      Cambridge University
         Discussion

12:30 pm Lunch Presidential Dining Room, Prospect

2:00 pm Position Statements of Participants

5:00 pm Cocktails Prospect

6:00 pm Dinner Room E, Prospect

7:30 pm Adjournment

Fig. 9.2. The beginning of the anthropic principle. On 21 February 1970 Brandon Carter gave his first presentation on large numbers in nature, interpreting them in terms of the existence of human observers.

Source: Courtesy B. Carter.

In 1973, the year of Carter's Cracow lecture, Collins and Hawking used what they called the 'Dicke-Carter idea' to come up with a 'most attractive answer' to the question of why the universe has such a high degree of isotropy. They reasoned that anisotropic universes evolved towards being highly anisotropic and that this would preclude the formation of galaxies. Since the existence of galaxies is presumably a necessary precondition for the development of intelligent life, our universe must be isotropic. To turn this observation into a kind of explanation they adopted the anthropic approach with its hypothesis that 'there is not one universe but a whole infinite ensemble of universes with all possible initial conditions'. Collins and Hawking concluded that life would be possible only in a tiny subset of the ensemble of universes. 'The fact that we have observed the universe to be isotropic is therefore only a consequence of our existence', they wrote. The two physicists repeated the provoking conclusion at the end of their paper: 'The answer to the question "why is the universe isotropic?" is "because we are here".'[39] Hawking gave a shorter presentation of the paper at the Cracow IAU symposium in which he repeated the formulation, but made it clear that what he really meant was that 'the isotropy of the Universe and our existence are both results of the fact that the Universe is expanding at just about the critical rate'.[40] Clearly, there is a great deal of difference between the two formulations.

The anthropic principle made its entrance into physics and cosmology in the late 1970s, at first without much fanfare and not with a great deal of controversy. In an influential review of anthropic structural aspects of the universe, published in *Nature* in 1979, Bernard Carr and Rees brought together all the anthropic arguments known at the time. The article did much to make scientists acquainted with the anthropic principle as a possible tool of science. However, although Carr and Rees found the principle to be greatly interesting, they expressed themselves cautiously. 'From a physical point of view, the anthropic "explanation" of the various coincidences in nature is unsatisfactory', they wrote, repeating Carter's evaluation from five years earlier. They added that it 'may never aspire to being much more than a philosophical curiosity'.[41] In his Milne Lecture of 1980 (Fig. 9.3), Rees expressed himself in a similar hesitant way about the scientific nature of explanations based on the anthropic principle: 'At best it can offer a stop-gap satisfaction of our curiosity regarding phenomena for which we cannot yet obtain a genuine physical explanation.'[42]

The anthropic principle, in one or more of its several versions, soon appeared also in more extensive reviews and popular books, such as the English physicist Paul C. W. Davies' *Other Worlds* from 1980 and *The Accidental Universe* from 1982. The new way of thinking about the universe was further disseminated by articles in popular science journals, including *Scientific American* and *Sky and Telescope*.[43] The publication in 1986 of John Barrow and Frank Tipler's comprehensive *The Anthropic Cosmological Principle*, still unsurpassed in detail and perspective, as well as speculation, marked the maturity of the anthropic principle and its widespread acceptance in parts of the scientific community. Collaborating with John Gribbin, an astrophysicist and science writer, in *Cosmic Coincidences* of 1989 Rees covered at a popular level many of the same subjects dealt with by

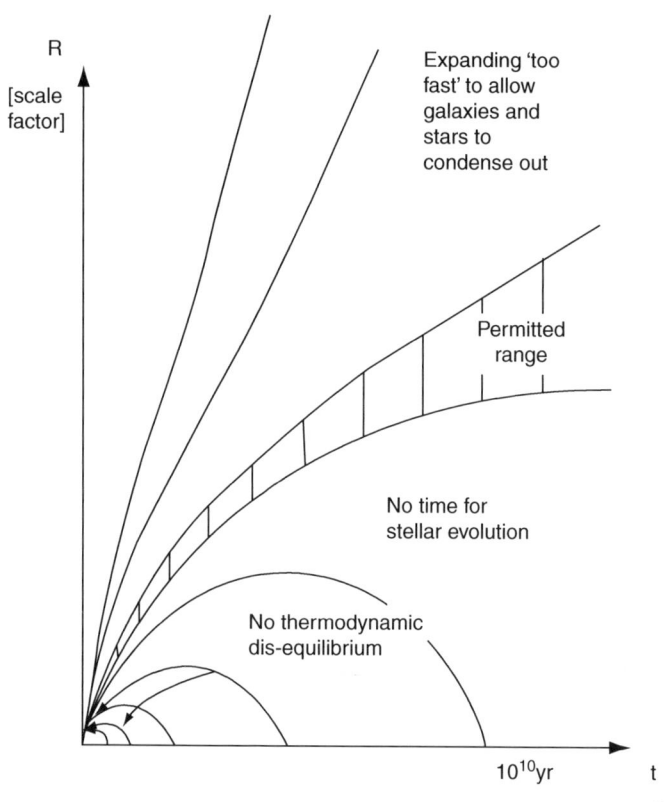

Fig. 9.3. A simple illustration of weak anthropic anthropic reasoning, as offered by Martin Rees in his Milne Lecture of 1980. If the initial expansion rate of the universe had been slow, it would start contracting before stars could be formed or perhaps even atoms could be formed. If the expansion is much faster than the critical rate, matter would have receded at such a high speed that it would not have condensed into stars and galaxies. Only for a range of initial conditions lying close to the critical value $\Omega_0 = 1$ will it be possible for complex structures to form and hence for life to evolve: in this respect (as in many others), the universe seems to be fine-tuned for life.
Source: Rees 1998.

Barrow and Tipler. The two authors summarized they key anthropic questions as follows:

> The conditions in our Universe really do seem to be uniquely suitable for life forms like ourselves, and perhaps even for any kind of organic complexity. But the question remains—*is* the Universe tailor-made for man? Or is it ... more a case that there is a whole variety of universes to 'choose' from, and that by our existence we have selected, off the peg as it were, the one that happens to fit? If so, what are the other universes, and where are they hiding?[44]

In 1988, the first international conference devoted to the study of the anthropic principle and its implications took place in Venice with participation of Carter, Sciama,

Hoyle, Ellis, Barrow, and others.[45] The following year another international meeting took place in St Petersburg (then still Leningrad), where a wide range of aspects related to the anthropic principle was discussed by Russian and invited foreign scientists.[46] By that time anthropic reasoning had become part of cosmology, if still a small and definitely a controversial part.

It is of course possible to appreciate the sensitivity of structures of matter to small changes in the numerical values of the fundamental parameters without accepting the anthropic principle. Fine-tuning is not necessarily an argument for the special position of intelligent life forms. This is what the Russian astrophysicist Iosif Rozental argued in a review of 1980 concerned with the effects that a hypothetical change of the constants would have. Covering much of the same ground as Carr and Rees, he found it unjustified to conclude in favour of the anthropic principle or otherwise to highlight complex biological structures. As Rozental emphasized, the connection between the constants of nature and the phenomenal world of physics occurs already at the lower levels, such as in chemical compounds, atomic structure, and the stability of nuclear matter. Consequently he advocated a kind of non-anthropic anthropic principle, what he called the 'principle of effectiveness' (and which Leibniz would have appreciated). The idea was that 'our basic physical laws, together with the numerical values of the fundamental constants are not only sufficient but also necessary for the existence of ground states'.[47] These ground states could range from atomic nuclei to galactic clusters, but Rozental saw no reason to single out biological or neurological structures as particularly interesting.

Parts of Rozental's criticism received support from a different perspective by P. J. Hall, a British philosopher who pointed out that many anthropic explanations assumed a strong form of reductionism. In addition, they disregarded the possibility of other life forms. 'To argue that the existence of life is impossible if the electronic charge were 1.603 [$\times 10^{-19}$] coulomb instead of 1.602 [$\times 10^{-19}$] coulomb is as absurd as deducing the value of the electronic charge from a study of biology.'[48] According to Hall, the Dicke–Carter anthropic principle had no proper explanatory power.

As Carter pointed out in 1974, anthropic explanations have force only in the context of the multiverse (a term still to be coined), the hypothesis of a whole range of hypothetical universes with varying properties. The association of the anthropic principle with the multiverse is not a necessary one, however, and it was not generally accepted in the 1970s and 1980s.[49] Only later did it become common to see an intimate connection between the two concepts, with the multiverse explaining anthropic fine-tuning (which is discussed further in Chapter 11). Many scientists were and are uncomfortable with postulating many universes to explain some of the properties of the one we live in. It does not strike them as a very economic approach. Rather than thinking in terms of many universes, one could think of a single oscillating universe and conceive the anthropic principle as related to its various cycles or temporally following universes. This idea, which had been briefly mentioned by Wheeler in 1972, did not appeal to most anthropically minded physiciststs. It did, however, appeal to Dicke, who had been interested in cyclic models of the universe for a long time. In 1982 he explained how he saw the connection between the cyclic universe and the anthropic principle:

> Suppose we have just one universe, but one that oscillates. It could be very nearly flat, just barely closed.... After many oscillations, the universe might contain many particles. It might then expand to the size of a walnut and collapse in about a millisecond. But it would bounce, and on each oscillation there would be a new and bigger universe. And somewhere down the line you would finally get to a universe big enough, with a long enough time scale, and we would exist.[50]

Another problem, apart from the many universes, that turned up at an early date and helped making the anthropic principle controversial, was its implicit teleological nature and apparent connection to religious modes of thinking. Physicists were not the only ones to take the anthropic principle seriously; so did philosophers and theologians. As Paul Davies pointed out in a review article of 1983, the strong anthropic principle represents a radical departure from the conventional concept of scientific explanation. It claims that the universe is somehow constructed as if living organisms, and more specifically intelligent life forms such as humans, were its very *purpose*. 'In this respect the strong anthropic principle is akin to the traditional religious explanation of the world: that God made the world for mankind to inhabit.'[51]

The same controversial association between teleology and the anthropic principle was made by Carr in a talk to a conference of 1982 on 'Cosmos and Creation' arranged by the Science and Religion Forum at the University of Surrey. Unusually for an astrophysicist, the title of his address referred to the 'purpose of the physical universe'. After a careful and sympathetic survey of the cosmic coincidences making up the evidence for the anthropic principle, Carr mentioned possible explanations in terms of either a many-worlds or a many-cycles universe, but found these to be 'rather bizarre'. It might very well be, he suggested, that no ordinary physical explanation could be found for the coincidences. Then what? 'One would have to conclude either that the features of the universe invoked in support of the Anthropic Principle are *only* coincidences or that the universe was indeed tailor-made for life. I will leave it for the theologians to ascertain the identity of the tailor!'[52]

As multiverse ideas were not initially seen as strongly connected to the anthropic principle, so the inflationary universe was not generally considered to constitute support for the principle. One can argue, as some cosmologists did, that inflationary cosmology tends to make anthropic explanations of, for example, the flatness and isotropy of the universe redundant. The mechanisms that inflation theory invoked to explain the flatness in no way relied upon anthropic reasoning. (But, of course, an anthropic-minded physicist might argue that these mechanisms will only work if the values of the associated parameters are anthropically constrained.)

The anthropic principle, so different from most other principles of science, was not welcomed by the majority of astronomers and physicists. William Press, an astronomer at Harvard University, saw the principle as a 'resurgence of teleological belief in science' which was 'threatening to the modern scientific enterprise'.[53] Malcolm Longair, a mainstream astrophysicist at Cambridge University, referred briefly and critically to the principle in his Halley lecture of 1985: 'I dislike this theory profoundly and regard it as an absolute last resort if all other physical arguments fail. The whole essence of the argument seems to run counter to everything one aspires to achieve as a scientist.'[54] Other physicists objected to the anthropic principle because it was too ambitious, claiming to be able to answer why-questions instead of the how-questions with which scientists are traditionally occupied. The Chicago astrophysicist David Schramm did not think that anthropic reasoning was within the purview of proper science: 'There is a circularity in this sort of reasoning and it would be premature to try to attach anything physical to the coincidences', he said. 'Physics tries to answer the "how" questions, and in some sense it is a philosophical rather than physical undertaking to have a go at these "why" questions, since they are unanswerable by the techniques of physics.'[55]

The generally sceptical attitude toward the anthropic principle in the late 1980s, at a time when inflation was accepted by a majority of cosmologists, can be further illustrated by *The Early Universe*, a monograph written by two distinguished theoretical astrophysicists at the Fermi National Accelerator Laboratory (Fermilab), Edward Kolb and Michael Turner. The authors dealt extensively with inflationary models, but characteristically they mentioned the anthropic principle only in a footnote: 'Since it is possible that the realization of physical law is different in different inflationary regions, inflation may, God forbid, provide some rational basis for the anthropic principle, as inflation provides a multitude of "Universes" from which to choose.'[56]

Alan Guth would later become an advocate of the multiverse, considering it intimately linked to inflation and providing a much needed justification of the anthropic principle. But this was not yet the case in 1988, when he was interviewed about questions of cosmology. Asked about his opinion of the anthropic principle, he responded as follows:

> Emotionally, the anthropic principle kind of rubs me the wrong way. I'm even resistant to listening to it. Obviously, there are some anthropic statements you can make that are true. If we weren't here then we wouldn't be here. As far as the anthropic principle as a way of approaching things, I find it hard to believe that anybody would ever use the anthropic principle if he had a better explanation for something.... I tend to feel that the physical constants are determined by physical laws that we can't understand now, and once we understand those laws we can make predictions that are a lot more precise.[57]

Guth further distanced himself from the view that life has any special role in the physical world or that the laws of nature were contrived to allow life to exist. 'It is a rather poor way to try to determine the laws using the fact that life exists', he said. 'The anthropic principle is something that people do if they can't think of anything better to do.' There is little doubt that at the time most of Guth's colleagues in astrophysics and cosmology would have agreed with him.[58]

9.2 ANTHROPIC REASONINGS

A principle or law may be of scientific value by leading to a deeper understanding of a field, for example by presenting knowledge in a unified and coherent manner, irrespective of whether or not it leads to specific predictions. This was the strength of Darwin's early theory of evolution, of Mendeleev's periodic system, and also of Lyell's uniformitarian geology from about the same period. When quantum mechanics arrived in 1925–26, it did not include novel predictions. But of course scientists prefer that a general principle also results in testable predictions (as it did in the case of Mendeleev, who predicted several new elements). Since the beginning of the anthropic principle scientists in favour of it sought to produce such predictions in order to demonstrate the scientific value of the principle. Ever since the Collins–Hawking paper of 1973 a large number of prediction claims have been made. I shall mention only a few of the better-known cases, mainly to illustrate the nature and diversity of anthropic arguments. But first an example from the past: what was allegedly an anthropic prediction made many years *avant le mot*.

Among the least convincing so-called anthropic predictions is the claim of Barrow and Tipler that the age-of-the-Earth debate in the latter part of the nineteenth century involved such a prediction. This debate, thoroughly investigated by historians of science, focused on William Thomson's argument, based on thermodynamics and the assumption of a cooling Earth, that the Earth was only 20–100 million years old. His preferred value of 20–40 million years was unacceptable to both geologists and natural historians (including Darwin), who needed a much longer timescale to explain geological and biological evolution processes. Among Thomson's critics was the Chicago geologist and astronomer Thomas Chrowder Chamberlin, who pointed out that Thomson had ignored the possibility that unknown atomic processes, involving the transformation of atoms, might occur under the extreme conditions in the interior of the Sun. From this, Barrow and Tipler not only infer that Chamberlin made an anthropic prediction, but also that in principle his argument 'could have led to the discovery of nuclear fusion reactions much earlier'![59]

One case which appears over and again in the anthropic literature is Fred Hoyle's famous prediction in 1953 of a resonance state of the carbon-12 atomic nucleus at 7.68 MeV, at the time not known experimentally.[60] It is frequently stated that here we have an early and genuine prediction based on the anthropic principle, if of course before the principle was formulated. According to Gribbin and Rees, Hoyle's insight is 'the only genuine anthropic principle prediction' and the best 'evidence to support the argument that the Universe has been designed for our benefit'.[61] A later author repeats that

'Hoyle was rigorously applying what would later become known as the anthropic principle', asserting that 'this was the first and only time that a scientist had made a prediction using the anthropic principle and had been proved right'.[62] There are numerous statements of a similar kind in both the scientific and popular literature. Let us take a closer look at this alleged anthropic prediction.

To put it briefly, the story is that physicists at the Massachusetts Institute of Technology had looked for a resonance level in the area about 7.5 MeV without finding one, and so it was assumed not to exist. (Evidence for an excited state near 7 MeV had been reported as early as 1940, but without being confirmed.) Now Hoyle argued that only if this state existed would it be possible for three helium nuclei to fuse into a carbon-12 nucleus under the physical conditions governing some stars. As shown by the Austrian–American physicist Edwin Salpeter in 1952, the triple-alpha process occurs in two steps, first with two alpha particles forming a beryllium-8 nucleus. This nucleus is highly unstable, but fortunately with a lifetime long enough that under the right conditions it can combine with another alpha particle and produce carbon. In addition, gamma photons are produced. Not only was the carbon resonance a lucky accident, so was it that the 7.12 MeV level of oxygen-16 lies slightly below the sum of the masses of carbon-12 and helium-4. Without these two lucky accidents the carbon produced in the triple-alpha process would immediately react with another helium nucleus and transform into oxygen:

$$3\,^4\text{He} \rightarrow {}^8\text{Be} + {}^4\text{He} \rightarrow {}^{12}\text{C} + 2\gamma, \text{ followed by } {}^{12}\text{C} + {}^4\text{He} \rightarrow {}^{16}\text{O}$$

Although Hoyle's theoretical argument was at first met with scepticism, experiments made by the Caltech physicist Ward Whaling and collaborators soon confirmed the predicted resonance state, which was found to be at 7.68 ± 0.03 MeV and is now known to be 7.644 MeV.[63] It is worth pointing out that Hoyle did not originally present his argument for the 7.68 MeV level as an important prediction. In his article of 1954, where he first called attention to it, he only mentioned it briefly and as a small part of a much longer investigation of the formation of elements in stars. His argument became a famous prediction, but this is not how it was seen originally.

In the early publications on the carbon resonance neither Hoyle nor others mentioned it as a case of fine-tuning, nor did they refer to the existence of life in the universe. A lecture of 1957 on the relationship between science and religion might have provided an opportunity for Hoyle to make the connection, but in fact he did not. Hoyle discussed the possibility that 'the laws of nuclear physics are designed to promote the origin of the complex atoms, ... [and] have also been deliberately designed to promote the origin of life'.[64] Although he found the hypothesis appealing, he did not clearly support it and he did not refer to his earlier calculation of the carbon-12 resonance as a case in point. As far as I know, Hoyle first referred to life in connection with the nuclear processes generating carbon and oxygen in 1965, when he noted that had the energy levels been just slightly different, 'it is likely that living creatures would never have developed'.[65] In a textbook published ten years later he repeated the comment, adding the speculation that the balance between the electromagnetic and nuclear forces (and hence the energy levels) might 'vary from one region of the universe to other, very distant regions'.[66] In that case, life as we know it would only form in some cosmic regions and possibly only in ours. In neither of the publications did Hoyle connect his work of 1953 with anthropic considerations.

The prediction of 1953 may first have been used as evidence (*post hoc*) of anthropic fine-tuning by Carr and Rees in their paper of 1979. In 1982 Hoyle apparently gave it his support when he said about the energy levels necessary to produce carbon and oxygen: 'A common sense interpretation of the facts suggests that a superintellect has monkeyed with physics, as well as chemistry and biology, and that there are no blind forces worth speaking about in nature.'[67] Since then the case has appeared routinely as evidence of the predictive power of the anthropic principle.

Whether or not Hoyle himself came to believe that he had found evidence for anthropic fine-tuning in 1953, he did not originally see it in that way. Hoyle *might* have reasoned something like this: since life is known to exist, and life as we know it is carbon based, there must exist a 7.68 MeV resonance. In that case it would have counted as an anthropic prediction. But this was not the way Hoyle argued in 1953. In his autobiography Hoyle said that the prediction caused him to contemplate the question of whether the existence of life might be due to coincidences in nuclear physics. Perhaps 'life would perforce exist only where the nuclear adjustments happened to be favorable, removing the need for arbitrary coincidences, just as one finds in the modern formulation of the weak anthropic principle'.[68] We are not told when he began thinking along these lines, but there is no evidence to suggest that such anthropic thoughts motivated his prediction. At any rate, it is far from clear that the resonances are of anthropic significance. Thus, in a study of 1989 Mario Livio and collaborators examined the consequences of carbon production by assuming the 7.644 MeV resonance level to be different. Their computer simulations showed that a 60 keV increase 'does not significantly alter the level of carbon production in stellar environments', and that a corresponding reduction in the energy difference 'leads to a significantly greater carbon production'. The fine tuning is not very precise. Admitting that the implications for the anthropic principle 'are not entirely free from subjective feelings', Livio and his coauthors concluded that their result weakened the anthropic significance of Hoyle's old prediction.[69]

Although Hoyle was intensely occupied with the nature and origin of life, he never endorsed the anthropic principle in any of its ordinary meanings and did not find it to be of much use for cosmology. In an address at the first Venice conference on cosmology and philosophy in 1987, he gave a review of his ideas of the relations between cosmology and biology, emphasizing that the key problem was how to explain the origin of life. According to Hoyle's reasoning, it was extremely implausible that life on Earth could have occurred by chance. As to the anthropic principle, he turned it upside down: 'Until we understand it [the origin of life], much, I believe, will remain to be discovered about cosmology, for surely the occurrence of life is the largest problem of which we are aware. It is not so much that the Universe must be consistent with us as that we must be consistent with the Universe. The anthropic principle has the problem inverted, in my opinion.'[70] At the following Venice conference, dedicated to the anthropic principle, Hoyle apparently adopted the strong principle, but it was in a version quite different from the usual one, namely that 'our existence leads to a potentially falsifiable prediction in the sense of Popper'.[71] That Hoyle used the anthropic principle in his own way, is further shown by his argument that the principle was in harmony with, and could indeed be used as support of, the steady-state theory of the universe.

Among the early and more remarkable attempts to demonstrate the predictive power of the anthropic principle was a paper of 1983 in which Carter applied weak anthropic reasoning to biological evolution in the universe. Of course, he only had one such example, the evolution of life on Earth, where life has existed for nearly four billion years, a period of the same order as the lifetime of the Sun. From various arguments, allegedly based on the anthropic principle, he arrived at some remarkable conclusions. One of them was that evolution is intrinsically likely to take far longer than it has taken on our planet, another that 'civilizations comparable with our own are likely to be exceedingly rare ... so that not much credibility can be attached to the exciting fiction scenarios involving reception of extraterrestrial communications, not to mention visitations'.[72] The 'fiction scenarios' he was referring to were probably the SETI programmes that assumed the existence of advanced life in our galaxy (see Chapter 12). Apart from the practical impossibility of testing Carter's claims, it is difficult to see how they relate to the anthropic principle and justify its predictive power. Yet Carter used his arguments for just this: to demonstrate that the weak anthropic principle has 'genuinely predictive power', that is, leads to non-trivial and testable predictions. More than twenty years later he continued to think in this manner, apparently considering the so-far unsuccessful search for extraterrestrial civilizations to be evidence supporting the prediction.[73]

A much discussed case of anthropic reasoning, and one of a quite different nature, refers to the cosmological constant Λ or, equivalently, the energy density of vacuum ρ_V. In the 1980s it was generally assumed that the cosmological constant is zero or very small, but estimates based on high-energy physics failed completely to give the small value indicated by observations. The result of the calculations indicated a vacuum energy density about 120 orders of magnitude larger than the energy density of the universe! The disaster, known as the cosmological constant problem, remained after the discovery of the acceleration of the universe. No one had a clue of how to calculate from quantum theory the energy density of vacuum so that it agrees with observational bounds. This is still the situation.

In 1987 Weinberg suggested that the smallness of the cosmological constant could be understood anthropically, 'because otherwise there would be no scientists to worry about it'.[74] His argument was that the value of the cosmological constant will have a dramatic effect on the formation of large structures in the universe, such as stars and galaxies. These structures can only be formed if gravitation is not dominated by the repulsive Λ–force. If the cosmological constant is too great relative to the attractive gravitational force, the universe will accelerate and the vacuum energy will increasingly dominate over matter energy, making large gravitational structures impossible. By developing this line of reasoning Weinberg found an upper bound of $\rho_V \leq 550\,\rho_0$, where ρ_0 is the present mass density. Two years later he suggested that 'we would expect a vacuum energy density $\rho_V \sim (10\text{--}100)\,\rho_0$, because there is no anthropic reason for it to be any smaller'.[75] This supposedly qualifies as an early anthropic prediction, especially in regard of the fact that at the time there was no observational evidence for such a high value of the cosmological constant. Weinberg actually predicted from the anthropic principle that 'this [cosmological] constant should be rather large, large enough to show up before long in astronomical observations'. Well, ten years later it did turn up in astronomical observations!

So, does Weinberg's impressive argument, later developed and sharpened by Alexander Vilenkin and many other physicists, qualify as a successful anthropic prediction? A prediction of sorts it is, but according to Smolin and other critics the problem is that it is not anthropic.[76] After all, although Weinberg did refer to intelligent life, his argument has nothing to do with lifeforms, intelligent or not. He merely made a deduction from an accepted cosmological model under the empirical constraint that galaxies and stars are plentiful. If there were no life in the universe it would not have changed his argument one iota, hence it cannot reasonably be called anthropic. The general structure of Weinberg's argument was quite similar to the earlier one adopted by Collins and Hawking: from the existence of galaxies one can infer that the universe must have a high degree of isotropy, or that there are bounds on the cosmological constant. Neither of the cases included a compelling argument that life follows from galaxies.

Another problem with anthropic determinations of the cosmological constant and similar quantities is that they are statistical in nature. The arguments rely on the probability that a 'typical' observer measures a certain value of the constant in a given universe, which is supposed to be a member of an ensemble of universes. However, there seems to be no one correct way to obtain the weighting factors, and different weighting schemes give widely different results for the peak of the probability function. A study of 2006 made by Glenn Starkman and Roberto Trotta, at the Case Western Reserve University and Oxford University, respectively, found that the anthropically predicted probability of measuring ρ_V as equal to or greater than the observed value was extremely small (about $1:10^5$). The result differs dramatically from the one found by Weinberg. The two cosmologists concluded that anthropic reasoning cannot be used to explain the value of the cosmological constant and that a similar conclusion probably holds for other constants as well.[77] They expressed their critique as follows:

> In its usual formulation, the anthropic principle does not offer any motivation—from either fundamental particle physics or probability theory—to prefer one weighting scheme over

another,.... Lacking either fundamental motivations for the required weighting, or other testable predictions, anthropic reasoning cannot be used to explain the value of the cosmological constant. We expect that similar statements apply to any conclusions that one would like to draw from anthropic reasoning.[78]

While it is generally agreed that established theory does not allow a calculation of the cosmological constant in agreement with observations, far from all physicists accept probabilistic explanations of the anthropic kind. Ideas of deriving an exact value of the cosmological constant in terms of fundamental parameters, much in the spirit of Eddington, are still being considered by a few physicists.[79]

A different way of approaching the problem of an anthropic determination of the cosmological constant has been proposed by Abraham Loeb, a Harvard University astronomer, who focuses on the possibility of intelligent life on planets formed at a time when the mean density of matter was about a thousand times greater than at present. Such planets may conceivably be found in nearby dwarf galaxies. Loeb argues that, if it turns out that planets are common in these galaxies, it will considerably weaken the anthropic argument that life depends on a value of the cosmological constant close to the one observed. The test would not confirm or refute the anthropic principle, but it would change its status: 'Any such test should be welcomed by proponents of the anthropic argument, since it would elevate the idea to the status of a falsifiable physical theory', Loeb says. 'At the same time, the test should also be welcomed by opponents of anthropic reasoning, since such a test would provide an opportunity to diminish the predictive power of the anthropic proposal and suppress discussions about it in the scientific literature.'[80]

A theoretical argument of a somewhat similar kind has been proposed by the Stanford physicist Roni Harnik and two colleagues, who in 2006 studied a hypothetical universe where the weak interactions were switched off.[81] In the standard model of particle physics it is hard to understand why the weak interactions are so strong relative to gravity (the Planck scale), namely 10^{32} times stronger, a serious 'hierarchy problem' that has challenged theorists since the 1980s. One possible solution is the anthropic argument that if the weak force was weaker, observers could not evolve. For example, neutrons would decay to protons inside nuclei, making chemistry beyond hydrogen impossible. The approach of Harnik and his colleagues was different, meant to challenge this kind of anthropic explanation. By adjusting the parameters of standard cosmology and particle physics they obtained a 'weakless universe' that looked surprisingly similar to our own and in which life could presumably thrive. In such a universe the usual hydrogen-to-helium stellar processes would not work, and the periodic system would stop at iron, but an alternative organic chemistry would nonetheless be possible. Harnik and his collaborators considered their study to provide a concrete counterexample to anthropic selection of the strength of the weak force. If a weakless universe can be habitable, it would seem to imply that the weak force cannot be explained by anthropic reasoning alone.

Work within the same line of thinking has recently been performed by varying the masses of the three light quarks and investigating how drastic the changes can be while still allowing for the evolution of intelligent life. It appears that, even with considerable changes, stable nuclei of atomic numbers 1 and 6 ('hydrogen' and 'carbon') will be formed, which can form the basis of a kind of organic chemistry and potentially life. Because fine tuning is not necessary, these studies 'cast some doubt on the usefulness of anthropic reasoning'.[82] It has likewise been pointed out that the 'vacuum angle' θ, one of the unexplained parameters of the standard model, is not related to the requirement that life exists. The value is known experimentally to be $|\theta| < 10^{-10}$, but there is no reason to assume that it is anthropically bounded.

Antagonists of the anthropic principle consider work such as the studies mentioned above (by Starkman and Trotta, Loeb, Harnik, and others) undermines anthropic arguments for the values of fundamental constants. On the other hand, these and similar critiques make little impression on the protagonists, who either criticize them or tend to ignore them.

It has been suggested that the strong anthropic principle rules out almost all cosmological theories that include an infinite past, such as the steady-state theories of Hoyle, Narlikar, and others. Had this anthropic objection been raised in the 1950s it would conceivably (but not realistically) have refuted the classical steady-state model years before the cosmic microwave background did the job. The argument was first used in 1978 by Paul Davies, who pointed out that if intelligent life must evolve somewhere in the universe—and this is the contention of the strong anthropic principle—then, assuming a static universe, it is a complete mystery why life appeared in some particular era rather than before or after that era.[83] Based on this premise, Davies suggested that a universe which is static in the large is contradictory.

In a more elaborate analysis of 1982 Frank Tipler generalized the argument, concluding that it ruled out cosmologies of the Hoyle–Narlikar type as well as most other models with an infinite past. What has been called the Davies–Tipler argument is, however, extremely speculative even by the standards of modern cosmophysics. Thus, it operates with 'an intelligent, self-reproducing robot rocket probe which is capable of travelling to another stellar system and there making copies of itself, copies which then travel to other stellar systems.' (Tipler calls such a creature a 'von Neumann probe', after the mathematician John von Neumann who had argued that machines of this kind were theoretically possible.) Moreover, it is taken for granted that at least some of these probes 'have the motivations of a living being, that is to expand and reproduce without limit'.[84] One wonders if these are really motivations of all advanced life forms, including humans.

Obviously, this kind of prediction, or rather *postdiction*, is very different from most other predictions based on the anthropic principle and hardly a convincing argument against modern cosmological models with an eternal past. As we have seen, there are several models of this kind, including the quasi-steady-state model, the Steinhardt–Turok cyclic model, the pre-big-bang model, and loop quantum cosmologies. Yet there are physicists who think that the Davies–Tipler argument 'serves as a powerful counterexample to those criticisms of anthropic principles which effectively reject them as shallow and uninformative tautologies'.[85] A more conventional anthropic explanation concerns the important discovery made by the Cosmic Background Explorer (COBE) satellite in the 1990s that the microwave background is not completely uniform but characterized by density fluctuations $\Delta\rho/\rho$ of the order 10^{-5}. Although this was in agreement with the inflationary scenario, the value did not follow from fundamental theory. So, in this sense it lacked an explanation. According to Max Tegmark and Rees, an explanation might be achieved by appealing to the anthropic principle.[86] By investigating cosmological scenarios with values of $\Delta\rho/\rho$ different from 10^{-5}, they found that life would be unable to evolve outside a range of the parameter between 10^{-4} and 10^{-6}. Anthropic selection seemed to favour 10^{-5}, which served as a kind of explanation of the measured value.

In a paper of 2001, the anthropic arguments of Linde, Weinberg, Tegmark, Rees, and others were critically examined by Anthony Aguirre at the Institute of Advanced Study in Princeton. For example, he pointed out that changing only one cosmological parameter, say the amplitude of density fluctuations, arbitrarily assumes that the other parameters will remain unaffected. However, if more than one parameter is changed, it may lead to effects in which one change is counteracted by another. Aguirre challenged the explanatory force of the weak anthropic principle in the form that 'observers will never measure a parameter to have a value which would preclude the existence of observers'. To illustrate his objections against the anthropic arguments, he constructed a class of cosmological models with parameters that differed widely from those known observationally. While the photon-to-baryon ratio is known to be about 10^9, Aguirre showed that anthropic arguments did not rule out universes with a much smaller ratio. His analysis led him to suggest that even if all the basic cosmological parameters varied by several orders of magnitude from their known values, it would not preclude in any obvious way the existence of intelligent life forms. 'This greatly complicates, and reduces the explanatory power of, anthropic arguments in cosmology', he concluded.[87]

Most of the examples mentioned so far date from before the discovery of the accelerating universe and the dark energy component. Of course, physicists and astronomers have continued to explore the anthropic principle and apply it to the more recent discoveries. To mention but one example, it has been argued that anthropic selection effects may be able to explain the small mass of the neutrinos that was first discovered in experiments from 1998.[88] The cosmic neutrino density is of the same order as the density of baryonic matter, both about $\Omega \cong 0.05$, a 'coincidence' that invites anthropic explanation. Generally speaking, modern physicists endorsing the anthropic principle deny that it is unpredictive. Although it does not lead to precise predictions of the kind known from quantum electrodynamics and other theories of conventional physics, they emphasize that it does allow predictions of a statistical nature. These can be confirmed or refuted at some specified confidence level, which shows that anthropic arguments are, after all, scientific in essentially the same sense as arguments based on conventional physics.

Among the more curious anthropic 'predictions' is one suggested by the philosopher John Leslie of the University of Guelph in Ontario, who has written widely about the anthropic principle and the concept of many universes. Leslie's prediction goes as follows: 'The notion that the visible universe must be typical of Reality as a whole, and that no observational selection effect could possibly be involved here, will be laughed at in introductory courses in philosophy of science.'[89] Surely, what students are being taught in future philosophy classes is a matter of sociology or of philosophical whims of fashion, and not of how nature works.

9.3 A CONTROVERSIAL PRINCIPLE

The anthropic principle was born controversial and has remained controversial throughout its more than 30 years of existence. As Rees formulated it in an address celebrating the 60th birthday of Hawking, the anthropic principle 'makes some physicists foam at the mouth, because they hate to feel that we may end up not being able to explain everything by some unique equations'.[90] This is one reason for the controversial nature of the principle, but not the only one.

The physicists who pioneered the principle were well aware of its unusual nature, not least in its strong version, and did what they could to present it as a useful scientific tool. Other physicists resisted it vigorously and sometimes emotionally, seeing it as a betrayal of the true spirit of science. The controversy concerning the status of the anthropic principle has changed over the years, but has far from ceased. Contrary to most other controversies in the physical sciences it has attracted much interest from philosophers, which is only natural. After all, one of the central questions is whether the anthropic principle is scientific in nature or merely a 'philosophical curiosity', as Carr and Rees expressed it in 1979. Does it have the capability of leading to such consequences that, if they are contradicted by experience, the principle must fall? From both camps in the controversy it is recognized that it is concerned with a potential major shift in the methodology of science. 'The anthropic controversy is about more than scientific facts and philosophical principles', says Leonard Susskind, a prominent string theorist. 'It is about what constitutes good taste in science.'[91] To speak of 'taste' is, however, an understatement: the controversy is in part about what constitutes *legitimate science*.

Carter realized that the anthropic principle did not go well with established standards of physics, and especially not with the standards assumed by many philosophers of science. Somewhat unusual among physicists, on several occasions Carter has dealt with the nature and aims of science in a general sense. In 1983 he criticized the view that the object of science is to find universal truths and he similarly criticized 'the doctrine that scientific theories are never verifiable but only falsifiable'.[92] 'In reality', he said, 'science is more modestly concerned with providing simple, coherent and comprehensible descriptions of natural phenomena.' This was a view which fitted much better with the anthropic mode of doing science:

> Scientific theories should not be judged as true or false, but rather should be evaluated as relatively good or bad on the basis of criteria such [as] degree of accuracy, range of applicability etc. The best theories can predict results in advance, but even partial historical explanations or mere botanical classification of previously known results should not be dismissed as valueless. Applications of the 'strong anthropic principle' should be judged by the standards of this humbler, merely explicative rather than predictive category.[93]

In a later article of 1989, Carter gave a more elaborate exposition of his views of the nature and methods of science. As to the overall aim of science, he defined it as being 'to obtain a logical description of as much as possible of what has been or may be observed on the basis of as few independent assumptions as possible', a formula that most physicists would probably agree with. More controversially Carter repeated and amplified his earlier attack on 'the widespread misunderstanding that has led to undiscriminating insistence on the requirement that a theory should satisfy the requirement of "refutability"'.[94] He found it a 'logical absurdity' always to consider falsifiable theories superior to non-falsifiable theories. Because, if a consequence of a theory is confirmed, and thus turned from hypothesis into a fact, it ceases to be refutable and yet the strength of the theory increases. Carter's critique of falsificationism repeated some of the points that were made during the earlier controversy related to the steady-state theory by McVittie and others (see Chapter 5).

Without referring to either Popper or other philosophers, Carter formulated his philosophy of science in the following, somewhat convoluted way:

> In so far as 'refutability' means 'verifiable predictive output' it is certainly better than unverifiable output and much better than non-deductive assertions (not to mention deductions that are empirically false) but, far from being indispensable, such 'refutability' is definitely less satisfactory (scientifically) than an equal amount of (irrefutable) 'verified postdictive output' provided the latter is deduced logically from the same amount of independently hypothesised input information.[95]

The suggestion that physics, or some part of it, is really nothing more than 'botanical classification' was not welcomed by the majority of physicists for whom botany is not a model science. Many years later a similar analogy between natural history and anthropic science was suggested by Craig Hogan at the University of Washington, Seattle. Hogan proposed a 'principle of humility', namely that there are some properties of the world that cannot be explained deductively from a fundamental theory of physics. These properties can only be understood in some other way, say in terms of an anthropic argument. Having reviewed a number of possible anthropic predictions, he concluded: 'One is reminded of Darwin's theory, which is a powerful explanatory tool even though some question its predictive power. Anthropic arguments are vulnerable in the same way to "Just So" storytelling but may nevertheless form an important part of cosmological theory.'[96]

Many of the standard arguments against the anthropic principle appeared for the first time in an article of 1985 written by the American particle physicist Heinz Pagels, executive director of the New York Academy of Sciences. For one thing, Pagels accused the anthropic principle of being a product of 'anthropocentrism', meaning that it tacitly presupposed intelligent life to resemble humans.[97] But why should this be the case? One can easily imagine very different life forms, even some that are not based on carbon chemistry. As far as methodology was concerned, Pagels argued that the principle was neither genuinely predictive nor testable in any real sense. Without mentioning Popper by name, he claimed that the anthropic principle was 'immune to experimental falsification—a sure sign that it is not a scientific principle'.[98] Unaware that the new theory of the inflationary universe would soon become a close ally of the anthropic principle, Pagels suggested that inflation theory made many of the anthropic claims redundant. For example, the theory provided an explanation, based on laws of nature, as to why the universe is uniform, thereby leaving the Collins–Hawking anthropic argument superfluous.

Pagels clearly disliked the anthropic principle, considering it reactionary and pseudoscientific. He believed, and certainly hoped, that 'the anthropic principle will soon be relegated to its proper role: as a museum piece in the history of science, gathering dust'. Not only was the anthropic principle scientifically and methodologically questionable, it was also potentially dangerous from a wider social and political perspective. Why had anthropic reasoning, in spite of its obvious shortcomings, become so popular? According to Pagels, the reason might be found in socio-psychological rather than scientific contexts.

> The anthropic principle's simplicity accounts for some of its appeal, particularly to the growing number of scientists who write for a popular audience. It is easier to convey a simple redundancy—that we can only see what we can see—than to grapple with the abstract mathematical arguments following from the unified field theories. In many respects, the anthropic principle is the lazy man's approach to science.

Why spend years of hard labour seeking for a fundamental explanation of the nature of things, uncertain if the search will ever succeed, if anthropic arguments might produce an easy answer? As if to insult believers in the anthropic principle, Pagels suggested that they were unknowingly participating in a quasi-theistic project. After all, was there any essential difference between the anthropic principle and the teleological argument for a divine creator of the universe? The anthropic principle, so Pagels suggested, 'is the closest that some atheists can get to God'.

Many of the critical points raised in Pagels' article were repeated and elaborated by later authors, both physicists and philosophers. For example, more than 20 years later Steinhardt and Turok objected to the use of anthropic arguments in basically the same way. Pagels' prophecy of 1985, that the anthropic principle would end up as a dusty museum piece, turned out to be wrong. Quite the contrary, with eternal inflation and landscape string theory anthropic physics became more popular and closer to scientific respectability than ever. In spite of the changes, many of the basic objections against the anthropic principle have remained the same. 'Science', Steinhardt and Turok wrote, 'should remain based on the principle that statements have meaning only if they can be verified or refuted. Ideas whose assumptions can never be tested lie outside the realm of science.'[99] Steinhardt and Turok admitted that anthropic arguments could be used to produce rough predictions of physical quantities, such as was the case with Weinberg's estimate of the cosmological constant, but did not see these predictions as crucially connected to the anthropic principle. If they turned out to disagree with observations, would the anthropic principle then be dismissed as wrong? 'The anthropic principle, with its malleability and reliance on untestable predictions, is never at risk of being proved wrong.'

Like Pagels (and also like Guth and Penrose), Steinhardt and Turok thought that anthropic-style physics was a 'lazy' approach compared to the traditional way of doing physics and one that created an 'unfair competition'. Moreover, they were worried that in the long run the anthropic approach might endanger the privileged position of science in society:

> It is not possible to draw a clean line between the anthropic principle, with its reliance on untestable assumptions, and other untestable beliefs and superstitions. The long-term effect of basing theories on anthropic reasoning could be to undermine the role of science in enlightening humankind, and in steering society away from poor decisions based on myths and fallacies.

It is worth mentioning, as will be elaborated in Chapter 11, that at about the same time the South African cosmologist George Ellis spoke out against the multiverse approach in much the same way, warning against its potential effects on the public credibility of science. There is a further analogy to the situation in the 1930s, when Herbert Dingle attacked the 'cosmythology' of Eddington, Dirac, and others (see Chapter 4). Concerned about 'the general intellectual miasma that threatens to envelop the world of science', Dingle considered the aprioristic tendencies of the new cosmophysics to represent 'a mental atmosphere in which the ideas fittest to survive are not those which stand in the most rational relation to experience, but those which can don the most impressive garb of pseudo-profundity'.[100]

The current controversy over the anthropic principle is scarcely distinguishable from the one concerned with the multiverse and the landscape interpretation of string theory that will be reviewed in Chapter 11. At least some of the protagonists of anthropic reasoning only adopted the principle in the light of the landscape multiverse, as did Susskind and Guth. Likewise, some of the antagonists of this collection of ideas are primarily motivated by their dislike of the anthropic principle. This is the case with Steinhardt, according to whom the anthropic string connection is 'an act of desperation'. In 2005 he stated his view as follows:

> I don't have much patience for the anthropic principle. I think the concept is, at heart, non-scientific. A proper scientific theory is based on testable assumptions and is judged by its predictive power. The anthropic principle makes an enormous number of assumptions—

regarding the existence of multiple universes, a random creation process, probability, distributions that determine the likelihood of different features, etc.—none of which are testable because they entail hypothetical regions of spacetime that are forever beyond the reach of observation. As for predictions, there are very few, if any.... Decades from now, I hope that physicists will be pursuing once again their dreams of a truly scientific 'final theory' and will look back at the current anthropic craze as millenial madness.[101]

Stefan Hollands and Robert Wald, two theorists at Chicago University, expressed themselves more diplomatically, but they agreed that the anthropic principle scarcely qualifies as scientific. Concerning anthropic explanations, they wondered 'whether arguments of this nature should be considered as belonging to the realm of science'.[102]

At a meeting in 2003 at the Case Western Reserve University, the physicist and science writer Lawrence Krauss characterized the anthropic principle as 'a way of killing time' when physicists had no better ideas.[103] (Compare this with Guth's statement of 1988, quoted above, that anthropic physics is 'something that people do if they can't think of anything better to do'.) David Gross similarly complained that anthropic predictions are inherently vague and imprecise and that they reflected a defeatist attitude to the difficult problems of physics. Present at the meeting was also Weinberg, who supported the anthropic principle, if somewhat reluctantly and only because it appeared to be necessary for the time being. He mused that, 'Those who favor taking the anthropic principle seriously don't really like it, and those who argue against it recognize that it may be unavoidable.'[104]

The dubious reputation that the anthropic principle has in a large part of the scientific community is to some extent due to unfortunate terminology and the silly, even outrageous, formulations that occasionally appear in both the popular and more scholarly literature. To present Carter's strong anthropic principle as the deep insight that 'from the fact that we observers do exist *it follows necessarily* that observers *are not impossible* in this universe' is little more than ridiculous.[105] The catchy name that Carter coined in 1974 has proved highly successful, but it is generally recognized, and admitted by Carter himself, that it invites associations that are both unfortunate and unintended. In 1989 he proposed as a more informative and less ethnocentric name 'observer self-selection principle', but without using the term himself.[106] Human beings are not really of crucial importance to most applications of the anthropic principle, which only refer to intelligent and conscious life forms in general, if to life at all. It may even be artificial life, such as advanced self-reproducing robots, or it may be imagined creatures like intelligent amoebae based on silicon or germanium rather than carbon. Yet in many cases it is tacitly assumed that our known biochemistry is the only basis for life, an assumption sometimes called 'carbon chauvinism'.

Especially outside the physical sciences the anthropic principle is often taken to imply that the universe was teleologically designed for our kind of life, an idea that Carter and most other physicists using the principle have never endorsed. Some authors prefer to speak of the 'biophilic' or 'biotropic' principle, but the names are no more widely adopted than 'observer self-selection principle'. Nor is this the case for the more awkward alternative 'cognizability principle', a name which stresses the importance of quantities that are observable by beings with cognitive abilities. As mentioned, this name was used by Dyson as early as 1972.

Names apart, the claim that we live in a bio-friendly universe is hardly more than a postulate, an unfounded extrapolation of the bio-friendliness of the Earth. To some advocates of the anthropic principle, but far from all, bio-friendliness is largely limited to the Earth, meaning that the principle becomes anthropocentric in a strong sense. Notably, this is the conclusion of Barrow and Tipler, who in their book of 1986 argued that advanced extraterrestrial life probably does not exist and that we are alone in the Milky Way. Carter arrived at a similar conclusion. In many cases life is not really important to anthropic or biophilic arguments, which are often of a quite trivial nature. A not particularly uncharitable formulation of the weak anthropic principle can be boiled down to the statement that the world as we know it must have evolved consistently with its past. This is indeed a fairly trivial observation, but it is neither a tautology nor devoid of scientific content.[107] The universe and its history must be consistent with the existence of human life, as it must be consistent with the existence of meteors, quasars, earthworms, and superconducting metals.

The anthropic principle claims to offer explanations of certain quantities and phenomena in nature, from the neutron-proton mass difference to the age of the universe. What kind of explanation is it? When it is said, for example, that the Hubble time is explained by the existence of human observers, it is not supposed to be either a teleological or a causal statement. No physicist, however warm to the anthropic principle, has seriously claimed that the age of the universe is *caused* by our existence in the ordinary sense of causation. Many anthropic writers use the term 'explanation' in unreflective and unconventional ways, which adds to the confusion about what the anthropic principle is all about. What they mean is, most often, that our existence imposes a selection effect on what we observe, which is not what is ordinarily meant by an explanation in science.[108]

The conclusion of the Collins–Hawking paper of 1973—that the answer to the question 'why is the universe isotropic?' is 'because we are here'—suggests a causal explanation, but this is not what is intended. Nor is it a teleological explanation, a claim that the universe is isotropic for the sake of humankind. In another slightly later paper on the same subject, Hawking explained how the isotropy could be explained by the present rate of expansion being nearly equal to the critical value required to avoid recollapse. This explanation, which is completely non-anthropic, would have sufficed, but Hawking saw it fit to add human observers to his conclusion: 'Since we could not observe the Universe to be different if we were not here, we can say, in a sense, that the isotropy of the Universe is a consequence of our existence.'[109] To say that X is a consequence of Y normally means that Y is caused by X, although Hawking did not mean it in this way. It is rather to be understood in a logical sense, like 'saying that being a woman is a logical consequence of being a wife'.[110]

9.4 FINE-TUNING WITH OR WITHOUT DESIGN

The original formulations of the anthropic principle quickly multiplied and became endowed with new meanings, some of which went much beyond the scientific domain. The many and varied instances of apparently fine-tuned cosmic coincidences may suggest that in some sense the fundamental features of nature have been designed with a purpose. Intelligent life may not be a product of blind chance but the outcome of a masterplan whose details we only understand dimly and incompletely. Although this kind of teleological thinking may seem to fit nicely with the anthropic principle, scientists have generally rejected it. The large majority of physicists and astronomers who recognize the significance of anthropic fine tuning do not associate it at all with the argument of design. On the contrary, it is common to see the core of anthropic reasoning as a naturalistic and *counter-teleological* explanation of cosmic parameters and coincidences, a scientific alternative to natural theology. Advocates of the anthropic principle repeatedly insist that 'anthropic "coincidences" or "fine-tunings" do not imply (intelligent) design ... [and] serious anthropic thinking has nothing to do with anthropocentrism'.[111] The modern version of the anthropic principle, says Susskind, 'offers a *wholly scientific* explanation of the apparent benevolence of the universe'.[112]

According to this line of thinking, anthropic arguments show that what appears to be a carefully designed nature is really due to self-selection effects imposed on our observations by our own existence. In this way anthropic selection works in much the same way that Darwinian selection was seen as a counter-teleological explanation in the biological sciences. Richard Dawkins, the prominent science writer, biologist, and advocate of atheism, considers the anthropic principle to be strong evidence for his message of a world without God. For some strange reason, 'religious apologists love the anthropic principle', he notes. But they are wrong, for 'The anthropic principle, like natural selection, is an *alternative* to the design hypothesis.'[113] It completes the job that Darwin started.

However, not all thinkers agree with this interpretation of the broader meaning of the anthropic principle. Numerous scientists, philosophers and theologians have discussed the spiritual and religious implications of the principle, in some cases embracing it and in other cases reinterpreting it. Lee Smolin considers the strong anthropic principle to be just a scientific version of the old 'God of the gaps' argument, that is, the idea of invoking God as an explanation for natural phenomena that cannot be, or have not as yet been, answered scientifically.[114] In *The Life of the Cosmos*, he charged that the anthropic principle in its strong version was 'explicitly a religious rather than a scientific idea. It asserts that the world was created by a god with exactly the right laws so that intelligent life could exist.'[115]

According to William Lane Craig, a philosopher and theologian, it is possible to understand the anthropic principle in a way compatible with divine design, to see cosmic fine tuning as evidence for a creator. Properly understood, Barrow and Tipler's *The Anthropic Cosmological Principle* 'becomes for the design argument in the twentieth century what Paley's *Natural Theology* was in the nineteenth'.[116] On the other hand, another theistic philosopher, Richard Swinburne, finds the anthropic principle to be both unnecessary and obfuscating when there is a much better explanation at hand: 'The peculiar values of the constants of laws and variables of initial conditions are substantial evidence for the existence of God, which alone can give a plausible explanation of why they are as they are.'[117]

While not all theologians are comfortable with anthropically based arguments, these arguments are not generally seen as a problem for a divinely created world. John Polkinghorne, an Anglican priest and former professor of mathematical physics, denies that the anthropic principle is an alternative to design. Quite the contrary: like Craig he thinks it provides ammunition for a 'new natural theology' that serves as a complement to science rather than a rival to it. 'Anthropic considerations are but part of a cumulative case for theism', he concludes. 'I believe that in the delicate fine-tuning of physical law, which has made the evolution of conscious beings possible, we receive a valuable, if indirect, hint from science that there is a divine meaning and purpose behind cosmic history.'[118] It seems that theologians can have it the way they prefer.

Other Christian writers have argued that the ordinary, naturalistic version of the anthropic principle is incomplete. By itself it is unable to explain why our universe, since its very beginning, has been arranged in such a delicate way that it ended up accommodating observers. Some additional assumptions are needed, either in the form of the multiverse hypothesis or in the form of a purposeful design. A third possibility would be to leave it to pure chance, and merely accept that this is the way things are. In that case there is no explanation, although one can always hope that one day the coincidences will be explained by a deeper theory and thus disappear. According to George Ellis, anthropic fine-tuning is neither due to chance nor to a selection principle combined with an ensemble of universes. It is the result of a purposeful design of the universe. Ellis goes as far as to propose a 'Christian anthropic principle' that combines design with the theological notion of God. Of course, this principle is not of a scientific nature, but something like it is held to be necessary to obtain an ultimate understanding of the universe. Ellis believes that such an understanding cannot be purely scientific, but must be founded on a combination of modern science and Christian theology. This view 'has the possibility of giving a much more profound basis for the anthropic principle than obtainable from a purely scientific view, with all the restrictions on mode of argumentation that that entails'.[119]

One does not have to be an orthodox Christian (or Muslim, or Jew) in order to find the anthropic principle appealing from a spiritual and religious perspective. Among the earliest advocates of anthropic reasoning, Dyson speculated in the late 1980s that scientific and philosophical design arguments might justify the existence of a kind of universal mind. As evidence for such a mind, operating on three levels, he mentioned phenomena from quantum mechanics and neurophysiology. The third level was 'the argument from design' as related to observations of nature on a cosmic scale:

> There is evidence from particular features of the laws of nature that the universe as a whole is hospitable to the growth of mind. The argument here is merely an extension of the anthropic principle up to a universal scale. Therefore, it is reasonable to believe in the existence of a third level of mind, a mental component of the universe. If we believe in this mental component and call it God, then we can say that we are small pieces of God's mental apparatus.[120]

The design argument is alive and well in the early part of the twenty-first century. It is not restricted to the kind of argument found in intelligent design, but can also be found in connection with anthropic reasoning, although such a connection was not part of the original anthropic programme. However, design arguments, with their unmistakably theological association, are largely confined to theological and philosophical contexts. They hardly ever appear in the growing scientific literature that makes use of anthropic reasoning in an attempt to answer questions that may not be answerable within the framework of traditional physics. The large majority of scientists undoubtedly agree with Weinberg that it is scientifically unfruitful to appeal to the argument from design, a doctrine they find to be foreign not only to traditional scientific reasoning but also to anthropic reasoning. 'You don't have to invoke a benevolent designer to explain why we are in one of the parts of the universe where life is possible: in all other parts of the universe there is no one to raise the question.'[121]

Notes for Chapter 9

1. The literature on the anthropic principle is forbiddingly large. For a useful bibliography that covers many of the works up to 1990, see Balashov 1991. Much of the later literature, especially of a philosophical nature, is listed in the online bibiliography at http://www.anthropic-principle.com.

2. Review of Barrow and Tipler 1986 in *Centaurus* **39** (1987), 191–94. Although impressed by the book, my attitude was critical: 'Under cover of the authority of science and hundreds of references Barrow and Tipler, in parts of their work, contribute to a questionable, though fashionable mystification of the social and spiritual consequences of modern science. This kind of escapistic physics, also cultivated by authors like Wheeler, Sagan and Dyson, appeals to the religious instinct of man in a scientific age. Whatever its merits it should not be accepted uncritically or because of the scientific brilliancy of its proponents.'

3. Quoted in Kragh 2008, p. 10. See also Ćirković 2003b. On the history of design arguments and anthropic-like ideas, see Barrow and Tipler 1986.

4. Ćirković 2002.

5. Wallace 1903, p. 411.

6. Preston 1879, p. 462.

7. Jeans 1926, p. 40, a lecture delivered at University College, London, on 9 November 1926.

8. Bondi and Gold 1948, p. 255. Emphasis added.

9. Idlis 1982, p. 357.

10. For Idlis' claim, see Idlis 2001 and http://www-philosophy.univer.kharkov.ua/Idlis1_eng.pdf. I have not seen Idlis' paper, but according to Zel'dovich 1981 the source is *Proceedings of the Astrophysical Institute of the Kazakh SSR Academy of Sciences* 7 (1958), 39–54.

11. Biographical introduction by D. Rabounski in Zelmanov 2006, p. 8, a translation of Zelmanov's dissertation of 1944. Rabounski says that the statement is Zelmanov's principle, as given 'in his own words', but does not provide any source or documentation. See also Rabounski 2006. According to Idlis 2001, Zelmanov formulated a version of the anthropic principle in a paper of 1970.

12. Dicke 1957b, p. 375, and similarly in Dicke 1957a.

13. Dicke 1959, p. 33. Reprint of paper originally published in the *Journal of the Washington Academy of Sciences* in July 1958.

14. Dicke 1961, p. 441. Reprinted in Leslie 1990, pp. 121–24.

15. See the papers included in Dicke 1964, especially p. 80. Together with his student Carl Brans, Dicke developed a new theory of gravitation which implied a progressive weakening of gravity, but at a slower and less determinable rate than the one proposed by Dirac.

16. Forty years later Carter placed a transcript of the notes on the arXiv website together with a postscript (Carter 2007). See also Bettini 2004, which provides a useful history of the anthropic principle.

17. In Davies 1982, p. 132, Rees, Ruffini and Wheeler 1974, p. 425, and a few other sources there are references to Carter's unpublished and provisional preprint of 1968.

18. Carter 2007, p. 1.

19. E-mail from Carter to the author of 18 February 2010.

20. Carter 1989, p. 190.

21. Longair 1974, p. 289. The session chaired by Wheeler was on 'The Structure of Singularities', a theme in which Carter's address did not fit at all.

22. Carter 1989, p. 189. It is unclear when Carter reacted against Dirac's argument for a varying constant of gravitation, but it was presumably at some time about 1970. When Carter started his work on the numerical coincidences in nature, Dirac was still in Cambridge as the holder of the Lucasian Chair. Like Carter, he was a member of DAMTP, but he had no office in the building and rarely came to the department. There seems to have been no interaction between Dirac and Carter, except that Carter followed Dirac's undergraduate lectures on quantum mechanics. He recalls having tried to put questions to Dirac, but with no more luck than most other people who addressed him. Dirac was notoriously taciturn. (Personal communication from Carter, E-mail of 7 February 2010.)

23. At least from a historical point of view, the term 'Copernican principle' is unfortunate. Although Copernicus removed the Earth and hence humans from the centre of the universe in a geometrical sense, his world system was far from uniform or without privileged parts. It has been argued that the weak anthropic principle is not really contrary to what is generally known as the Copernican principle, but can be considered an instance of Copernicanism (Roush 2003).

24. Carter 1990, p. 126. The paper was originally published in Longair 1974, pp. 291–98.

25. Carter 1990, p. 126.

26. Ibid., p. 129.

27. Gell-Mann 1994, p. 212; Weinberg 2001, p. 238 and also at other occasions.

28. Tipler 1989, p. 32.

29. Bostrom 2002.

30. Gardner 1986, a spirited and critical review of Barrow and Tipler 1986. The final anthropic principle plays an important role in so-called physical eschatology, as reviewed in Chapter 12.

31. Barrow 1983, p. 149.

32. Carter 1990, p. 133. For the Everett doctrine, or many-worlds interpretation of quantum mechanics, see Chapter 11. Whereas all the worlds of the many-worlds interpretation are often claimed to be real, Carter considered as real only those worlds which can accommodate observing organisms of some kind.

33. Ibid.

34. Ibid., p. 130.

35. Dyson 1972, p. 235. Dyson was receptive to anthropic arguments. As he wrote the year before: 'As we look into the Universe and identify the many accidents of physics and astronomy that have worked together to our benefit, it almost seems as if the Universe must in some sense have known that we are coming.' Dyson 1971, p. 59.

36. Mehra 1973, p. 58. In an address at a symposium at the Smithsonian Institution in 1973, commemorating the 500th anniversary of the birth of Copernicus, Wheeler asked: 'Has the universe had to adapt itself from its earliest days to the future requirements for life and mind?' He suggested that although this was a question 'stranger than science has ever met before', it should be taken seriously and not be dismissed as meaningless. Wheeler 1975, p. 283.

37. Rees, Ruffini, and Wheeler 1974, p. 307. Wheeler's chapter entitled 'Beyond the End of Time' was adapted from two lectures delivered in 1971. The term 'anthropic principle' did not occur in the book.

38. Personal communication (e-mail from Carter of 7 February 2010).

39. Collins and Hawking 1973, p. 319 and p. 334. See also Barrow and Tipler 1986, pp. 422–30.

40. Hawking 1974, p. 285.

41. Carr and Rees 1979, p. 612. For later and more comprehensive discussions of apparently fine-tuned physical and cosmological parameters, see Barrow and Tipler 1986 and Hogan 2000.

42. Rees 1998, p. 66, the Milne Lecture of 1980. First published in *Quarterly Journal of the Royal Astronomical Society* **22** (1981), 109–24.

43. Gale 1981, a thoughtful and philosophically informed account, helped to make the anthropic principle known to a broad audience.

44. Gribbin and Rees 1989, p. 269.

45. Abramowicz and Ellis 1989; Bertola and Curi 1993.

46. Balashov 1990 is a review of the St Petersburg meeting. The great Russian cosmologist Yakov Zel'dovich adopted the 'anthropogenic principle' in a paper of 1981 in which he defended the priority of his compatriot Grigory Idlis (Zel'dovich 1981).

47. Rozental 1980, p. 296. In Rozental 1988, he dealt more fully with the anthropic principle.

48. Hall 1983, p. 446.

49. Deakin, Troup, and Grant 1983 argued that the concept of a world ensemble was unnecessary to the anthropic principle and that it only added to the speculative nature of the principle.

50. Quoted in Simmons 1982, p. 22. Dicke's rediscovery of the big bang universe in 1963–64 was based on the idea of a bouncing universe with many big bangs and big squeezes.

51. Davies 1983, p. 33.

52. Carr 1982, p. 253. Other speakers at the conference included John Polkinghorne, Michael Shallis, and Stanley Jaki.

53. Press 1986, a critical review of Barrow and Tipler 1986.

54. Longair 1985, p. 187.

55. Quoted in Simmons 1982, p. 20.

56. Kolb and Turner 1994, p. 315 (originally published 1990).

57. Interview in Lightman and Brawer 1990, p. 479.

58. Roger Penrose did. According to him, the strong anthropic principle 'tends to be invoked by theorists whenever they do not have a good enough theory to explain the observed facts' (Penrose 1990, p. 561). See also the interview in Lightman and Brawer 1990, where Penrose characterized the anthropic principle as 'a way of stopping and not worrying any further' (p. 430).

59. Barrow and Tipler 1986, p. 165, who suggest that 'Chamberlain's [sic] argument amounted to an Anthropic Principle prediction.' On the controversy over the age of the Earth, see Burchfield 1975. Nuclear fusion processes in the interior of the Sun were first proposed by Eddington in 1920, but only turned into a proper theory in the 1930s. The Chamberlin case is not the only doubtful example of early anthropic predictions mentioned by Barrow and Tipler. Another one is the claim related to the philosopher Charles Hartshorne with regard to global time (Barrow and Tipler 1986, pp. 194–95).

60. Hoyle 1954.

61. Gribbin and Rees 1989, p. 247.

62. Singh 2004, pp. 395–96. See also Chown 2003, a newspaper article commemorating the 50th anniversary of Hoyle's prediction: 'The [resonance] state had to exist, reasoned Hoyle, because life existed and life was based on carbon... To this day, Hoyle is the only person to have made a successful prediction from an anthropic argument in advance of an experiment.'

63. Hoyle et al. 1953, a brief preliminary announcement given at the meeting of the American Physical Society in Albuquerque in September 1954. Dunbar et al. 1953: 'Hoyle (private communication) explains the formation of elements heavier than helium by this process [$^8B + \alpha \rightarrow {}^{12}C + \gamma$] and concludes from the observed abundance ratios of $O^{16}:C^{12}:He^4$ that this reaction should have a resonance at 0.31 MeV or at 7.68 MeV in C.'

64. Untitled chapter in Mott 1959, p. 65.

65. Hoyle 1965b, p. 147. In 1980 Hoyle referred to the anthropic principle in the context of nucleosynthesis, but without endorsing the principle (Hoyle 1980a, p. 55).

66. Hoyle 1975, p. 402.

67. Hoyle 1982, p. 16 and slightly differently in Davies 1982, p. 118. For a critique of the anthropic interpretation of the carbon resonance case, see Klee 2002 and Weinberg 2001, pp. 235–37, who do not consider the resonance an instance of fine-tuning.

68. Hoyle 1994, p. 266. Hoyle spoke of the prediction as 'an early application of what is known nowadays as the anthropic principle' (p. 256).

69. Livio et al. 1989.

70. Hoyle 1991, p. 518.

71. Hoyle 1993, p. 85. Popperian falsifiability is not a methodological virtue commonly associated with the anthropic principle.

72. Carter 1983, p. 354. See also the detailed critique in Wilson 1994. A somewhat similar critique of the existence of extraterrestrial intelligent life was made in Tipler 1980.

73. Carter 2006, p. 177, speaking of 'the prediction that the occurrence of anthropic observers would be rare, even on environmentally favorable planets such as ours'.

74. Weinberg 1987b, p. 2607. See also Weinberg 1989. Anthropic bounds on the cosmological constant were earlier considered by the Israeli physicist Tom Banks. The models discussed by Banks 'can generate many universes and provide a basis for the application of the anthropic principle to the observed value of the cosmological constant' (Banks 1985, p. 354).

75. Weinberg 1989, p. 8.

76. Smolin 2007.

77. Starkman and Trotta 2006.

78. Trotta and Starkman 2006, p. 328.

79. Beck 2009 derives a formula for Λ in terms of h, G, m, and α that Eddington would undoubtedly have appreciated.

80. Loeb 2006, p. 3.

81. Harnik, Kribs, and Perez 2006. For a popular presentation, see Chown 2006. The study of hypothetical worlds with fundamental parameters different from what they are known to be is a well-established tradition in theoretical physics. For example, it was cultivated by Eddington, Gamow, and Hoyle and can be found farther back in time. More recently Robert Cahn at the Lawrence Berkeley National Laboratory studied worlds where electrons are replaced by muons, or protons are heavier than neutrons (Cahn 1996). Such worlds are *a priori* possible within the framework of the standard model, so why do they not exist? Cahn did not mention either the anthropic principle or the multiverse solution.

82. Jenkins and Perez 2010, p. 48. See also Jaffe, Jenkins, and Kimchi 2009.

83. Davies 1978.

84. Tipler 1982, p. 38 and earlier in Tipler 1980. See also Barrow and Tipler 1986, pp. 601–608. As mentioned above, Hoyle saw no conflict between the anthropic principle and steady-state cosmology.

85. Ćirković 2000, p. 33.

86. Tegmark and Rees 1998.

87. Aguirre 2001, p. 11.

88. Tegmark, Vilenkin, and Pogosian 2005.

89. Leslie 1989, p. 148.

90. Rees 2003a, p. 36.

91. Susskind 2006, p. 111. The nature of scientific controversies has been studied extensively by historians and sociologists of science. According to the definition suggested by Ernan McMullin, the ongoing dispute concerning the anthropic principle easily qualifies as a controversy. It is what he calls a controversy of principle (McMullin 1987).

92. Carter 1983, p. 352, who further argued that the Popperian doctrine implied that 'all existing theories are not only falsifiable, but may safely be assumed in advance to be false'. Carter was familiar with Popper's falsifiability criterion since his days as a student, but only in a simplified folklore version and without having read Popper. He was not aware of Thomas Kuhn's objections to Popper's views at the time. (E-mail to author, 7 February 2010.)

93. Carter 1983, p. 352.

94. Carter 1989, p. 185 and p. 194.

95. Ibid., p. 195.

96. Hogan 2000, p. 1160.

97. Scientific anthropocentricism has been suggested from a quite different perspective by the Israeli philosopher Mark Steiner in an attempt to explain the increasing mathematization of the physical sciences. According to Steiner, because modern physics is based in mathematics it inherently follows an anthropocentric strategy. The reason, he says, is that mathematics relies on human standards of beauty and convenience, which is an anthropomorphic policy. For the detailed argument, see Steiner 1998, who does not relate his proposal to the anthropic principle.

98. Pagels 1985, reprinted in Leslie 1990, pp. 174–80. The following quotations are from the same source.

99. Steinhardt and Turok 2007, quotations from pp. 234–37.

100. Dingle 1937b, p. 1012.

101. Edge World Question Center (http://www.edge.org/q2005/q05_print.html).

102. Hollands and Wald 2002, p. 2046.

103. Quoted in Overbye 2003.

104. Ibid.

105. Leslie 1990, p. 14.

106. Carter 1993, p. 38.

107. Roush 2003 has argued that, far from being a tautology, the weak principle is not even true, because what we observe is not necessarily restricted by the conditions required for our existence.

108. See, for example, Deltete 1993. Philosophically based critique of anthropic explanations and predictions is extensive. See Mosterin 2004 for an overview and literature.

109. Hawking 1974, p. 286.

110. Leslie 1989, p. 137.

111. Ćirković 2002, p. 252.

112. Susskind 2006, p. 11. Emphasis added.

113. Dawkins 2006, p. 136.

114. Smolin 2001, p. 197. According to Horgan 2006, p. 59, the anthropic principle is 'cosmology's version of creationism'.

115. Smolin 1997, p. 203. The accusation is unfair: neither Carter nor other physicists have explicitly advanced the strong anthropic principle as a religious idea.

116. Craig 1988, p. 393. *Natural Theology*, a book published in 1802 by the English philosopher and theologian William Paley, was hugely successful and the period's most influential attempt to prove the existence of God from the design of nature. Contrary to most earlier works in natural theology, Paley focused on biological rather than astronomical phenomena.

117. Swinburne 1990, p. 164.

118. Polkinghorne 1996, p. 92.

119. Ellis 1993, p. 398.

120. From Dyson, 'Science and Religion', first published in D. Byers, ed., *Religion, Science and the Search for Wisdom* (Boston: National Conference of Catholic Bishops, 1987). Here quoted from Worthing 1996, p. 40. In his Gifford Lectures of 1985, Dyson described himself as 'loosely attached to Christian beliefs by birth and habit but not committed to any particular dogma' (Dyson 2004, p. 5).

121. Weinberg 2001, p. 238. The essay 'A designer universe?' was first published in *The New York Review of Books* (21 October 1999), 46–48.

10

The Multiverse Scenario

Alice laughed. 'There's no use trying' she said: 'One can't believe impossible things.'
'I daresay you haven't had much practice', said the Queen. 'When I was your age
I always did it for half-an-hour a day. Why, sometimes I've believed as many as six
impossible things before breakfast.'

Lewis Carroll, *Alice in Wonderland*, 1865

Among the recent physical theories that have attracted much attention both in scientific circles and in the public arena, the theory of the 'multiverse' stands out as a particularly interesting example of a higher speculation.[1] The basic claim of this theory—that there exists a multitude of other universes with which we have no contact and never will have contact—is not new in itself; but it is new that the claim has become part of scientific discourse and won acceptance in a not insignificant part of the community of theoretical physicists and cosmologists. What used to be a philosophical speculation is now claimed to be a new paradigm in cosmological physics, meant to replace the traditional ideal of explaining the universe and what is in it in a unique way from first principles. Perhaps, says one theorist, 'we are facing a deep change of paradigm that revolutionizes our understanding of nature and opens new fields of possible scientific thought'. Another theorist confirms that 'We are in the middle of a remarkable paradigm shift in particle physics'.[2] The new multiverse physics invites a different style of science where strict predictability and ordinary testability are abandoned or given low priority. Probabilistic reasoning based on the anthropic principle is an important part of the new style.

The universe-versus-multiverse debate is an interesting case of a contemporary controversy that concerns foundational issues. Part of the controversy is philosophical in nature insofar that it deals with the definition of science, but the participants are nonetheless the physicists themselves rather than professional philosophers. It opens a window on what might be thought to be a phenomenon of the past, namely, physicists acting as natural philosophers. At the same time, bandwagon effects and other sociological mechanisms are clearly at play. 'The smart money will remain with the multiverse and string theory', announces *New Scientist*, quoting the prominent string theorist Brian Greene of Columbia University as a convert to the new kind of physics: 'I have personally undergone a sort of transformation, where I am very warm to this possibility of there being many universes, and that we are in the one where we can survive.'[3] The multiverse style of doing physics has gained momentum, but it is too early to say if it will be the framework for tomorrow's theoretical cosmology. In spite of its undecided and peripheral status (as seen from the perspective of mainstream cosmology), it is worthwhile to examine it critically within a historical context.

10.1 EARLY IDEAS OF MANY WORLDS

Speculations concerning multiple worlds, conceived in either a spatial or temporal sense, can be traced back to the pre-Socratic philosophers, when such ideas were first discussed by Anaximander and Anaximenes. Epicurus, the Greek philosopher who lived about 300 BC and who is also known for his version of atomism, believed that 'there are infinite worlds both like and unlike this world of ours'. He argued as follows: 'For the atoms being infinite in number ... have not been used up either on one world or on a limited number of worlds, nor on all the worlds which are alike, or on those which are different from these. So that there nowhere exists an obstacle to the infinite number of worlds.'[4] Much later, the fascinating idea of many worlds reappeared in a debate among scholars in the Middle Ages, who usually assumed the hypothetical other worlds to be identical or nearly identical to ours. One of the most prominent of the scholastic thinkers, the fourteenth-century Parisian philosopher and mathematician, Nicole Oresme, entertained ideas not only about worlds within worlds, but also about worlds that exist beyond our world and are concentric with it.[5] He considered the scenario a logical possibility, although he admitted that it could neither be proved by reason nor by evidence from experience. Medieval philosopher-theologians might conclude that the omnipotent God could have created a multitude of universes (since the notion was logically allowed), but they also concluded that God in his fathomless wisdom had chosen not to do so.

In the late renaissance, the notion of many worlds figured prominently in the cosmology of Giordano Bruno and since then the idea has been a standard ingredient in cosmological speculations. Bruno was convinced that the universe at large was infinite and that it contained an infinity of complete 'worlds' or solar systems, some of which he thought were entirely separated from our own world. A century later Leibniz suggested his famous hypothesis of 'possible worlds', with which he referred to the infinity of worlds that God could have created but had chosen not to actualize. Leibniz argued that God must have chosen this universe out of a multitude of other possibilities, 'for this existing world being contingent and an infinity of other worlds being equally possible, and holding, so to say, equal claim to existence with it, the cause of the world must needs have had regard or reference to all these possible worlds in order to fix upon one of them'.[6] In general he subscribed to the so-called principle of plenitude, the metaphysical idea that all that can exist actually does exist. Or, in an alternative formulation, what is not forbidden is compulsory.

The many possible worlds hypothesized by Leibniz were logically self-consistent but not as perfect as the existing one, the 'best of all possible worlds' (note that he did not refer to the Earth). Since Leibniz distinguished between possible worlds and the one and only real world, he cannot be considered a precursor of the multiverse. At any rate, when considering the early ideas of multiple worlds, such as were suggested in very different ways by Bruno and Leibniz, one should keep in mind that they were not, and were not meant to be, scientific contributions to astronomy. They were philosophical speculations, usually serving a moral and theological purpose. Yet they were commonly known and discussed, both in the contexts of theology and natural philosophy. One example among many is the famous Scottish philosopher David Hume, who in the posthumously published *Dialogues Concerning Natural Religion*, an attempt to undermine the generally accepted belief in natural theology, has one of his characters say:

> Many worlds might have been botched and bungled, throughout an eternity, ere this system was struck out; much labour lost; many fruitless trials made; and a slow and continued improvement carried on during infinite ages in the art of world-making. In such subjects, who can determine where the truth, nay, who can conjecture where the probability lies, amidst a great number of hypotheses which may be proposed, and a still greater which may be imagined.[7]

It is as if Hume anticipated the much later controversy over the multiverse. Another Enlightenment natural philosopher who speculated about the possibility of many worlds with different properties was Roger Boscovich, a contemporary of Hume, who was mentioned in Chapter 1.

Speculations about others worlds or dimensions were as common in the Victorian era as they were at the time of Leibniz and Hume. Some of them were mathematical pastimes, some were science fiction, and others again were associated with a spiritual meaning. Stewart and Tait's *Unseen Universe* of 1875, describing a world connected by bonds of etherial energy to the visible universe, belonged to the latter category (see Chapter 2). Louis-Auguste Blanqui, a French revolutionary activist and utopian communist, subscribed to the idea of an infinite and eternal universe consisting only of matter moving in space. He argued that, in a materially homogeneous and infinite universe, atoms must combine in identical structures and do so an infinite number of times. Therefore, at any given moment in time there would be exact replicas of any number of humans elsewhere in the universe, all of them performing the same actions and thinking the same thoughts. These doubles, he wrote, 'are of flesh and blood, or in pants and coats, in crinoline and chignon. These are not phantoms: they are the now eternalized'. Far from admitting that he was speculating, Blanqui claimed that his conclusions were 'a simple deduction from spectral analysis and from Laplace's cosmology'.[8] Replace 'spectral analysis' with 'string theory' and 'Laplace's' with 'inflationary', and we have a claim close to that of modern multiverse physicists.

The famous American astronomer Simon Newcomb, a professor of mathematics and astronomy at Johns Hopkins, was among the scientists who toyed with ideas of many universes, but without taking them very seriously. 'Right around us', he wrote, 'but in a direction which we cannot conceive, . . . there may exist not merely another universe, but any number of universes.'[9] Newcombe maintained that even if a fourth space dimension existed, the other universes or 'hyperspaces' associated with it would forever remain unknown to us and therefore not belong to the realm of true science.

Among the many pre-1900 speculations of multiple worlds, the one of greatest scientific impact was probably Boltzmann's idea of how the universe might locally escape the heat death predicted by the second law of thermodynamics. In the 1890s the Austrian pioneer of statistical physics argued that while the universe as a whole, implicitly supposed to be infinitely old, would be in an equilibrium state corresponding to maximum entropy, this might not be the case with all the parts of the universe. Basing his argument on the probabilistic notion of entropy he had introduced in 1877, in 1895 he developed a remarkable scenario of anti-entropic pockets in an infinite or perhaps just exceedingly large universe:

> If we assume the universe great enough we can make the probability of one relatively small part being in any given state (however far from the state of thermal equilibrium) as great as we please. We can also make the probability great that, though the universe is in thermal equilibrium, our world is in its present state. . . . Assuming the universe great enough, the probability that such a small part of it as our present world be in its present state, is no longer small. If this assumption were correct, our world would return more and more to thermal equilibrium, but because the whole universe is so great, it might be probable that at some future time some other world might deviate as far from thermal equilibrium as our world does at present.[10]

In Boltzmann's many-worlds scenario the 'worlds' were just different parts of the universe—he might have thought of different stellar systems or nebulae—not causally separated areas as in later multiverse ideas.

To jump ahead in time, relativistic cosmology changed the notion and way of thinking of many separate universes. Shortly after the introduction of the expanding universe, Eddington pointed out that the accelerated expansion of the closed Lemaître-Eddington universe with a positive cosmological constant would eventually lead to a situation with many distinct universes, although these were located in the same cosmic space: 'Objects separating faster than the velocity of light are cut off from any causal inference on one another, so that in time the universe will become virtually a number of disconnected universes no longer bearing any physical relation to one another.'[11] Here we have for the first time a scientifically sound prediction of a simple multiverse. Incidentally, in the same paper Eddington introduced the famous balloon analogy of a closed expanding universe, asking his audience to 'imagine the nebulæ to be embedded in the surface of a rubber balloon which is being inflated'.

As mentioned in Chapter 8, it is also from this period that we have the first relativistic models of a temporal multiverse, namely a cyclic universe of the kind contemplated by Richard Tolman in particular. In an investigation of inhomogeneous solutions of the cosmological field equations, Tolman was brought to consider also a different kind of multiverse, the spatial version. He observed that in such a universe, known as the Lemaître–Tolman model, there is the possibility that the universe can be open in one part of spacetime and closed elsewhere. Although inhomogeneous as a whole, the Lemaître–Tolman universe may contain independent homogeneous regions of different density and curvature. 'Some of these regions', he wrote, 'might be contracting rather than expanding and contain matter with a density and stage of evolutionary development quite different from those with which we are familiar.'[12]

Related ideas of what became known as bubble universes (the name seems to be due to Eddington) were proposed by a few astronomers and physicists in the 1960s, but without attracting much attention. The earlier mentioned Hoyle–Narlikar steady-state theory is an example. In the development of this cosmological theory the two physicists were led to consider separate and continually forming bubble universes of which our own was just one bubble among others. Hoyle even speculated that the empirically known physical constants, such as the mass ratio between the proton and the electron, might reflect the size of the particular bubble universe we inhabit and thus not be constant on the largest possible scale. 'The particular values we find for the dimensionless numbers of physics, or of some of these numbers, could conceivable belong to our locality', he said. 'If their values were different in other localities the full range of the properties of matter would be incomparably richer than it is usually supposed to be.'[13] Some forty years later a new generation of multiverse physicists would repeat Hoyle's speculation that the environment determines the physical laws and constants.

Of course, philosophers had long been familiar with the notion of other universes, if usually taken in a metaphysical rather than physical meaning. 'There is nothing necessary about a physical universe', Lewis Feuer, a young American philosopher, pointed out in 1934. 'There is no formal contradiction in supposing that the physical relationships might be other than they are. One may legitimately speculate on the possible existence of regions where different laws obtain.'[14]

10.2 THE MODERN CONCEPT OF THE MULTIVERSE

Although the term 'multiverse' can be found in the early part of the twentieth century, and possibly earlier, in a scientific context it is quite new.[15] The first time it appeared in the title of a scientific publication seems to have been in 1998 (according to the Web of Science). However, the name itself is unimportant and there are other, broadly synonymous but less catchy names such as pluriverse, megaverse, and parallel worlds. More important is the meaning of the term, which in a general sense refers to 'worlds' with which we have no causal contact and possibly never will have contact. We cannot communicate with them, nor can we receive physical signals of any kind from them. In other words, we cannot establish their existence by direct empirical means.

There are different versions or levels of the multiverse and several ways to classify multiverse theories. Although some are more popular than others, none has won general recognition. According to a simple classification, which has some merit from the point of view of history of science, one can distinguish between (1) temporal multiverse models, (2) spatial multiverse models, and (3) models with other-dimensional universes. To these may be added (4) hierarchical or fractal universes.[16] The first class comprises the cyclic models dealt with in Chapter 8.

The simplest spatial multiverse is not very exotic as we only have to refer to our own universe, assuming that it is flat and infinite and satisfying the cosmological principle of uniformity. The classical Einstein–de Sitter universe of 1932, which expands critically as $R \sim t^{2/3}$, might be an example. In this model, where the expansion decelerates, one can in principle travel to arbitrarily distant regions, which is not possible in a universe with a growing expansion. As pointed out by Eddington, the universe does not have to be open to evolve into a multiverse, as does the closed and accelerating Lemaître–Eddington model. If the universe is both open and accelerating, as we now have strong reason to believe, there will be an infinity of causally disjoint regions or subuniverses. Not only are they inaccessible, but because there is an infinity of them one meets a number of conceptual problems that appear with a realized infinite ensemble. Still, this kind of multiverse is relatively uncontroversial. It constitutes the first level in the hierarchy of multiverse models proposed by the Swedish-born MIT physicist Max Tegmark, which includes universes increasingly more exotic and different from the one we know.[17] Whereas there is only one big bang in class I, in Tegmark's class II there are many big bangs, leading to a multitude of different universes governed by low-energy 'effective laws' with different dimensionality, particle content, constants of nature, etc. On the other hand, the truly fundamental laws are the same. Level III universes are essentially those associated with the Everett many-worlds interpretation of quantum mechanics, while the ultimate multiverse of level IV comprises an infinity of 'Platonic' universes governed not only by different laws of physics but also with different mathematical structures (see below).

Several historical roots can be identified for the idea of the modern multiverse such as emerged in the early years of the twenty-first century. For one thing, although ideas of many universes were not highly regarded, they had been discussed ever since the 1930s and were known to most cosmologists. Based on very different arguments, namely that anthropic coincidences could be explained on the basis of a 'world ensemble' hypothesis, Brandon Carter introduced in 1974 the modern formulation of the anthropic principle (see Chapter 9). His argument included the hypothesis of 'an ensemble of universes characterised by all conceivable combinations of initial conditions and fundamental constants'. Carter's anthropic principle came to play a very important role in later multiverse physics, but the connection was only established in the 1990s (although it was suggested earlier). Until then the multiverse and the anthropic principle were rarely seen as naturally connected. Of greater importance for the increased interest in the multiverse was the later recognition of what is known as the many-worlds interpretation of quantum mechanics. In fact, Carter thought that the latter interpretation supported the idea of a world ensemble: 'Although the idea that there may exist many universes, of which only one can be known to us, may at first sight seem philosophically undesirable, it does not really go very much further that the Everett [many-worlds] doctrine to which one is virtually forced by the internal logic of quantum theory.'[18]

In 1957 the young American mathematician and physicist Hugh Everett, at Princeton University, came out with a radically new 'relative state' interpretation, which challenged the orthodox Copenhagen interpretation due to Bohr, Heisenberg, Rosenfeld, and others. With Wheeler as his supervisor, he stated the new theory in the form of a PhD dissertation, and later the same year a much shorter version was published in *Reviews of Modern Physics*.[19] According to the Copenhagen view, a physical system described by a superposition of wave functions cannot be assigned reality before it is measured or observed. What happens at the moment of measurement is that the quantum state collapses to the one corresponding to the observed result. The other possible outcomes—and in general there are many of these—remain unrealized: reality is brought into existence by the collapse of the wave function caused by an observation. Everett's picture was entirely different, as he argued that the puzzles of quantum mechanics (such as Schrödinger's cat) could be explained by denying the collapse. Wave functions, he said, are real and continue to be described by the Schrödinger equation; they describe events that are real irrespective of observation and intervention of human consciousness. The post-measurement states that are unobserved are no less real than the measured one. Each outcome of a possible quantum event exists in a real world, if not in ours. In his published paper Everett faced the objection that no kind of such a 'splitting' process is known empirically:

> Arguments that the world picture presented by this theory is contradicted by experience, are like the criticism of the Copernican theory that the mobility of the earth as a real physical fact is incompatible with the common sense interpretation of nature because we feel no such motion. In both cases the argument fails when it is shown that the theory itself predicts that our experience will be what it in fact is.[20]

According to the Everett interpretation, by every subatomic process the world splits, branches, or multiplies, with our own bodies and brains being parts of the ceaseless multiplication. The other worlds are not 'failed' worlds, potential but non-realized worlds in some Leibnizian sense, they are every bit as real as the one we live in. Schrödinger's famous cat is not dead *or* alive, it is dead *and* alive, if not at the same place and time. Unfortunately the other worlds are strictly disconnected from ours. Everett showed in his work of 1957 that this strange picture, based on the postulate that all quantum possibilities are real, results in the very same experimental predictions as the

Copenhagen interpretation of quantum mechanics. Thus, from an instrumentalist point of view the two interpretations were equivalent.

The amazing alternative to understanding quantum mechanics that Everett proposed only became widely known in 1970, after it was described and popularized by Bryce DeWitt, a physicist at the University of North Carolina who at the time specialized in problems of quantum gravity.[21] For this reason it is also known as the Everett–DeWitt interpretation or sometimes the Everett–DeWitt–Wheeler interpretation. (There are differences between the versions of Everett and DeWitt, but these are unimportant in the present context.) DeWitt gave more emphasis to the splitting and the many 'maverick worlds' arising from the infinities of measurements:

> This [our] universe is constantly splitting into a stupendous number of branches, all resulting from the measurementlike interactions between its myriads of components. Moreover, every quantum transition taking place on every star, in every galaxy, in every remote corner of the universe is splitting our local world on earth into myriads of copies of itself.[22]

As he remarked with an understatement: 'The idea of 10^{100+} slightly imperfect copies of oneself constantly splitting into further copies, which ultimately become unrecognizable, is not easy to reconcile with common sense.' DeWitt also intimated that the many-worlds interpretation might have testable implications for big bang cosmology, but otherwise he did not relate the new picture of quantum mechanics to the universe as studied by astrophysicists and cosmologists. Nor did Everett in his works of 1957.

By the 1980s the many-worlds interpretation was receiving increasing attention, not least among physicists trying to describe the universe in terms of quantum theory. Although it was still a minority view, as it supposedly still is, it induced physicists and cosmologists to think about other worlds as something more than just a philosophical speculation. There is no doubt that the more sympathetic response to the many-worlds interpretation was a factor in the new multiverse cosmology that emerged shortly after the turn of the century. But it is unclear to what extent this was the case and it is also not very clear if the universes of the multiverse are to be thought of in the same way as the worlds of the Everett interpretation of quantum mechanics. In a series of works starting in the 1980s Viatcheslav Mukhanov argued that the multiple quantum worlds are all real, exhibiting all possible values of physical constants and other parameters. He maintained that all conceivable processes, including those violating the second law of thermodynamics, take place in reality. Not all proponents of the Everett interpretation are willing to go that far or even to admit the physical reality of the other worlds.

According to some enthusiasts of multiple worlds, the many-worlds interpretation is the *only* logical interpretation of quantum mechanics and the two are really just different sides of the same coin. 'The discovery of quantum mechanics', says Mukhanov, 'was in fact the discovery which gave a solid scientific basis to the "Multiverse versus Universe" debate.'[23] Tegmark too believes that the existence of many universes follows as a prediction of quantum mechanics: 'Accepting quantum mechanics to be universally true means that you should also believe in parallel universes.'[24] So does Tipler, who considers the many-worlds interpretation a misnomer, since it is the only possible interpretation: 'More precisely, if the other universes and the multiverse do not exist, then quantum mechanics is objectively false. This is not a question of physics. It is a question of mathematics.'[25] In spite of such bombastic statements, far from all physicists endorse the Everett interpretation. Nor do they agree that acceptance of quantum mechanics leads to the multiverse or that the two are intimately connected or connected at all. According to critics, the many-worlds interpretation has nothing to do with either the multiverse or the anthropic principle. Besides, many physicists see no reason to adopt the interpretation of Everett and DeWitt instead of the standard Copenhagen interpretation.

At about the same time that physicists warmed to hypotheses of many universes, philosophers discussed somewhat similar ideas. However, there was little connection between the two groups. In a book of 1981, the American philosopher Robert Nozick at Harvard University introduced the 'fecundity assumption', without referring to Carter's anthropic principle or the many-worlds interpretation of quantum mechanics. According to the principle of fecundity, the answer to the question 'why X rather than Y'—for example, why do we live in the universe X rather than in some other universe Y?—is that both X and Y exist but that we happen to experience X. All possibilities are realized and the actual world is merely the world we inhabit, because it makes our existence possible. 'The hypothesis of multiple independent worlds', Nozick said, 'enables us to avoid leaving something as a brute fact, in this case, that there is something.'[26] Nozick's principle of fecundity was a multiverse idea, but apparently it had no impact on the works of the cosmologists.

Although ideas of many universes were well known about 1980, the majority of physicists and cosmologists tended to consider them heterodox, weird, and speculative. The successful inflationary theory of the early universe proposed by Alan Guth in 1981 did much to change the situation, if not immediately. This first happened with the versions of 'chaotic' and 'eternal' inflation introduced by the Russian-born physicists Andrei Linde and Alexander Vilenkin a few years later. (Vilenkin was actually born in Ukraine, then part of the Soviet Union; he emigrated to the United States in 1976 and was followed by Linde in the late 1980s.) Linde, then at the Lebedev Physical Institute in Moscow, concluded early on that after the brief inflationary phase the universe became divided into an infinity of bubble- or subuniverses. At the Nuffield Workshop on the Very Early Universe that convened in Cambridge, England in the summer of 1982, he gave a brief description of the new bubble-universe scenario and also related it to the anthropic principle:

> In the scenario suggested above the universe contains an infinite number of mini-universes (bubbles) of different sizes, and in each of these universes the masses of particles, coupling constants etc. may be different due to the possibility of different symmetry breaking patterns inside different bubbles. This may give us a possible basis for some kind of Weak

Anthropic Principle: There is an infinite number of causally unconnected mini-universes inside our universe, and life exists only in sufficiently suitable ones.[27]

Linde likened what he called the chaotic inflationary scenario to an infinite chain reaction with no end and possibly no beginning either. It followed as a consequence of the scenario that 'the universe is an eternally existing, self-reproducing entity, that it is divided into many mini-universes much larger than our observable portion, and that the laws of low-energy physics and even the dimensionality of space–time may be different in each of these mini-universes'.[28] Here we have the multiverse fully expounded. Moreover, in an important paper of 1986 he made it clear that he thought of the universe as a multiverse, or rather a tiny part of it, and, moreover, that he saw the idea associated with and supported by string theory:

> *All* types of mini-universes in which inflation is possible should be produced during the expansion of the universe, and it is unreasonable to expect that our domain is the only possible one or the best one. From this point of view, an enormously large number of possible types of compactification which exist, e.g. in the theory of superstrings should be considered not as a difficulty but as a virtue of these theories, since it increases the probability of the existence of mini-universes in which life of our type may appear. The old question why our universe is the only possible one is now replaced by the question in which theories the existence of mini-universes of our type is possible.[29]

In a book of 1987, celebrating the tercentenary of Newton's *Principia*, Linde repeated his idea of an anthropic multiverse. He concluded by stressing the difference between the new approach and the traditional one based on deductions from a fundamental theory: 'The line of thought advocated here is an alternative to the old assumption that in a "true" theory it must be possible to compute unambiguously all masses, coupling constants, etc. . . . This assumption is probably incorrect; in any case it is not necessary.'[30]

By 1990, then, there existed a variety of ideas of how multiple universes might be generated. Some of them were based on inflation theory, others on hypotheses of cyclic universes, and others again were motivated by the many-worlds interpretation of quantum mechanics.

According to the self-reproducing or eternal inflationary scenario, bubble universes will be produced constantly from regions of false vacuum and the universe as a whole, meaning the multiverse, will regenerate eternally. Vilenkin claims that inflationary cosmology, at least in the favoured eternal version, makes the multiverse 'essentially inevitable',[31] a claim supported by some other advocates of the multiverse. Guth, the primary originator of inflation theory, came to share the belief of Linde, Vilenkin, and others that inflation means multiplicity of universes. In *The Inflationary Universe*, a popular book that appeared in 1997, Guth wrote:

> If the ideas of eternal inflation are correct, then the big bang was not a singular act of creation, but was more like the biological process of cell division. . . . Given the plausibility of eternal inflation, I believe that soon any cosmological theory that does not lead to eternal

reproduction of universes will be considered as unimaginable as a species of bacteria that cannot reproduce.[32]

As Guth and others have pointed out (or claimed), in a multiverse consisting of an infinity of universes anything that can happen will happen, and it will happen infinitely often. Anything is possible, unless it violates some absolute conservation law. According to the analysis of Guth and his collaborators, inflation will go on forever in the future, but it is probably not eternal in the past. In that case a primary big bang is still part of the picture, just as it is in cyclic theories with a limited number of past cycles.

The possibility of an infinite universe, whether connected with multiverse ideas or not, has caused concern among some modern cosmologists, as it did among philosophical cosmologists in the past. In a paper of 1979, Ellis and his collaborator G. B. Brundrit pointed out that in a low-density expanding universe it is highly probable that there exists an infinity of 'worlds' with an infinity of 'populations of beings identical in number and genetic structure with that on the Earth'.[33] Although we will never be able to observe these other worlds or populations, we can be 'reasonably confident' of their existence. As a way out of the dilemma, they mentioned the possibility that the cosmological principle might not hold true for the universe at large. The same problem, but now in the context of inflationary cosmology, was examined several years later by Vilenkin and Jaume Garriga of the Autonomous University of Barcelona. Apart from its basis in inflation, the conclusion of Garriga and Vilenkin was largely the same, namely that there is an infinite number of 'unquestionably real' cosmic regions with histories identical to ours.[34] The kind of universe considered by Ellis and Brundrit, an open and uniform one, was what Tegmark later called a level I universe. By counting the number of quantum states that a Hubble volume can have, Tegmark concluded again that in some far away galaxy there will exist identical copies of you and me. In fact, there will be an infinite number of such copies, the closest one (don't worry) some $(10^{10})^{29}$ m away.[35]

The problem of an infinite universe, either in a spatial and material sense or in a temporal sense, has been considered more seriously by Ellis and his two colleagues U. Kirchner and William Stoeger, who argue forcefully against an actually existing infinite set of anything: electrons, universes, or time units. The objections against such actual or realized infinities are old and of a conceptual and logical rather than scientific nature. David Hilbert was only one of many who objected: in a lecture of 1925, he concluded that 'the infinite is nowhere to be found in reality; it neither exists in nature, nor does it provide a basis for rational thought'.[36] Finding the situation in modern cosmology, with its apparent justification of an infinity or worlds, to be 'distinctly uncomfortable', Ellis and his coauthors suggest that a cosmologically flat space is probably an abstraction that does not hold physically.[37] Most other cosmologists seem to be undisturbed by the infinity problem and consider it to be of no scientific relevance.

One should not believe that the ghost of infinity is something discovered by modern cosmologists. On the contrary, it is hard to think of an older problem. It has been discussed since the time of Aristotle, the consensus view being that actual infinities cannot exist. In many cases the rejection of real or physical infinities was theologically based, namely that infinity is a quality that belongs to God alone. For example, this was one of the reasons the brilliant mathematician (and devout Catholic) Augustin Cauchy gave for rejecting the possibility of an actual infinity. Later in the nineteenth century the question was re-examined in an original way by the German mathematician Georg Cantor in his theory of transfinite numbers. It also became part of the controversy over the heat death, which was generally seen as a realistic scenario of the future only if the universe was of finite size. One of the standard objections to the heat death was to postulate an infinite universe to which the law of entropy increase presumably did not apply. During the period of controversy, roughly 1870–1910, the possibility of actual infinities was scrutinized by many scientists, philosophers and theologians. The Catholic philosopher Constantin Gutberlet spoke for most of them when he concluded that 'it is absolutely impossible that the number of celestial bodies or atoms can be infinite'.[38] The concern of Ellis and his colleagues can to some extent be seen as a continuation of this nineteenth-century historical tradition.

But let me return to the multiverse of the modern period. Since the beginning of the twenty-first century there has been a marked change in the interest in and attitude to the multiverse scenario. Some eminent physicists have 'converted' from the idea of a single universe to the possibility of many universes, and more find it at least worthwhile to discuss the multiverse. Part of the reason has been the increased focus on the cosmological constant as a source of the dark energy that followed the discovery of the accelerated universe.

Another very important reason, apart from the inflation model, is that theoretical advances in string theory (or M-theory) have inspired confidence in the multiverse. As mentioned, the idea of relating the multiverse to concepts of string theory was suggested by Linde in 1986, but it only appeared in detailed form several years later. Similarly, the connection between string theory and the anthropic principle was largely ignored in the two last decades of the twentieth century, when string theorists tended to shun anthropic ideas. As the Dutch theorist A. N. Schellekens observed, 'The number of string papers before 2000 containing the "A-word" can be counted on the fingers of one hand.'[39] The situation only began to change with a paper of 2000 in which two string theorists, Raphael Bousso at Stanford University and Joseph Polchinski at the University of California, Santa Barbara, demonstrated mathematically that a very large number of string vacuum states might explain the size of the cosmological constant without direct appeal to fine-tuning.[40] The Bousso–Polchinski theory indicated a way to create new bubble universes somewhat similar to that of eternal inflation, but based on the fundamental string theory.

String theorists have traditionally hoped that the theory, when sufficiently developed and understood, would result in a unique compactification or in a 'vacuum selection principle' from which the one and only vacuum state describing the universe would emerge (see further Chapter 11). It now seems that this ambitious hope has to be abandoned for good. Apparently there is no unique way in which string theory can predict all the constants of nature by compactifying the six extra dimensions that are additional to the ordinary four dimensions of space-time. Each of these compactifications corresponds to a distinct vacuum state of space–time with a particular set of physical parameters, interactions, and types of particles. Such a vacuum is taken to represent a possible low-energy world with its own laws and constants of physics. This theory of a 'landscape' of universes has since 2002 been promoted and developed by many physicists, in particular by Leonard Susskind of Stanford University, one of the founding fathers of string theory.[41] Susskind's popular book *The Cosmic Landscape* appeared in 2006, significantly subtitled *String Theory and the Illusion of Intelligent Design*.

Like most advocates of the multiverse, Susskind believes that the anthropic principle plays an important and legitimate role in cosmology, but in a version that does not in any way indicate a benevolent creator. It is a general feeling among physicists that the multiverse is intimately connected with anthropic reasoning. Some believe that if it can be proved that the landscape follows from string theory, the anthropic principle will become an almost inevitable part of physics. 'The combination of inflationary cosmology and the landscape of string theory gives the anthropic principle a scientifically viable framework', Guth says.[42]

Linde has spelled out the connection from the landscape to the anthropic principle as follows:

> If this scenario [the landscape] is correct, then physics alone cannot provide a complete explanation for all properties of our part of the Universe.... According to this scenario, we find ourselves inside a 4-dimensional domain with our kind of physical laws, not because domains with different dimensionality and with alternative properties are impossible or improbable, but simply because our kind of life cannot exist in other domains.[43]

In another comment Linde has summarized the twin ideas of the multiverse and the anthropic principle, at the same time indicating the controversy these ideas have brought with them: 'In some other universe, people there will see different laws of physics. They will not see our universe. They will only see theirs. They will look around and say, "Here is our universe, and we must construct a theory that uniquely predicts that our universe must be the way we see it, because otherwise it is not a complete physics." Well, this would be a wrong track because they are in that universe by chance.'[44]

The string landscape provides the possibility of an enormous number of universes, and eternal inflation provides a mechanism for generating them. To describe the very large but discrete set of states, the word 'discretuum' has been coined. The number of different vacuum states or possible universes that come out of string theory is a staggering 10^{500} or more.[45] Sometimes the figure 10^{1000} is quoted, but what matters is that the number of universes is huge beyond comprehension. Since no reasons have been found that any of the vacua are preferred over others, it is assumed that each of the vacua is *a priori* a valid candidate for a universe. The many universes are claimed to really exist or to be parts of the 'populated' landscape. However, it is not obvious what such a reality claim implies, just as it is not obvious what the reality of the many worlds in the Everett theory means. Susskind explains: 'What physicists ... mean by the term *exists* is that the object in question can exist *theoretically*. In other words, the object exists as a solution to the equations of the theory. By that criterion perfectly cut diamonds a hundred miles in diameter exist. So do planets made of pure gold. They may or may not actually be found somewhere, but they are possible objects consistent with the Laws of Physics.'[46] Apart from being a modern reincarnation of the principle of plenitude, this is an unorthodox and problematic notion of existence which assumedly is not shared by the majority of physicists.

The meaning and scientific relevance of the string landscape remain highly controversial, and it is not even certain that the landscape exists as more than the possibility of a multitude of stable or metastable vacuum states. In 2004 three theorists at the Santa Cruz Institute for Particle Physics concluded that 'the possibility that the real world is described by one of these states appears somewhat dim'.[47] They pointed out that whereas some parameters can be tightly constrained by anthropic arguments, other parameters cannot. For example, life is believed to be consistent with a lower limit of 10^{16} years for the proton's lifetime, whereas it is known experimentally that the lifetime is greater than 10^{32} years. For this and other reasons the three physicists hesitated in accepting the 'new paradigm for scientific explanation', that is, the anthropic-multiverse paradigm.

The universes generated by eternal inflation have a common causal origin and share the same space–time, for which reason they do not form a completely disconnected multiverse. The multitude of other 'domain universes' are not accessible to observers located in our universe but are nonetheless connected, as a leaf on an oak tree is connected to all the other leaves of the tree. (However, while the leaves are largely of the same shape and size, the universes may differ widely.) This distinguishes this kind of multiverse from the more radical notion of a multiverse made up of strictly disjoint universes as proposed by Tegmark and others. It is only in the first case that regular properties across the ensemble of universes can be expected. In the absence of such regularity it seems hard or perhaps impossible to say anything about the universes on a scientific basis. As Ellis and others have argued, there is a great deal of difference between the two kinds of multiverse. The hypothesis of domain universes can be ruled out if it turns out that there never was an inflationary era in the history of the cosmos. There is no similar way to rule out the hypothesis of strictly separate universes.

Although the domain or bubble universes making up the multiverse are separate, they need not always have been entirely separate. According to some physicists, including string theorist Laura Mersini-Houghton of the University of North Carolina at Chapel Hill, there is the possibility that neighbouring universes may interact gravitationally. Perhaps our universe collided with another one shortly after the big bang, or perhaps one or more other universes exert a gravitational pull on large lumps of matter in our universe. The recent proposal of a 'dark flow', a phenomenon based on observations of the motion of galactic clusters, is seen by some theorists as a signature of the impact of another universe. Mersini-Houghton and her collaborator Richard Holman at the Carnegie Mellon University consider the dark flow to be evidence for a particular model of the landscape universe.[48] Generally, many physicists believe that cosmological observations (rather than laboratory experiments) may provide the first tests of 'new physics' such as the string landscape.

Multiverse physics, in its widest sense, leads to a surprising and entirely new conception of laws of nature and the relationship between law-bound and contingent phenomena. Physicists are used to thinking, and have reasons to think, that the fundamental laws, whether we know them or not, are the unique and first principles from which natural phenomena can in principle be calculated. This is the Einsteinian view of physics. But according to multiverse thinking there is nothing particularly elevated about the laws that govern *our* universe. They may be merely local and anthropically allowed by-laws, that is, consistent with life as we know it. From the grander perspective of the multiverse they are contingent and so are the values of at least some of the physical parameters. According to Martin Rees: 'The entire history of our universe could be just an episode in the infinite universe; what we call the laws of nature (or some of them) may be just parochial by-laws in our cosmic patch.'[49]

In principle, then, Newton's law of gravitation does not have a status any more dignified than that we assign to the 'law' that the inner planets are small and have no or few moons. Rather than accepting that the environment is determined by the laws of nature, multiverse physicists suggest that the laws are determined by the environment. This is a most radical suggestion. The idea that some physical quantities are in this sense environmentally selected and thus to be explained by means of anthropic reasoning appear in several works of modern physics, if not necessarily in connection with the multiverse hypothesis.[50]

10.3 AN ULTIMATE THEORY OF EVERYTHING?

The legendary 'theory of everything', a term that may first have turned up in the scientific literature as late as 1985,[51] is a putative theory that fully explains all natural phenomena within a unified framework of all fundamental but particular physical theories, that is, theories with a limited domain. Reflecting a significant reductionistic bias, it is taken for granted that the theory of everything (TOE) belongs to physics and not to any other science—to speak of a 'botanical theory of everything' presumably does not make sense outside botany. Of course, the very notion of a theory of everything is problematic, as a theory, in the ordinary meaning of the term, must necessarily rest on some assumptions concerning entities and concepts, these assumptions being physical, mathematical, or philosophical. It would seem that there will always be some properties that are inexplicable. In spite of such general arguments against the existence of a theory of everything, physicists have happily pursued the goal of an ultimate theory, if mostly in a somewhat less ambitious version than a truly final theory.[52]

The dream of being able to represent all fundamental knowledge about nature by a single theoretical system, or perhaps even by a single master equation, is reflected in the history of unification that started with Newton and took pace with Maxwell and his generation. We have met several attempts at establishing a unified physical framework in this book, from the vortex theory of the 1870s to Heisenberg's 'world equation' of the 1950s. However, the modern and more successful attempts at unification only occurred a little later, starting with the unified quantum theory of electromagnetic and weak interactions established by Steven Weinberg, Abdus Salam, and others about 1970. The electroweak theory was an important milestone in the unification programme, and another one was its extension to an electronuclear 'grand unified theory' some years later (Sheldon Glashow, Howard Georgi, Weinberg, and others).[53] Yet, impressive as these theories are, they are far from qualifying as theories of everything. Not only do they contain several free parameters, more seriously they do not incorporate the gravitational force. A theory of everything must, as a minimum, unify the grand unified quantum field theories with the general theory of relativity.

The most discussed and developed version of such a unified theory, and consequently the best candidate for a theory of everything in the traditional sense, is the theory of superstrings or M-theory (see Chapter 11). As we have seen, recent developments in string theory have led to the idea of a string landscape which, in conjunction with eternal inflation, has been interpreted as a theory of multiple universes. However, theories of the multiverse are not usually seen as candidates for a theory of everything, although they may turn out to be consequences of such a theory.

In a series of works starting in 1998, Max Tegmark has developed a most ambitious theory of everything with strong links to the multiverse. His idea of a mathematical multiverse is not really a theory in the traditional sense, but should rather be considered a framework for a future theory or a meta-theory, or perhaps as a philosophical claim for a meta-theory (Fig. 10.1). The basis of Tegmark's theory (which I shall nonetheless call it) is not the hypothesis of multiple universes, which is presented as a consequence of the theory and which did not appear explicitly in his first communication of 1998, written before the inflation-landscape multiverse was introduced. In this work Tegmark discussed in general terms the nature of a theory of everything and in particular the relation between what exists mathematically and what exists physically.

Tegmark's basic postulate is that the world is purely mathematical and that in the strongest possible sense, namely that physical reality *is* a mathematical structure and nothing but. This ontological claim he later called the mathematical universe hypothesis or MUH. He realizes of course that this is an unconventional view but dismisses the alternative—that the world is not completely mathematical—as 'somewhat of a resignation' when it comes to predictive power.[54] The standard view is that mathematics is

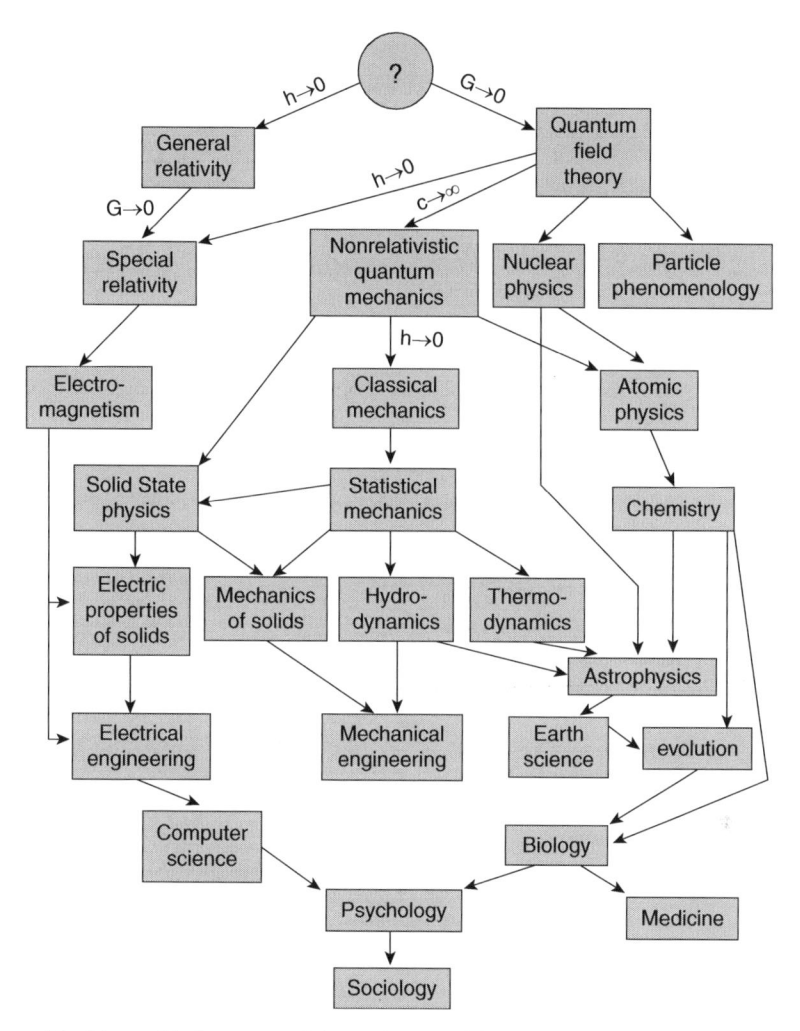

Fig. 10.1. Max Tegmark's illustration of the mathematical theory of everything from which all other theories are derivable. The degree of fundamentality is highest at the top and lowest at the bottom.
Source: Max Tegmark, 2008 'The mathematical universe' *Foundations of Physics* **38** (2008), 101–150.

merely a tool that describes and can be useful in understanding aspects of the physical world, but this is a view far from Tegmark's. His view, 'a form of radical Platonism' that he considers a candidate for a theory of everything, is that *everything* that exists mathematically is also endowed with physical existence. One's immediate response to such a claim is presumably that it is either nonsense or a freewheeling philosophical speculation that may or may not be intellectually interesting.[55] Tegmark argues, however, that it qualifies as a scientific theory because it makes non-trivial statistical predictions. Moreover, he suggests that it is superior to views that deny the mathematical nature of the world or only accept that some mathematical structures have physical existence.

Feynman found it 'quite amazing that it is possible to predict what will happen by mathematics, which is simply following rules which really have nothing to do with the original thing'.[56] At the same time Eugene Wigner, reflecting on the same amazing fact, famously problematized what he called 'the unreasonable effectiveness of mathematics in the natural sciences'. According to Tegmark, the problem raised by Feynman and Wigner (and many others) receives a natural explanation within a theory according to which the world is a mathematical structure. Within such a theory, 'our successful theories are not mathematics approximating physics, but mathematics approximating mathematics'.[57]

In later works Tegmark has developed these ideas and related them to the multiverse. As Geoffrey Chew advocated 'nuclear democracy' in the 1960s, so Tegmark advocates 'mathematical democracy' in the early twenty-first century. The general idea of reducing physical reality to mathematics is not new, of course, but it may never before have appeared in such an extreme form as in Tegmark's theory of everything. Many years earlier James Jeans described God as a pure mathematician, arguing that reality was basically mathematical. 'All the concrete details of the [physical] nature', he said, 'the apples, the pears and bananas, the ether and atoms and electrons, are mere clothing that we ourselves drape over our mathematical symbols—they do not belong to Nature, but to the parables by which we try to make Nature comprehensible.'[58] Tegmark argues similarly, but in much greater detail, that in an ultimate theory all human 'baggage'—the words and concepts we use to make sense of the equations and to relate symbols to measurements—will disappear, leaving only the bare mathematics. Not only is this baggage (or Jeans' 'clothing') redundant, so is the entire empirical domain. The theory of everything of the twenty-first century differs from Laplace's omniscient demon, but it expresses the same dream and the same kind of unrestricted hubris. 'All properties of all parallel universes...could in principle be derived by an infinitely intelligent mathematician.'[59]

In Tegmark's later expositions of his mathematical universe he argues that the commonly accepted assumption of an external physical reality independent of humans *implies* the mathematical universe hypothesis. Moreover, the multiverse is seen as following from the hypothesis because, if the mathematics is immensely richer than what we know from the physics of our universe, the majority of mathematical structures must be realized in other universes. *Ex hypothesis* there cannot be mathematical structures without physical existence. But can the mathematical universe hypothesis ever be tested? According to Tegmark it can, at least in the weak sense that there is evidence in favour of it. One piece of evidence is held to be the increasing mathematization of physics and the uncovering of new mathematical regularities in nature. 'I know of no other compelling explanation for this trend than the physical world really is mathematical.'[60]

While Tegmark argues that physics is in essence mathematical, other scientists have suggested that mathematics originates in the physical multiverse rather than the other way around. Inspired by computer and information science they speculate that the true theory of everything, and indeed reality itself, must be understood in computational terms, say as a universal quantum computer that can simulate everything.[61]

One may suspect that the mathematical universe with its infinity of structures and corresponding worlds is so rich that it amounts to an anything-goes universe. But this is not quite the case for any of the classes of the multiverse. There may well be universes with tartan elephants, if such creatures are mathematically allowed (as they presumably are), but none of the kinds of multiverse imply that all *imaginable* universes or objects exist. One can easily imagine a flat-space universe where the circumference of a circle differs from $2\pi r$ (at least I can), but no such mathematical universe exists. Most multiverse models, including the string landscape, do not even claim to represent all *possible* universes but only a tiny subset of them. The reason is that they rely on a number of assumptions, such that the universes are described by quantum mechanics and have only one time dimension. Although 10^{500} universes are a lot, the number is infinitesimal compared to the number of possible universes. Not even the mathematical universe includes all imaginable universes, 'only' those which can be mathematically defined.

In his argument for a mathematical universe Tegmark naturally draws upon the ideas of other physicists who have suggested a greater role for mathematics in physics than the one ordinarily accepted. For example, in an important paper of 1931 in which he predicted the existence of the positron and the magnetic monopole on the basis of quantum mechanics and relativity theory, Dirac offered the following advice, quoted by Tegmark:

> The most powerful method of advance that can be suggested at present is to employ all the resources of pure mathematics in attempts to perfect and generalize the mathematical formalism that forms the existing basis of theoretical physics, and after each search in this direction, to try to interpret the new mathematical features in terms of physical entities.[62]

Even more relevant, but less well known, is Dirac's 1939 address mentioned in Chapter 7, with its suggestion that 'the whole of the description of the universe has its mathematical counterpart'. In this work Dirac questioned the traditional separation in physical theories between laws of nature and contingent initial conditions. This idea reappears in multiverse physics, where there is no fundamental difference between the two parts. As Tegmark expresses it: 'A TOE with a landscape and inflation reclassifies many of the remaining "laws" as initial conditions, since they can differ from one post-inflationary region to another, but since inflation generically makes each such region infinite, it can fool us into misinterpreting these environmental properties as fundamental laws.'[63]

Tegmark's multiverse theory is not the only modern theory of everything that relies on the notion of physical reality being essentially mathematical. Frank Tipler entertained somewhat similar ideas in an ambitious paper of 2005 in which he suggested that the standard model of elementary particles conjoined with a theory of quantum gravity is, in a sense, the correct theory of everything. The TOE already exists! While Tegmark equates mathematical and physical reality, Tipler regards physical reality to be a subset of a much larger mathematical reality. His Pythagorean–Platonic universe rests on the assumption that 'physical reality is not "real" ultimately; only number—the integers comprising the true ultimate reality—is actually real'.[64] Tipler, too, operates with a multiverse consisting of an indefinite number of universes. In the multiverse fashion he explains the values of the free parameters of the standard model by appealing to a claim that 'all possible values of the constants are obtained in some universe of the multiverse at some time'.[65]

According to Tipler, 'The multiverse is forced upon us by observation'.[66] By this he means that since, in his view, quantum mechanics leads necessarily to a multiverse, and since observations tell us that quantum mechanics is true, then observations also tell us that there is a multitude of other universes. There are an awful lot of them, but not quite as many as in Tegmark's mathematical multiverse. The multiverse argued by Tipler does not involve all logically possible universes, but only those that are consistent with the fundamental laws of physics as we know them today. Moreover, his theory of everything differs from many other proposals of unified cosmophysics by not involving ideas such as superstring theory and cosmic inflation. These ideas he does not believe in. First and foremost, what distinguishes Tipler's ideas from most other theories of everything is the crucial and controversial role he assigns to intelligent life, a topic which will be taken up in Chapter 12.

Whether associated with string theory or some other theoretical framework, ideas of theories of everything are controversial and not generally taken very seriously. Some scientists consider them to be innocent pastimes, mere mathematical games, while for others they represent an unhealthy and speculative trend in theoretical physics.

Many physicists specializing in condensed matter physics and related fields object to the arrogant reductionism of particle physics and the entire philosophy underlying the notion of a theory of everything. They argue that the quest for an ultimate theory in terms of elementary particles and fields is fundamentally misguided because there is no deductive link between such a theory, should it exist, and most phenomena of real physics. These phenomena are emergent, that is, they depend on organizing principles and collectives states that cannot be reduced to simpler states. Following up on earlier criticism of this kind by the eminent solid-state theorist Philip Anderson, Robert Laughlin, a Nobel laureate at Stanford University, has attacked what he considers the dominant fundamentalism of theoretical physics. In a paper co-authored by David Pines of the Los Alamos National Laboratory he criticized the 'imperative of reductionism', which 'requires us never to use experiment, as its objective is to construct a deductive path from the ultimate equations to the experiment without cheating'. In sharp contrast to the ideas of Tegmark and Tipler, the two physicists concluded:

Rather than a Theory of Everything we appear to face a hierarchy of Theories of Things, each emerging from its parent and evolving into its children as the energy scale is lowered. The end of reductionism is, however, not the end of science, or even the end of theoretical physics. How do proteins work their wonders? Why do magnetic insulators superconduct? Why is ^{3}He superfluid?... The list is endless, and it does not include the most important questions of all, namely those raised by discoveries yet to come. The central task of theoretical physics in our time is no longer to write down the ultimate equations but rather to catalogue and understand emergent behavior in its many guises, including potentially life itself.[67]

While Laughlin and Pine spoke of theoretical physics being in 'the midst of a paradigm change', particle and string theorists such as Gross and Polchinski would have nothing of it. They denied that the fundamental laws of nature are emergent phenomena or that complexity is to be epistemically preferred over simplicity. 'To me, the history of science seems to be a steady progression toward simpler and more unified laws', Polchinski commented in the *New York Times*. 'I expect to see this continue and to contribute to it. Things may take many surprising twists and turns, but we reductionists are still quite happily and busily reducing.'[68]

10.4 WORLDS, UNOBSERVED AND UNOBSERVABLE

The increasing popularity of multiverse cosmology and anthropic arguments has caused a great deal of debate in the physics community, although it is a debate that the large majority of physicists and astronomers probably tend to find irrelevant or may even be unaware of. The overarching question is whether multiverse cosmology is a science or not. Have physicists in this case unknowingly crossed the border between science and philosophy, or perhaps between science and theology? Almost all physicists agree that a scientific theory has to speak out about nature in the sense that it must be empirically testable, but they do not always agree what testability means or how important this criterion is relative to other criteria. In this section I consider some of the arguments for and against the multiverse as a scientific proposal.

The eminent astrophysicist Dennis Sciama, a former student of Dirac and for a period an enthusiastic advocate of the steady-state theory, was an early convert to the multiverse. In the late 1980s he suggested that the existence of many worlds was necessary not only to explain the fine-tuning of natural constants but also to explain why the possibility of many other universes did not, apparently, correspond to the physical realities of these possibilities. During the 1989 Venice conference he argued strongly in favour of a multiverse consisting of many universes, each governed by its own fundamental theory. He did not need either inflation or string theory to reach the desired conclusion, but only the doctrine that 'everything which is not forbidden is compulsory', that is, the principle of plenitude.[69] In an essay of 1993, well before the controversy had gained momentum, Sciama considered some of the potential objections against the hypothesis of many worlds:[70]

(i) The hypothesis is much too extravagant and bizarre to be credible.
(ii) It violates the well-established tradition in theoretical physics to explain phenomena deductively from a fundamental theory.
(iii) The multiverse hypothesis has no real predictive power.
(iv) It is unscientific to postulate the existence of other universes, many of which are unobservable even in principle.

The objections listed by Sciama are among those that entered the controversy over the multiverse that only started in earnest after his death in 1999.

With regard to the first point, although the multiverse is certainly bizarre, this cannot in itself be a valid argument against the hypothesis. Since Copernicus claimed that the Earth rotates about its axis and further whirls around the Sun (a most bizarre theory by the standards of the time), scientists have learned that weird consequences of a theory do not necessarily imply that the theory is false. Besides, there are other theories that can compete with the multiverse in weirdness, the Everett interpretation of quantum mechanics being one of them and the wormholes of general relativity perhaps another. There are even physicists who apparently consider weirdness a sign of health rather than a problem. As Mukhanov says about the many-worlds interpretation, it is 'crazy enough to be true'.[71] Nonetheless, arguments based on epistemic or ontological weirdness do play a role and often contribute to the overall assessment of a theory, as they did in the case of the steady-state theory in the 1950s.

The stupendous number of parallel universes may seem to be wasteful and hence to violate the principle of simplicity accepted by most physicists. However, as a response to this objection it may be enough to recall that principles of simplicity and economy are notoriously ambiguous and offer little practical guidance for choosing between theories. Moreover, it can be and has been argued that the multiverse does not really run counter to Ockham's famous razor and in an explanatory sense is in fact a simpler concept than a single universe.[72]

The distinguished astrophysicist Martin Rees, a supporter of the multiverse, has on several occasions argued that there are good reasons to believe in the many unobservable universes or at least to take the hypothesis seriously. His 'slippery slope' argument is: 'From a reluctance to deny that galaxies with redshift 10 are proper objects of scientific enquiry, you are led towards taking seriously quite separate spacetimes, perhaps governed by quite different laws.'[73] With this he means that we have no problem in accepting that galaxies that have crossed the visible horizon are still real parts of the universe. They are unobservable, but it would take a naïve empiricist view to deny their existence. Nor have we serious problems with conceiving galaxies passing beyond the horizon corresponding to an infinite redshift; they disappear, but remain real. From this there is but a small step to accept the existence of galaxies that disappear at an ever-increasing rate, although these are and forever will be unobservable in principle. We may now compare these causally disjoint regions, held to be real, with other disjoint regions that emerge from the big bang according to the eternal inflation scenario. Rees's point is that if we have confidence in the reality of the first class of regions, why not believe in the reality of the second class as well?

The slippery slope argument may be more seductive than compelling, especially because it does not recognize the drastic difference between earlier observed regions of the universe and the infinity of disjoint regions. It is an argument that does not convince Paul Davies: 'As one slips down that slope, more and more must be accepted on faith, and less and less is open to scientific verification.'[74]

There is another kind of argument that is not specific to the multiverse, namely, that if a theory can explain phenomena that cannot otherwise be explained, then this theory is likely to be true. Linde offers the following version of the argument:

> We don't have any other alternative explanation for the dark energy; we don't have any alternative explanation for the smallness of the mass of the electron; we don't have any alternative explanation for many properties of particles.... These are experimental facts, and these facts fit one theory: the multiverse theory. They do not fit any other theory so far. I'm not saying these properties necessarily imply the multiverse theory is right, but you asked me if there is any experimental evidence, and the answer is yes. It was Conan Doyle who said, 'When you have eliminated the impossible, whatever remains, however improbable, must be the truth.'[75]

It is quite remarkable that Linde considers unexplained facts such as the proton–electron mass ratio to be 'experimental evidence' in favour of the multiverse. This is about the same logic that Boscovich adopted in the eighteenth century, when he suggested that phenomena such as cohesion and chemical affinity amounted to evidence for his theory of matter. Again, by the same logic Eddington's fundamental theory of the 1930s should have been readily accepted as true by his contemporaries since it was solidly supported by experimental evidence that other theories could not account for.

Yet another defence of the existence of a multitude of unobservable universes relies on historical analogy. There have been, so the argument goes, other cases in the history of science of predictions of unobservable entities and phenomena, and we have confidence in some of them. A theory cannot be considered scientific if *all* its predictions concern unobservable entities, but if some of them are observable and testable things are different. It is not a valid objection against a theory that it makes untestable predictions, for almost all scientific theories do that. Newton's theory of gravity predicts how the moon of some distant planet beyond the horizon revolves around the planet, but since the planet can never be observed the prediction is untestable. Yet we do not dismiss Newton's law because the prediction cannot be tested. Nor did scientists in the late eighteenth century question the scientific nature of the law of gravity because it, in conjunction with the emission theory of light, predicted the existence of invisible 'dark stars', a kind of classical black hole.[76] What matters is that a theory must make *some* testable predictions, as Newton's does abundantly, and it is on the basis of these testable predictions that it will be judged.

A well-established theory with empirical successes may include predictions which cannot be tested, and in such a situation one can argue that we have reason to believe in them in spite of their hypothetical nature. This is a very common argument in support of the multiverse scenario. We believe in quarks and gluons because they are predicted by the reliable quantum theory of strong interactions, and we believe in properties of black holes because they are predicted by the reliable theory of general relativity. As Tegmark phrases it:

> Because Einstein's theory of General Relativity has successfully predicted many things that we *can* observe, we also take seriously its predictions of things we cannot observe, e.g., that space continues inside black hole event horizons and that (contrary to early misconceptions) nothing funny happens right at the horizon. Likewise, successful predictions of the theories of cosmological inflation and unitary quantum mechanics have made some scientists take more seriously their other predictions, including various types of parallel universes.[77]

The argument presupposes that the physics behind the multiverse hypothesis, mainly eternal inflation and string theory, has the same credibility and epistemic authority as the theory of relativity, which may well be questioned. At any rate, historical analogies only carry limited epistemic force.[78] Without singling out any particular theory, Rees imagines a highly successful physical theory of the future that explains, for example, why there are three families of leptons and why the proton is about 1836 times as massive as the electron. 'If the same theory, applied to the very beginning of our universe, were to predict many big bangs, then we would have as much reason to believe in separate universes as we now have for believing inferences from primordial nucleosynthesis about the first few minutes of cosmic history.'[79]

Whereas critics of the multiverse claim that predictions of many universes escape testing, proponents of the idea argue that it is testable, albeit not in the ordinary sense. A multiverse theory may of course be trivially falsifiable if it is specific enough, say that it predicts that all universes are devoid of water. More generally, a multiverse theory can be ruled out if it predicts that none of the universes in its ensemble have properties observed in our world. Unfortunately, real multiverse theories are anything but specific and cannot be tested in this way. It is generally agreed that theories of the universe cannot result in definite predictions of the kind known from other parts of physics. Nonetheless, proponents of the multiverse insist that testable predictions are possible, but that the predictions will appear in the form of probability distributions. For example, it should be possible to determine what fraction of an immense and possibly infinite number of universes includes a cosmological constant of a size within a certain range.

The problem of how to define and compute probabilities in multiverse physics, that is, to calculate from a multiverse theory the probability that we should observe (in our universe, of course) a given value for some physical property, is known as the 'measure problem'. This problem is a hot topic in current research, but in spite of much work it has not led to a solution. It involves comparison of one infinity with another, which in general leads to an undefined expression. Aurétien Barrau, a French physicist and advocate of the multiverse, admits that 'except in some favourable cases, ... it is hard to refute explicitly a model in the multiverse'.[80] Physicists seem to agree that although it is possible to derive some probability predictions from a multiverse theory, this can be done only if certain strict conditions are satisfied. These conditions do not hold if the laws of physics vary from universe to universe, in which case no predictions of any kind appear to be possible.

The strongest and most articulate critic of the multiverse is possibly George Ellis, who in several works has not only raised technical objections to the theory but also questioned its scientific nature. Is the multiverse a scientific concept, a reality which follows nearly inevitably from fundamental physics? Or is it an interesting speculation whose proper place is in philosophy departments and science fiction literature? Whereas Susskind and Linde supports the first claim, Ellis is more in favour of the second one, maintaining that the existence of a multiverse 'remains a matter of faith rather than proof'. In an illuminating discussion with Bernard Carr, he stresses that the very nature of science is at stake in the current discussion about the multiverse:

> Its advocates propose weakening the nature of scientific proof in order to claim that the multiverse hypothesis provides a scientific explanation. This is a dangerous tactic. . . . There has been an increasing tendency in theoretical physics and cosmology to say it does not matter whether a proposal is testable. . . . [But] can one maintain one has a genuine scientific theory when direct and indeed indirect tests of the theory are impossible? If one claims this, one is altering the meaning of science. One should be very careful before so doing. There are many other theories waiting in the wings, hoping for a weakening of what is meant by 'science'. Those proposing this weakening in the case of cosmology should be aware of the flood of alternative scientific theories whose advocates will then state that they too can claim the mantle of scientific respectability.[81]

That is, if multiverse cosmology is admitted as a science, how can scientists reject pseudosciences such as astrology, intelligent design, and crystal healing on methodological grounds?

The current cosmophysical debate, in some measure reminiscent of the debate in the 1930s, is in part about the legitimate standards of physical science and the role of speculations. Both parties accept that speculative proposals have an important part in science, and in cosmology in particular, but they disagree whether the multiverse proposal is speculative or not, and if the multiverse is admitted as a speculation, whether it is a scientific or philosophical speculation. The critics argue that in the strong sense of an ensemble of totally disconnected universes, the multiverse theory definitely belongs to the latter class. Protagonists of the mutiverse are more inclined to accept even this extreme idea as a scientific speculation. Admitting that the multiverse is 'highly speculative', Rees maintains that the existence or non-existence of other universes is nonetheless a scientific question. As seen from his perspective it belongs to science, not metaphysics. Characteristically, he appeals to the traditional standard of demarcation, namely that the multiverse hypothesis results in 'some claims about other universes [which] may be refutable, as any good hypothesis in science should be'.[82]

Many physicists with a career in the more traditional approaches to particle physics have their misgivings about string theory in general and the landscape scenario in particular (see also Chapter 11). As one example, consider Burton Richter, an eminent particle physicist, Nobel laureate, and former director of SLAC, the Stanford Linear Accelerator Center. Jointly with the Brookhaven physicist Samuel Ting, in 1976 Richter received the Nobel Prize for the work that led to the discovery of the J/Ψ meson and which was an important part of the so-called November revolution in particle physics. Richter followed the development in multiverse physics with a mixture of wonder and disgust. This was not his kind of physics, if physics at all. In 2006, 30 years after having been awarded the Nobel Prize, he offered the following advice to the 'landscape gardeners':

> Calculate the probabilities of alternative universes, and if ours does not come out with a large probability while all others with content far from ours come out with negligible probability, you have made no useful contribution to physics. It is not that the landscape model is necessarily wrong, but rather that if a huge number of universes with different properties are possible and equally probable, the landscape can make no real contribution other than a philosophic one. That is metaphysics, not physics.[83]

10.5 BETWEEN SCIENCE AND PHILOSOPHY

Directly or indirectly, many of the questions discussed in the multiverse controversy are of a philosophical nature, not least when it comes to the proper standards of science on which the multiverse scenarios should be evaluated. Interestingly, the questions are discussed mostly within the scientific community, whereas philosophers so far have shown little interest in them. In a sense there is nothing new in the present situation. Cosmology has always been a field where metaphysical and other philosophical considerations have played a role. In spite of the great progress that has occurred during the last century, parts of cosmology may still be more philosophical than scientific in nature. The two fields cannot be easily and cleanly separated. This became clear during the steady-state controversy in the 1950s, and it appears no less clearly in the current controversy over the multiverse.

One of the problems is the infinite number of universes in some multiverse theories, or more generally the appearance of realized infinities in cosmology, as mentioned previously. When such infinities turn up, they inevitably raise problems of a philosophical nature that cannot be solved by scientific methods alone. Another issue concerns the relationship between prediction and explanation, two qualities which are highly regarded by practically all physicists. Indeed, most will agree that a theory must be able to predict as well as explain parts of nature, a theory which neither predicts nor explain just does not qualify as science. However, the two qualities do not always go easily together. There are theories that have great explanatory power but rate poorly when it comes to testable and specific predictions. As Ellis and other critics have argued, multiverse theories are extreme in this respect since they offer no specific predictions and yet are able to explain about everything. A theory which operates with 10^{500} or 10^{1000} universes can accommodate almost any observation; and should the observational result be revised, it will have no problem with explaining that either. As Ellis objects, 'The existence of universes with giraffes is certainly predicted by many multiverse proposals, but universes where giraffes do not exist are also predicted'.[84]

Ellis recognizes of course that accepted norms of science are not static and what has passed as legitimate science has changed over time and from one science to another. Nonetheless he insists that there is a core feature of science that must be retained at all cost, namely that scientific theories are empirically testable. Leave this criterion, and you have left science. Lee Smolin is no less adamant in his advocacy of falsifiability as a *sine qua non* of science. Referring to the lack of testability of the string landscape, he deplores that 'some of its proponents, rather than admitting that, are seeking leave to change the rules so that their theory will not need to pass the usual tests we impose on scientific ideas'.[85] On the other hand, physicists sympathetic to the multiverse call attention to the methodological changes that have occurred throughout the history of science, and they are more willing to accept softened versions of the principle of testability. Only very few will dispense with it altogether.

Among the antagonists of the multiverse and anthropic reasoning are also Steinhardt and Turok, who argue that their own model of an infinite cyclic universe is methodologically superior to the inflationary multiverse. 'Science should remain based on the principle that statements have meaning only if they can be verified or refuted', they say, concluding that the multiverse fails miserably on this count. The two theorists note with regret the trend towards accepting anthropic reasoning, which 'seems likely to us to drag a beautiful science towards the darkest depths of metaphysics'.[86] The Steinhardt–Turok cyclic universe is itself a kind of (temporal) multiverse, but in a sense very different from the one currently discussed in the universe–multiverse debate.

In a situation where the very standards of science are at stake one might expect the scientists to appeal to philosophical notions and demarcation criteria of science. Although this has happened in the modern cosmological debate, it is only on rare occasions and without much effect. Generally speaking, physicists have little respect for or are plainly uninterested in the opinion of philosophers (not to mention sociologists and theologians). They see it as part of their job to expand the domain of physics at the expense of philosophy and other branches of knowledge, to turn vague philosophical doctrines into precise and operational scientific concepts. And they believe they have succeeded in doing so on many occasions. For example, the semiphilosophical anthropic principle has now moved into physics and thus, because it is no longer philosophical, become scientifically acceptable. This kind of imperialist rhetoric is far from new, but it is particularly common in the modern debate over string theory and the multiverse.

About the only philosopher of science who is widely known among physicists is Karl Popper, famous for his falsificationist criterion of science which he stated in his classic work *The Logic of Scientific Discovery* of 1958. (The book first appeared in German in 1934, as *Logik der Forschung*, but without being much noticed.[87]) Whether or not they make reference to Popper's philosophy, many physicists (including Ellis and Smolin) feel that testability and falsifiability are indispensable for a scientific theory. Mario Livio, a physicist who is to some extent sympathetic to the multiverse and anthropic explanations, has emphasized that a theory that cannot be tested even in principle can hardly be counted as scientific. It 'goes against the principles of the scientific method, and in particular it violates the basic concept that every scientific theory should be falsifiable', he says.[88] As mentioned, Rees is another advocate of the multiverse who has defended its scientific nature by arguing that it does lead to falsifiable claims. But not all physicists have the same reverence for the falsifiability criterion.

As early as 1989, during the Venice conference on the anthropic principle, Carter discussed the principle and its implications in relation to Popperian philosophy of science, which he criticized in a general way for its 'negativist' attitude.[89] Carter objected to Popper's 'refutability principle', with which term he referred to the one-sided emphasis on refutation of predictions at the expense of confirmation. Still, although he rejected the 'folklore version of the Popper principle', he maintained that the anthropic principle was falsifiable and thus not in direct conflict with Popper's criterion of science. Other physicists advocating the anthropic principle and the multiverse have been more openly hostile to philosophers in general and Popperianism in particular. After all, have philosophers any right to dictate to the physicists the norms of science and hence to decide whether their theories belong to science or not?

Provoked by the charges against the multiverse of being unfalsifiable, Barrau insists that science can only be defined by the scientists themselves: 'If scientists need to change the borders of their own field of research, it would be hard to justify a philosophical prescription preventing them from doing so.'[90] Indeed, the multiverse controversy has highlighted the fundamental question of the very definition of science and the demarcation criteria that distinguishes science from non-science. Perhaps philosophers have no 'right' to decide which criteria are valid, but do the scientific experts have such a right? What if the experts disagree? Should the question then be decided by a vote? Or by a court of justice? Inspired by the multiverse controversy, the American physicist Robert Ehrlich argues that 'decisions as to what constitutes a legitimate scientific theory are simply too important to be left to the practitioners of that field, who obviously have vested interests in it, such as a desire to keep the funding coming'.[91]

Susskind is another multiverse advocate who has little patience with armchair philosophy and philosophical demarcation criteria. As to Popper's falsificationism he has no confidence at all, only scorn:

> Throughout my long experience as a scientist I have heard unfalsifiability hurled at so many important ideas that I am inclined to think that no idea can have great merit unless it has drawn this criticism.... Good scientific methodology is not an abstract set of rules dictated by philosophers. It is conditioned by, and determined by, the science itself and the science who create the science.... Let's not put the cart before the horse. Science is the horse that pulls the cart of philosophy.[92]

The multiverse theory clearly has problems with testability, as usually understood, and even greater problems with falsifiability. And so what?—is the response of some multiverse physicists. They either argue that science needs no formal norms except those that scientists agree upon, or they point to other criteria that may overrule the one of testability. An acceptable physical theory has to lead to statements that can be compared with observations and experiments, but it is generally agreed that there are other factors at play than mere empirical testing. We can have good reasons for believing in a theory even though it does not lead to directly testable consequences. Almost all physicists agree that a satisfactory theory, in addition to being testable, must also be simple and internally consistent, it must show explanatory power, and it must connect to the rest of science. Where the waters divide is when it comes to the priority given to these criteria. Is empirical testability absolutely necessary? And, if this is granted, how should testability be understood?

It is quite clear that some of the multiverse physicists have no respect at all for philosophers of science in general and for the 'Popperazi' in particular, to use Susskind's nickname for the modern followers of Popper. 'As for rigid philosophical rules', he says, 'it would be the height of stupidity to dismiss a possibility [such as the string multiverse] just because it breaks some philosopher's dictum about falsifiability'.[93] On the other side, Smolin and Ellis subscribe to Popperian standards of science, falsificationism included, if not in quite the dogmatic sense of Bondi, according to whom 'There is no more to science than its method, and there is no more to its method than Popper has said' (see Chapter 5). Given the abstract nature of Tegmark's theory of the mathematical universe one might expect him to side with Susskind, but this is not the case. He refers positively to Popper's criterion of falsifiability, arguing that both the multiverse and the mathematical universe theory of everything satisfy the criterion. On the other hand, Tegmark also seems to agree with Carter's point that science is not so much about proving theories wrong as it is about accumulating positive evidence for theories. 'What we do in science isn't falsifying, but "truthifying"—building up the weight of evidence', he says.[94] Tegmark's 'truthifying' is commonly known as verification or confirmation.

As pointed out by Michael Heller, a Polish cosmologist, philosopher, and Catholic priest, multiverse physicists often refer to Popperian falsifiability in ways that are vague and scarcely legitimate.[95] Most likely, few of them have ever read Popper or looked into the philosophical literature concerned with falsificationism. Whereas Popper held, to express it briefly, that it is a necessary condition for a scientific theory that it must be falsifiable in principle, of course he never claimed that it is a sufficient condition. There are evidently falsifiable statements that do not qualify as scientific. A new theory of gravitation from which it follows that planets necessarily move in circular orbits (such as was thought in pre-Keplerian astronomy) is certainly falsifiable, but this alone does not make it scientific.[96] Likewise, Tegmark's appeal to Popperian falsifiability presupposes that because a theory leads to consequences that can be proved wrong by observation, and indeed are wrong, it is of a scientific nature. But this is plainly a misrepresentation of what falsifiability means.

As philosophy is involved in the debate over the multiverse, so is religion, if mostly indirectly and between the lines. It is common among supporters of the multiverse to conceive it as an alternative to a divinely created world and to demarcate the theory from ideas of natural theology. The multiverse appeals to the anthropic principle, but without any trace of intelligent design. 'If there is only one universe', Bernard Carr says, 'you might have to have a fine-tuner. If you don't want God, you'd better have a multiverse.'[97] Richard Swinburne, the eminent theistic philosopher, agrees that the multiverse is as contrary to Christian belief as the anthropic principle. 'To postulate a trillion trillion other universes, rather than one God in order to explain the orderliness of our universe, seems the height of irrationality,' he comments.[98]

On the other hand, there is no one-to-one correspondence between the multiverse and belief in a divine creator. Several physicists with a dislike of the multiverse idea have called attention to what they conceive as the religious or quasi-religious elements it contains. To Ellis, belief in the multiverse is essentially based in 'faith' (if not religious faith), and Paul Davies find multiverse explanations to be 'reminiscent of theological discussion'. To his mind, they are effectively reintroducing divine explanations in cosmology: 'Far from doing away with a transcendent Creator, the multiverse theory actually injects that very concept at almost every level of its logical structure. Gods and worlds, creators and creatures, lie embedded in each other, forming an infinite regress in unbounded space.'[99]

The views of Don Page, a theoretical physicist at the University of Alberta and former collaborator with Stephen Hawking, may illustrate that the multiverse does not necessarily contradict Christian belief. By his own account a conservative Christian, Page believes that the divinely created universe has a purpose, and that whether there is a single universe or a lot of them. 'I do believe the Universe was providentially created by God', he says, and 'I also strongly suspect that the Universe is a multiverse, with different parts having different values of the physical parameters'.[100] Page finds the multiverse theory simpler, more elegant, and with greater explanatory force than the single-universe theory, and he sees no reason why God should not have decided to create a multiverse instead of a universe. At a symposium in 2008 Page gave a presentation entitled 'Does God love the multiverse?' in which he argued that the multiverse is not an alternative to design by God. He tended to answer his question affirmatively.

It has even been suggested that multiverse physics and intelligent design share some methodological ground, a suggestion nearly all advocates of the multiverse will vehemently deny. Yet, one of the standard arguments for dismissing creationism and intelligent design is that they cannot be tested and provide no room for falsification. This was essentially the argument of the US Supreme Court when it decided in 1986 that scientific creationism is religious and not scientific in nature. In this case the opponents of creationism, including a large number of American physicists, emphasized that 'An explanatory principle that by its very nature cannot be tested is outside the realm of science'.[101] But, so it has been argued, intelligent design is hardly less testable than many multiverse theories. To dismiss intelligent design on the ground that it is untestable, and yet accept the multiverse as an interesting scientific hypothesis, may come suspiciously close to applying double standards.[102] As seen from the perspective of some creationists, and also by some non-creationists, their cause has received unintended methodological support from multiverse physics.

Notes for Chapter 10

1. Kragh 2009d is a historically oriented introduction to the subject.
2. Barrau 2007. Schellekens 2008, p. 1. Kuhnian metaphors such as 'revolution' and 'paradigm' abound in the literature on superstrings and multiverse physics, but almost never with references to Kuhn's philosophy of science.
3. Ananthaswamy 2009. A few years earlier, Greene expressed his reservations with respect to anthropic multiverse ideas, which he felt were too easy. 'Maybe', he said in an interview, 'you just needed five more years of hard work and you would have answered those unresolved questions, rather than just chalking them up to, "That's just how it is".' *Scientific American Special Edition* **15**, issue 3 (2005), 50–55.
4. Quoted in Crowe 1999, p. 3.
5. Grant 2007, p. 228.
6. Quoted in Crowe 1999, p. 28. The source is *Théodicée* from 1710, one of Leibniz's most important works.
7. Hume 1980, p. 36.
8. Blanqui 1872, p. 47.
9. Newcomb 1906, p. 164. On the interest in hyperspaces and other universes, see Bork 1964.
10. Boltzmann 1895, p. 415. For more than a century Boltzmann's argument has attracted the interest of physicists and philosophers alike. For a recent review, see Ćirković 2003a.
11. Eddington 1931b, p. 415.
12. Tolman 1934b, p. 175. Reprinted in *General Relativity and Gravitation* **29** (1997), 935–43. The Lemaître–Tolman model is also known as the Tolman–Bondi model because Bondi investigated it in 1947.
13. Hoyle 1965b, p. 131.

14. Feuer 1934, p. 346.

15. The American novelist and historian Henry Adams used the term 'multiverse' a couple of times in his autobiographical *The Education of Henry Adams* first published in 1918. Writing in the third person, he said: 'The child born in 1900 would, then, be born into a new world which would not be a unity but a multiple. Adams tried to imagine it ... He could not deny that the law of the new multiverse explained much that had been most obscure....' The first time 'multiverse' appeared in a scientific journal (the *Irish Astronomical Journal*) may have been in 1982, but then in relation to Adams. See Jaki 1982.

16. Gale 1990. See also the historical review in Trimble 2009.

17. Tegmark 2003.

18. Carter 1990, p. 131 and p. 133 (first published in Longair 1974, pp. 291–98).

19. The article in *Reviews of Modern Physics* (Everett 1957) is reproduced in Wheeler and Zurek 1983, pp. 315–23. On Everett's life and work, see Byrne 2007.

20. Everett 1957, p. 460.

21. DeWitt 1970. DeWitt's work on quantum gravity led him to suggest that quantum mechanics applies to the entire universe at all times, such as Everett had earlier proposed. According to the Copenhagen interpretation, only an open system interacting with a classically describable measuring device (an observer) can be treated quantum mechanically, for which reason there can be no quantum mechanics of the universe. In 1967 DeWitt proposed a wave equation for the universe, known as the Wheeler–DeWitt equation, and expressed the hope that mathematical consistency alone would lead to a unique wave function.

22. DeWitt 1970, p. 33.

23. Mukhanov 2007, p. 270.

24. Tegmark 2007, p. 23, a tribute to Everett's theory on the occasion of its 50th anniversary.

25. Tipler 2007, p. 95. See also Tipler 1994, pp. 483–88. In the early 1950s, David Bohm's causal interpretation of quantum mechanics was rejected as metaphysical nonsense by leading quantum theorists, including Pauli, Heisenberg, and Rosenfeld. According to them, there was only one interpretation, namely the Copenhagen interpretation. In a letter to Rosenfeld of 16 April 1958, Heisenberg said: 'I agree with you that the expression *Copenhagen interpretation* is not really fortunate, in so far that one could think that there are also other interpretations, such as claimed by Bohm, for example. Of course, we both agree that these other interpretations are nonsense...' Quoted in Pauli 1996, pp. 342–43.

26. Nozick 1981, p. 129. The fecundity assumption is related to the principle of plenitude, but the latter refers only to the realization of possibilities in the actual world we live in. Whereas Nozick did not mention the anthropic principle, he did refer to Wheeler's speculations of many universes with different laws of physics.

27. Linde, 'Nonsingular regenerating inflationary universe', reprint of July 1982. Accessible online as http://www.stanford.edu/%7Ealinde/1982.pdf.

28. Linde 1990, p. 29.

29. Linde 1986, p. 399.

30. Linde 1987, p. 628. In the same volume, Hawking made use of the weak anthropic principle to explain why we live in an expanding rather than a contracting universe.

31. Vilenkin 2007, p. 163.

32. Guth 1997, pp. 251–52. See also the self-creating universe scenario in Gott and Li 1998.

33. Ellis and Brundrit 1979, p. 38. This is pretty much the same conclusion that Blanqui reached back in 1872, although in his case on a purely speculative basis (see above).

34. Garriga and Vilenkin 2001.

35. Tegmark 2003. The contention, made or implied by Ellis, Tegmark, Guth, and others, that in an infinite universe with an infinity of things there will be objects with any combination of those things, is wrong. Infinity alone is obviously not a sufficient condition for these combinations to occur.

36. Hilbert 1925, p. 190. Hilbert referred both to quantum theory (the impossibility of infinitely small quantities) and cosmology (the impossibility of infinitely large quantities).

37. Ellis, Kirchner, and Stoeger 2004.

38. Quoted in Kragh 2008, p. 84, where further information on the controversy can be found.

39. Schellekens 2008, p. 8.

40. Bousso and Polchinski 2000. This paper did not refer to the 'A-word'.

41. Susskind 2006. See also Bousso and Polchinski 2004 and Carroll 2006. A comprehensive review appears in Douglas and Kachru 2007. The landscape multiverse hypothesis was quickly taken up by journalists and disseminated to the public. See, for example, Overbye 2002.

42. Guth and Kaiser 2005, p. 888.

43. Linde 2007, p. 134.

44. Quoted in Folger 2008.

45. For an explanation of the number 10^{500} that does not require expert knowledge of string theory, see Conlon 2006.

46. Susskind 2006, p. 177.

47. Banks, Dine, and Gorbatev 2004, p. 2.

48. Mersini-Houghton and Holman 2009. See also Gefter 2009.

49. Rees 2007, p. 66.

50. For example Jaffe, Jenkins, and Kinchi 2009, who prefer to speak of environmental constraint rather than selection.

51. Schwarz 1985, p. v: 'Superstring theory... could be a "theory of everything"'. Ellis 1986 and Rújula 1986 also referred to the possibility of a theory of everything, both of them in connection with the new theory of superstrings.

52. On the final theory, see Weinberg 1992, especially pp. 211–40. See also Taylor 1993, who concludes that any theory of everything contains the seeds of its own destruction.

53. For a historical perspective on different views on unity and disunity in the period, see Cat 1998.

54. Tegmark 1998, p. 2.

55. A critique of Tegmark's Platonic multiverse hypothesis appears in Heller 2009, pp. 107–13. Heller concludes that the hypothesis may at best 'serve as an inspiration for science fiction novelists'. It should be noted that the mathematical universe hypothesis is not Platonic in the real or authentic sense, according to which the mathematical objects (which alone are real) do not exist in any physical universe but in some transcendent non-spatial and non-temporal world.

56. Feynman 1992 (originally published 1965), p. 171.

57. Tegmark 2003, p. 50. Wigner 1960. On a different and more modern perspective on the mathematics-physics relationship, see Jaffe and Quinn 1993. See also the original analysis in Steiner 1998.

58. Jeans 1934, p. 356.

59. Tegmark 2008, p. 125.
60. Ibid., p. 141.
61. Inspired by the ideas of David Deutsch and others, Karthik 2004 speculates that we live in a level III multiverse which self-replicates an infinite number of times. What is traditionally called a theory of everything (TOE) is for him merely a theory of something (TOS). Deutsch, a physicist at the University of Oxford, has long promoted his own ideas of the multiverse, the theory of everything, simulated worlds, physical eschatology, and the like. See e.g. Deutsch 1997.
62. Dirac 1931, p. 60. On Dirac's view on the relation between mathematics and physics, see Kragh 1990 and Bueno 2005. Dirac's scientific practice was not consistent with the praise of the power of pure mathematics, as he expressed it in 1939 and on some other occasions. He basically conceived mathematics as 'only a tool' (Dirac 1958, p. viii) and emphasized physical ideas at the expense of mathematical formalism.
63. Tegmark 2008, p. 117.
64. Tipler 2005, p. 905.
65. Ibid., p. 960.
66. Tipler 2007, p. 15.
67. Laughlin and Pines 2000, p. 31. Laughlin received the 1998 Nobel Prize for his co-discovery of the fractional quantum Hall effect. On Anderson's early criticism of reductionistic particle physics, see Anderson 1972 and Cat 1998.
68. Quoted in Johnson 2001.
69. Sciama 1993b, p. 108. Like many other protagonists of the multiverse, Sciama felt a need to legitimate the scientific nature of the hypothesis. He argued that, 'by making a testable prediction, the hypothesis that there exist many disjoint universes is a physical hypothesis'. In his case, the prediction was that the Penrose–Hawking singularity theorem was wrong.
70. Sciama 1993a.
71. Mukhanov 2007, p. 272.
72. For example Sciama 1993b, p. 108 and Tegmark 2007, p. 123. Victor Stenger, a physicist, science writer, and advocate of atheism, has argued that 'The cosmology of many universes is more economical if it provides an explanation for the origin of our universe that does not require the highly nonparsimonious introduction of a supernatural element that has not heretofore been required to explain any observations'. (Stenger 1995, p. 236). Roush 2003 defends the opposite view.
73. Rees 2007, p. 63 and similarly in Rees 2003b.
74. Davies 2003.
75. Interview with Linde in Folger 2008. What philosophers of science sometimes call the Sherlock Holmes strategy refers to a passage in one of Conan Doyle's most famous short stories, *Silver Blaze*. It goes without saying that the strategy is questionable. How can one know that all alternative explanations have been eliminated?
76. The invisible dark stars of very high density discussed by John Michell, Laplace, and others were not, strictly speaking, undetectable. It was realized that in principle their existence might be revealed by their gravitational perturbations on visible stars or planets. On the history of the idea of dark stars and its relation to black holes, see Israel 1987.
77. Tegmark 2008, p. 124, and very similar in Livio and Rees 2005.

78. The use and misuse of historical analogies can be followed through most of history of science. In a discussion of the methods of cosmology, Thomas Gold warned against relying on such analogies: 'The most valuable lesson to be learnt from the history of scientific progress is how misleading and strangling such analogies have been, and how success has come to those who ignored them' (Gold 1956, p. 1722). Gold's argument was directed against Dingle's critique of the steady-state theory and other modern models of the universe.

79. Rees 2003b, p. 385.
80. Barrau 2007.
81. Carr and Ellis 2008, 2.33.
82. Rees 2003b, p. 388.
83. Richter 2006, p. 9.
84. Carr and Ellis 2008, p. 2.35.
85. Smolin 2006, p. 170.
86. Steinhardt and Turok 2004.

87. However, *Logik der Forschung* was noticed by Einstein, who read it soon after it appeared. He liked it very much, so as he told Popper in a letter of 15 June 1935: 'Your book has pleased me in many ways. Rejection of the "inductive method" from an epistemological standpoint. Also the falsifiability as determining property of a theory of reality.' Quoted in Van Dongen 2002, p. 39.

88. Livio 2000, p. 187.
89. Carter 1993.
90. Barrau 2007.
91. Ehrlich 2006, p. 86.
92. Susskind and Smolin 2004.
93. Susskind 2006, p. 196.
94. Quoted in Matthews 2008.
95. Heller 2009, pp. 88–89.

96. This example was used by McVittie against Bondi's frequent claims that the steady-state theory, being easily falsifiable, was more scientific than the relativistic rival models of the universe. See Kragh 1990, p. 250.

97. Quoted in Folger 2008.

98. Swinburne 1996, p. 68. Weinberg (2007, p. 39) has called attention to an article in the *New York Times* in which Christoph Schönborn, archbishop of Vienna, grouped together 'neo-Darwinism and the multiverse hypothesis in cosmology' and attacked them for trying 'to avoid the overwhelming evidence for purpose and design found in modern science'. The source is *New York Times* of 7 July 2005, p. A23.

99. Davies 2003.

100. Page 2007, p. 412. See also the interview in Lightman and Brawer 1990, where Page expressed his sympathy for a multiverse. While in favour of the multiverse, he rejects Tegmark's mathematical (level IV) multiverse because it involves contradictions.

101. The documents from the 1986 court case can be found online as http://www.talkorigins.org/faqs/edwards-v-aguillard/amicus1.html.

102. Luskin 2006. The suspicion of double standards is also discussed by Robert Ehrlich, a physics professor with no sympathy for intelligent design and no sympathy for the multiverse either (Ehrlich 2006).

11

String Theory and Quantum Gravity

> *So it seems God's faithful*
> *angels have tired of dancing*
> *on pinheads to the horns*
> *of creation and are wracking*
> *their branes to tune*
> *the superstrings*
> *of their harps' fundamental.*

J. Radke, 'Life × 10^{-33}' in Miller and Verma, *Riffing on Strings*, 2008, p. 147

Of all the grand attempts to establish an ultimate theory of the particles and forces of nature, none has been as popular, ambitious, and controversial as the modern theory of superstrings 'living' in a space–time of many dimensions. Not only is the theory considered attractive by a large part of the theoretical physics community, it has also received a great deal of press coverage and been disseminated to the public in the form of numerous popular books, magazine articles, and internet discussion sites. It has even inspired authors to write a substantial number of poems and short stories. At least one drama about superstrings has been performed on stage.[1]

The string theory or research programme proclaims a revolution in our understanding of nature on its most basic level, yet the proclamations and results of the extended work have had curiously little impact outside mathematics and string theory itself. After 40 years and several 'revolutionary' breakthroughs it is still an open question whether string theory is speculative mathematical physics or a theory of the real world. In any case, the theory of superstrings is generally recognized as the best offer of a theoretical framework encompassing quantum mechanics and general relativity, and hence the fulfilment of the old dream of unification. It has even proved possible to establish cosmological models on the basis of the theory, as we have seen in Chapter 8. But it is far from the only theory of quantum gravity, nor is it the only quantum-gravitational theory with cosmological consequences. A very different approach, known as loop quantum gravity, has been developed over some 20 years and is sometimes presented as the main alternative to string theory, an alternative that in a sense is more classical and deviates less from the traditional conception of physics.

String theory has in common with other theories of quantum gravity that it is mathematically complex but empirically weak. The problem of connection to the low-energy world where experiments can be performed is, in the eyes of most physicists, even more serious than the problems of internal consistency with which most researchers in quantum gravity are occupied. Yet string theorists, along with physicists cultivating other approaches to quantum gravity, are keenly aware that ultimately a physical theory needs to lead to testable consequences. In spite of its excessive mathematical machinery and worrying distance from experiment, string theory is not an attempt to establish physics on an *a priori* basis in the manner of Descartes or Eddington.

Although the debate over string theory may be seen as a fight between two paradigms, one based on theory assessment by experiment and another by intratheoretical progress, string theory does not imply an entirely new way of assessing scientific theories. It is only recently that claims of a widely different methodology have emerged, and then in connection with the landscape interpretation of string theory and its connection to multiverse speculations. The standards of evaluation in this branch of scientific speculation differ more from the classical standards than is the case in traditional string theory.

11.1 MANY-DIMENSIONAL SPACE

The early attempts to establish a theory that integrated gravitation, electromagnetism, and quantum phenomena in a single framework took many directions. Interesting work on quantization of gravity was done by Léon Rosenfeld, Matvei Bronstein, and a few others in the 1930s, but I shall here deal only with another and earlier approach.[2] This is the approach based on the hypothesis that there are more spatial dimensions than the three we perceive and on which physical theories are normally based. Whether this hypothesis turns out to be true or not, it has exerted a profound influence on the development of modern theories of fundamental physics.

As mentioned in Chapter 2, the general idea of hidden or extra dimensions can be found in nineteenth-century speculations of 'hyperspaces', often in connection with the etherial world view that was so popular in the late part of the century. Among the scientists and philosophers who entertained such ideas, if only as hypotheses, was the respected astronomer Simon Newcomb. In an address to the American Mathematical Society of 1896 he discussed the possibility of atoms vibrating in some extra dimension. 'We have no right to say that those motions are necessarily confined to the three dimensions', he said. 'Perhaps the phenomena of radiation and electricity may yet be explained by vibrations in a fourth dimension'.[3] However, Newcomb and his contemporaries never developed the hypothesis of a fourth spatial dimension beyond the level of uncommitted speculation. It was only in the 1920s that the hypothesis entered physics in a scientifically meaningful way. Not only was the idea a scientific novelty, it also served as an exemplar for later many-dimensional theories of quantum gravity, including the string theories of the last part of the century.

With the presentation in 1916 of Einstein's general theory of relativity, the force of gravity was understood in terms of a basic theory of physics. The other familiar force of nature, electromagnetism, was accurately described by Maxwell's equations, and hence it became desirable to unify the two very different theories into a single supertheory. The unification programme pursued over the next couple of decades was primarily aimed at producing a single theory of electrogravity from which general relativity and electrodynamics would appear as special cases. Leading scientists within the first phase of this programme, such as Einstein, Weyl, Eddington, and the French mathematician Elie Cartan, were of course aware of the new quantum theory of atoms and radiation, but as mentioned in Chapter 3 they did not seriously attempt to incorporate it in their theories.

Einstein—who was not only the founder of general relativity but also a pioneer of quantum theory—realized early on that somehow the strange quanta had to be taken into account. As early as 1916, in the communication to the Prussian Academy of Sciences in which he introduced the notion of gravitational waves, he noted that the revolving electrons in atoms must emit a tiny amount of gravitational radiation. Since no such radiation was known to exist, 'it appears that not only must quantum theory lead to a modification of Maxwellian electrodynamics, it must also modify the new theory of gravitation'.[4] He reiterated the suggestion in his second paper on gravitational waves, which appeared two years later. Hermann Weyl, too, recognized the importance of quantum theory. In the fourth edition of his celebrated *Raum, Zeit, Materie*, published in 1921, he presented his own candidate for a unified theory about which he said: 'I feel convinced that it contains no less truth than Einstein's theory of gravitation—whether this amount of truth is unlimited or, what is more probable, is bounded by the quantum theory.'[5]

The first physicist who not only talked about integrating quantum theory and general relativity, but actually did something about it, was the young Swede Oskar Klein, a collaborator of Bohr who was originally trained in physical chemistry. Although he came to the hypothesis of a fifth dimension independently, he was anticipated by the German mathematician Theodor Kaluza, an unknown privatdocent from Königsberg.[6] In fact, Kaluza's general idea of embedding space–time into a five-dimensional world can be found in a little known work by the Finnish theoretical physicist Gunnar Nordström as early as 1914, that is, before the full formulation of general relativity. Nordström's introduction of an extra dimension occurred in an attempt to unify Maxwell's equations with the equations of gravitation. 'The five-dimensional world has a characteristic axis', he wrote, 'with the four-dimensional space–time world being perpendicular to it.' Nordström considered his unification to be purely formal, which may have been the reason why it made no impact on contemporary physicists and mathematicians. 'Of course, there is no new physical content in the equations. Still, it is not impossible that there may be a deeper reason for the formal symmetry I have found.'[7] These were not words likely to attract the attention of the physicists. Although the Finnish theorist later did important work on general relativity theory, including an early contribution to black hole physics, his five-dimensional theory was ignored.

Kaluza, apparently unaware of Nordström's earlier work, presented his idea of a new way of unifying electromagnetism and gravitation to Einstein in 1919, but it took two more years before it appeared in print. His theory was based on the postulate of a new fifth dimension in addition to the one time dimension and the three space dimensions of ordinary space–time. The physical meaning of the new dimension was unclear. Kaluza's theory included 14 potentials—10 gravitational and four electrodynamical—and it was constructed in such a way that the physical effects of the extra dimension were unobservable. What mattered was that both the field equations of general relativity and Maxwell's equations fell out of the five-dimensional equations in a natural way. In this sense it unified the two theories.

Einstein initially expressed admiration for Kaluza's five-dimensional theory. 'I have great respect for the beauty and the boldness of your thought', he wrote to the Königsberg mathematician in 1919.[8] Although Kaluza did not take into account quantum phenomena, in agreement with Einstein and others he recognized that 'any hypothesis that lays claim to universal significance is threatened by the sphinx of modern physics—quantum theory'.[9] The reference to quanta remained on a rhetorical level, as was the case with the references of Einstein and Weyl.

Klein's approach of 1926–28 appeared at a time when Schrödinger's wave mechanics had just become known. In fact, it was meant as a contribution to the new wave theory of atoms rather than yet another attempt of unification. The main result of Klein's work was a generalization of the Schrödinger equation that agreed with the special theory of relativity. (This equation, being of the second order in both space and time, became known as the 'Klein–Gordon equation' because it was independently discovered by the German physicist Walter Gordon.) In Klein's theory, the fifth dimension was non-observable even in principle, something he did not see as a problem. On the contrary, he emphasized that observability is not a valid criterion for physical meaningfulness and that the new dimension was far from superfluous.

According to Klein, Kaluza's fifth dimension related to the elementary electrical charge. In this way he hoped to explain the atomicity of electricity as a quantum law and also to account for the then known basic building blocks of matter, the electron and the proton. Klein conjectured that what we think of as a point in three-dimensional space is really a tiny circle going round the fifth dimension in a loop with a certain period λ. The loop is not in ordinary space, but in a direction that extends it. As he explained, 'the origin of Planck's quantum may be sought just in this periodicity in the fifth dimension'. As to the period and its relation to the quantum of action, he suggested

$$\lambda = \frac{hc}{e}\sqrt{2\kappa} = \frac{4h}{e}\sqrt{\pi G} \cong 0.84 \times 10^{-32} \text{m}$$

'The small value of this length... may explain the non-appearance of the fifth dimension in ordinary experiments as the result of averaging over the fifth dimension', he wrote.[10] That is, he assumed that the extra dimension was rolled up to a less than microscopic size—compactified, as it was later called. The quantity λ, sometimes known as the Klein length, can be written in terms of the fine-structure constant and the Planck length as

$$\lambda = 4\pi\sqrt{\frac{2}{\alpha}}\ell_{Pl}$$

The energy or mass associated with this scale is of the order 10^{17} GeV, far beyond anything which can be produced in the laboratory. In a more general way, and much as Einstein in 1916 and Weyl in 1921 had already done, Klein argued that general relativity could not remain uninfluenced by quantum theory. It is to be expected, he said in a paper of 1928, 'that the general theory of relativity stands in need of revision in the sense of the quantum postulate, as also results from the fact that various of its consequences stand in contradiction to the requirements of the quantum theory'.[11]

The five-dimensional Kaluza–Klein theory attracted considerable attention in the interwar years, when it was examined and developed by Einstein, Fritz London, Vladimir Fock, Schrödinger, Pauli, and others. Klein returned to it in an important paper on gauge theory of 1939, which was, however, little noticed at the time. Einstein, who in 1926 had described Klein's paper as 'beautiful and impressive', took up the Kaluza–Klein approach from time to time, but in 1931, in a brief report on unified field theories, he concluded that although it was superior to the earlier theories of Weyl and Eddington, it was 'not acceptable'. His reason was this: 'It is anomalous to replace the four-dimensional continuum by a five-dimensional one and then subsequently to tie up artificially one of these five dimensions in order to account for the fact that it does not manifest itself.'[12] Yet Einstein continued to show some interest in the Kaluza–Klein theory. Only in 1943 did he abandon it for good, having found that localized particles were incompatible with the five-dimensional field theory. Nor did other physicists find the five-dimensional approach convincing, among other reasons because it seemed remote from physical testing and application. Like so many candidates for a unified theory, it remained on the periphery of physics for many years, considered to be somewhat speculative and more interesting from a mathematical than a physical point of view.

What Abdus Salam called the 'Kaluza–Klein miracle'[13] in his Nobel lecture of 1979 was resuscitated about 1970, now with more than one extra dimension. Higher-dimensional unified theories became almost fashionable when physicists found ways to express gravity in supersymmetric ways. The concept of supersymmetry implies many more particles than those known experimentally, for example in the case of gravity, several 'gravitino' superpartners to the graviton. Generally, supersymmetry implies a correspondence between fermions and bosons, meaning that to each fermion there has to exist a bosonic partner particle with the same mass, and vice versa. Apart from the known electrons and quarks there have to exist hypothetical spin-zero 'selectrons' and 'squarks', which are not governed by the Pauli exclusion principle. Similarly, the photon has to have a partner in the form of the 'photino' with spin one-half. A world populated with supersymmetric partner particles would be completely different from the one we know. Therefore, if supersymmetry is valid at all, it must be a spontaneously broken symmetry that secures that electrons have mass while selectrons do not, etc.

The original theory of supergravity, proposed in 1976 by Daniel Freedman, Peter van Niewenhuizen, and Sergio Ferrara, was four-dimensional but was soon generalized to higher dimensions in the style of Kaluza–Klein.[14] Whereas the original theory of Kaluza and Klein operated with only five dimensions, the most popular of the supergravity models of the period was formulated in 10 space dimensions and one time dimension. When taking over the Lucasian Chair of Mathematics in Cambridge in 1979 (the chair that Newton had once held, and later Dirac), Stephen Hawking addressed the possibility 'that the goal of theoretical physics might be achieved in the not too distant future, say, by the end of the century'. By this he meant, 'that we might have a complete, consistent and unified theory of the physical interactions which would describe all possible interactions'. A main reason for Hawking's 'cautious optimism' was the progress that had recently been made in the 11-dimensional theory of supergravity, which he considered a serious candidate for a theory of everything.[15]

The theory of supergravity aroused great excitement (Fig. 11.1), but excitement waned when various flaws were discovered. At about the same time much work was done in developing many-dimensional Kaluza–Klein theories where the electromagnetic, weak, and strong forces appeared as adjuncts of the gravitational force. Although it turned out that the 11-dimensional supergravity and Kaluza–Klein theories did not work, the mathematical efforts were not wasted. The theories were developed in interaction with the new theory of superstrings and by the 1990s they were regarded as features of the 10-dimensional string theory rather than alternatives to it. At any rate, Kaluza's nearly century-old idea of space having more than three dimensions has proved more fruitful than he might have dreamed of.

11.2 A BRIEF HISTORY OF STRING THEORIES

Currently the best offer of a theory of everything, the 40-year-old string theory has a remarkable if largely unexplored history.[16] This much-discussed theory did not start out as an attempt to establish a unified theory, but, more modestly, in the context of the S-matrix-inspired programme for understanding the strong interactions (see Chapter 6). In 1968 the Italian theorist Gabriele Veneziano, at the time only 26 years old, suggested

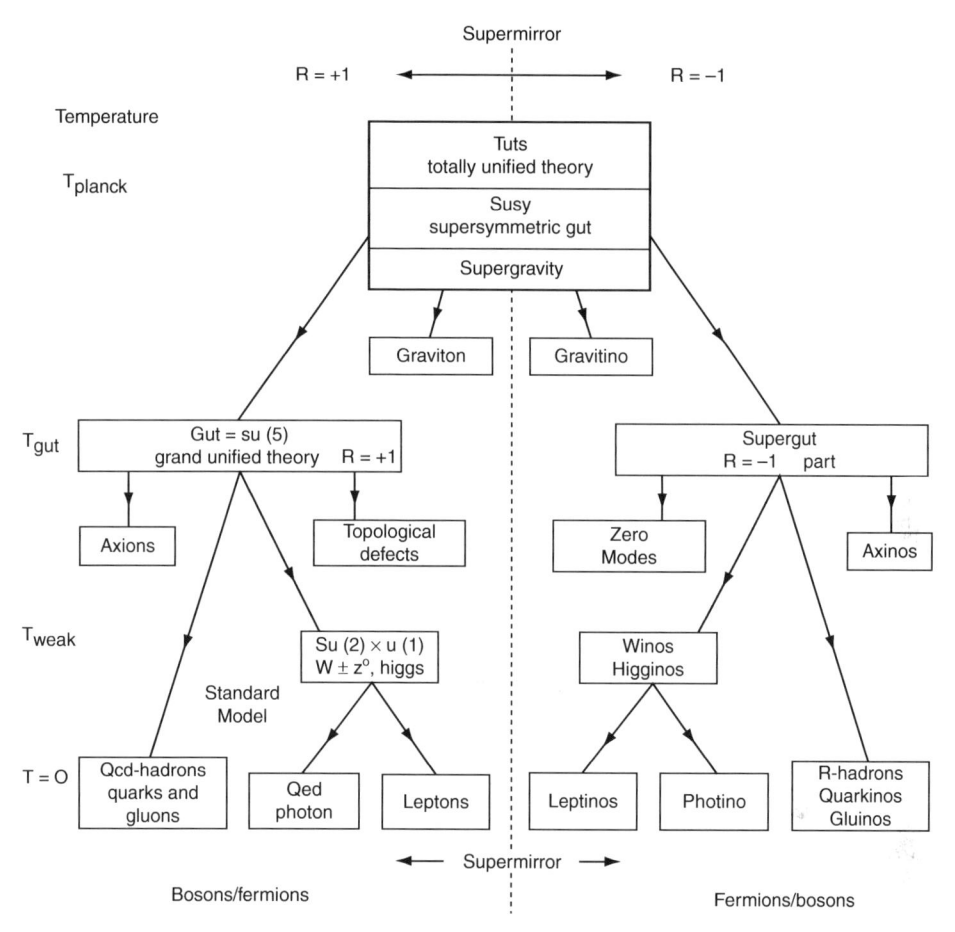

Fig. 11.1. Heinz Pagel's attempt to picture the unity in particle physics as known by the early 1980s. At the beginning of the universe (above), when the temperature was nearly infinite, all particles and forces are assumed to be governed by a single TUT, a 'totally unified theory'. With the cooling of the universe, a series of spontaneous symmetry breakings begin, which at low temperatures result in the particles known today. These and the theories that describe them are given on the left side of the picture (R = +1); on the right (R = −1) are shown the assumed supersymmetric counterparts. In his review of unification physics, Pagels did not include string theory.

Source: Heinz Pagels, 'Microcosmology: new particles and cosmology,' *Annals of the New York Academy of Sciences* **422** (1984), 15–32.

a 'dual resonance' model of hadrons in order to explain the phenomenology of certain scattering processes involving strongly interacting particles. He did not think of strings, nor did he anticipate them in his paper, which only pioneered string theory in an indirect sense. Two years later, Yoichiro Nambu of the University of Chicago, Holger Bech Nielsen of the University of Copenhagen, Tetsuo Goto of Nihon University, Japan, and Leonard Susskind had proposed a physical interpretation of Veneziano's theory by representing the strong force as one-dimensional vibrating quantum strings. Susskind recalled that he was greatly intrigued when he learned about Veneziano's model:

> I immediately set to work on trying to construct a model of hadrons consisting of a qq [quark-antiquark] pair interacting by a harmonic force that would scatter according to the Veneziano Model. By the spring of 1969 Yoichiro Nambu and I had independently discovered the string model.... According to this theory, a meson was a qq pair joined by an elastic string with an energy that increased linearly with the length of the string.[17]

This first string picture of hadrons allowed quarks to exist and explained why they could never appear isolated from the hadrons of which they were parts. Indirectly it suggested to the experimentalists that they should abandon their searches for free quarks.

However, it quickly turned out that the early string model was unphysical, as it only described bosons and not fermions such as protons and neutrons. Moreover, in order to be consistent with special relativity and quantum mechanics it required a 26-dimensional space–time and predicted a massless spin-2 particle as well as a tachyon as a ground state of the mass spectrum. These looked more like predictions from a toy model than from a theory representing the real world of physics.

Tachyons, hypothetical particles that travel faster than light, may seem to be precluded by the theory of relativity. Indeed, in his seminal paper of 1905 Einstein declared that such particles could have no existence, a verdict that probably dissuaded most later physicists from looking closer at the matter.[18] Only in 1923 did an unknown Russian physicist by the name of L. Strum demonstrate that the assumptions of special relativity do not rule out tachyons as long as they remain superluminal. Strum's argument went unnoticed and only reappeared in the 1960s when Gerald Feinberg, George Sudarshan, and a few others made what was effectively the same argument. In 1967 Feinberg coined the name tachyon, derived from the Greek 'tachys', meaning swift. Based on the principle of plenitude and little else it was suggested that, since tachyons can exist, they actually do exist. During the following decade, tachyon physics attracted considerable interest, including several attempts to detect the elusive particles (and, as might be expected, a few discovery claims). After a brief period of discussion it became the established view that although tachyons are acceptable from the point of view of relativity alone, they are inconsistent with theories based on quantum mechanics. The kind of tachyons that entered string theory were not really particles that moved faster than light, but particle states with an imaginary rest mass, meaning $m^2 < 0$. (For a particle with $v > c$, this follows from the special theory of relativity.)

After this brief digression, back to the early 1970s, when a more realistic 10-dimensional string theory, accommodating fermions as well as bosons, was found in different ways by Pierre Ramond at Yale University and by John Schwarz at the California Institute of Technology and the French theorist André Neveu. String theorists at that time pictured hadrons as consisting of quarks bound together by elastic strings of the order of 10^{-15} m, roughly the radius of an atomic nucleus. The inevitable occurrence of a massless spin-2 particle was originally seen as a serious flaw, for no such particle could be of any relevance to hadron physics. However, in 1974 Schwarz, in collaboration with a young French theorist, Joël Scherck, turned the difficulty into a virtue by realizing that the particle might be a graviton, the particle of the quantized gravitational field.[19] Until then, string theory had been a theory of strong interactions alone, with no reference to gravitation. The theory was now reinterpreted as a unified theory of all the fundamental forces, perhaps a framework for a theory of everything. It turned out that if the constant of gravitation was to have the correct value, the length scale of the strings needed to be close to the Planck length of 10^{-35} m, that is, immensely smaller than the original hadronic strings. On the other hand, strings do have a finite size, however small, which is an important difference from the particles of quantum field theory.

The objects of the theory, the strings, were very different from the lepton and quark point particles appearing in the standard model of particle physics. They were one-dimensional curves, either closed or open, living in a 10-dimensional space–time. The tensions of the strings were enormous, corresponding to energies of the Planck level of 10^{19} GeV. This tension, a measure of how much energy is contained per unit length of string, was seen as the basic defining constant of the theory. It corresponds to an absolute minimum uncertainty in length of the order of 10^{-34} m. The excitations or vibrations of a string were interpreted as giving the spectrum of elementary particles, hopefully including those known empirically but also including a multitude of other hypothetical particles. In superstring theories there are many particles corresponding to a given type of vibration and all of the known particles are supposed to be described by ground states of the string. The necessary 10 dimensions do not, of course, correspond to the world we experience, but the six extra dimensions of the 'internal' space were thought to be curled up or compactified, and hence be unobservable, in a manner similar to the fifth dimension in the original Kaluza–Klein theory.

Interest in string theory was limited in the 1970s, in part because of the successful development of quantum chromodynamics that followed the discovery of charmed quarks and which signalled the triumph of the standard gauge model of strong interactions. Still, a small number of physicists continued working on string theory, in part because they found it to be mathematically attractive. As Schwarz recalled, paraphrasing Dirac's philosophy of physics, 'string theory was too beautiful a mathematical structure to be completely irrelevant to nature'.[20]

By 1980 string theory had become superstring theory, thus incorporating the fermion-boson correspondence characteristic for supersymmetric theories. The introduction of supersymmetry was originally proposed by Ramond as early as 1971 and during the next few years independently by other physicists in the Soviet Union and the United States. The 10-dimensional superstring theories, as they were by the early 1980s, had no tachyonic ground state and were seen as mathematically attractive, but not likely to be of much physical relevance. Even on the theoretical level there were severe problems, namely that the theories were plagued by infinities and what are technically known as anomalies. Anomalies are terms that violate the symmetries or conservation laws when the theory is quantized, and therefore make the theory inconsistent. There were other consistency problems, for example that the 10 dimensions seemed incompatible with the correct form of superstring theory. For these and other reasons, string theory was not considered particularly attractive by the majority of theoretical physicists. In the early 1980s, the annual output of papers dealing with strings and superstrings was about 50 (Fig. 11.2).

Drawing on work by Edward Witten and his collaborator Luis Alvarez-Gaumé, in the summer of 1984 Schwarz and his British collaborator Michael Green at Queens College, London, brought new life to string theory by showing that all of the anomalies would cancel each other out if the theory was governed by one of two particular symmetry groups.[21] In one of the resulting theories, known as SO(32), charge conservation arose as a result of including gravity, and it was this theory that Schwarz and Green first developed. Shortly after this breakthrough, Witten suggested how this version of superstring theory could be compactified to get a four-dimensional theory. The Schwarz–Green paper initiated what in the string community is known as the 'superstring revolution', or the first of such revolutions (Fig. 11.3).

Another important component of the 1984 revolution was the development of yet another new version of superstring theory, known as the $E_8 \times E_8$ heterotic theory, by David Gross and his collaborators at Princeton University. (The team, whose other members were Emil Martinec, Jeffrey Harvey, and Ryan Rohm, was jokingly known as the Princeton string quartet.) Being a mixture of the older 26-dimensional boson string theory and the new 10-dimensional supersymmetric theory, the heterotic theory was considered promising with respect to the low-energy world. As Gross and his coauthors noted, 'The heterotic $E_8 \times E_8$ string is perhaps the most promising candidate for a unified field theory. One can easily contemplate physically interesting compactifications

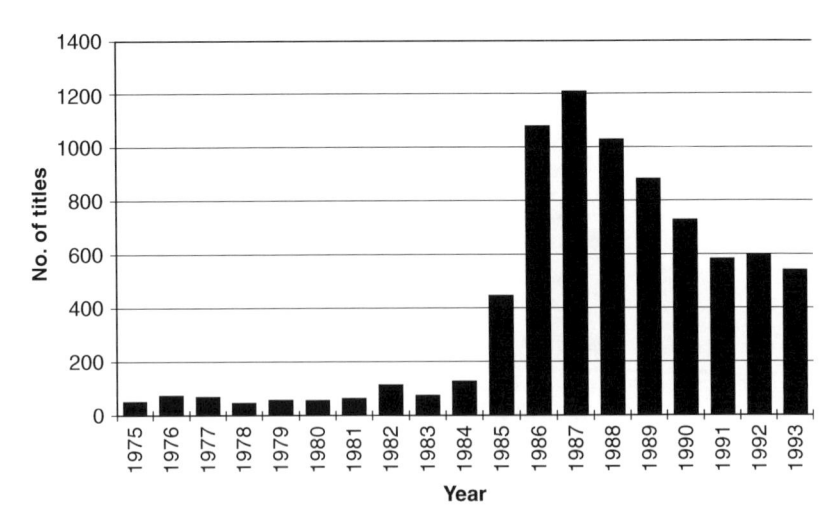

Fig. 11.2. Number of papers on strings and superstrings 1975–93, showing the impact of the 1984–85 revolution.

Source: Galison 1995, in *Laws of Nature. Essays on the Philosophical, Scientific and Historical Dimensions*, ed. Friedel Weinert, pp. 369–408. Berlin: Walter de Gruyter.

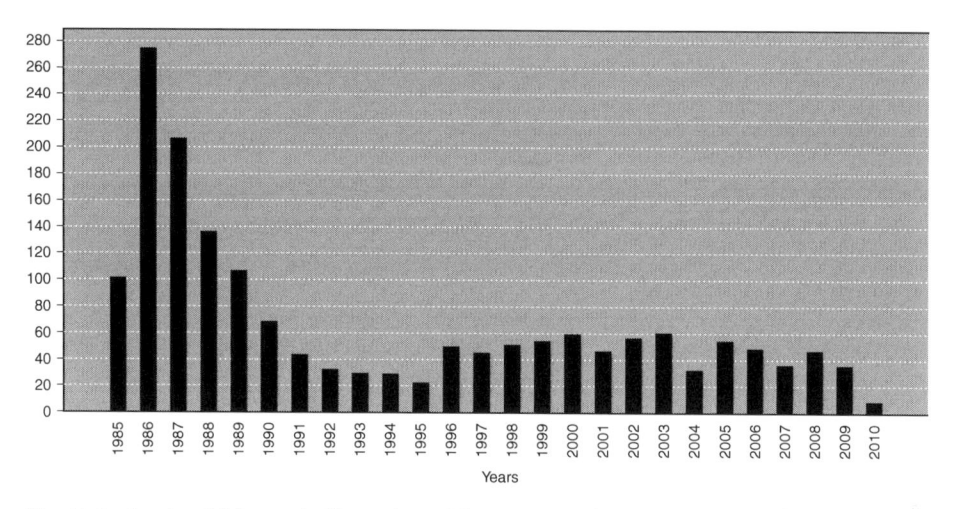

Fig. 11.3. Another bibliometric illustration of the string revolution in 1984–85. The graph shows the number of citations to the 1984 paper by Green and Schwarz. While after the peak in 1986 the number of citations begins to decline, following the normal pattern, from 1995 it slowly grows. The rise probably reflects the increased interest in string theory caused by the M-theory formulation in 1995.

Source: Data from ISI Web of Science.

of this theory to four dimensions.'[22] The paper of the Princeton theorists was followed by one dealing with the compactification problem written by Philip Candelas, Gary Horowitz, Andrew Strominger, and Witten.[23] They showed that, given certain conditions, the six extra dimensions of space–time take the form of a six-dimensional so-called Calabi–Yau manifold. This type of space is named after two American mathematicians, Eugenio Calabi and Shing Tung Yau, who studied it prior to string theory and for purely mathematical reasons. It is one more example of an exotic mathematical idea which unintentionally proved useful in (perhaps equally exotic?) physical theory.

The remarkable thing about the three string theories known by the late 1980s was that they were unique and completely free of adjustable parameters.[24] The mathematical structure was so tightly knit that that it could not be changed without destroying the theories. This was a feature that appealed strongly to many theoretical physicists and contributed to making the new superstring theory the favoured framework for an ultimate theory of physics.

The reborn superstrings of 1984–85 changed the fate of string physics and made supersymmetric theories, whether in the context of strings or other theories, very popular. While only 16 papers were published on supersymmetry in 1983, two years later the number had increased to 316, and in 1986 no less than 639 scientific papers included supersymmetry as a keyword. Following the much-publicized string revolution, expectations ran high. In an address of 1986, Weinberg embraced the new superstring theory as a likely candidate for a final theory of nature. Admitting that 'the phenomenological success of superstring theory is not part of its justification', which obviously was an understatement, he suggested that the lack of connection to experiments might be temporary and at any rate not an important problem. Because, 'the real reason so far for being interested in superstring theories is that they are mathematically consistent [and] finite... relativistic quantum theories, of which there are pretty few, and these are the only ones we know that contain gravity'.[25]

Developments in string theory were and still are purely theoretical, but of course some physicists stopped their frenetic calculations from time to time, asking themselves if the theory they worked with was testable physics. Writing in 1987, Schwarz was optimistic, not because the theory had delivered any testable predictions but because he thought it was moving in the right direction. 'The theory should enable us to calculate the properties of elementary particles at ordinary energies', he said, adding that 'with all the brainpower to bear, there is no reason to be pessimistic about the eventual testability of the theory.'[26] Twenty years later string physicists still appealed to the 'eventual testability' of the theory. Alvaro De Rújula, a physicist at CERN, was not so easily carried away by the spirit of optimism following the breakthrough in 1985: 'Such a gregarious fascination for theories based almost exclusively on faith has never before charmed natural philosophers, by definition.' Superstrings, he noted, might have nothing to do with physical reality, and yet 'they have the irresistible power of addiction'.[27]

The addiction metaphor was not out of place: A kind of bandwagon effect occurred in the years following the 1984 revolution. Whereas the annual number of papers dealing with string theory had been fewer than 100 during the years 1975–83, in 1987 alone about 1200 papers were published on the subject, after which publications slowly declined. Superstring theory was hailed as the near-fulfilment of a century-old dream, the stepping stone to a new physics, the sought-after holy grail of a quantum theory of gravity. At least, this was how the string theorists themselves looked upon the situation and it was also the picture they conveyed to the public. Witten expected that a proper understanding of string theory would be obtained within the near future. When this happened, it would 'involve a revolution in our concepts of the basic laws of physics—similar in scope to any that occurred in the past'.[28] He prophesied that the theory 'will dominate the next half century, just as quantum field theory has dominated the previous half century'.[29]

However, the enthusiasm was moderated when the new theory of superstrings was more closely investigated, with the result that problems arose. About 1990 there existed five competing theories, an embarrassingly large number for a supposedly unique theory of everything. No unique theory had been found, no unique way of curling up the six extra dimensions. All the theories predicted gravitons, which was a good thing. But they also predicted particles that had not been detected and might not exist, such as extremely heavy particles (of masses of the order of the Planck mass of 10^{-5} g) with fractional charges that might or might not have been produced in the big bang. In spite of the many predictions, the proper predictive power of the theories was close to zero. In an important mathematical paper of 1986, Andrew Strominger, at the time at the Institute for Advanced Study in Princeton, ended with a few comments on the physical implications of the theory, something which was not usual in the string literature. Noting that 'the class of supersymmetric superstring compactifications has been enormously enlarged', he said: 'While this is quite reassuring, in some sense life has been made too easy. All predictive power seems to have been lost.'[30]

Yet by the end of the decade optimism and excitement had returned to the string community. Some theorists had begun to study other types of string-like dynamical objects, including two-dimensional surfaces that might be pictured as membranes. These kinds of surfaces or 'branes' were introduced by Polchinski in 1995. In an address of the same year at the University of Southern California, Witten put the five consistent string theories together in an 11-dimensional umbrella theory called M-theory. According to Witten, 'M stands for magic, mystery, or matrix, according to taste', and he foresaw that 'physicists and mathematicians are likely to spend much of the next century trying to come to grips with this theory'.[31] His key idea was that the five theories in 10 dimensions were somehow the same theory or different limits of an underlying meta-theory. What this theory was, more precisely, neither he nor other string physicists could tell.

This second superstring revolution, as it is often called, brought a new wave of optimism and confidence to string physics. 'For the first time', said Cumrun Vafa, a string theorist at Harvard University, 'there's a hope that we can solve string theory in our lifetime.'[32] With M-theory and what followed, new recruits were attracted to the field. It is estimated that the string community amounts to some 1500 scientists worldwide, a remarkably large number given the abstract and purely theoretical nature of string physics.[33] Incidentally, this means that there are as many string theorists today as the total number of academic physicists in the world about 1900, all countries and fields of physics included.[34] The comparison not only speaks to the popularity of string theory but also to the phenomenal growth of physics in the twentieth century.

The renewed optimism did not mean that string physics at the turn of the century had entered a state of smooth progress. Low-energy physics in the form of the standard model still could not be derived from string theory, and the mechanism for breaking supersymmetry (which is necessary since the observed world is not supersymmetric) remained a problem. Another problem was the discovery of the accelerating universe in the late 1990s, which is generally taken to imply the existence of a dark energy due to a positive cosmological constant. Dark energy came as an unwelcome surprise to the string physicists, who were not prepared for it at all. Although no definite value of the cosmological constant was derived from string theory, it was agreed that the constant could not be positive, which meant that the theory was faced with a serious problem. Later work done by a group of Stanford theorists, known as KKLT (Shamit Kachru, Renata Kallosh, Andrei Linde, and Sandip Trivedi), solved the problem by producing a string theory consistent with a positive cosmological constant. Receiving some 900 quotations within a few years, the KKLT paper was seen as very important in the string community.

But there was a price for it. The theory, as developed in the early years of the new century, implied a huge number of distinct string theories—a landscape of them. As early as in the late 1980s it appeared that there were many more Calabi–Yau spaces than expected, and hence a large number of compactifications, but it took several years until the physical implications became clear and it was realized that string theory might be likened to a hydra with 10^{500} heads.[35] The recent landscape interpretation of string theory marked a drastic change in the thinking of string theorists. It also led to the landscape multiverse, as discussed in Chapter 10.

The historical outline presented above is a brief standard version of the main steps in the development of string theory over a period of forty years. There are other, more sociological ways to present and illustrate the history, for example by means of bibliometric publication and citation data. Scientometric analyses support the view of the dominant position of string theory in modern theoretical physics and also the importance of the two string revolutions of 1984 and 1995.[36] Thus, in the list of ScienceWatch in 1997, four of the ten most-cited physics papers were about string theory. The corresponding list in 1999 was even more dominated by contributions from string physicists. As another indication of the impact of string theory, consider the citations of the papers of Witten. In the period 1981–2010, his research papers received a total of no less than 51,200 citations and in part of the period Witten was the most cited of all physicists. Given that his papers are purely theoretical, this is quite remarkable. One of his early and important papers, 'The dynamical breaking of supersymmetry' of 1981, has been cited more than 2000 times. Even 26 years later, in 2008, it was cited 94 times. As mentioned, initially string theory did not attract much attention in the physics community: the two papers that Schwarz and Scherck published in 1974 received only a total of 53 citations over the next decade. The changes from about 1985 can be illustrated by the citations of a paper on 'Superstring theory' that Schwarz published in 1982 and in which he presented a detailed review of current string theory.[37] This paper was cited 46 times in 1983–84, while in 1985–86 it received 246 citations.

Whether seen from a sociological or internal scientific perspective, the history of string theory has been strikingly different from the histories of relativity theory and quantum mechanics, the two most dramatic changes in modern theoretical physics. The theory of relativity was largely the work of Einstein, who based it on physical ideas he conceived in 1905 and extended over the following years. Quantum mechanics was not the result of a single genius, yet only a handful of physicists contributed crucially to its creative phase: Heisenberg, Schrödinger, Born, Jordan, and Dirac. The new theory of atoms and quanta developed with astonishing speed and was quickly disseminated to the physics community at large. Less than three years after Heisenberg's breakthrough in the summer of 1925, quantum mechanics was complete and had been extended to the realm of special relativity. In both cases, relativity theory and quantum mechanics, the theories made predictions that were confirmed experimentally.

String theory has developed in an entirely different way. It was not based on any new principle of physics, and there still is no such principle, no equivalent to the principle of relativity or the indeterminacy principle. String theory has gradually if unevenly developed over a period of 40 years, based on the work by an army of physicists. In spite of some outstanding string physicists, the field has no equivalent of an Einstein or a Heisenberg. Last but not least, although great progress has been made in the theory, solid contact with experiments is still missing.

As mentioned, string theory is often singled out as the premier candidate for a theory of everything. This is, however, a hype more commonly used by science writers and journalists than by string theorists themselves. It is a hype that rarely turns up in the scientific literature and with which many physicists are unhappy. Green and Witten are just two of the more prominent string physicists who have objected to calling string theory a theory of everything. In *The Elegant Universe*, a popular and widely read account of the marvellous theory of superstrings, Brian Greene enthusiastically wrote that 'string theory provides the promise of a single, all-inclusive, unified description of the physical universe: a theory of everything (T.O.E.)'. But he also emphasized that string theory, or some future development of it, will at most be a theory of everything in principle, never in a practical reductionistic sense: 'Almost everyone agrees that finding the T.O.E. would in no way mean that psychology, biology, geology, chemistry, or even physics had been solved or in some sense subsumed.'[38] Yet there are those who are willing, at least in the context of popularization, to go further. One of them is Michio Kaku at the City University of New York, an expert in string theory and author of popular science books:

> If string theory is sound, it should allow us, mathematically, to compute basic properties of the universe from first principles. For instance, it should explain all the properties of familiar subatomic particles, including their charges, mass, and other quantum properties. The periodic table of elements that students learn in chemistry class should emerge from the theory, with all the properties of the elements precisely correct. If the computed properties do not fit the known features of the universe, string theory will immediately become a theory of nothing. But if the predictions accurately match reality, that would represent the most significant discovery in the history of science.[39]

Such an unrestricted reductionism, which at any rate is in principle only, is also part of Tegmark's idea of a theory of everything. However, this idea is very different from the string theory version.

11.3 THE TROUBLE WITH STRING PHYSICS—OR STRING PHYSICISTS?

String theory was not initially seen as controversial, but after the 1985 revolution and the drastic rise in popularity of the theory it began to attract methodological and other criticism. This was not only because of the theory's glaring lack of connection to experiments, but also because of the way enthusiastic string theorists spoke of and promoted it as 'the only game in town' when it came to unifying the forces of nature. One is reminded of Chew's promotion of the bootstrap theory in the 1960s. It can be difficult to distinguish clearly between epistemic and sociological elements in the controversy over string theory, which started in the 1980s and has continued to this very day. Understandably, many high-energy experimentalists received the superstring revelation with indifference, distrust, or hostility. And they were not alone in opposing a theory which seemed to signal a new style of theoretical physics and which could apparently be tested only by means of mathematics. As pointed out by historian of science Peter Galison, in the string controversy of the late twentieth century (continuing into the early twenty-first century) we witness a profound and contested shift in the position of theory in physics.[40]

According to Richard Feynman, physics is and should be essentially an interaction between experiment and theoretical calculation. Based on his equations, the theorist will calculate some physical property, which, directly or indirectly, can then be confronted with experiments. If the experiments agree with the prediction, the theory need not be correct but the physicist will have increased confidence in it. If they disagree, he will have to reconsider the theory or possibly scrap it. According to Feynman, physics is about precise and testable predictions, such as Dirac's prediction of the positron on the basis of his relativistic wave equation, or the calculations from quantum electrodynamics of the electron's magnetic moment. By the 1980s, the magnetic moment of the electron, measured in Bohr magneton units, μ_B, was determined experimentally as:

$$\mu = (1.00115965221 \pm 0.00000000003)\mu_B$$

whereas calculations gave:

$$\mu = (1.00115965246 \pm 0.00000000020)\mu_B$$

This was Feynman's kind of great physics.

In 1987, a few months before he died at the age of 69, Feynman objected in an interview that string theory was unable to come up with answers to unsolved problems. For example, why was the muon 207 times as heavy as the electron? As he saw it, string theory moved in a wrong direction because it did not result in calculations of measurable quantities. 'I don't like that for anything that disagrees with an experiment, they [the string theorists] cook up an explanation—a fix-up to say "Well, it still might be true".... So the fact that [string theory] might disagree with experience is very tenuous, it doesn't produce anything; it has to be excused most of the time. It doesn't look right.'[41] In another interview, conducted by the physicist and historian of science Jagdish Mehra, Feynman said of the 11 dimensions of space–time postulated by string theory: 'The world doesn't have eleven dimensions, so it rolls up seven. Why not six, why not four? It's a hell of a theory, isn't it? One can't even check the number of dimensions.'[42]

Another Nobel laureate and distinguished theorist, Sheldon Glashow, was no less adamant in his opposition to string theory and the style of physics it represented. Together with Paul Ginsparg, a Harvard colleague, he complained about the arrogance of string theorists and their lack of concern with contact with experiments. 'In lieu of the traditional confrontation between theory and experiment, superstring theorists pursue an inner harmony where elegance, uniqueness and beauty define truth.' Glashow and Ginsparg feared that mathematics and aesthetics were on their way to supplanting experiment and turning theoretical physics into a sterile intellectual game, a scholastic exercise.

> Contemplation of superstrings may evolve into an activity as remote from conventional particle physics as particle physics is from chemistry, to be conducted as schools of divinity by future equivalents of medieval theologians. For the first time since the Dark Ages, we can see how our noble search may end, with faith replacing science once again. Superstring sentiments eerily recall 'arguments from design' for the existence of a supreme being.... How satisfying and economical to explain everything in one bold stroke of our aesthetic, mathematical and intuitive sensibilities, thus displaying the power of positive thinking without requiring tedious experimentation! But *a priori* arguments have deluded us from ancient Greece on.[43]

Glashow's caustic allusion to theology was not accidental. He repeated it two years later, in words that were no milder. 'Until the string people can interpret perceived properties of the real world, they simply are not doing physics. Should they be paid by universities and be permitted to pervert impressionable students?... Are string thoughts more appropriate to departments of mathematics or even to schools of divinity than to physics departments?'[44] To Glashow, the string theorists were 'kookie fanatics following strange visions' and the superstring fashion—or fad—a disease 'far more contagious than AIDS'.[45] In a comment of 2006 another veteran of high energy physics, Burton Richter, attacked string theory and similar advanced theories for being more like 'theological speculation' than science with testable consequences. Referring to Popper, Richter maintained that testability and falsifiability were the hallmarks of science, and he could find neither of them in string physics.[46]

Of course, string theorists are neither mathematical theologians nor intelligent designers. To suggest any link between string theory and intelligent design is an insult, which may be the reason why the suggestion has nonetheless been made from time to time. For example, the physicist and cosmologist Lawrence Krauss (who is strongly opposed to intelligent design) has on some occasions associated string theory with intelligent design, if only from a methodological point of view:

> Look at string theory, how can you falsify that? It's no worse than intelligent design. I do think there are huge differences between string theory and intelligent design. People who are doing string theory are earnest scientists who are trying to come up with ideas that are viable. People who are doing intelligent design aren't doing any of that. But the question is, is it falsifiable? And do we do a disservice to real theories by calling hypotheses of formalisms theories? Is a multiverse—in one form or another—science?[47]

The questions raised by Krauss were discussed more fully by Robert Ehrlich, a professor of physics at George Mason University, Virginia. In broad agreement with Krauss, Ehrlich concluded that while string theory was 'marginally within the realm of science', intelligent design was not scientific at all.[48]

Glashow subsequently softened his view, among other reasons because of a lack of progress in conventional field theories of grand unification, a branch of physics he had himself helped to establish. Some 15 years after his attack on string theory, he was willing to admit it as physics, even useful physics. 'Still', he said, 'It does not make predictions that have anything to do with experiments that can be done in the laboratory or with observations that could be made in space from telescopes.' He further complained that string theory broke with the cumulative nature of science, in the sense that it did not incorporate the details of earlier knowledge, such as the standard model of elementary particles. 'Until it does that, it is not yet physics in a conventional form. It is a perhaps promising corner of physics that may some day say things about the world. But today they're saying things about string theory to one another.'[49]

Attacks on string theory have continued. Most of the critical comments have focused on the theory's lack of testability and its failure to produce results concerning the world as it is experienced. Daniel Friedan, an American physicist who was active in the early phase of string theory, was more critical than most. Defining physics as 'reliable knowledge of the real world, based on experiment', he stated the traditionally accepted credo of physicists, namely that a theory 'must be capable of making definite statements than can be checked'. If judged by these standards, string theory failed miserably:

> The long-standing crisis of string theory is its complete failure to explain or predict any large distance physics.... String theory is incapable of determining the dimension, geometry, particle spectrum and coupling constants of macroscopic spacetime. String theory cannot give any definite explanations of existing knowledge of the real world and cannot make any definite predictions. The reliability of string theory cannot be evaluated, much less established. String theory has no credibility as a candidate theory of physics.... More broadly, string theory, as it stands, is incapable of generating the variety of large characteristic spacetime distances seen in the real world.[50]

Not only did Friedan suggest that string theory did not really explain or predict gravity, he also saw no reason to base a fundamental theory on supersymmetry, which after all is not observed in nature. Nor was he impressed by string theorists' frequent appeals to beautiful mathematics: 'History suggests that it is unwise to extrapolate to fundamental principles of nature from the mathematical forms used by theoretical physics in any particular epoch of its history, no matter how impressive their success. . . . Mathematical beauty in physics cannot be appreciated until after it has proved useful.'[51]

Most physicists in the large string community ignore objections of the kind made by Glashow, Friedan, and Peter Woit, which they feel are irrelevant and unfair, not to mention that they are 'philosophical' (or, even worse, 'sociological'). But some have responded to them, either directly or indirectly. The standard attitude to the allegation that string theory is divorced from experiment is not to deny the ultimate role of experiment, but to ask for more time to meet the requirement. Superstring theory is terribly complex and its structure and underlying principles not even understood very well, so it may take many years until we can come up with results that refer to the low-energy world we live in. But they will come, and meanwhile there is no reason to worry. (There is here another analogy to the vortex atom theory described in Chapter 2, as in when Tait said in 1876 that the mathematical problems of the theory would occupy the next generation or two of the world's mathematicians.)

In an interview of 1987, Weinberg was asked if and how string theory could explain the properties of the electron. 'I find your question an awkward one', he responded. 'It's like asking "how in general relativity do you work out the shape of a suspension bridge?"'.[52] More than 20 years later Schelleken made a similar point, namely, that one cannot reasonably expect string theory to provide precise answers to the problems of particle physics. While Feynman would have string theory predict the correct value of the muon–electron mass ratio, Schelleken found it irrelevant: 'Rejecting a theory of gravity that makes no particle physics predictions may be like rejecting the theory of continental drift because it does not predict the shape of Mount Everest.'[53]

Apart from Feynman and Glashow, several other Nobel Prize winning physicists have expressed their dislike of string theory and its associated culture of physics. The Dutch–American theorist Martinus Veltman ended a popular book on particle physics by justifying why it did not mention the theory of superstrings: 'This book is about physics, and this implies that theoretical ideas must be supported by experimental facts. Neither supersymmetry not string theory satisfy this criterion. They are figments of the theoretical mind.'[54] Philip Anderson, too, found string theory to be futile as physics and nothing more than an interesting specialty of mathematics. 'String theory is the first science in hundreds of years to be pursued in pre-Baconian fashion, without any adequate experimental guidance', he said in 2005. 'It proposes that Nature is the way we would like it to be rather than the way we see it to be; and it is improbable that Nature thinks the same way as we do.'[55] This was a variation of an older theme, similar to the criticism that McVittie in 1940 levelled against the cosmology of Milne, whom he accused of 'telling Nature what she ought to be like' (see Chapter 4).

Almost all physicists, string physicists included, agree that testability and predictability are important epistemic values. They may not be relevant to all aspects or in all phases of the development of a theory, but ultimately they cannot and should not be ignored. What Gross called 'the dream of all string theorists' was 'to predict a phenomenon, which would be accessible at observable energies and is uniquely characteristic of string theory'.[56] Alas, 35 years after the birth of string theory it remained a dream. In the absence of such unique predictions, string theorists have had to fall back on other kinds of prediction or to interpret the concept in manners that differed from the usual ones and were more adapted to what string theory can actually produce. In an interview of 2007, the Indian theorist Ashoke Sen, at the Harish-Chandra

Research Institute in Allahabad, was pressed on the question of whether string theory is falsifiable. Can it ever be proved wrong? Sen's answer was characteristically vague: 'One way that would happen, obviously, is if somebody else finds a theory that does better at describing particle physics, the universe we live in.... Another way it could turn out to be wrong is if suddenly there should be some unexpected experimental phenomenon that would lead us to a different concept or which could not be explained by string theory. Although what such a development might be, I wouldn't know.'[57] Not all string theorists are equally evasive.

If direct empirical testing is so difficult, can it not be avoided by replacing it with a criterion that is purely theoretical? Suggestions of this kind are sometimes made, if rarely as a serious alternative to empirical testing. Brian Greene has suggested that if 'unification nirvana' is reached with a *unique* unification of quantum mechanics and general relativity, then there is no need to require novel testable predictions from the theory. 'After all, a wealth of experimental support for both quantum mechanics and general relativity already exists, and it seems plain as day that the laws governing the universe should be mutually compatible', he points out. 'If a particular theory were the unique, mathematically consistent arch spanning the two experimentally confirmed pillars of twentieth-century physics, that would provide powerful, albeit indirect, evidence for the theory's inevitability.'[58] Of course, the problem is that it seems impossible to know that the unified theory in the form of some future version of the theory of superstrings is the only possible one. If this cannot be proved, tests remain necessary to convince the sceptic of the truth of the theory.

It is commonly argued by string physicists that their theory does lead to predictions and that some of them are actually verified. According to Witten and others, if we did not already know about gravity, we could discover it from string theory. We are invited to imagine a hypothetical history of physics without Newton and Einstein, with string theory somehow emerging on the scene and, by implication, general relativity. It is indeed remarkable that string theory includes gravitation, especially because it was not designed to do so, but it hardly counts as a prediction in the ordinary sense of the term. Would one say that Newton predicted Galileo's laws of falling bodies or that Maxwell predicted the laws of optics? Probably not. On the other hand, Witten insists that 'supersymmetry . . . is a *genuine prediction* of string theory'.[59]

Although supersymmetry has neither been confirmed nor refuted by experiments, it is certainly a testable hypothesis. Supersymmetric partner particles might turn up in high-energy experiments, which would undoubtedly be hailed as a triumph of string theory. However, it would not confirm the theory in any definite sense, only increase confidence in it, for supersymmetry is not exclusive to string theories. Conversely, if accelerator experiments fail to reveal any trace of supersymmetry, it does not falsify string theory; for the failure can be explained by arguing, for example, that the energies of the experiments were not high enough. Suppose, said Schwarz in 1998, that high energy experiments failed to find evidence of supersymmetry. Should we then shelve string theory? Not at all:

I believe that we have found the unique mathematical structure that consistently combines quantum mechanics and general relativity. So it must almost certainly be correct. For this reason, even though I do expect supersymmetry to be found, I would not abandon this theory if supersymmetry turns out to be absent.[60]

According to critics, superstring theory seems to have established a kind of problematic self-protection against empirical control. This is what Paul Halpern, a theoretical physicist at the University of the Sciences in Philadelphia, charges. Calling supersymmetry 'one of the most audacious proposals in the history of modern scientific thought', he says: 'No other physical theory has won so many supporters with so little experimental support, surviving instead on the basis of its own mathematical beauty and internal consistency.'[61] As we have seen, this is a view shared by several other physicists.

Another possible candidate for making contact with the world of experiments, apart from supersymmetry, are cosmic strings. These are string objects that have supposedly been blown up to large scales during the brief inflationary era and therefore, if they still exist, might be detected in distant regions of the universe. Being very massive, cosmic strings may reveal their presence by gravitational lensing or the emission of gravitational waves. Some versions of them may behave like superconducting wires, in which case they should produce dramatic cosmological effects. Much theoretical work has been done on cosmic strings, but none have been observed. Moreover, these hypothetical entities are not exclusively connected to the strings of superstring theory, and so it is uncertain if signals from them would amount to a confirmation of string theory.

There are some other ways, apart from accelerator experiments and cosmic strings, in which string theory may be tested experimentally, but again it is doubtful if they are specific enough to either confirm or refute the theory.[62] As briefly mentioned in Chapter 10, there is the possibility that signatures of the string world may turn up in cosmological observations, either in the cosmic background radiation or in new phenomena such as the 'dark flow'. Another possibility is to search for string theory's extra dimensions by checking the validity of Newton's inverse-square law of gravity at very small distances. Erich Adelberger at the University of Washington, as well as other teams of physicists, have for several years pursued this line of high-precision experimental research, in part motivated by the possibility that deviations from Newton's law at very small distances (less than a millimetre) may reveal the existence of the extra dimensions required by string theory.[63] Neither these nor other attempts to provide string theory with experimental support have yielded definite results in favour of the theory. They are not of a kind that can falsify the theory or otherwise test it in a crucial way. Generally speaking, falsification and string theory do not seem to go well together.[64]

Weinberg, one of the fathers of quantum chromodynamics, did not participate actively in the early development of string theory, but unlike Glashow he found the theory to be very appealing and he continues to do so. Weinberg's reasons for supporting the programme of string theory were basically meta-theoretical. 'It has the smell of inevitability about it', he said in 1987. The theory of superstrings 'cannot be altered without messing it up...[and] for this reason, quite apart from the fact string theory incorporates gravitation, we think we have more reason for optimism now about approaching the final laws of nature than we have had for some time'.[65] Weinberg recognized of course the difficulty in associating superstring theory with a physical picture that could be tested experimentally, but did not consider it a serious flaw. He thought it was the price to be paid for a fundamental theory:

> The final theory is going to be what it is because it's mathematically consistent. Then the physical interpretation will come only when you solve the theory and see what it predicts for physics at accessible energies. This is physics in a realm which is not directly accessible to experiment, and the guiding principle can't be the physical intuition because we don't have any intuition for dealing with that scale. The theory has to be conditioned by mathematical consistency. We hope this will lead to a theory with solutions that look like the real world at accessible energies.[66]

Thirteen years later Weinberg was ready to consider string theory, or rather some future development of it, as a possible candidate for the legendary final theory of physics:

> When at last we have a simple, compelling, mathematically consistent theory of gravitation and other forces that explains all the apparent arbitrary features of the standard model, it will be a good bet that this theory really is final. Our description of nature has become increasingly simple. More and more is being explained by fewer and fewer fundamental principles. But simplicity can't increase without limit. It seems likely that the next major theory that we settle on will be so simple that no further simplification would be possible.[67]

Although most physicists working with string theory believe that the theory can and will be tested in the traditional, empirical way—and that it will pass these tests of the future—there are also those who consider mathematics to be no less important than experiments. The view of Michio Kaku is that 'verification of string theory might come entirely from pure mathematics rather than from experiment'. In agreement with Weinberg and Kaku, Schelekken similarly suggests that 'Consistency may be the only guiding principle we have'.[68]

In his best-selling *Dreams of a Final Theory*, Weinberg offered some further reflections on the status of string theory and the negative responses from some of his colleagues. Fully aware of the difficulties of the theory, he said that even if these difficulties were solved and a single consistent theory established, there would be one overarching problem left: why would that particular string theory be the one that applied to the real world? A possible answer might be found in 'a principle with a dubious status in physics', namely the anthropic principle.[69] What Weinberg in 1992 only mentioned in passing as a somewhat speculative possibility would soon enter as a popular ingredient in string theory and cosmological models related to it.

Among the arguments in favour of string theory as physics, not just fancy mathematics, is the sociological observation that the theory actually is investigated by a large number of physicists and taken seriously by them. In the self-understanding of a large part of the physics community, the theory of superstrings is an impressive and exciting physical theory. It is a theory which is too good, or too beautiful, to be wrong. This is the view of Alan Guth, according to whom the lack of direct experimental tests is partly compensated for by consistency tests. As he said in 2004:

> If the goal of string theory is to build a quantum theory that's consistent with general relativity, that's a very strong constraint, and so far string theory is the only theory that seems to have convinced a lot of people that it satisfies that criterion. Just from a sociological point of view, theoretical physicists have been looking for a consistent quantum theory of gravity for at least 50 years now, and so far there's really only one theory that has reached the mainstream—string theory.[70]

The string controversy took a new turn in the early twenty-first century when the plethora of different string theories began to be recognized as unavoidable. As a consequence, string theory came to enter an unexpected alliance with the anthropic principle (see Chapter 10). While grand unification theories contain many free parameters, such as the masses of leptons and quarks, and therefore enough freedom to be a candidate for anthropic reasoning, string theory has no free parameters and therefore apparently allows no room for fine tuning. According to Gordon Kane at the University of Michigan, if all low-energy physics could be derived uniquely from string theory in terms of compactification and processes breaking supersymmetry, how could anthropic reasoning possibly enter? This is what he and his collaborators Malcolm Perry and Anna Zytkow argued in a paper of 2002, somewhat dramatically entitled 'The beginning of the end of the anthropic principle'. The usual anthropic arguments, they said, 'are unlikely to be relevant to understanding our world if string theory is the right approach to understanding the law(s) of nature and the origin of the universe'.[71] Kane and his coauthors realized the possibility that there might be many different ways of compactifications, corresponding to a large number of universes, and ended with the more cautious conclusion that string theory 'removes the possibility of making any anthropic arguments beyond those of choosing from a discrete set of possible vacua'.

With the landscape version promoted by Susskind and others, it turned out that the discrete set of possible vacua, also known as the 'discretuum', was much larger than expected. The embarrassing existence of a huge number of vacuum states soon came to be seen as a virtue rather than a disaster. Some of the leading string theorists resisted the idea, so different from the Einsteinian ideal of one ultimate and unique theory that had driven string theory so far. After all, the original aim of string theory had been to find a theory which consistently *and* uniquely described all of nature and was controlled only by requirements of self-consistency. And now it seemed that the laws of physics discovered in our universe are not the only ones possible. Would uniqueness have to be abandoned as an epistemic desideratum? Was it time to admit that it was 'nothing other than wishful thinking'?[72]

Nobel laureate and leading string theorist David Gross did not think so, and he was not happy at all about the landscape scenario and the kind of physics it exemplified.[73] In 2004 he described the appearance of an enormous number of stable vacuum states as 'one of the great mysteries of the theory, and the greatest stumbling block to its predictive power'. Gross did not reject the landscape idea because it relied on string theory, but because of its element of anthropic arguments:

> I find this kind of reasoning premature and defeatist. In the absence of a single consistent string cosmology (one where the initial conditions are determined and the initial singularity is resolved) and in the absence of a fundamental formulation of string theory, it is premature to come to such strong conclusions. Anthropic reasoning should, at best, be the last resort of physical theory. I think that at this stage of our understanding it is silly to give up on the hope that string theory will live up to its potential predictive power.[74]

Rather than accepting the anthropic principle as part of physics, Gross thought that some new concepts, probably of a very radical nature, might solve the problem and yield a unique vacuum state from which all physical parameters could in principle be calculated. Among the possible changes in a future theory might be to replace the space–time continuum with something more fundamental—but what? Keeping closely to the spirit of Einstein's dream, Gross says that 'the unified and completely predictive theory remains the ultimate goal of physics, and a guiding principle'. Moreover, referring to physics in the style of the anthropic landscape: 'A theory that contains arbitrary parameters or, worst of all, arbitrarily fine-tuned parameters, is deficient.'[75]

The more recent string controversy, closely relating to the controversy over the multiverse, was fuelled by popular anti-string books written by scientists, especially Lee Smolin's *The Trouble With Physics* and the mathematician Peter Woit's *Not Even Wrong*.[76] From the other side of the battlefield, Susskind defended the string landscape idea in his book *The Cosmic Landscape*. According to Smolin, string theory did not live up to established standards of physics. In the absence of empirical progress, string theorists have created their own criteria of success and tailored them to meet what string theory can offer. As Smolin saw it, this was to manipulate the criteria of what constitutes good science.

On the other hand, it is generally recognized that a research programme—and this may be the best way to characterize string physics—can progress in other ways than along the empirical dimension. For example, explanatory and unifying power are recognized methodological virtues, and it can be argued that with regard to these virtues string theory has indeed shown signs of progress. As pointed out by Nancy Cartwright and Roman Frigg, two philosophers of science, a research programme that makes progress only in some dimensions, but fails to do so empirically, does not count as being progressive. At least, this is not the case according to the methodology of scientific research programmes developed by the Hungarian–British philosopher Imre Lakatos in the 1970s. Cartwright and Frigg conclude that, 'as it stands, string theory is not yet progressive because it has made progress only along a few of the many dimensions that matter to a research programme's success'.[77] Significantly, the two philosophers do not conclude, as some have done, that because string theory is de facto unfalsifiable it is not scientific.

Although physicists do not generally listen to what philosophers say, some philosophers do listen to what physicists say and follow what they do. They may even be willing to change their views on the nature of science under the impact of important new developments in the sciences. Dudley Shapere, professor at Wake Forest University and a highly respected philosopher, believes that 'physics is in fact approaching, or perhaps has reached, the stage where we can proceed without the need to subject our further theories to empirical test'.[78] Referring to superstrings, compactified hidden dimensions, and theories of 'other regions of the universe, or even other universes, which are forever unconnected with us', he finds it reasonable to ask: 'Could empirical enquiry, which has guided science up to a certain point in its history, lead at that point to a new stage wherein empiricism itself is transcended, outgrown, at least in a particular domain?' Evidently, such a view is in much better accord with string theory, whether in the landscape version or not, than the traditional view associated with Popper's philosophy of science and its emphasis on crucial empirical tests.

The theory of superstrings is remarkable and impressive in many respects, mathematically as well as physically. It is equally remarkable from the perspective of sociology and history of science. During a period of 40 years a large number of physicists, and also a considerable number of mathematicians, have cultivated this approach to an ultimate theory of nature, and yet the thousands of articles and hundreds of conferences have not produced a solid bridge to the laboratory. Never before in the history of science has so many resources been devoted to a purely theoretical research programme over such an extended period of time. String theorists believe that their theory will establish contact with the world of experience if only they are given the necessary time to develop the connection. As two leading string theorists say: 'The primary goal of superstring compactification is to find realistic or quasirealistic models. Real-world physics . . . is rather complicated, and it should not be surprising that this goal is taking time to achieve.'[79] Contrary to what some critics have charged, string theorists have not abandoned the commonly accepted view of experiment as the final arbiter of physical theory. Although they place much emphasis on purely theoretical results, this is not because they assign higher priority to theory than experiment. In the end they value the connection to the empirically known world no less than their colleagues in other branches of physics.

11.4 AN ALTERNATIVE: LOOP QUANTUM GRAVITY

The term 'quantum gravity' generally refers to physical theories aimed at accounting for gravitational interactions in which matter and energy are described by quantum mechanics. In most of these theories, including string theory and loop quantum gravity, gravity is quantized, but 'quantum gravity' may also refer to theories in which this is not the case. Indeed, some physicists and philosophers of science have questioned the general assumption that a quantum theory of gravity is necessary. They point out that it is not necessary in order to solve empirical problems and is largely based on a desire for theoretical unification.[80]

It is a routinely stated argument for superstring theory that it is the only known consistent quantum theory of gravity. However, although string theory is the most developed and far the most popular theory of quantum gravity, it is not the only game in town. There exist several other theories of or approaches to quantum gravity, all of them tentative to varying degrees. Among these proposals, so-called loop quantum gravity (LQG) is the most elaborated alternative and the only one which has attracted a substantial group of researchers. According to a count of articles on quantum gravity in the late 1990s, the field was roughly divided into several approaches dominated by string theory and loop quantum gravity. The distribution, as given by the average number of articles per month, was as follows:[81]

String theory	69	(56.1%)
Loop quantum gravity	25	(20.3%)
QFT in curved space	8	(6.5%)
Lattice approaches	7	(5.7%)
Euclidean quantum gravity	3	(2.4%)
Non-commutative geometry	3	(2.4%)
Others	8	(6.5%)

Emerging on the scene later than string theory, much work has recently been done in loop gravity, which is considered by some physicists to be a serious contender for a consistent theory of quantum gravity and a preferred substitute for string theory. Physicists within this research programme sometimes refer to it as 'post-string physics'.

The first papers in loop gravity appeared 1987–90, when Lee Smolin and Carlo Rovelli, an Italian physicist at the University of Trento, found a formulation for a quantum theory of gravitation that was independent of the space–time background. As mentioned in Chapter 8, this work was based on a novel way of expressing general relativity discovered in 1986 by the Indian theorist Abhay Ashtekar at Syracuse University, which again built on earlier work by Amitabha Sen. Ashtekar's mathematical reformulation of general relativity greatly simplified Einstein's equations and also had the advantage that it expressed the theory in a language that was close to that used in quantum chromodynamics. Moreover, the new formulation was background-independent, meaning that there is no fixed background of space–time over or on which physical events occur; instead the geometry itself is treated as a dynamical quantity. String theory, on the other hand, assumes an existing space–time in which the strings move.

The theory of quantum gravity introduced by Smolin and Rovelli in 1988 incorporated fully relativistic space–time into quantum field theory, which they interpreted in terms of 'loop variables', the loops being quantum excitations of the lines of force of the gravitational field. Over the next years loop quantum gravity developed into a mature research programme that was continually improved and led to a number of applications and results not originally anticipated.[82] It was particularly important when Smolin and Rovelli realized in 1994 that a concept called 'spin networks' fitted beautifully into their new quantum picture of space–time. The idea of spin networks went back to the 1970s, when it was introduced by the British mathematician Roger Penrose in an attempt to understand the quantum origin of space.

From loop quantum gravity emerged a new picture of space, which at a very small scale was no longer seen as continuous but as having a discrete structure. Space is atomistic, not with atoms floating in space but with space itself being made up of discrete units, a kind of space-atom. To the extent that quantum space can be pictured at all, it may look like a fine fabric woven by loops. These loops are in a sense the atoms of space. The theory predicts minimal physical areas and volumes, the minimum size of an area being about 10^{-70} m^2 and for a volume about 10^{-105} m^3. All volumes and areas must be multiples of these very small units. The existence of minimal units is not an assumption, but a prediction of loop gravity. Because of the existence of these minimal quantities, corresponding to a smallest length of 10^{-35} m (the Planck length), the infinities in quantum field theories are eliminated. Thus, something like the old idea of a smallest length advocated by Heisenberg, Ivanenko, and others in the 1930s, and with roots further back in time, reappears in the modern theory of loop quantum gravity. It may seem that a minimal observable length (or area and volume) is incompatible with ordinary special relativity, where all lengths depend on the state of motion of the observer, but it turns out that there is no problem.

The almost counterintuitive idea of renouncing the space–time continuum had been considered from time to time by none other than Einstein. But his speculations of a possible 'algebraic physics' led to nothing. In a letter of 1954 to his old friend Michele Besso, he wrote: 'I consider it entirely possible that physics cannot be based upon the field concept, that is on continuous structures. Then *nothing* will remain of my whole castle in the air including the theory of gravitation, but also nothing of the rest of contemporary physics.'[83] Einstein consequently abandoned the idea of a discrete space–time.

As physicists working within the framework of loop quantum gravity like to point out, this theory differs radically from string theory, both when it comes to technical and to conceptual issues. Loop gravity is mathematically rigorous and ambitious, but physically it is more modest than string theory and methodologically it is more conservative. It aims at establishing a complete unification of gravity and quantum mechanics, not a unification of all the interactions of nature. Thus, it has nothing directly to say about elementary particles and their interactions, and in its spirit it is quite foreign to the ideal of a theory of everything. However, recent results suggest that loop quantum gravity may nonetheless be developed into a theory that generates the particles of the standard theory. The theory is in a sense conservative, since it is a quantum theory of space–time based on well established physical principles: those underlying quantum mechanics and general relativity. Contrary to most other approaches to quantum gravity, the loop gravity approach does not consider modifications to general relativity or to the basic principles of quantum mechanics. Smolin considers this methodological conservatism to be a virtue and loop quantum gravity an example of fundamental physics which obviates the need to revise the scientific method. It is 'science done the old-fashioned way', he says.[84]

Compared with string theory the theory of loop quantum gravity is parsimonious, including neither supersymmetry nor a space–time with more than the four known dimensions. Contrary to string theory and conventional quantum field theory, loop gravity does not rely on a given background space–time. The two theories or research programmes are also very different from a sociological point of view, a difference that some loop gravity physicists are fond of pointing out. Although the two programmes are not incommensurable, they focus on different questions, speak in different languages, and invent different imaginary worlds. According to Rovelli: 'String theory and loop gravity differ not only because they explore distinct physical hypotheses, but also because they are expressions of two separate communities of scientists, which have sharply distinct prejudices, and who view the problem of quantum gravity in surprisingly different manners.'[85] From a media and public perspective the difference is great as well: while string theory has attracted much controversy and massive press coverage, loop quantum gravity, if far from invisible in the public arena, generates much less press.

There is no direct experimental support for loop quantum gravity, just as for string theory. Yet physicists working in the loop gravity programme are no less eager than the string theorists to emphasize that there are, after all, connections from theory to experiment. These connections are mostly indirect and relate to possible future experiments, but they are nonetheless real. 'The theory makes definite quantitative predictions', maintains Rovelli, referring to the discreteness of space on the Planck scale, a prediction which unfortunately seems to be testable only in principle.[86]

There are other predictions, some of them following strictly from the theory while others are mere possibilities or expectations. To the latter category belongs the prediction that the speed of light will depend slightly on the wavelength: the higher the energy of a photon, the slower it travels (but the difference in speed is minute). The predicted effect has never been observed, but it may possibly be tested in satellite observations of high-energy photons from distant gamma-ray bursts. In fact, in the autumn of 2009 a team of researchers concluded from an analysis of data from a short gamma-ray burst recorded by a detector onboard the Fermi Gamma-Ray Space Telescope that there was no photon dispersion. Photons in the range between 10 keV and 31 GeV, after travelling a distance of 7.3 billion light years, were found to arrive at almost the same time. The data strongly disfavour some theories of quantum gravity that predict a discrete space–time, although not all such theories.[87] 'Einstein's theory prevails', as the *New York Times* announced.

Another kind of test relates to the thermodynamics of black holes and the radiation from them predicted by Hawking in 1974. Calculations based on loop quantum gravity have reproduced the so-called Bekenstein–Hawking formula according to which the entropy of a black hole is proportional to the surface area A of its event horizon (the relationship is $S = kc^3 A/2h$). It has also proved possible to reproduce the Hawking radiation and further to predict a discrete fine structure in its spectrum. This is counted as one of the 'definite quantitative predictions' arising from loop gravity, but it is worth recalling that the prediction concerns a phenomenon for which there is not the slightest experimental evidence. Finally, as string theory has given rise to models of the universe, so has loop quantum gravity. As mentioned in Chapter 8, the theory known as loop quantum cosmology has resulted in predictions that may be tested in the cosmic microwave background.

When it comes to predictions and testable consequences it is probably fair to say that there is not a great deal of difference between string theory and loop quantum gravity. Both theories lead to a few results that are testable, some in principle only but others more realistically. Although work in the loop gravity programme is purely theoretical, as it is in the string programme, physicists in both camps insist that what they are doing is physics. As such their theories must at some stage result in predictions that can be confronted with experimental or observational results. 'Because this is science, in the end experiment will decide', says Smolin.[88] But when is the end? Noting that so far experimental support for either of the two theories is lacking, Rovelli admits that presently a comparison needs to be based on non-empirical standards: 'Waiting for experiments, a theory must be evaluated and compared with alternatives only in terms of its consistency with what we know about Nature, internal coherence, and its capacity to produce unambiguous novel predictions. But sound scientific standards demand that no definitive conclusion be drawn.'[89]

In spite of the great differences between string theory and loop quantum gravity, the two theories may not be contradictory. It is possible, as some physicists have suggested, that they are complementary approaches, two different sides of the same underlying theory. According to this vision, the two theories stand in about the same relationship to the super-theory of the future as Galileo's theory of falling bodies and Kepler's theory of planetary orbits did to Newton's mechanics in the seventeenth century. The characteristic feature of loop quantum gravity, that space is conceived as discrete and consisting of minimal portions of the order of the Planck volume, does not in itself disagree with string theory. This theory, too, suggests that there is no operational meaning to sub-Planck distances and that continuous space may turn out to be an emergent concept. If the two different approaches could be made to meet, the result might be a stronger and more complete theory that avoids the weaknesses of each of the two approaches. But it is uncertain if the two theories can ever be conjoined.

Notes for Chapter 11

1. See the collection of writings presented in Miller and Verma 2008.
2. For a condensed history of early quantum gravity, see Stachel 1999.
3. Newcomb 1896, p. 195.
4. Einstein 1996, p. 356.
5. Weyl 1922, p. vi.
6. Information about Kaluza and his influence on Einstein's research programme can be found in Wünsch 2005.
7. Nordström 1914. On Nordström and his theory of gravitation, see Norton 1992. The unfortunate fate of his five-dimensional theory is considered in Halpern 2004a.
8. Wünsch 2005, p. 282.
9. Quoted in Vizgin 1994, p. 158.
10. Klein 1926. On Klein and five-dimensional quantum theory, see Kragh 1984 and Halpern 2007.
11. Klein 1928, p. 188.
12. Einstein 1931, p. 439. A similar critique would later be voiced with regard to the many compactified dimensions of string theory.
13. A. Salam, 'Gauge unification of fundamental forces'. See http://nobelprize.org/nobel_prizes/physics/laureates/1979/salam-lecture.html.
14. Freedman and Nieuwenhuizen 1978.
15. Hawking 1980, pp. 1–2.
16. A condensed account of the development of string theory appears in Kragh 1999a, pp. 415–19 and Schwarz 1996. A more detailed history, covering the development up to the late 1980s, is presented in Galison 1995. See also Smolin 2006, pp. 101–200. Aspects of the early phase are dealt with in Shapiro 2007.
17. Susskind 1997, p. 234.
18. On early ideas of superluminal particles, both before and after special relativity, see Fröman 1994. Arnold Sommerfeld was among the physicists who considered the possibility of superluminal electrons. In 1904 he thought for a brief while that there was experimental evidence for such particles.
19. Scherck and Schwarz 1974. A Japanese physicist, Tomiaki Yoneya, came independently to the same conclusion.
20. Schwarz 1996, p. 698.
21. Green and Schwarz 1984.

22. Gross *et al.* 1985, p. 504.
23. Candelas *et al.* 1985.
24. For a contemporary review, see Green 1985.
25. Quoted in Galison 1995, pp. 372–73.
26. Schwarz 1987, p. 38.
27. Rújula 1986.
28. In Davies and Brown 1988, p. 97.
29. 'Anomalies cancellation launches superstring bandwagon', *Physics Today* **38** (July 1985), 17–20, on p. 20.
30. Strominger 1986, p. 284.
31. Witten 1998, p. 1129. Other meanings for the 'M' have been proposed, such as 'messy', 'maybe' and 'mother' (of all theories). Magueijo (2004, p. 238) suggests 'masturbation' as befitting.
32. Quoted in Taubes 1995.
33. The number is given in *Physics World*, September 2007, p. 43.
34. The number of academic physicists in 1900 was probably more than 1200 and less than 1500. See Kragh 1999a, p. 13.
35. Conlon 2006, who explains 'how the dream of a unique world consistent with string theory turned into a nightmare of 10^{500} such worlds'.
36. Chen and Kuljis 2003. It should be kept in mind that the number of citations, being a measure of impact in the scientific community, is sociological and not epistemic in nature. It does not necessarily indicate the quality, originality, or truth of the reported research.
37. The citation history of Schwarz's paper is analyzed in Budd and Hurt 1991.
38. Greene 1999, p. 146. To use the apt phrase of John Barrow, 'We cannot expect everything of a Theory of Everything' (Barrow 1992, p. 121).
39. Kaku 2005a.
40. Galison 1995, p. 372.
41. In Davies and Brown 1988, p. 194.
42. Interview of January 1988, in Mehra 1994, p. 507. Although Feynman had neither interest in nor respect for philosophers' views of science, his ideas of the methodology of physics were close to those of Popper and his followers. This is clearly seen from passages in Feynman 1992, originally published 1965.
43. Ginsparg and Glashow 1986, p. 7.
44. Quoted in Galison 1995, p. 399.
45. In Davies and Brown 1988, p. 191.
46. Richter 2006.
47. 'The energy of empty space that isn't zero: A talk with Lawrence Krauss', in http://www.edge.org/3rd_culture/krauss06/krauss06.2_index.html.
48. Ehrlich 2006.
49. 'Viewpoints of string theory', http://www.pbs.org/wgbh/nova/elegant/view-glashow.html (2003). In the September 2007 issue of *Physics World*, Glashow is quoted: 'String theory is different to religion because of its utility in mathematics and quantum field theory, and because it may someday evolve into a testable theory.'
50. Friedan 2003, p. 8.

51. Ibid., p. 11. Friedan's comment agrees with my analysis of mathematical beauty in relation to Dirac's view of physics (Kragh 1990, pp. 275–392) and also with James McAllister's more comprehensive investigation of beauty in science (McAllister 1996).

52. Davies and Brown 1988, p. 216.

53. Schelleken 2008, p. 12.

54. Veltman 2003, p. 308. Martinus Veltman shared the 1999 Nobel Prize in physics with his former PhD student Gerardus 't Hooft.

55. *New York Times*, 4 January 2005. Philip Warren Anderson received the Nobel Prize in 1977 for his work on solid-state theory.

56. Gross 2005, p. 104.

57. ScienceWatch, May–June 2007 (http://www.sciencewatch.com).

58. Greene 2004, p. 378.

59. Witten 1998, p. 1128. Emphasis added.

60. Schwarz 1998, p. 2. See also Hedrich 2007.

61. Halpern 2004b, p. 231.

62. See Kaku 2005a for an accessible survey of ways to testing string theory.

63. Adelberger, Heckel, and Nelson 2003.

64. Distler *et al.* 2007 suggested a way to falsify models of string theory, but critics have argued that the proposal is not really about string theory. Whereas the first versions of the paper, as posted on the arXiv website, were entitled 'Falsifying string theory through WW scattering', in the published version 'string theory' in the title was substituted with 'new physics'. The first versions promised a way to 'falsify string theory', which in the published version had become a way to 'falsify generic models of string theory'.

65. Weinberg 1987a, p. 105.

66. Davies and Brown 1988, p. 221.

67. Weinberg 2000.

68. Kaku 2005b, p. 282. Schellekens 2008, p. 12.

69. Weinberg 1992, p. 220.

70. Interview of 12 August 2004. See http://sciencewriter.org/alan-guth-interview/.

71. Kane, Perry, and Zytkow 2002 p. 52, partly in opposition to Hogan 2000.

72. Schelekken 2008, p. 12, in a section entitled 'Against uniqueness'.

73. Gross was awarded the 2004 Nobel Prize (sharing it with David Politzer and Frank Wilczec) for the discovery in 1973 of asymptotic freedom, a concept that gave strong support to the validity of quantum field theory and paved the way for the quark gauge field theory. Although there have presumably been many nominations, no Nobel Prize has been awarded for work in string theory. Leading string theorists Michael Green, John Schwarz, and Edward Witten have been named as likely Nobel Prize winners (Chen and Kuljis 2003, p. 438), but so far none of the predictions have been unsuccessful. See also http://science.thomsonreuters.com/news/2005-08/8289814. On the other hand, several of the leaders of string theory have received other prestigious prizes. Thus, the Fields Medal, often described as the Nobel Prize of mathematics, was awarded to Witten in 1990. The Dirac Medal, issued by the Abdus Salam International Centre for Theoretical Physics, has been awarded to Witten (1985) and Schwarz (1989), and also to Juan Maldacena, Joe Polchinski, and Cumrun Vafa (2008).

74. Gross 2005, p. 105.

75. Gross 2008, p. 288.

76. Woit 2006. Smolin 2006. The publication of the two books gave rise to much debate and media coverage. For example, an article in the *Wall Street Journal* of 23 June 2006 suggested that string theory betrayed science because it 'abandoned testable predictions'.

77. Cartwright and Frigg 2007. A more detailed discussion of string theory from a Lakatosian perspective appears in Johansson and Matsubara 2009, who conclude that according to Lakatos' criteria string theory is a degenerating research programme.

78. Shapere 2000, p. 161. Other quotations from p. 153 and p. 159.

79. Douglas and Kachru 2007, p. 789.

80. Wüthrich 2005 and Zinkernagel 2006. James Mattingly, a physicist at Georgetown University, concludes that 'Standard arguments from physics as well as general... arguments from the theory of science do not compellingly indicate that gravitation theory should be quantized' (Mattingly 2005, p. 337).

81. Rovelli 1998, who based his count on the electronic archives hep-th (theoretical high energy physics) and gr-qc (general relativity and quantum cosmology). Most of the string papers were classified in the first group, most of the others in the second group.

82. For a popular introduction to loop quantum gravity, as of the early 1990s, see Bartusiak 1993.

83. Einstein to Besso, 10 August 1954, quoted in Stachel 1993, p. 286.

84. Smolin 2006, p. 254.

85. Rovelli 2008, p. 7.

86. Rovelli 2003, p. 3.

87. Abdo *et al.* 2009. *New York Times*, 29 October 2009: '7.3 billion years later, Einstein's theory prevails.'

88. Smolin 2003, p. 70.

89. Rovelli 2008, p. 5.

12

Astrobiology and Physical Eschatology

It is better to be too bold than too timid in extrapolating our knowledge from the known into the unknown.

Dyson *Time without end.* 1979, p. 449

Physics based on the anthropic principle is not the only approach in modern physics in which humans, or other intelligent beings more or less like them, are considered of crucial importance. Life in the universe outside the Earth has traditionally been seen as more like science fiction than science, but since the 1960s the subject has been eagerly studied by groups of physicists, astronomers, and other scientists. 'Astrobiology' may be thought to be a curious scientific discipline, in so far as it deals with a subject that is not known to exist, but it is nonetheless a thriving branch of modern science.

One corner of the broad and interdisciplinary field of astrobiology is concerned with the fate of intelligent life in a cosmological context, meaning the remote future as it follows from cosmological models. 'Physical eschatology' and related studies are, however, not primarily interested in *human* beings, but in intelligent life understood in a more general and sometimes quite abstract way. Humans are of interest mainly because we are the only known member of the class of intelligent and conscious beings. Present-day humans can definitely not survive the conditions in a universe where the stars no longer shine, or where galaxies have turned into black holes, but we can at least imagine that our descendants can. These imagined descendants may be advanced biological species or they may be self-reproducing robotic supercomputers.

This kind of astro- or cosmobiology is necessarily speculative and parts of it are *very* speculative indeed. However, it is claimed that it deals with scientifically based speculations insofar that the scenarios and hypotheses of the far future are constrained by the known laws of physics and also by knowledge from information and computer science. In this sense the scenarios are held to be scientific, although of course they cannot be tested in any direct sense. Much of the work in this area can be conceived of as thought experiments, theoretical explorations of an unknown cosmic future by means of scenarios involving intelligent life forms more or less of one's own making. Other works are so far away from ordinary science that they are probably better understood as a peculiar version of a philosophical exercise. At least some of the publications dealing with physical eschatology and related subjects are controversial because they may seem to border on pseudoscience. They may be seen as the products of smart people investigating weird things and perhaps even believing weird things.[1] In some of the earlier chapters we have encountered theories and ideas proposed by scientists that perhaps belong to some area of thought other than science. In this chapter we shall meet other and no less remarkable examples of higher speculation.

12.1 PLURALISM AND EXOBIOLOGY

Belief in the existence of intelligent life outside the Earth, a concept known as 'pluralism', has been held since antiquity and for most of the time has not been regarded as particularly controversial.[2] The ancient Greeks were well aware of the idea, which continued to be discussed during the Middle Ages, usually in a theological context. During the scientific revolution and the subsequent Enlightenment era, pluralism was championed by an array of prominent authors, including many of the most distinguished scientists of the time. For example, in the posthumously published *Kosmotheoros* of 1698, the leading physicist and astronomer Christiaan Huygens described in great detail the characteristics of the inhabitants of the other planets. Lack of even the slightest trace of evidence in no way prevented respected scientists and popularizers of science from making proposals of the nature of extraterrestrials and their distribution throughout the universe. William Herschel, generally recognized as one of the greatest astronomers ever, was not only convinced that there must be inhabitants on the Moon, but also suggested that the Sun was inhabited by beings whose constitution was specially adapted to the immense heat of their habitat. The idea of solar beings was far from new. For example, it was seriously discussed by some of the medieval philosophers.

What are probably the most detailed calculations ever of the number of extraterrestrial beings, and possibly the most unfounded as well, were due to Thomas Dick, an Irish-born clergyman and writer. In 1838 he published *Celestial Scenery*, a classic in pluralist literature which ran into six editions and numerous reprintings.[3] According to Dick, the small planet Mercury housed a population of 8.96 billion intelligent beings, and Saturn's ring system alone no less than 8142 billion. In another work, *The Sidereal Heavens* of 1840, he went even further, now calculating the population of the entire universe to be an astounding 6.0573×10^{22}. Reverend Dick was unacquainted with the value of Avogadro's number, or the number of molecules in 24 litres of a gas, which was only determined later in the century. Had he known the value of the constant, 6.0221×10^{23}, he would undoubtedly have noticed the remarkable coincidence that the number of extraterrestrials equals one tenth of Avogadro's number. Can this be just a coincidence, or does it require an explanation, say in terms of the anthropic principle?

About the turn of the nineteenth century extraterrestrial life was as hot a topic as ever, nourished in particular by the notorious claims of the astronomer and convinced pluralist Percival Lowell and others that observations of Mars revealed that there was or had been an advanced civilization on the planet. At about the same time the Swedish chemist and physicist Svante Arrhenius, a Nobel chemistry laureate of 1903, advocated a theory of 'panspermia' according to which primitive life in the form of bacteria was propelled through space by the pressure of stellar radiation. He described the hypothesis, which was not quite new, most fully in his book *Worlds in the Making* from 1908. Although not widely accepted by his contemporaries, Arrhenius held the idea of panspermia to his death in 1927. This and similar ideas did not explain the origin of life, it merely put it off. After all, where did the panspermia come from? Yet it did offer an argument for the pluralist belief in nearly universal life, a belief Arrhenius fully subscribed to and did much to promote.[4]

Much later, in the 1970s, Fred Hoyle and his former student, the Cardiff University astronomer Chandra Wickramasinghe, put forth a different version of the panspermia hypothesis while at the same time acknowledging its affinity to Arrhenius' old hypothesis. Based upon spectroscopic measurements they argued that interstellar dust clouds carried with them biomolecules such as proteins and nucleic acids. These molecules, Hoyle and Wickramasinghe suggested in their book *Lifecloud* from 1978, were assembled in space to simple living cells. Their basic idea was that cells in the form of freeze-dried bacteria were prevalent everywhere in the universe. According to what they called 'cosmic biology', biological processes operate throughout the universe, which they saw as an alternative to the idea of primitive life being a chance product of a Darwinian biochemical evolution process that once happened on Earth. In a book entitled *The Relation of Biology to Astronomy*, Hoyle concluded that the universe is fundamentally biological. 'Is the biological control over astronomy to be an intelligent control or is it to be a product of blind evolutionary processes signifying nothing?' he asked. 'My personal speculation would be that the control is intelligent.'[5]

Other forms of astrobiological speculations related to the possibility of life in the far future. In so far that these ideas were discussed on the basis of the thermodynamically based heat death, as they often were, they referred to cosmology and not only to the more narrow fields of planetary or stellar astronomy. They were about the end of life and also about the end of the universe. We find such physics-based speculations about endless life in the early twentieth century, in connection with the controversy over the heat death. Some of the German scientists involved in this controversy argued that life might persist even in the very high-entropic environment of the far future. As the physics teacher Caspar Isenkrahe wrote, 'An end of human descendants does in no way imply the end of life, and it is still an open question if there is a limit to the adaptability of life to changed external conditions.'[6] In his eschatological physics Isenkrahe considered the effects on future life of such modern knowledge as intra-atomic vibrations, the nature of molecular forces, radioactivity, and light pressure. His approach did not differ qualitatively from that of Dyson and late-twentieth-century physical eschatologists.

As another and later example, the young evolutionary biologist John B. S. Haldane argued in 1927 that although the Sun would eventually burn out and thus destroy life on Earth, this did not imply an absolute end of life in the universe. He thought that on other planets, revolving around other suns, biological evolution would go on, possibly in the form of descendants of humans. The following year, in a response to James Jeans' conclusion that the universe would necessarily end in an irreversible heat death, Haldane considered the end of the universe in a cosmological context. Evidently inspired by Boltzmann (but without mentioning him), he suggested that we live in a huge low-entropic fluctuation, an improbable bubble universe. Although such a state is exceedingly unlikely, 'in the course of eternity any event with a finite probability will occur'. An eternal universe would certainly be 'dull', except for the infinitesimal period of time in which life happened to exist.

> But during most of eternity there can be no living creatures at all resembling ourselves to be bored. For since all organisms live by the utilisations of processes involving increase of entropy, they can presumably only exist during the aftermath of a very large fluctuation. This is why we are witnesses of this excessively unusual occurrence.[7]

No wonder that Haldane is sometimes mentioned as an early anthropic thinker.

Ernest William Barnes was not only Bishop of Birmingham, but also an accomplished mathematician whose works in pure and applied mathematics had made him a Fellow of the Royal Society. In a remarkable book of 1933, *Scientific Theory and Religion*, he discussed in technical detail recent developments in relativistic cosmology, including Lemaître's new theory of the exploding universe (which he dismissed as 'a brilliantly clever *jeu d'esprit* rather than a sober reconstruction of the beginning of the world'.) Two of the sections of this wide-ranging work, which was based on his Gifford Lectures of 1927–29, carried the titles 'Is the whole cosmos the home of intelligent beings?' and 'The final state of the cosmos'. Barnes reasoned that life was abundant throughout the universe and 'in many places its development has reached stages immeasurable in advance of that attained by man upon the earth'.[8] As to mind, he speculated that it could take forms entirely different from what we know. There might even be 'highly developed organisms . . . which could only "live" when the matter of which they were composed was in the state in which it exists in the bright stars'. Barnes further considered the final state of the universe, which he conceived as a thin soup of radiation or perhaps a collection of extinct and very cold stars. As a Christian he had no problems with the end of the world, yet he felt it pertinent to point out that 'our galactic universe will not reach this state until many millions of millions of years have elapsed'. In an earlier paper he speculated, as a few others had previously done, that it might be possible to establish contact with extraterrestrial civilizations by means of radio communication.[9]

A rationalist cosmologist if there ever were one, Edward Arthur Milne was not foreign to speculations. In his Cadbury Lectures for 1950, published posthumously as *Modern Cosmology and the Christian Idea of God*, he suggested that God was constantly busy with creating opportunities for organic evolution throughout the universe. He considered 'the infinity of galaxies as an infinite number of scenes of experiment in biological evolution'. Milne had for long argued that the notion of entropy increase was invalid for the universe at large and that the second law of thermodynamics did not necessarily imply an end of all life and order. As a Christian pluralist he was fully aware of the theological problems of extraterrestrial life, such that the incarnation being a unique event. Milne came up with a science-fiction like solution which he thought was 'not fantastically improbable'. He believed that in principle, and perhaps even in practice, the new science of radio astronomy might solve the problem by securing communication with inhabitants of other planets. With interplanetary communication might follow interplanetary salvation:

> There is no prima facie impossibility in the expectation that first of all the whole solar system, secondly our own group of galaxies, may by inter-communication become one system. In that case there would be no difficulty in the uniqueness of the historical event of the Incarnation. For knowledge of it would be capable of being transmitted by signals to other planets and the re-enactment of the tragedy of the crucifixion in other planets would be unnecessary.[10]

Although Milne's speculations were extraordinary, at about the same time the possibility of detecting extraterrestrials by means of radio signals began to be seriously discussed in the physics community. If intelligent life is common in the universe, such as there are reasons to believe, how is it that we have no signs of it at all? Why have we detected no radio transmissions from the supposedly abundant civilizations outside the solar system? This question is sometimes referred to under the dignified name 'Fermi's paradox' because it was raised by the great physicist Enrico Fermi during a casual lunch conversation he had in 1950 while working at the Los Alamos National Laboratory. 'Where is everybody?' Fermi is supposed to have asked.[11] Of course, one answer was the 'melancholy theory', as Shklovskii and Sagan called it, namely that we are alone in the universe. This was also the answer of Michael Hart, who in 1975 argued that other explanations, whether of a physical or sociological nature, were much less likely.[12] But this was an answer most scientists, not to mention the public, found unappealing. Within the more recent tradition of cosmobiology the Fermi paradox and related arguments are often taken to imply that although life may be plentiful on a cosmological scale, humans are likely to be the only highly developed civilization in the Milky Way system.

Whatever the depth of Fermi's innocent question and his role in the development, the 'paradox' came to serve as an important motivation for the later Search for Extra Terrestrial Intelligence (SETI) programmes that were pioneered by Philip Morrison, Frank Drake, Giuseppe Cocconi, Carl Sagan, and a few others. A landmark work of the new research field was *Intelligent Life in the Universe* by the Russian astrophysicist Iosef Shklovskii, which in 1966 appeared in a revised translation co-authored and extended by Sagan. This book gave a wide-ranging and thorough, and often quite speculative, survey of the possible existence of extraterrestrials and the means of communicating with them.[13] In agreement with the unbounded technological optimism and the no-limits-to-growth philosophy of the 1960s—a time when global warming was not yet on the agenda—the two authors were convinced that productivity and consumption must increase continually. 'Even the slightest progressive decrease would, after thousands of years reduce the technological potential to essentially nothing.'[14] Shklovskii and Sagan dealt extensively with the astrophysical and technical aspects of SETI problems, but did not place them in a proper cosmological perspective. It should be recalled that in 1966 cosmology was in a state of transition, with the new hot big-bang model just beginning to develop into a standard picture of the universe.

In an attempt to systematize the many probabilities that are involved in the question of communicating with alien life, Drake formulated in 1960 a simple equation that states the likelihood of the existence of civilizations in our galaxy. Drake worked at the time at the National Radio Observatory in Green Bank, West Virginia, for which reason the equation is sometimes referred to as the Green Bank equation. The Drake equation, which neatly summarizes the main problems of SETI programmes, exists in several versions. For example, the number of communicative civilizations N at any given time can be stated as

$$N = R \times F_p \times n_e \times F_l \times F_i \times F_c \times L$$

where R is the rate of formation of suitable stars, F_p is the fraction of those stars with planets, n_e is the number of Earth-like planets per planetary system, F_l is the fraction of those Earth-like planets where life actually develops, F_c is the fraction of planets on which radio communications technology develops, and L is the lifetime of communicating civilizations.

Although the Drake equation is a useful framework for discussion, it is generally recognized that it is scarcely more than a way of organizing scientists' ignorance of civilizations outside the solar system. For one thing, some of the factors (such as F_l and F_c) are completely unknown and can be given almost any value one fancies. For another thing, the equation does not take into account the possibility of interstellar colonization, a possibility that some scientists believe is important. For these and other reasons the Drake equation leads to no definite results and has, as one might expect, been used to estimate wildly different values for N. For example, whereas Shklovskii and Sagan suggested that $N \cong 10^9$, Barrow and Tipler argued in 1986 that $N \cong 1$. Most other estimates lie within these limits, but there are also scientists who have argued for $N < 1$ (meaning that the chance for a successful comprehensive SETI programme is less than one). It is generally assumed that the Drake equation refers to stars and planets in our own galaxy, the only part of the universe from where we can realistically hope to receive radio signals. (Our nearest neighbour, the Andromeda galaxy, is located some 23 million light years away.) It is of course possible to extend the equation to cover other galaxies as well, but this is rarely done. By far the majority of SETI investigations consider areas in the Milky Way, while for understandable reasons extragalactic searches for intelligent life have received much less attention.

SETI-related speculations came from science fiction writers and astronomers in about equal measures. In 1964, a Russian astrophysicist and former student of Shklovskii at the Sternberg Astronomical Institute in Moscow, Nicolai Kardashev, proposed dividing advanced extraterrestrials according to their capability to restructure large objects. The most highly developed of Kardashev's civilizations would be able to restructure entire galaxies. By extrapolating the annual increase in energy consumption on Earth, Kardashev found that in 3200 years from now the energy consumption would equal the power of the Sun (4×10^{33} erg s^{-1} or 4×10^{26} watt); 5800 years into the future it would be equal to the output of 10^{11} suns. Noting that these figures might seem 'inordinately high', Kardashev nonetheless suggested that they were realistic.[15] His ultimate civilization would master an energy of the scale of its own galaxy, with an energy consumption estimated at about 4×10^{44} erg s^{-1}, which could be used for interstellar communication. He speculated that some of the newly found radio sources (quasars) might be artificial systems created by a technologically advanced civilization.

In 1971 NASA funded a SETI programme to look for transmissions from civilizations on distant planets and since then SETI and related initiatives have developed into a major research field, as fascinating as it is controversial. Most astrobiological research is, however, more limited in scope, dealing with possible primitive life forms in our planetary system and not with advanced civilizations outside it. Parts of astrobiology deal with life on Earth under extreme conditions, such as organisms living in hot springs or at the bottom of the deep oceans. In 1979 'Extraterrestrial Life' became a separate category in the volumes of *Astronomy and Astrophysics Abstracts*, and three years later the Boston University astronomer Michael Papagiannis formed a commission on 'bioastronomy' under the International Astronomical Union. The field consisted essentially of the study of and search for extraterrestrial life and civilization (Fig. 12.1).[16]

As one might expect, far from all scientists welcomed the new field defined by the SETI programmes. Among them was Frank Tipler, who in a paper of 1980 argued that we are the only intelligent species in the Milky Way. He estimated the probability of extraterrestrial intelligence to be less than 10^{-10}. From the anthropic principle he inferred that the universe must contain 10^{20} stars in order to contain a single intelligent species, hence 'We should not therefore be surprised if indeed it contains only one'.[17] Tipler thought that the anthropic principle should be given more weight and that it effectively ruled out the existence of advanced extraterrestrial civilizations. As mentioned in Chapter 9, this was also what Carter suggested in 1983. Using the Popperian-like argument that 'no experiment will ever convince the ETI believers that we are alone', Tipler grouped bioastronomy together with parapsychology and other pseudosciences.[18] Considering that the extensive radio search for extraterrestrials had failed so far, he suggested that it was time to expel bioastronomy from the true sciences covered by the International Astronomical Union.

But this did not happen. Astrobiology is today a flourishing research area with its own institutes, societies, and communication structures. Results from astrobiological research are published either in the traditional journals devoted to physics, astronomy,

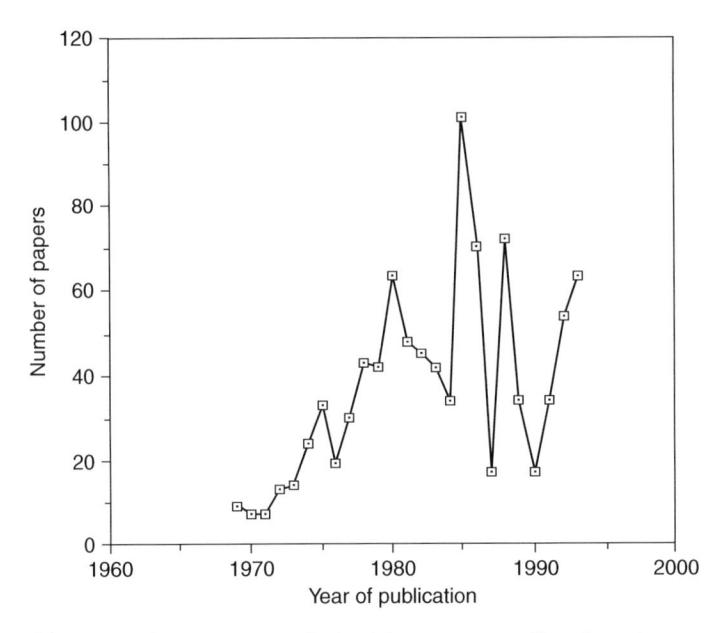

Fig. 12.1. Publications in bioastronomy, exluding Mars, 1969-1993. Data from *Astronomy and Astrophysics Abstracts*. The peaks reflect proceedings from IAU conferences in 1979, 1984, 1987 and 1993.

Source: Steven J. Dick, *The Biological Universe: The Twentieth-Century Extraterretrial Life Debate*. Cambridge: Cambridge University Press, 1996, p. 493. Reproduced by permission of Cambridge University Press.

biology, and the earth sciences, or in new and more specialized journals such as *Astrobiology, International Journal of Astrobiology,* and *Astrobiology Magazine*. The field is interdisciplinary, covering not only aspects of the physical and biological sciences, but also parts of psychology, neurology, sociology, and computer science. When the *International Journal of Astrobiology* was launched in 2002, its editor, David Wynn-Williams, defined the field as 'the study of the origin, evolution, adaption and distribution of past and present life in the Universe'. Surprisingly, he did not include future life in his definition, although this was a main concern of several physicists and astronomers who would publish many papers on this subject in the new journal. According to the 2006 Astrobiology Science Conference, 'Astrobiology is the scientific study of the living universe, how it arrived at this point in time, and where it is heading.'[19]

12.2 THE FAR, FAR FUTURE OF THE UNIVERSE

It has always been a major aim of evolutionary cosmology to account for the past history of the universe, from the very earliest times to the present (say, from 10^{-43} s after $t = 0$, or how close one can come to that magical moment). How did the universe come to be as it is? While this question forms an essential part of cosmological research and has been investigated in great detail since the late 1940s, when it was first considered by Gamow, much less work has been devoted to the future of the universe. And yet Einstein's equations of relativistic cosmology are symmetric in time, telling us not only about the past but also the future. When Freeman Dyson in 1978 gave the 'James Arthur Lectures on Time and its Mysteries' at New York University, he used them to present a broad-ranging study of the physical and biological conditions of the universe in the far future. Dyson could not help noting the glaring lack of symmetry between physicists' studies of the early universe and the few works devoted to the other extreme of the timescale. Not only were papers on the latter topic few and scattered, Dyson also felt that 'they are written in an apologetic or jocular style, as if the authors were begging us not to take them seriously'.[20]

In the early days of modern cosmology the future state of the universe was rarely a subject of scientific interest. When it was considered, it was mostly in connection with the heat death, the controversial prediction that the universe was bound to end in 'death and annihilation', as Jeans phrased it. In a lecture of 1928 given at the University of Bristol, Jeans explained that life and matter would disappear in the future, being destroyed by the effects of high-energy radiation. His view of the end of the universe was this:

> There can be no creation of matter out of radiation, and no reconstruction of radioactive atoms which have once broken up. The fabric of the universe weathers, crumbles, and dissolves with age, and no restoration or reconstruction is possible. The second law of thermodynamics compels the material universe to move ever in the same direction along the same road, a road which ends only in death and annihilation.[21]

This was one of the first cases in which speculation about the future state of the universe was lent authority by what seemed to be exact scientific arguments, an early example of scientific eschatology. In another lecture of the same year Jeans considered life in the universe, which he tended to see as a rare and accidental by-product of cosmic processes. 'It does not at present look as though Nature had designed the universe primarily for life', he said. 'Life is the end of a chain of by-products; it seems to be the accident, and torrential deluges of life-destroying radiation the essential.' Jeans warned against drawing wide-reaching inferences about the universe being either friendly or unfriendly to life, inferences of the kind that later became formulated in the various versions of the anthropic principle. He illustrated his point as follows:

> Each oak in a forest produces many thousands of acorns, of which only one succeeds in germinating and becoming an oak. The successful acorn, contemplating myriads of acorns lying crushed, rotten, or dead on the ground, might argue that the forest must be inimical to the growth of oaks, or might reason that nothing but the intervention of a special providence could account for its own success in the face of so many failures. We must beware of both types of hasty inferences.[22]

With the recognition of the expansion of the universe in the early 1930s, forecasts about the end of the world got a new dimension. Eddington was possibly the first to discuss astronomical eschatology within the framework of the expanding universe, which he did in his presidential address of 5 January 1931 to the Mathematical Association. Assuming a spatially closed ever-expanding universe with no beginning in time (the Lemaître–Eddington model), he agreed with Jeans that nothing could prevent the entropic heat death. 'It is widely thought', he ended his lecture, 'that matter slowly changes into radiation'. According to Eddington,

> If so, it would seem that the universe will ultimately become a ball of radiation growing ever larger, the radiation becoming thinner and passing into longer and longer wavelengths. About every 1500 million years it will double its radius, and its size will go on expanding in this way in geometrical progression for ever.[23]

Neither Jeans nor Eddington speculated about the fate of life in the far future.

Only from about 1970, after the hot big bang had become the established framework for cosmological research, did a few astronomers and physicists take a closer look at the fate of the universe in the far future. These studies, with their suspicious air of science fiction, were not always considered respectable science. After all, they built on extravagant extrapolations and involved future scenarios that could not be tested in any realistic sense. As the British astrophysicist Malcolm Longair remarked in his Halley Lecture for 1985, 'The future of our Universe is a splendid topic for after-dinner speculation.'[24] Yet, some scientists thought it was more than that.

Studies of the state of the universe in the remote future, meaning zillions of years from now, may or may not include considerations of the continuation of intelligent life. Some of the studies do not mention life explicitly, while others do it only briefly and casually (in what Dyson called 'an apologetic or jocular style'). Other works again are specifically concerned with the survival of humans and similarly advanced if hypothetical beings. It is these works that belong to the new field of astrobiology, or perhaps better *cosmobiology*. It should be noted that the term 'cosmobiology' is sometimes used in a very different sense, namely to designate a branch of or approach to astrology founded by Reinhold Ebertin, a twentieth-century German astrologist and physician. Needless to say, this is not what I have in mind when I speak of cosmobiology.

A new subfield of astrophysics and cosmology has emerged during the last few decades, sometimes called 'physical eschatology' because it is concerned with, among other things, the final state of life and everything else. The term is used from time to time, but has not been generally or officially adopted.[25] (It is not to be found in either the *Oxford English Dictionary* or the *Encyclopedia Britannica* online editions.) According to Barrow and Tipler, physical eschatology is 'The study of the survival and the behaviour of life in the far future' and its basic problem is 'to determine if the forms of matter which will exist in the far future can be used as construction materials for computers that can run complex programs, if there is sufficient energy in the future environment to run the programs, and if there are any other barriers to running a program'.[26]

Other scientists occupied with this kind of theoretical research have somewhat different opinions of what it is about. What is common to them is that they conceive the field as 'a respectable astrophysical discipline'[27] and eagerly stress that, in spite of its obvious speculative elements, it builds on known physics and established scientific methodology. The papers on life in the far future 'have shown the progression required of physical science', Barrow and Tipler assured in 1986. Because, 'The papers subsequent to Dyson's first article built on, improved, and corrected their predecessors, and the discussion is now based entirely on the laws of physics and computer theory.'[28] Critics of physical eschatology are not so sure that it belongs to the physical sciences, or that it counts as scientific at all.

The Serbian astrophysicist Milan Ćirković and other promotors of physical eschatology stress that the idea or research programme has no religious connotations at all. Perhaps it has not, but theologians and religious authors have nonetheless found it interesting and taken it up, just as they have taken up the multiverse and the anthropic principle—and about everything else. The biologist and theologian Arthur Peacocke has emphasized that there is a world of difference between 'scientific (and pseudoscientific) futurology' and proper eschatology, first of all because the latter is about the ultimate destiny and goal of man. Wolfhart Pannenberg, a distinguished Evangelical theologian, likewise notes the apparent conflict between the two scenarios of the future, asking 'Is the Christian affirmation of an imminent end of this world that in some sense invades the present even now, reconcilable with scientific extrapolations of the continuing existence of the universe for billions of years ahead?'[29]

In spite of these and other cautions, it can be argued that physical eschatology has a quasireligious element or is at least compatible with Christian eschatology. According to William Lane Craig, physical eschatology furnishes grounds for taking seriously the hypothesis of a transcendent creative and omnipotent agent, also known as God.[30] From the cosmologists' point of view, should such an agent exist—and this is line with some versions of physical eschatology—it must be immanent, part of physical nature and not outside it. The agent must be a pantheistic God or, to use a term from Whitehead's process philosophy, a panentheistic God. (According to panentheism, God and the world stand in a mutual and perpetual relationship: God creates the world, and the world creates God. Contrary to the standard conception, panentheists do not consider God to be eternally changeless.)

In 1969, in what he called an 'eschatological study', Martin Rees investigated the fate of a closed universe during its phase of contraction. As the universe decreases in size the radiation temperature rises, with the result that the stars will eventually be destroyed. Before that happens, the galaxies will be squeezed together and merge with one another. And these are only some of the processes that will make the end of the contracting universe a most unfriendly place.[31] Although the road towards the big crunch takes time (and a lot of it), the final crunch is by all measures a violent catastrophe. It easily compares with the eschatological scenarios in the Bible, as described in 2 Peter 3:10–13: 'The heavens shall pass away with a great noise, and the elements shall melt with fervent heat, the Earth also and the works that are therein shall be burned up.'

Rees did not mention either humans or other forms of life in his early contribution to eschatological astrophysics. He speculated (as others had done before him) that the collapse might be followed by a new expanding phase and that the cyclic behaviour might proceed endlessly. In spite of the reference to 'eschatology' in the title of his paper, it was a fairly conventional piece of theoretical research, an examination of the long-term behaviour of a cosmological model that built on established physics. Incidentally, Rees was not the first scientist to refer to eschatology in an astronomical or cosmological context. In his posthumously published biography of James Jeans, Milne discussed Jeans' unshakable belief in the heat death as the effective end of the universe. Milne took his distinguished colleague to task for having advocated this kind of 'astronomical eschatology'.[32]

Many years earlier, in a speculative world system published as *Kosmologie* in 1880, a German amateur scientist by the name Hermann Sonnenschmidt included a chapter on the eschatological aspects of cosmology. Sonnenschmidt's concern with the end of the world was shared by many scientists at the turn of the century. Among German authors who explicitly approached eschatological subjects were Johann Rademacher (*Der Weltuntergang*, 1909) and Caspar Isenkrahe (*Energie, Entropie, Weltanfang, Weltende*, 1910). Some of their scenarios were quite close to what physicists have come up with in the twentieth century. The best-known French example of the genre is probably the astronomer and prominent pluralist Camille Flammarion's widely read *La fin du monde* from 1894, although in this case it was a science-fiction novel mixed with popular science.[33]

Four years after Rees' eschatological study of 1969, Paul Davies published a brief paper on the thermal future of ever-expanding cosmological models in which he examined how various physical effects would affect the heat death of the future.[34] He mentioned life only at the end of the paper, as if it were an afterthought. In reality his paper was no more concerned with life than Rees' earlier paper. The same kind of added reference to life processes appeared in a cosmic forecast published by four American physicists in *Scientific American* in 1983. They argued that when the universe, supposed to be open, had existed for about 10^{32} years, matter would consist of black holes and a highly diluted soup of electrons, positrons, neutrinos, and photons. At the much higher age of 10^{100} years even massive black holes would have evaporated via the Hawking process. 'With the evaporation of black holes there will be a cosmic energy shortage, ... [and] any constant rate of energy consumption by life forms will in due course become untenable.'[35] Characteristically, in their more technical paper published in the *Astrophysical Journal* of the previous year, a paper which focused on the effects of proton decay, they did not mention life at all.

Other works from the two last decades of the century reached roughly similar conclusions, depending on what physical processes were considered, the assumptions of the chosen cosmological model, and the details of the calculations.[36] For example, in addition to the hypothetical proton decay the effects of massive neutrinos, gravitational entropy, exotic particles of dark matter, and magnetic monopoles could be taken into account.[37] In a contribution to the new field of eschatological studies of 1978, Barrow and Tipler arrived at a picture of eternity in which the ever-expanding universe would end in radiation, electrons, and neutrinos. But contrary to earlier pictures of physical eternity, their scenario included the idea that space–time would become more and more irregular. 'Eternity is unstable', they concluded.[38] Among the hypothetical processes considered by Barrow and Tipler was the annihilation of protons and electrons into photons ($p + e \rightarrow 2\gamma$), the very same process that Jeans, Eddington, and others had discussed in the 1920s.

The new physical eschatologists were generally careful to stress that their scenarios built on standard physics in the sense that the presently known laws of physics were assumed to remain valid in the indefinite future. This assumption had the status of an 'article of faith', as two authors expressed it.[39] The physicists prided themselves of doing science in the ordinary way, both epistemically and sociologically. They had established a small research tradition where new work built on and improved earlier results and thereby secured the progress that they held to be a defining feature of science. Whereas the early contributions to physical eschatology were based on estimates and rough calculations, later works aimed at more detailed calculations and use of advanced physical theory. 'Our aim', wrote the two Michigan physicists Fred Adams and Gregory Laughlin in an important paper of 1997, 'is to proceed in as quantitative a manner as possible. We apply known physical principles to investigate the future of the universe on planetary, stellar, galactic, and cosmic scales.'[40] Perhaps to keep their investigation as close to respectable physics as possible, they explicitly avoided considering the issue of life.

The picture of the material evolution of the future cosmos presented by eschatological physicists depended on whether the model of the universe was open or closed. The favoured scenario was the open, continually expanding case, where the picture could start with the extinction of stars and their later transformation into neutron stars or black holes, a phase to be reached in some 10^{14} years from now.[41] After a time of 10^{25} years, dead stars would have evaporated from the galaxies, leaving a material composition of 90% dead stars, 9% black holes, and 1% hydrogen and helium. Then, due to the decay of protons, dead stars and planets would disappear, with the result that at 10^{34} years all matter would have degenerated to a cold soup of black holes, electrons, positrons, neutrinos, and photons. But this is not the ultimate end, for according to Hawking black holes do not live for ever. They slowly evaporate, first the light ones and then the more massive ones. (The time for radiating away a black hole by emission of blackbody radiation varies with the cube of the mass of the hole.) At a time of about 10^{100} years even black holes of the mass of a galaxy will have evaporated. Should there exist any black holes of the mass of clusters or superclusters of galaxies it will take a little longer, perhaps 10^{118} years. All that is left in this ultimate future is an exceedingly thin electron–positron plasma immersed in a cold radiation of neutrinos and photons. Not a happy prospect.

While most work in physical eschatology has been concerned with the remote future, there are also studies of the nearer future of humankind in the form of probability arguments that claim to predict the future lifetime of the human species. This kind of study, covering the middle ground between philosophy and speculative astrophysics, is known as the 'doomsday argument'. The Princeton astrophysicist J. Richard Gott III was among the first to contribute to the doomsday literature and connect the argument to the anthropic principle, which he did in a paper of 1993. His reasoning was based on what he called the Copernican anthropic principle, namely 'that the location of your birth in space and time in the Universe is privileged (or special) only to the extent implied by the fact that you are an intelligent observer, that your location among intelligent observers is not special but rather picked at random'.[42] Based on this principle Gott calculated the near-future prospects of humans, concluding that our descendants can expect to survive at most a few million years. As to the possibility of colonizing space, he found it to be completely unlikely. The probability that humans would colonize our galaxy came out as less than 10^{-9}, and that of establishing a civilization using the energy output of the entire galaxy (as envisaged by Kardashev) as less than 10^{-17}. According to Gott, the future of humankind was limited and bleak.

Gott's doomsday argument is only one paper in an extensive literature, which is predominantly philosophical rather than scientific. Gott stressed, however, that his reasoning, based on the hypothesis of present humans being random observers, was fully scientific because the hypothesis is falsifiable. It could of course be proved wrong if we received radio messages from extraterrestrials, but also 'if more than 2.7×10^{12} more human beings are born'. This falsifiable prediction may give an idea of what scientists in this area of research mean when they speak of testing a hypothesis. Other scientific contributions to the doomsday discussion have dealt with the end of the world as it is supposed to be in a universe dominated by dark energy. For example, if dark energy is made up of phantom energy, in which the sum of the pressure and energy density is negative ($w < -1$), the universe will suffer a spectacular death in a 'big rip'.[43] Nor is this a happy prospect.

12.3 SPECULATIONS OF INDEFINITE INTELLIGENT LIFE

The series of lectures that Dyson gave in New York in 1978 were not the first mathematically informed speculations of a physicist on the survival of intelligent life in the far future. But, given by an eminent theoretical physicist at the right time and with the right mix of mathematics and popular philosophy, they turned out to be very influential. Dyson had been interested in space travels, intelligent extraterrestrials, and similar SETI topics of a partly speculative nature for many years. For example, in 1960 he wrote a paper in *Science* in which he suggested that a technologically advanced civilization might be able to enclose their star with an artificial shell (a 'Dyson sphere') that allowed a maximal utilization of the star's radiation energy.

Dyson's lecture of 1978 on 'Time without end' was published the following year in the prestigious *Reviews of Modern Physics* and is today recognized as a founding paper in the field of speculative astrobiology. It was certainly a remarkable paper—by many different standards. Thus, Dyson openly questioned the 'taboo against mixing knowledge with values' formulated, for example, by the Oxford philosopher George Edward Moore in the so-called naturalistic fallacy. The consensus view among both philosophers and scientists undoubtedly is that science has no place for values. For example, Einstein was adamant that science and ethical values should be kept apart, as he emphasized in an address of 1939 at the Princeton Theological Seminary: 'Science can only ascertain what is, but not what should be, and outside its domain value judgments of all kinds remain necessary.'[44] With regard to this question, Dyson disagreed with Einstein.

Incidentally, Dyson was not the first modern physicist to question the 'taboo'. Some years earlier the American physicist Robert Bruce Lindsay at Brown University did the same in an article dealing with the moral and social implications of the second law of thermodynamics. According to Lindsay, it was a moral imperative that humans should 'combat the natural tendency for order in the universe to be transformed into disorder', for only then would 'life [have] a meaning in the face of the remorseless natural increase in entropy of the universe'.[45]

As for Dyson, he did not hide his own ethical values and preference for eternal survival, if not of humans then of intelligent life in a more general sense. Whereas Weinberg in his popular book on the history of the universe, *The First Three Minutes*, had famously concluded that 'The more the universe seems comprehensible, the more it also seems pointless', Dyson's conclusion was diametrically opposite. He believed he had found 'a universe growing without limit in richness and complexity, a universe of life surviving forever and making itself known to its neighbors across unimaginable gulfs of space and time'.[46] Not only did he argue by means of mathematical equations that such a universe was possible, he also very much *wanted* it to be true and apparently saw no problem in letting his value judgment influence his conclusion, indeed dictate it.

Considerations of perpetual life were not new among post-World War II cosmologists. For example, Hoyle and Sciama saw the possibility as one more reason to prefer the steady-state model over models of the big-bang universe. At a meeting of the Royal Astronomical Society in late 1948, Hoyle pointed out that the new theory of the universe recently proposed by Bondi, Gold, and himself included the possibility of 'physical evolution, and perhaps even of life, may well be without limit'.[47] Sciama was attracted to the Hoyle–Bondi–Gold model in part 'because it's the only model in which it seems evident that life will continue somewhere'.[48] And Herman Zanstra and a few other supporters of the cyclic universe admitted that eternity was an additional reason to favour this kind of model (see Chapter 8).

To mention another and more interesting example, when Dirac was confronted with Dicke's alternative to a varying gravitational constant based on the large number hypothesis, he replied that on this assumption planets 'could exist indefinitely in the future and life need never end.... I prefer the [assumption] that allows the possibility of endless life'.[49] Although Dirac only made this one public remark on eternal life and, unlike Dyson, did not engage in speculations about the future fate of intelligent life, it was more than just a casual comment. In private notes of 1933 he stated as his belief that 'the human race will continue to live for ever and will develop and progress *without limit*'. What he characterized as an 'article of faith' was 'an assumption that I must make for my peace of mind'.[50] However, the references to endless life by Hoyle, Sciama, Dirac, and a few other physical cosmologists were typically brief and without foundation in calculations of the future conditions of the universe. This kind of careful quantitative consideration related to the possibility of life only appeared in the late 1970s.

Dyson hesitatingly admitted that in a closed and supercritical universe there was probably no chance for life to survive, and *for this reason* he focused on the open case. It was as if the mere idea of a closed universe gave him claustrophobia. What distinguished Dyson's arguments from most other works in the same genre was that they were based on detailed quantitative estimates, if of course combined with many assumptions of a somewhat arbitrary nature. Among the more basic assumptions adopted by Dyson and also by most other scientists in this area of research was that the laws of physics do not vary in time. Another basic assumption was that the fundamental laws known today will remain the relevant laws of the far future. 'New physics' is not considered, although scientists of course realize that such physics may arise and most likely will arise. In that case the calculations will have to be changed, perhaps drastically.

Apart from choosing to disregard the possibility of proton decay, at the time a new and untested hypothesis from grand unified theory, Dyson's reasoning was very broadly of the same kind as that adopted by later scientists such as Adams and Laughlin. What is more, to Dyson the conditions of life could not be separated from cosmic development. As he saw it, biology in a generalized sense was part and parcel of the problem, and that included non-physical aspects like consciousness and memory. As to consciousness, or what Eddington and writers of his generation called mind, Dyson adopted the non-materialistic viewpoint that it was based on structure rather than matter. On the other hand, he did not consider mind without a material structure. Not that he gave any reasons for this position, but it was the only way that he, by his own account 'a philosophical optimist', could maintain his desire of an indefinite existence of conscious life in the cosmos.

The nature of concepts such as mind, consciousness, and memory have been investigated in thousands of works written by philosophers, psychologists, and theologians. Not only did Dyson ignore this tradition in scholarship, so have most later physicists occupied with similar scenarios of the future prospects for intelligent life.

There is some contact between physicists, computer scientists, and philosophers in this area of speculative research, but it is very limited.

In a paper of 1977, the physicist Jamal Islam, of University College, Cardiff, had reached the pessimistic conclusion that it was extremely 'unlikely that civilization in any form can survive indefinitely'.[51] This was a conclusion that strongly disagreed with Dyson's philosophical desires and which he therefore felt he had to counter. As a way to avoid the conclusion that the total amount of energy in the universe is insufficient for eternal survival, Dyson suggested that life might 'hibernate' during repeated immense intervals of time. During the periods of hibernation, metabolism stopped while radiation of waste energy continued. At extremely low temperatures the time to think a single thought might be trillions of years, but Dyson argued that with respect to 'subjective time'—the scale of time experienced by the thinking being—the mental act would proceed normally. By making use of the hibernation hypothesis, subjective time, and some other assumptions, he concluded that intelligent life might indeed survive for ever, even after the temperature had irreversibly dropped to zero.

And this was not all, for Dyson also argued that communication by radio might bring a sense of biological unity and dominance to the whole universe: 'Every society in the universe could ultimately be brought into contact with every other society.' Dyson seems to have thought about life in an abstract biological sense of extremely adaptable intelligent beings, whereas he did not include advanced robots and similar machines in his considerations. Nor did he reflect on the biochemical basis of life or possible life forms based on elements other than carbon, such as silicon. (Silicon-based life has been considered by a few astrobiologists, but the speculation is not taken very seriously.)

The hibernation strategy for eternal life that Dyson advocated was later shown to be untenable by Lawrence Krauss and Glenn Starkman at the Case Western Reserve University, Cleveland. According to their analysis, the recently discovered evidence for a positive cosmological constant made the case against eternal intelligent life even stronger. Contrary to the result obtained by Dyson in 1979, the scenario outlined by Krauss and Starkman in 2000 was announced as 'not optimistic'.[52] On the other hand, by assuming dark energy to be given by an energy density different from the cosmological constant (such as quintessence) Katherine Freese and William Kinney argued that the optimistic conclusion of indefinite life could be maintained. Should their argument fail, there was no reason to worry too much. The two Columbia University physicists calmed readers of their article in *Physics Letters* by stating that, 'Perhaps someday one can find a way to create and use wormholes.'[53] Dyson found the Krauss–Starkman argument to be clever, but not a fatal blow against eternal life. Arguing that it was based on the assumption of material systems being digital, he rebutted the idea that if life were analogue it would still be able to survive for ever in a cold expanding universe. In 2001 78-year-old Dyson commented:

> Perhaps this implies that when the time comes for us to adapt ourselves to a cold universe and abandon our extravagant flesh-and-blood habits, we should upload ourselves to black

clouds in space rather than download ourselves to silicon chips in a computer center. If I had to choose, I would go for the black cloud every time.[54]

Dyson and Islam had to assume that cosmological knowledge, in so far it related to the future conditions of life, was basically complete. This is a rule of the game for this kind of long-term prediction, but it is a rule that may well turn out to be violated by new theories and discoveries. When Dyson wrote his paper he could not know of the inflationary scenario that burst upon the scene of cosmology a few years later, and he could not know either that by the close of the century the expansion of the universe was found to be accelerating. The new developments did not overthrow the earlier speculations of physical eschatology, but they implied revisions and in general offered new opportunities for speculating about life in the far future, as illustrated by some of the examples mentioned above.

Linde, one of the founding fathers of the new inflation scenario, thought that 'One of the main purposes of science is to investigate the future evolution of life in the universe'.[55] As he pointed out, in the chaotic inflationary scenario life can in principle exist eternally because it will appear again and again in new domains. In this respect the situation resembled the one of the steady-state universe. Unfortunately the many bubble universes do not promise eternity to *our* kind of life or to other kinds of life in *our* universe. Our domain will either evolve into a nearly empty universe or perhaps into a black hole, neither of which possibilities will be able to sustain life. Further considering the role of a very small cosmological constant, Linde concluded that it would only make things worse. Undeterred, he proposed as a possible strategy of survival that our descendants might be able to travel to new cosmic domains where conditions were more hospitable than in our dying domain; and when the new host domain became inhospitable for life, they could move to another one, and so on *ad infinitum*. Should it turn out that our universe evolves into a black hole that swallows up everything, matter and space included, he comforted his readers that it might take $10^{10\,000}$ years before the black hole formed. 'Hopefully, within this time we will be able to elaborate a successful strategy of survival or a more philosophical attitude towards our future.'[56]

Speculations of a scale and kind similar to Linde's were later entertained by Vilenkin, who, with Jaume Garriga, Viatcheslav Mukhanov, and Ken Olum, studied the possibility of sending messages to future civilizations in an eternally recycling universe. 'We would then become a branch in an infinite "tree" of civilizations, and our accumulated wisdom would not be completely lost', they philosophized.[57]

Frank Tipler of Tulane University, New Orleans, disagreed with Linde's conclusion that in a chaotic inflationary universe life would in principle be able to survive for ever. In 1992, defining life in a reductionistic way that might be more appealing to physicists than biologists—'an entity which codes information that is preserved by natural selection'—Tipler examined the possibility of eternal life in a collapsing universe arbitrarily close to the singular end-state.[58] If this definition of life appears to be reductionistic, it is probably because it was intended to be so. 'I am an uncompromising reductionist', Tipler admitted two years later; 'everything, including human beings, can be completely described by physics'.[59] Like Dyson in 1979, he made use of a 'subjective time', which is the time it takes for an individual or a civilization to process a certain amount of information, or, amounting to nearly the same thing, to think a certain number of thoughts. On the basis of this measure of time he found that life could exist for an infinite span of time: 'It is in principle consistent with known physics for life to continue forever in a closed universe arbitrarily close to the final singularity, but such survival is not possible in an eternal chaotic inflation cosmology.'[60]

The idea that intelligent life exists eternally in the future conforms with the final anthropic principle. According to Barrow and Tipler, this principle is a testable hypothesis because it requires matter to have some non-trivial properties. Moreover, they argue that the final anthropic principle has predictive power insofar that it rules out closed models with a decreasing entropy during the contracting phase of the universe.[61] Elsewhere in their book they explain that although the principle, which they insist is a statement of physics, has no direct implications for moral values, it nevertheless is closely connected to such values: 'The validity of the FAP is the physical precondition for moral values to arise and to continue to exist in the Universe: no moral values of any sort can exist in a lifeless cosmology.'[62] The last part of the claim is certainly correct, if trivially so. The first part is, it seems to me, a empty postulate.

Whether associated with the final anthropic principle or not, it is clear that the allegedly scientific speculations of Dyson, Tipler, and some other physicists are motivated in the age-old philosophical or religious desire of eternal life. They find it emotionally unbearable that intelligent life should some day disappear, and for this reason they develop scenarios more acceptable to their taste. This is what Tipler's argument for the final anthropic principle boils down to: 'I once visited a Nazi death camp; there I was reinforced in my conviction that there is nothing uglier than extermination. We physicists know that a beautiful postulate is more likely to be correct than an ugly one. Why not adopt the Postulate of Eternal Life—FAP, that the extinction of everything we could possibly care about is not inevitable—at least as a working hypothesis?'[63]

The reductionism of Tipler and similar spirits implies that all other sciences (and not only natural sciences) can be fully understood in terms of physics and ultimately are nothing but branches of physics. This kind of reductionism, or perhaps better physical imperialism, is also known from other physicists and other eras. It has a depressingly long history. Dirac expressed it in a moderate form in 1929, when he suggested that chemistry was merely applied quantum physics (see Chapter 6). Nine years earlier Max Born described the relationship between physics and chemistry in a more militant rhetoric: 'We realise that we have not yet penetrated far into the vast territory of chemistry, yet we have travelled far enough to see before us in the distance the passes that must be traversed before physics can impose her laws upon her neighbour science.'[64] Sommerfeld and Bohr likewise wrote about chemistry as a somewhat immature science to be conquered by the physicists. Tipler's reductionism follows a historical tradition, only it is much more extreme in its unrestrained imperialism. Describing the domain of physics as 'the whole of reality', he writes: 'An *invasion* of other disciplines is inevitable, and indeed the advance of science can be measured by the extent of the *conquest* of other disciplines by physics.'[65]

Most physicists (and, I believe, all philosophers) will have problems with the kind of reductionism espoused by Tipler. For one thing, no theory of physics is able to explain intentionally made objects such as teapots or aircrafts. As Ellis has pointed out, 'Even if we had a satisfactory fundamental physics "theory of everything",... physics would still fail to explain the outcome of human purpose, and so would provide an incomplete description of the real world around us.'[66] But this is precisely what Tipler and a few other hard-core reductionists deny. If the human mind is not only physically based, but can be understood in detail from principles of physics and computer science, intentionality can be explained as well.

12.4 COSMOBIOLOGY AT THE FRINGE, OR BEYOND

According to some cosmobiologists, far from being an evolutionary chance event, intelligent life is essential to the universe. Tipler is the major champion of this strong claim, which he has developed into a speculative theory of the 'omega point', a name borrowed from the French philosopher, theologian, and palaeontologist Teilhard de Chardin. In Tipler's version the omega point is the final state of the universe, which is reached after life has engulfed the entire universe and turned it into a sort of an all-encompassing organism or computer programme:

> At the instant the Omega Point is reached, life will have gained control of *all* matter and forces not only in a single universe, but in all universes whose existence is logically possible; life will have spread into *all* spatial regions in all universes which could logically exist, and will have stored an infinite amount of information, including *all* bits of knowledge which it is logically possible to know.[67]

In the years following the publication of *The Anthropic Cosmological Principle*, Tipler presented his theory of the omega point at a conference at the Vatican Observatory in Rome and also at a meeting of the Philosophy of Science Association held in Evanston, Illinois. Few of the theologians and philosophers listening to him were convinced.

In a controversial book of 1994, *The Physics of Immortality*, Tipler further developed some of these remarkable ideas, which now included allegedly scientific arguments for immortality and the resurrection of the dead.[68] According to the author, intelligent life—not organic, but in the form of computers to which our mind patterns have been transferred—will eventually gain control over the entire universe. The process will culminate at the omega point with the emergence of a computer superintelligence, which is as close to God as eschatological physicists can imagine. In fact, in a later work Tipler made it clear that the cosmic singularity *is* God, and the Judeo-Christian God at that.[69] At the omega point, where the final eternity has been reached, the temporal becomes atemporal. Contrary to Dyson's conclusion that life could not survive the collapse of a closed universe, Tipler argued that the collapse would not lead to a singularity with an infinite temperature. It would occur at different rates in different directions, and this would imply a temperature difference providing the power for life. Paradoxically, 'It is the very collapse of the universe itself which permits life to continue forever.'[70]

Not only did Tipler claim that the omega point scenario is consistent with the known laws of physics, he also claimed that the scenario follows from these laws, indeed that he had *proved* life to be eternal. However, his alleged proof rested on a Platonic claim of reality, namely this: 'Mathematical reality, the class of all logically consistent propositions, is regarded as the ultimate reality, and physical reality is a proper subclass of ultimate reality' (see also Chapter 10). As if to emphasize the scientific nature of the omega point theory, Tipler derived what he claimed were several testable predictions from it. Among these were his findings that the universe must be closed, have a density parameter in the range $1 < \Omega_0 < 1 + 10^{-6}$ and a Hubble parameter $H_0 \leq 45$ km s^{-1} Mpc^{-1}, and that the mass of the top quark, still undetected in 1994, is 185 ± 20 GeV. Tipler believed his theory to be 'very conservative from the physics point of view' because it did not rest on new equations or laws of physics.[71]

However, omega cosmology (which does not operate with extra dimensions) requires the universe to be closed and also implies that the cosmological constant must eventually decrease to zero. These were reasonable assumptions through most of the 1990s, but not after the discovery in 1998 of the accelerating universe. As to the Hubble constant, observations made in 1994 gave a value of 73 km s^{-1} Mpc^{-1}, which has remained fairly stable; no modern observations have resulted in the low value predicted by Tipler. So has the omega point theory been refuted by observations? Of course Tipler does not think so. Instead he suggests that the observed acceleration of the universe is a result of a temporary excess of matter over antimatter, not of a positive cosmological constant. In 2005 he repeated his conviction that the role played by continually progressive intelligent life in the universe is crucial, arguing that this is required both by the laws of physics and also for speeding up the annihilation of protons that will help in closing the universe. Not only is life in general of crucial importance, so are we humans of the twenty-first century:

> After all, we are now attempting to reproduce our ultimate biological ancestor, the first living cell from which all life on Earth descended. We would be the first rational beings from which all rational beings in the far future would be descended, so in reproducing us in these far future computers, life in the far future would just be learning about their history. So the laws of physics will not only be for us in the sense of requiring the biosphere to

survive, they are also for us in the sense that they will eventually allow every human who has ever lived have a second chance at life.[72]

Is this science, or merely some speculations disguised as science? Is Tipler a 'Dr Pangloss in disguise' or perhaps another Dr Zöllner?[73] Martin Gardner, writing in the *Skeptical Inquirer*, thought it fit to compare Tipler with L. Ron Hubbard. The omega point theory, he opined, could either be viewed as 'a new scientific religion superior to Scientology' or 'a wild fantasy generated by too much reading of science fiction'.[74]

Theologians have not been enthusiastic about Tipler's eschatological physics either. And no surprise, for according to Tipler theology is merely a branch of physics: 'From the physics point of view, theology is nothing but physical cosmology based on the assumption that life as a whole is immortal.'[75] This is not a view likely to arouse enthusiasm among theologians. Nor is his later conclusion that 'Christianity is not a mere religion but an experimentally testable science' (and one which has been verified).[76] Philosophers and theologians generally objected not only to the book's extreme positivism and reductionism, but also to its complete lack of understanding of theological concepts. In the opinion of one reviewer, Tipler's work had, however, one advantage: 'His attempt to transform theology into physics is so absurd that in the long run he will have no emulators.'[77]

Nor was the theory kindly received by Tipler's colleagues in physics and cosmology. George Ellis, for one, dismissed it as nothing but pseudoscience. Together with D. H. Coule he questioned several of the technical arguments and also objected to the general nature of the 'idiosyncratic' paper of 1992. For example, they found the definition of life chosen by Tipler to be arbitrary, inadequate, and in conflict with biological knowledge. More seriously, Ellis and Coule criticized Tipler for disregarding what was physically plausible and substituting it with what was merely imaginable, but doing so without any evidence. To Ellis and his coauthor, Tipler's elaborate speculations scarcely qualified as science: 'The question is how we distinguish this kind of speculation from science fiction or fantasy. Without the constraints of known physical theory, we can claim anything at all; but then a scientific paper is not the appropriate place for its publication.'[78] Ellis likewise dismissed *The Physics of Immortality* as 'a masterpiece of pseudoscience' based on 'a fertile and creative imagination unhampered by the normal constraints of scientific or philosophical discipline.'[79]

Some other writers have found the criticism of the omega point theory to be unfair and defended the theory as an interesting contribution to scientific speculation. The majority of cosmologists preferred to ignore it or at most regard it as 'a topic of after-dinner conversation' (as Longair described the anthropic principle). Although Tipler's ideas have not been kindly received, eschatological speculations of a less extreme kind are seriously discussed by many physicists, philosophers, and theologians who find them interesting in the ongoing dialogue between science and religion.[80]

Undeterred, Tipler has continued his exotic explorations in speculative cosmobiology. While many of the works in this branch of science have been concerned with the far future of the universe and the possibility of life in it, there are also works that do not belong to physical eschatology proper. One of the classical problems of astrobiology is the possible abundance of intelligent life. Are we alone in the universe? Or are we perhaps just one of a multitude of advanced civilizations? Although the latter position is sometimes assumed by cosmo- and astrobiologists, there is no compelling argument for either it or its alternatives. We just do not know. Yet, according to Tipler and some other scientists, we can know and do know. Based on a 'limited resources argument' he concludes that there is indeed intelligent life elsewhere in the universe, but it is very rare, perhaps of the order of one civilization per Hubble volume. Moreover, Tipler has repeated his earlier teleological claim that intelligent life will continue to exist forever: 'Intelligent life will survive until the end of time because the laws of physics require it. Or to put it another way, because such survival is one of the goals of the universe.'[81]

This message is also a leading theme in a more recent book, *The Physics of Christianity*, where Tipler deals more directly with his version of cosmotheology and modifies some of the arguments that appeared in *The Physics of Immortality*. He now gives priority to the laws of physics rather than to the eternity of life itself, but the difference is of little consequence since he concludes that the latter follows from the first. 'I do not *assume* life survives to the end of time', he says. 'Life's survival follows from the laws of physics. If the laws of physics be for us, who can be against us?' What matters is the development towards complete dominance of intelligent life in the universe:

> We can say that life must have become omnipresent in the universe by the end of time.... Therefore, if life is to continue guiding the universe—which it must if the laws of physics are to remain consistent—then the knowledge of the universe possessed by life must also increase without limit, becoming both perfect and infinite at the Final Singularity. Life must become omniscient at the Final Singularity.[82]

Although Tipler's excursions into cosmotheology are not highly regarded, it is noteworthy that somewhat similar physics-based speculations are seriously discussed by a small community of cosmobiologists in reputed scientific journals. The speculations may not be widely accepted, but in some circles of science they are recognized as interesting contributions to an extreme form of physical eschatology and cosmobiology. They are considered to be science, yet if so it is science of a very different kind than the traditional mode of understanding nature.

Tipler and some other cosmobiologists and SETI researchers are keen to warn against anthropocentric ways of thinking. We may tacitly assume that our technology exhausts what is possible using the known laws of physics, but this would be a gross mistake that greatly underestimates the power of future technologies. While it is undoubtedly true that the technology of the future will dwarf anything we can imagine today, cosmobiologists seem to be curiously confident that our present laws of physics will also be those of the future. They tacitly assume that our fundamental laws of physics are the true and final laws, that they are also nature's laws. Clearly, this assumption lacks credibility and is unwarranted from the perspective of history of science. This elementary lesson was emphasized by Hoyle in a lecture of 1957, when he reminded his audience:

> The laws of science are not inviolable.... The world of science is not identical with the physical world itself, with the real world if you like. Science is a model of the real world that we construct inside our own heads. The model is arranged by us to work according to a set of prescribed rules. These are the laws of science. And when we speak of comparing our scientific theories with observation we mean that a comparison is being made between our model and the events that comprise the real world.[83]

The kind of science (or, if one prefers, pseudoscience) exemplified by the omega point theory is exceptional, but there are other examples of a similar kind in the literature dealing with physical eschatology and speculative cosmobiology. Just to give a taste, a recent paper in the *International Journal of Astrobiology* discussed seriously and at length the idea of 'ancestor simulation', arguing that individuals may be re-animated in the computers of the far future. Sensing that this may sound absurd, the author remarks that 'the question of the requirement of some form of continuation of personhood is a possibility that many would see as a proper or appropriate eschatological conclusion to any FAP [final anthropic principle] or ELP [eternal life postulate]'.[84] Another paper in the same journal, modestly dealing with the theory of everything and the eternity of life, may illustrate the pseudoprofundity and the extreme reductionism and scientism that are characteristic features of this branch of speculative science. The theory of everything will be 'a discovery with titanic consequences for the human civilization' because it makes everything amenable to science. You may think that poverty, pollution, overpopulation, wars, and global warming are things to worry about, but not so if seen from the elevated perspective of the physical eschatologist:

> We must start taking physical eschatology more seriously right away, for the very future of all life in the Multiverse is at stake here. What is the point of our lives if none of it will exist tomorrow? What would be most exciting and rewarding is how humanity and other life forms might be able to save themselves and continue their survival indefinitely into the future. Can and will we?[85]

In the late 1980s Guth examined the question of whether it is in principle possible to create synthetic universes from a region of false vacuum. Although he was forced to conclude that 'it is rather difficult to create a new universe in the laboratory', he and some other cosmologists found it an intriguing possibility.[86] Guth's investigations inspired another American cosmologist of repute, Edward Harrison of the University of Massachusetts, to suggest a remarkably speculative theory of 'natural selection of universes'. This was neither the first nor the last time that Darwinian metaphors were used in a modern cosmological context. Linde had spoken of his chaotic inflation scenario as 'a kind of Darwinian approach to cosmology' as early as 1983.[87] About a decade later Lee Smolin developed a theory according to which the universe might be conceived as a living entity evolving in a strict Darwinian sense. As an alternative to the anthropic explanation of the values of the fundamental constants of physics, he suggested that the constants evolved in a series of collapses and expansions of universes controlled by the production of black holes. Relying on ideas of quantum gravity, he developed a speculative 'mechanism for natural selection in cosmology', which he argued was of a nature similar to biological evolution, including the elements of randomness and natural selection.[88] According to Smolin, his multiverse theory of cosmological natural selection, intended to be an alternative to the anthropically based multiverse, made genuine predictions, just as a scientific theory is expected to do.

Harrison's speculation of 1995 combined elements of the ideas of Guth and Smolin, which he supplied with the idea that our universe has been constructed by intelligent beings living in some other universe. These beings were supposed to be vastly more intelligent than us, but basically of the same kind. At some time in the not so distant future, Harrison suggested, our superintelligent descendants would themselves create new universes. 'If beings of our limited intelligence [such as Guth] can dream up wild yet seemingly plausible schemes for making universes, then beings of much higher intelligence might know theoretically and technically exactly how to do it.'[89] What advantages could such a science-fiction-like proposal possibly have? Harrison argued that it provided an explanation of the apparently fine-tuned constants of nature that was preferable not only to the teleological-theistic explanation but also to the usual anthropic explanation. The latter he found unacceptable because of its 'undeniable element of circularity' and also because of its uneconomical 'multiworld wasteland of mostly dark and barren universes'. On the assumption of Harrison's hypothesis it became understandable why the universe is both comprehensible and hospitable to intelligent life:

> The universe is comprehensible because it was created by comprehensible beings who had thought processes basically similar to our own.... Our universe is hospitable to life because it was created by beings of superior (but finite) intelligence who occupied a universe that was compatible with their physical existence and therefore was most likely similar to our own.[90]

Harrison apparently meant his proposal, published in the respected *Quarterly Journal of the Royal Astronomical Society*, as a contribution to the scientific understanding of the universe. However, it did not include any predictions, calculations or suggestions of how to test it. Nor did it explain how the first intelligent universe was created. Nonetheless, it was taken seriously by at least some cosmologists.[91]

Notes for Chapter 12

1. Shermer 2002, pp. 273–314.

2. The rich and fascinating history of beliefs and disbeliefs in extraterrestrials is covered in the authoritative works of Dick 1982, Dick 1996, and Crowe 1999. For a condensed account, see Tipler 1981.

3. On Dick's blend of pluralism and natural theology, see Crowe 1999, pp. 196–202.

4. Arrhenius 1908. For details on the history of the panspermia hypothesis, see Kamminga 1982.

5. Hoyle 1980a, p. 23.

6. Isenkrahe 1910, p. 45. A similar argument was made in Rademacher 1909. Isenkrahe's contributions to the controversy over the heat death and the entropic creation of the universe are considered in Kragh 2008. As noted in Chapter 4, J. J. Thomson speculated in 1905 that in the final state of the universe it would consist solely of free electrons.

7. Haldane 1928, p. 809. On Haldane's cosmological views, see Kragh 2004, pp. 221–25. On his biological eschatology, see Adams 2000. Shklovskii and Sagan 1966, a classic in the modern literature on extraterrestrials, was dedicated to Haldane.

8. Barnes 1933, p. 403. The two following quotations are from the same source, p. 402 and p. 404. For his rejection of Lemaître's primeval atom hypothesis, see p. 408. On Barnes' views on cosmology and its relation to Christianity, see Kragh 2004, pp. 113–17.

9. Untitled contribution to symposium on 'The question of the relation of the physical universe to life and mind' in *Nature* **128** (1931), 700–22, pp. 719–22.

10. Milne 1952a, p. 154. On Milne's cosmophysics, see Chapter 4. Later Christian cosmologists have considered the same issue as Milne, that is, whether separate worlds or civilizations need their own Christ. Ellis believed that what he called 'the many-Christ view' strengthened the case of a finite universe, for 'Surely an infinite number of Christ-figures must be too much, no matter how one envisages God.' Ellis 1993, p. 395.

11. Fermi's colleagues Emil Konopinski, Edward Teller, and Herbert York recalled the conversation. See Jones 1985b and also the more detailed Jones 1985a, which is online as http://www.fas.org/sgp/othergov/doe/lanl/la-10311-ms.pdf.

12. Hart 1975. The question is sometimes known as the Fermi–Hart paradox.

13. Shklovskii and Sagan 1966, based upon Shklovskii's *Vselennaia, Zhizn, Razum* (Moscow, 1963). The early history of the search for extraterrestrial intelligence is detailed in Dick 1996, pp. 399–445.

14. Shklovskii and Sagan 1966, p. 472.

15. Kardashev 1964. Shklovskii and Sagan 1966, pp. 394–95, 477–78. See also Kaku 2005b, pp. 307–18. As mentioned in Chapter 9, in 1990 Kardashev examined models of the cyclic universe, in part because he believed they made eternal life possible.

16. On the emergence of astrobiology and SETI research as a scientific discipline, see Dick 1996, pp. 473–501. Apart from the names 'astrobiology', 'exobiology', and 'bioastronomy', the name 'cosmobiology' is occasionally used. This name, which is appropriate in the present context, may first have been used by John D. Bernal in a lecture of 1952 to the British Interplanetary Society (ibid., p. 379). As mentioned in the text, 'cosmobiology' is sometimes used in a quite different, astrological context.

17. Tipler 1980, p. 279.

18. Tipler, review of Michael D. Papagiannis, ed., *The Search for Extraterrestrial Life: Recent Developments* (Boston: Reidel, 1985), in *Physics Today* **40** (1987), 92.

19. http://abscicon.ar.nasa.gov/index.html.

20. Dyson 1979, p. 447.

21. Jeans 1928a, p. 698.

22. Jeans 1928b, p. 470, a lecture delivered before the Royal Society of Arts on 7 March 1928.

23. Eddington 1931, p. 452. A slightly different version appeared in *The Mathematical Gazette* **15** (1931), 316–24.

24. Longair 1985, p. 188.

25. See the useful bibliography in Ćirković 2003c. Barrow and Tipler 1986, pp. 658–82, includes a survey of physical eschatology. Tipler 1994 speaks both of 'physical eschatology' and 'physical theology' (p. 116).

26. Barrow and Tipler 1986, pp. 658–59.

27. Ćirković 2003c, p. 122.

28. Barrow and Tipler 1986, p. 658.

29. Peacocke 2004, p. 332. Pannenberg 1981, pp. 14–15.

30. Craig 2008.

31. Rees 1969.

32. Milne 1952b, p. 164.

33. I have discussed this tradition in Kragh 2008.

34. Davies 1973.

35. Dicus et al. 1983, p. 85. See also Dicus et al. 1982 for technical details.

36. For example Frautschi 1982, Barrow and Tipler 1986, pp. 613–82, Rozental 1988, pp. 125–28, and Adams and Laughlin 1997.

37. According to the grand unified theories of the late 1970s, the proton should be unstable, decaying into lighter particles such as a positron and a neutral pion with a lifetime of 10^{31}–10^{32} years. Many experiments have looked for decaying protons, but none have been found. If the proton is radioactive, experiments tell that its half-life is greater than 7×10^{32} years.

38. Barrow and Tipler 1978, where the fate of life is not mentioned.

39. Adams and Laughlin 1997, p. 338.

40. Ibid.

41. Tables of the time scales of the universe, ranging from 10^{10} years to 10^{1500} years or more, can be found in Dyson 1979, Barrow and Tipler 1986, pp. 653–54, Adams and Laughlin 1997, and several other sources.

42. Gott 1993, p. 316. Other contributions to the doomsday argument are listed in Ćirković 2003c.

43. Caldwell et al. 2003. On phantom energy, see Chapter 8.

44. Einstein 1982, p. 45.

45. Lindsay 1959, p. 385, who was foreshadowed by Ostwald and his 'energetic imperative' in the early years of the twentieth century. See Hakfoort 1992.

46. Dyson 1979, p. 459. Other quotations are from the same source. Dyson repeated his belief in immortal life in his Gifford Lectures of 1985 (Dyson 2004, pp. 54–96).

47. Hoyle, in *Observatory* **68** (1948), 214–16.

48. Interview by Spencer Weart of 1978, quoted in Kragh 1996, p. 254.

49. Dirac 1961.

50. Farmelo 2009, p. 221. Tipler 1994 (p. 329) speaks of the 'Dirac/Dyson Eternal Life Postulate', which may be to make too much of Dirac's remark of 1961. Tipler (ibid., p. 11) further claims that Dirac was the first physicist to argue for the postulate of eternal life, which is definitely not the case.

51. Islam 1977. After Dyson's publication of 1979, Islam gave an extended exposition of his ideas which referred to Dyson's more 'optimistic' conclusions (Islam 1979) and four years later he

expanded his eschatological studies into a popular book. Possibly under the influence of Dyson, he speculated that 'a new form of intelligent life created artificially especially adapted to survive in extremely cold surroundings may be able to survive indefinitely' (Islam 1983, p. 109). Islam and Dyson agreed that in the case of a closed universe, indefinite survival of life was nearly impossible.

52. Krauss and Starkman 2000, p. 29.

53. Freese and Kinney 2003, p. 7. González-Diaz 2010 (p. 777) similarly concludes that 'even if all universes are filled with phantom energy, life could endure eternally in the realm of an infinite number of universes connected by worm holes'. A wormhole is a connection between two regions of space–time that is allowed according to some solutions to the equations of general relativity and which may permit time travel between the regions. Wormholes are probable not parts of the real universe.

54. Dyson, 'Is life analog or digital?' *Edge* online discussion of 13 March 2001, http://www.edge.org/documents/archive82/html. See also Dyson's article with the same title in Ellis 2002, pp. 140–57. The 'black cloud' refers to the science fiction novel *The Black Cloud* written by Fred Hoyle in 1957.

55. Linde 1988, p. 31.

56. Ibid. See also Linde 1989.

57. Garriga et al. 2000, p. 1892. The message or container sent to the future might be 'a device which reproduces our civilization in the new region, rather than waiting for new civilizations to evolve.... One can even imagine some individual members of the old civilization surviving in the container into the new region, perhaps by having their physical form and state of knowledge encoded in some compact and durable way for later reproduction' (p. 1899).

58. Tipler 1992 and earlier in Tipler 1980. According to Tipler 1994 (p. 124), life is 'a form of information processing, and the human mind—and the human soul—is a very complex computer program'.

59. Tipler 1994, p. 352.

60. Tipler 1992, p. 42.

61. Barrow and Tipler 1986, pp. 673–74. Barrow 1992, p. 160, repeats that the final anthropic principle (or conjecture) is to be considered a scientific statement about our world and not a philosophical speculation.

62. Ibid., p. 23.

63. Tipler 1989, p. 32.

64. Born 1920, p. 382.

65. Tipler 1994, p. 8. Emphasis added. Tegmark's reductionism is at about the same level as Tipler's. Even Dyson has lapsed into this kind of scientism, as when he comments that 'philosophy is nothing but empty words if it is not capable of being tested by experiment' (Dyson 2004, p. 96).

66. Ellis 2005.

67. Barrow and Tipler 1986, p. 677.

68. Tipler 1994.

69. Tipler 2007, p. 1.

70. Ibid., p. 117.

71. Interview conducted by Michael Shermer of 11 September 1995, in Shermer 2002, p. 260.

72. Tipler 2005, p. 905.

73. The reference to Dr. Pangloss, used by Sherman 2002, alludes to a figure in Voltaire's *Candide*, a professor of 'metaphysico-theology-cosmolonigology'. Karl Friedrich Zöllner, a brilliant and controversial German astrophysicist in the second half of the nineteenth century, advocated a 'transcendent physics'. In 1881 he published *Naturwissenschaft und Christliche Offenbarung*, a speculative attempt to integrate Christian theology and spiritualism with physical science.

74. Gardner 1991, p. 132.

75. Tipler 1994, p. 17.

76. Tipler 2007, p. 4.

77. Mutschler 1995, p. 480. On the other hand, the respected theologian Wolfhart Pannenberg considered Tipler's book to be an important contribution to the dialogue between theology and science.

78. Ellis and Coule 1994, p. 738.

79. Ellis 1994.

80. See, for example, Ellis 2002, where physical eschatology and similar subjects are discussed by, among others, P. Davies, G. Ellis, J. Barrow, M. Heller, and F. Dyson.

81. Tipler 2003, p. 143. Tipler 2005.

82. Tipler 2007, p. 62 and p. 93.

83. Untitled chapter in Mott 1959, p. 55.

84. Prince 2005, p. 223.

85. Karthik 2004, p. 324.

86. Farhi and Guth 1987, p. 154.

87. Linde 1983.

88. Smolin 1992. See also the elaborate exposition in Smolin 1997. For a critique of Smolin's speculations, see McCabe 2004.

89. Harrison 1995, p. 198.

90. Ibid., p. 200.

91. The article was cited 17 times until 2007. Gott and Li 1998 speculated that Harrison's super-intelligent civilization could have left us a message in the form of the dimensionless constants of nature.

13

Summary: Final Theories and Epistemic Shifts

I think the universe is all spots and jumps, without unity, without continuity, without coherence or orderliness or any of the other properties that governesses love. Indeed, there is little but prejudice and habit to be said for the view that there is a world at all.

Bertrand Russell, *The Scientific Outlook* 1931, p. 95.

13.1 CANDIDATES FOR A FINAL THEORY

Although this is not a book about theories of everything, or of the largely synonymous final or ultimate theories, several of the chapters have dealt with theories or theoretical frameworks that belong to this category.[1] In this area, there is a great deal of continuity from the days of Descartes to those of Witten. Descartes' dream of explaining all of nature deductively from a few principles or laws set a pattern that can be followed right up to the present, and so can his belief in the crucial role of mathematics. Boscovich's mathematical theory of matter allegedly led to 'explanations of all the chief properties of bodies, & of the phenomena of Nature', a claim repeated by several later theories.

The early research programmes associated with the vortex theory of atoms and the electromagnetic world picture included theories that aimed at finality. They were proclamations about nothing less than the ultimate aim of science and also forecasts of an ultimate end of empirical scientific research. When the final theory had been firmly established, 'all physical phenomena will be a branch of pure mathematics', as Hicks phrased it in 1895. In the same spirit, Mie, Weyl, and Hilbert later proposed certain world equations from which every physical phenomenon would be explicable. When the equations had been found, physics as usually understood would be over, for then, as Hilbert said, 'one needs only thinking and conceptual deduction to acquire all physical knowledge'. This was a conviction shared by Eddington, whose idiosyncratic fundamental theory promised a way to deduce unambiguously all the laws and constants of physics from epistemic considerations. But the grandiose proclamations were just that—proclamations. As Einstein (not a foreigner to such theorizing) pointedly remarked in relation to Mie's theory, it was a fine frame but unfortunately with no filling.[2]

The problem with these early unified and supposedly final theories was not only that they failed to reproduce most concrete phenomena of nature and predict new ones. It was also that the theories were necessarily established on the basis of what was known about the fabric of the world, and that this basis seriously restricted the power of the theories. A final theory proposed at any given time naturally reflects the knowledge about nature at the time, and for this reason it is ill-prepared for new discoveries that conflict with or radically extend that knowledge. Mie's unified electromagnetic theory presupposed that matter consisted of positive and negative electrons (protons and electrons), which was also the picture on which Eddington originally built his theory. Recall that the title of his great book of 1936 was *Relativity Theory of Protons and Electrons*. The new discoveries of particles and interactions in the 1930s just did not fit into his fundamental theory, and the way he nevertheless sought to make ends fit convinced nobody than himself. Chew's later bootstrap theory was different, not only because it was restricted to the strong interactions but also because it built on the requirement of self-consistency. This implied that if the *S*-matrix was known, 'any experiment can be predicted', as Chew proudly said in 1963. But the bootstrap theory could not, in fact, predict the outcomes of the high-energy experiments in the 1970s that led to the gauge field theory of quarks and leptons.

String theory, the most discussed of the modern attempts to establish a kind of final theory, shares with the old bootstrap programme the element of self-consistency. Contrary to all earlier unified theories, it contains no free parameters and can in principle account for new particle discoveries and much more. Alas, in principle only. The theory of superstrings does not differ qualitatively from its predecessors when it comes to concrete calculations of known phenomena and properties or to the prediction of new discoveries. For example, the discovery of the acceleration of the universe driven by dark energy should in principle have been predicted or at least expected by string theory, but in reality it took the string theorists by surprise.

Physicists only rarely announce a new fundamental theory to be 'of everything', a notion which is not only pretentious but also carries with it reductionism in a strong sense. Heisenberg's ill-fated unified theory of the 1960s came close, but it was neither reductionistic nor really of everything. Nor was this the case with Feynman's discovery from the same period of an all-encompassing world equation of the beautifully simple form $U = 0$.[3] If one wants a modern theory that pretends to be a theory of everything, and is in fact announced as such, Tegmark's mathematical theory of nature is a good candidate. Yet, with its basis in the so-called mathematical universe hypothesis it is of a very different kind to other attempts at a final theory. It is probably better described as a metascientific or philosophical system than a scientific theory. I would characterize Tipler's recent theory of the structure of the world in the same way.

The predominant role of mathematics is another feature common to unified or final theories. In these theories, mathematics is much more than merely a tool to establish theories and derive consequences from them. It is supposed to be the king of science, and in some sense is superior to physics. It is also a model towards which physical theories must aspire. In different versions and formulations, Hilbert, Minkowski, Eddington, and Milne likened the laws of physics to mathematical theorems. Hilbert looked forward to the day when 'physics in principle becomes a science of the type of geometry'. Milne's mathematical rationalism was no less absolute. He, too, aimed at 'a complete reconstruction of physics from the bottom up, on an axiomatic basis', and he described the laws of nature as 'no more arbitrary than geometrical theorems'.

Radical as these early conceptions were, they were not nearly as extreme as those suggested by Tegmark and Tipler at the beginning of the twenty-first century. According to their Platonic minds, there is no physical reality beyond or apart from mathematical reality. Everything physical is based on mathematics, because mathematics is all there is. The role assigned to mathematics in these conceptions is more extreme than in string theory, which is sometimes accused of being mathematics disguised as physics. It may be true that string theory is essentially guided and tested by mathematical consistency considerations (rather than experimental physics), but the theory does not claim to reduce physics to mathematics in any ontological sense.

At any rate, mathematics is hardly the final answer to everything. One problem may be that a theory of everything will necessarily have to be mathematically formulated, which seems to imply that it has to face the famous incompleteness theorem that the German logician Kurt Gödel first proved in 1931. According to this theorem, mathematics is essentially incomplete, meaning that it is always possible to construct true but unprovable mathematical statements. If this problem carries over into physics, it would seem to rule out a physical theory of everything in the strict sense. At least some scientists believe (or hope) that this is the case. Dyson is one of them:

> It is my hope that we may be able to prove that the world of physics as inexhaustible as the world of mathematics. Some of our colleagues in particle physics think that they are coming close to a complete understanding of the basic laws of nature.... But I hope that the notion of a final statement of the laws of physics will prove as illusory as the notion of a formal decision process for all of mathematics. If it should turn out that the whole of physical reality can be described by a finite set of equations, I would be disappointed.[4]

On a more mundane level, the mathematical symbols entering the equations of a theory of everything have to be interpreted in order to make physical sense, and the interpretation cannot come from the mathematical formalism itself. During the nineteenth-century controversy over the age of the Earth, Thomas Huxley criticized William Thomson for his blind confidence in mathematical calculations. Huxley's point is worth quoting: 'Mathematics may be compared to a mill of exquisite workmanship, which grinds you stuff of any degree of fineness; but, nevertheless, what you get out depends upon what you put in; and as the grandest mill in the world will not extract wheat-flour from peascod, so pages of formulae will not get a definite result out of loose data.'[5]

As pointed out in Chapter 9, the very concept of a theory or principle of everything is problematic for a number of reasons.[6] Thus, if an explanatory principle P is truly ultimate, there can be no deeper explanation that covers P. Either P must be left unexplained, in which case it is not about everything, or it must somehow explain itself, which is not possible for ordinary explanations. If a candidate for P is nonetheless found, how can we ever know that it is the ultimate explanation? To rely on the subjective notion of self-evidence is hardly a satisfactory answer, although this is what Wheeler did when he addressed the problem: 'Some principle uniquely right and uniquely simple must, when one knows it, be also so obvious that it is clear that the universe is built, and must be built in such and such a way and that it could not possibly be otherwise.'[7] Descartes would have agreed, but that does not make the argument any better.

The standard conception among philosophers and most scientists is presumably that an ultimate theory of everything is nothing but a chimera. It is a seductively attractive dream with a rich history, but it can never be other than a dream or a useful illusion. In a comment on Tipler's claim of having provided an ultimate explanation for the existence and nature of the world, the philosopher of science Lawrence Sklar expressed his reservations as follows:

> Should we even hope for explanations of the existence and nature of the universe which are unconditioned on any antecedent contingent posits? ... Conditioned by the ever increasing success of deeper and deeper explanations of phenomena, we hope that someday we will find the ultimate explanation, which, unlike all of its transient anticipators, requires no contingent posit at all. But such an unconditioned accounter of the conditioned is, according to Kant, an unobtainable ideal. The belief in its possibility is a useful illusion. Useful because the eternal hope that the final explanation will be found, like the carrot forever in front of the donkey, goads us ever onward in our search for the obtainable deeper and deeper relative explanations.[8]

13.2 FROM SCIENCE TO GOD, OR AWAY FROM GOD

It should come as no surprise that theories as speculative, abstract, and ambitious in scope as those referred to in this book attract religious and other non-scientific interest of a nature that can loosely be characterized as ideological. After all, if a theory is purportedly about everything it naturally relates to an area that is traditionally part of the religious domain. None of the theories reviewed here have been directly motivated by religion or designed to meet emotional needs, but in several cases they have been used for such purposes, either by the scientists themselves or, more commonly, by groups outside the scientific community. Thomson and Tait saw the vortex atom theory as congruent with their Christian sentiments, Milne drew religious consequences from his cosmophysics, and Tipler has identified God with the wave function of the universe. But they all insisted that their theories were scientific.

There is no law against using a physical theory in the service of religion or, for that matter, in the service of anti-religion. This can and has been done in a variety of ways, and there is absolutely nothing new in it. Just as Newtonian mechanics was an important part of the natural theology of the age of Enlightenment and thermodynamics was mixed up in the battle over religion in the nineteenth century, so the vortex theory and the electromagnetic world view were occasionally used in religious, spiritual, and political contexts. Occultists and spiritualists could use the physicists' ether, but so could the militant atheist and dialectical materialist, Lenin. A common way to 'apply' physics or some other science to non-scientific domains is by way of extrapolation, a particular form of scientism that does not respect the limited domain of physics. (Tipler explicitly denies that the domain of physics is limited, but his view is, after all, exceptional.) The extrapolation is sometimes to the religious realm, but it can be to any kind of moral, spiritual, or societal realm. One example is provided by Capra's use of the bootstrap programme in support of a whole new world view that unites nature with the human mind in a way that is in agreement with the philosophical insight of Eastern mysticism. Another and more isolated example is the emotional and sometimes quasi-religious appeal that cyclic conceptions of the cosmos has held on some scientists. It is one of history's many ironies that eternally cyclic models were traditionally seen as associated with atheism and materialism.

The most discussed and most controversial of the potential connections between religion and modern physical theories of the universe is currently to be found in the anthropic principle and its corollary (as it may sometimes be thought) the multiverse. Even before its association with multiverse philosophy, advocates of the anthropic principle had to face the apparently teleological nature of the principle and its relationship with classical arguments of divine design. While some theologians were quite happy with the anthropic multiverse,[9] others considered it part of an atheistic project. As far as anthropic physicists are concerned, the general attitude is to ignore or deny any connections to religion and design. Indeed, there are those who consider it not only to be counter-teleological but also a scientific alternative to belief in God. What matters for these physicists and cosmologists is not so much to combat religion as to present the anthropic multiverse as a respectable and purely scientific hypothesis that explains what other hypotheses cannot explain. There is hardly any 'correct' way of understanding the religious relevance or irrelevance of the multiverse, only ways that differ according to the interests of the actors.

Religious issues also turn up in connection with the new tradition of physical eschatology, a name that directly invites thinking about links or parallels between physics and theology. Yet by far the majority of the scientists contributing to this kind of extreme extrapolation of physics are eager to distance themselves and their field from any kind of religious association. (As usual, Tipler is an exception.) Just as in the case of anthropic reasoning, they consider it important to downplay the speculative and quasi-religious elements and present physical eschatology as a serious if somewhat unconventional science. In spite of their assurances, the connection to religion is too obvious to be ignored. What is going on in some works of physical eschatology is not apologetic physics but rather a scientistic version of it, in which speculative cosmological physics is introduced as an *Ersatzreligion*—a secular and scientifically respectable substitute for religious faith. There is more than a little similarity between these modern works and earlier attempts in the history of science to either base religion on science or substitute religion with a new gospel of science.[10]

13.3 EPISTEMIC SHIFTS AND NEW PARADIGMS

In the development of modern theoretical physics there have been several cases in which a new theory or principle has been proclaimed as ushering in a whole new paradigm in something like a Kuhnian sense, a new view of science radically different from and perhaps incommensurable with the old one. Such proclamations have almost always turned out to be more rhetoric than reality. Enthusiasts of the electromagnetic world view considered themselves revolutionaries on their way to building a new foundation of physics that would 'break the framework of the old physics to pieces' and 'overturn the established order of ideas and laws', as Langevin proclaimed in the early twentieth century. However, the electron and ether theories of the early twentieth century were not truly revolutionary. When it came to justifying the new theories, the physicists used well-established standards and methods, including as a crucial standard that the theories must pass experimental tests. Other standards of a transempirical nature appeared as well, and sometimes quite prominently, but the emphasis on epistemic values such as unification, simplicity, and explanatory breadth was neither new nor was it conceived as an alternative to empirical testability.

I take the concept of *epistemic shift* to imply proposals from scientists that traditional criteria of evaluation of scientific theories or practices are no longer adequate and should therefore be replaced by new criteria that better fit the problems under investigation. The proposed changes in epistemic standards may be so drastic that they challenge the traditional meaning of science and in effect introduce a new definition of what counts as science. This is the situation in some branches of contemporary fundamental physics, where a redefinition of science is openly discussed—not by philosophers, but by a minority of active scientists. A recent article in *New Scientist*, commenting on new theories of the universe, carried the title 'Do we need to change the definition of science?'[11] Of course, multiverse string physics and the like is not the only kind of new science that has been claimed as marking a break with the previous understanding of science and form the basis of a new paradigm. There are other self-proclaimed revolutions in modern science, proposals to change the 'definition' of science, but these are not or are only loosely connected with the subject of the present book.[12]

The notion of revolutions in the development of science may or may not be seen in the perspective of Thomas Kuhn's ideas as they appeared in his influential book *The Structure of Scientific Revolutions* of 1962. Although the terms 'revolution' and 'paradigm' are popular among scientists and science writers, they rarely use them in any specific or reflected sense. According to Kuhn's original ideas, revolutions or paradigm shifts are changes so radical that two competing paradigms are incommensurable and wholly different world views. Apart from during brief revolutionary phases, science develops 'normally' or monoparadigmatically. Although some authors have suggested that the string revolutions of the late twentieth century show a striking resemblance to Kuhn's description of paradigm shifts, in reality they were changes of a much lesser scale.[13]

Although there is no invariant definition of science, there are certain standards of evaluation that are relatively stable and shared by almost all physical scientists. These standards are not, of course, divinely given but the result of a long historical and social process with roots stretching back to the scientific revolution or even further. They have changed over time and are likely to continue to change.[14] On the other hand, the stabilized scientific standards are neither arbitrary nor merely social constructs.

The large majority of physicists undoubtedly consider testability to be more than just a desideratum that scientists happen to have agreed upon and which suited physics at a certain stage of development. They consider it an indispensable precondition for a theory being scientific, and in this sense a crucial and stable epistemic value. Yet, although there is a high degree of consensus with regard to this value, consensus does not necessarily extend to its translation into scientific practice. Scientists in the same field do not always interpret testability in the same way. For example, some may argue that testability in principle is the important thing, while others may insist on actual and instant testability. Again, should a theory result in precise and directly testable predictions, or will indirect testability and probability predictions do? Can one test a theory mathematically, or will a real test have to be empirical, a confrontation with nature in the form of an observation or a laboratory experiment? At what time in the development of a theory or a research programme is it reasonable to demand testability? Do predictions of novel phenomena count as more important than 'predictions' of phenomena that are already known to exist? None of these questions can be given a definite answer.

In spite of its almost sacred position in the physical sciences, experimental tests are only one way among several to assess a scientific theory, hypothesis, or model. It is generally agreed that there are other factors at play, such as explanatory power and breadth—epistemic values that do not necessarily agree with testable predictions. There is also a class of consideration that does not relate to experiment or observation, but more to the structure of the theory. Transempirical and sometimes aesthetic values such as simplicity, coherence, consistency, and 'mathematical beauty' play a role in modern science, as they have always done. Physicists sometimes find a theory inherently attractive even though it fares badly in empirical tests, or they may dislike a theory even though it has an excellent empirical record. They may refer to it in emotional terms, as we have seen with regard to cyclic models of the universe (e.g. Eddington, Zanstra, and Tolman). Another example is Dyson's dislike of closed universe models because they do not allow eternal life. In spite of such emotional bonds, scientists are careful not to let them influence their final evaluations, which they maintain should primarily be based on a comparison with observations.

There are very few examples of scientists who have seriously proposed that experiment and observation should not have the final word in theory evaluation. The most remarkable one is probably to be found in the cosmophysics of the 1930s, where Eddington advocated a drastic epistemic shift according to which knowledge claims of fundamental physics should be judged 'without turning up the answer in the book of nature'. He held that some of these claims, such as relating to the constants of nature, were *a priori* and therefore independent of measurements. Although Milne did not accept Eddington's selective subjectivism, his system was no less rationalistic. A view according to which laws of nature are nothing but 'the inevitable consequences of our having said exactly what we are talking about' does not leave much room for observational tests. When Milne spoke of his programme of kinematic relativity as 'a complete reconstruction of physics from the bottom upward', he meant it seriously. But neither Eddington's nor Milne's versions of rationalist cosmologies survived for long and were at any rate restricted to the British scene.

In the 1960s, when their views were largely forgotten, we hear somewhat similar tunes from Chew and a few other advocates of the hadronic bootstrap, who spoke of it as a 'precursor of a new science' or even a 'completely new form of human intellectual endeavor . . . one that not only will lie outside physics but will not even be describable as "scientific"'. The complete bootstrap would explain nature as it is because 'this is the only possible nature consistent with itself'. While these suggestions of a possible new paradigm remained on the rhetorical level, elements similar to them later became part of string physics.

Many of the ideas about the universe that have been forwarded during the last couple of decades are highly speculative. Yet it is a characteristic feature of such proposals that scientists defending them go to great lengths in order to argue that they connect with empirical reality. They maintain that, however speculative and arcane they may look, their views lead to testable predictions and *therefore* qualify as scientific ideas. They may be extreme, but they are nonetheless scientific and therefore respectable.

Among the many examples, recall the efforts of the cosmologists who have proposed new cyclic models of the universe to show crucially that these can be confronted with empirical data. Steinhardt and Turok are keen to emphasize that 'science is nothing without observational test' and that their cyclic model lives up to this basic standard. Defending an alternative model based on loop quantum gravity, Smolin agrees that 'in the end experiment will decide'. Also advocates of the string-based model of the pre-big-bang universe stress the importance of testability and that the final choice between rival theories will 'be made on the basis of experiment'. The experiments that the cosmologists appeal to are in many cases unrealistic for the time being, but what matters is their insistence on the possibility of testing, which to them counts as an indispensable criterion of science. A further example is provided by Penrose's highly speculative scenario of a conformally cyclic universe, about which he repeats the mantra that 'in principle at least, there should be clear-cut predictions which should be observable'.[15]

The high epistemic value assigned to testable predictions may be in principle only, but it is nonetheless taken seriously by nearly all modern physicists and cosmologists. The anthropic principle is in many ways a foreign element in science, yet scientists making use of anthropic reasoning do not see it as an alternative to predictions. On the contrary, they maintain that the anthropic principle does have predictive power and in this respect lives up to established standards in physics. On the other hand, the kind of predictions they come up with are often of an unusual nature and have failed to satisfy the critics. Does the apparent absence of extraterrestrials really qualify as an example of the 'genuinely predictive power' of the anthropic principle, as argued by Carter? If it follows from anthropic arguments that cosmologies with an infinite past are wrong (the Davies–Tipler argument), should the inference be accepted as a testable prediction? These kinds of predictions are surely of an altogether different nature to Hoyle's alleged anthropic prediction of 1953, in which he deduced an unknown energy level in the carbon-12 nucleus from astrophysical evidence.

The multiverse, with its close connection to anthropic reasoning, is probably the most radical of the modern attempts to establish fundamental science on a new footing. Indeed, terms such as 'revolution' and 'paradigm shift' are commonly heard in the controversy over the multiverse. If multiverse physics promises 'a deep change of paradigm that revolutionizes our understanding of nature' (Barrau), then it involves an epistemic shift as radical as anything in the past. It is more than just a new style of doing physics, for it questions the significance and meaning of some of the most important concepts of science, such as explanation, physical existence, and natural law.

When physicists predict that something exists, the prediction will be judged upon whether or not the predicted object is actually found in nature. Dirac's relativistic quantum theory predicted the existence of positrons, and when these were discovered in cosmic rays it amounted to a verification of the theory. The positron existed as 'a solution to the equations of the theory' (Susskind), but the true existence was to be found in nature and not in mathematics.[16] According to some multiverse and string theorists, nature is in this respect irrelevant. The existence of an object, they say, refers solely to the equations, meaning that the theoretical object is consistent with the laws of physics on which the equations build. Also, in the case of natural law, does the multiverse introduce a radically new perspective, since what are traditionally known as fundamental laws now appear as merely laws applying to our own universe? This local universe or sub-multiverse is privileged only by the fact that we happen to live in it.

Multiverse physicists do not really revolt against established standards of physics such as predictivity and testability, but they want to make use of new kinds of prediction and they also tend to deny that testability implies falsifiability. 'The multiverse remains within the realm of Popperian science', Barrau asserts. 'It is not qualitatively different from other proposals associated with usual ways of doing physics... [but] falsifiability is just one criterion among many possible ones and it should probably not be overdetermined.'[17] Whereas most physicists consider falsifiability to be of crucial significance, this is an epistemic criterion that does not fit very well with anthropic multiverse science. As we have seen, in the 1980s Carter objected explicitly to what he called the 'requirement of refutability' and suggested that anthropic science should be judged on different, more humble standards than those normally used in physics. Later, Susskind had accused the 'Popperazi' of strangling the progress of science by imposing unrealistic and irrelevant methodological restrictions on what science is about. The response of Susskind and some other multiverse physicists seems to be that, if their favoured cosmological theory does not live up to the standards of science defined by Popper, perhaps it is better to ignore or change those standards. If there is a need to appeal to some other standards of science, there are several alternatives on the philosophical market.

One such alternative is the Bayesian approach, which according to many philosophers offers a richer and more realistic picture of how science works.[18] Bayesianism, so named after the eighteenth-century English mathematician and priest Thomas Bayes, emphasizes that science is basically concerned with providing and weighing evidence in favour of a theory and its rival theories. Not only do many philosophers of science subscribe to Bayesianism, the approach has also commanded the attention of several physicists and cosmologists in their attempts to assess the plausibility of various cosmological ideas. For the last decade or so much work has been done in astrophysics and cosmology based on Bayesian methods, most of which is concerned with areas other than multiverse speculations. Although not all cosmologists are happy with the new 'cosmostatistics', an increasing number finds the approach to be both valuable and necessary. 'The application of Bayesian tools to cosmology and astrophysics is blossoming', says Roberto Trotta of Oxford University.[19]

With its basis in probabilistic reasoning of degrees of belief, Bayesian methodology corresponds much better to multiverse physics than Popperian methodology. On the other hand, from the point of view of a strict Bayesian analysis there seems no reason to prefer the multiverse explanation of cosmic fine-tuning over the explanation offered by a single universe hypothesis. According to a recent analysis, cosmic fine-tuning does not lend support to the notion of multiple universes.[20] The leading Bayesian philosopher Colin Howson has no problem in admitting the multiverse as an entirely legitimate scientific hypothesis. 'If Popper condemns it as pseudoscience because it is "unfalsifiable"—and it may not always be—then so much the worse for Popper.'[21] Yet the fact remains that Popper's philosophy of science has been and continues to be very influential in cosmology and related sciences. Whether they are aware of it or not, most physicists and cosmologists subscribe to the basic doctrines of Popperian falsificationism, or at least to the folklore version of it.[22]

As Ellis and a few other scientists have pointed out, the debate over the multiverse is not merely a question of disagreement about the proper methods of science. It goes deeper, by involving the very understanding of science and the ways to distinguish between science and non-science. To Ellis, Smolin, Richter, and probably the majority of physical scientists, actual testability (and not merely testability in principle) is a *sine qua non* for a theory that deserves to be called scientific. If this fundamental criterion is abandoned or just weakened, how can we justifiably claim that physical cosmology is a science while astrology is not? There is another and related issue at stake, namely to whom the 'right' to define science belongs. Should that be left entirely to the scientific experts, or should philosophers have a say? Or, why not the politicians? Under normal circumstances the question is hardly important, but in situations of controversy like the present one it is important. The question is not restricted to the multiverse hypothesis, for it may be equally or even more relevant in relation to those astrobiological speculations and exercises in physical eschatology that seem to defy any reasonable criterion of testability.

In an article in *Nature* of 2006, Susskind is quoted as follows: 'It would be very foolish to throw away the right answer on the basis that it doesn't conform to some criteria for what is or isn't science.'[23] He apparently suggests we forget about normative prescriptions and let the scientists, or perhaps the scientific community, determine by way of their practices what is and what is not science. Since many physicists and cosmologists take the multiverse seriously and investigate the hypothesis as part of their research, surely it must be a scientific idea. However, this view is troubling for a number of reasons. Not only is it circular reasoning to define science as what scientists do, it also presupposes that all scientists have roughly the same ideas of what constitutes science, which is definitely not the case. Moreover, it makes no sense to speak of a 'right answer' without appealing, explicitly or implicitly, to some criteria of science. To conclude that a theory is either valid or invalid necessarily involves certain standards of scientific validity. These standards need not be part of a philosophical system, nor do they need to be explicitly formulated, but it is hard to see how they can be avoided. Nature herself is not kind enough tell us when an answer is right.

String theory was controversial many years before it became associated with multiverse reasoning, primarily because it seemed to represent a new style of theoretical physics that broke with the traditional and eminently fruitful dialogue between theory and experiment. Many physicists of the older generation (and some of the younger) complained that it was bad science, and some denied that it was science at all. The main complaint was and still is that string theory is divorced from experiment and therefore cannot be tested by ordinary empirical means. Have we any reasons to believe in the 10-dimensional world of superstrings other than mathematics, aesthetics, and authority?

In fact, classical string theory does not really represent an entirely new kind of theoretical physics and it does not involve very radical changes in epistemic standards.

If the balance of justification is heavily oriented towards mathematical arguments, it is not because string theorists have decided to scrap the traditional empirical mode of physics. The predominant role of mathematics is forced upon string theory because of the difficulties in producing testable consequences from it. The 'mathematical tests' are to some extent seen as temporary alternatives to the empirical tests that so far cannot be performed but may be realized in the future. In this respect string theory is basically in the same situation as the vortex theory of atoms was in the late nineteenth century. But whereas we know that matter is not composed of ethereal vortices, we do not yet know whether or not elementary particles are made up of superstrings.

Critics of the string theory often say that it violates Popperian norms and therefore is not acceptable science. However, even if one keeps to a traditional Popperian perspective (and there is no strong reason why one should do so), this is not necessarily the case.[24] Popper's philosophy of science rules out a theory if it cannot be falsified even in principle, but it does not require instant falsifiability or falsifiability at any given time after the formulation of the theory. The lack of testable implications may not be a permanent feature, in which case the theory will be considered a legitimate research project that is likely to lead to a truly scientific theory. Of course, if string theory never evolves into a testable theory, the verdict is different. At some stage one will have to decide that long enough is long enough. Critics believe that this stage was passed many years ago, while supporters of the theory continue to develop string theory in the hope of establishing firm contact with the world of experiments.

String theorists share with other physicists a belief that testability and predictability are important epistemic values. 'The acid test of a theory comes when it is confronted with experiments', as two string theorists phrase it.[25] The problem is that these values are rather far from what physicists can produce from their calculations. They therefore have to rely, at least to a large extent, on mathematical consistency, not only as a guiding principle but also as a criterion of progress. Whereas string theory is in principle testable and capable of generating predictions about the real world, it is more questionable if it is also falsifiable in the sense that future experiments will force physicists to abandon it. It may be so in principle, but not necessarily so in practice. The situation is not unlike the one of Mie's unified theory in the 1910s. This theory resulted in a few predictions, but they were of such a kind that they were 'experimentally useless', as Mie admitted.

The epistemic problems of string theory have only aggravated in recent time, especially since the landscape interpretation was widely accepted in the string community. The traditional virtue of 'Einsteinian physics', that a fundamental theory should not only be consistent but also lead uniquely to predictions of what there is in nature, is clearly challenged by the new understanding of string theory. Some string and multiverse physicists have suggested that the new situation indicates the limits of traditional cosmological physics and the emergence of a new paradigm in which anthropic reasoning has a place that is as natural as it is necessary. 'We are in the middle of a remarkable paradigm shift in particle physics', says Schellekens, referring to the anthropic string landscape.[26] The landscape model does not and cannot lead to precise and unique predictions, but it can produce probability predictions that can at least be compared rationally with observed natural phenomena. Don Page of the University of Edmonton has recently provided numerical estimates of the fine-structure constant and the mass of the proton based on anthropic arguments. This may be seen as impressive, but the calculations are no more proper predictions than were Eddington's calculations of the 1930s.[27]

Some commentators have suggested that in particular domains of physics, such as cosmology and Planck-scale physics, empirical enquiry is no longer either needed or possible. Perhaps we have reached the stage 'where we can proceed without the need to subject our further theories to empirical test' (Shapere). We may already have entered a 'post-modern cosmology' guided by only aesthetic and mathematical considerations, a 'new branch of thought [that will] use the language of physics, but will not rely on experimental perspectives'.[28] Such post-empirical theories belong to the 'not-even wrong' category. They may also be said to belong to 'hyperphysical physics', to use the phrase coined by Schelling in the late eighteenth century.

It is hard to think of a more speculative area of science than the recent emergence of so-called physical eschatology, including attempts to evaluate the possibility of life in the far future by extrapolations of the currently known laws of physics. Some scenarios in this post-modern branch of science might not unreasonably be labelled science fiction had it not been for the fact that they are contemplated by scientists and appear in recognized scientific journals. Physical eschatology is conservative is the sense that it normally builds on established laws of physics. In all other senses it is anything but conservative. It is a field that almost by definition cannot be tested, and yet some physicists have suggested kinds of prediction from doomsday scenarios that they claim are testable.

Even though speculations and extrapolations have always been part of the physical sciences, it seems that they have reached new heights during the last couple of decades. They are not cultivated in parallel with and isolated from science, but have become increasingly integrated in the social and epistemic system of science. Publications by many physicists, some of them of high distinction, show a tendency to unrestrained extrapolation of physics into domains that according to the traditional view are inaccessible to the methods of physics.[29] The flights of fancy represented by some recent works in speculative cosmology, astrophysics, and physical eschatology are of no less interest from a sociological and psychological point of view than they are from a scientific point of view. Indeed, future historians of science may well conclude that they are to be explained primarily in terms of external factors, such as the socio-psychological mechanisms that are shaping some aspects of modern science.

Notes for Chapter 13

1. There is a considerable philosophical literature on the possibility of a unified theory of everything. See, for example, Kitcher 1981 and Rescher 1999. For general discussions, see Lindley 1993 and Scoular 2007.

2. Einstein's remark is valid for most of the unified theories, even for string theory. The Dutch theoretical physicist Gerardus 't Hooft, recipient of the 1999 Nobel Prize for his work on electroweak theory, argues that string theory is not really a 'theory' but 'just a hunch'. He says: 'Imagine that I gave you a chair while explaining that the legs are still missing, and that the seat, back and armrest will perhaps be delivered soon; whatever I did give you, can I still call it a chair?' Hooft 1997, p. 163.

3. Feynman's law of 'unworldliness' was a joke, intended to show that with some appropriate mathematical formulation one can unify all the laws of physics without the unification being physically meaningful. Feynman, Leighton, and Sands 1964, p. 25–11.

4. Dyson 2004, p. 53. In a talk of 2002 celebrating the centennial of Dirac's birth, Hawking likewise suggested that Gödel's theorem makes a final theory impossible (http://www.damtp.cam.ac.uk/strings02/dirac/hawking/). On the relevance of mathematical incompleteness theorems in physics and related problems, see Robertson 2000 and Barrow 1998.

5. Quoted in Hallam 1988, p. 115.

6. For these reasons, see e.g. Nozick 1981, pp. 116–19.

7. Rees, Ruffini, and Wheeler 1974, p. 297.

8. Sklar 1989, p. 54.

9. I cannot refrain quoting Bertrand Russell: 'Theologians have grown grateful for small mercies, and they do not much care what sort of God the man of science gives them so long as he gives them one at all' (Russell 1931, p. 110). The comment referred to James Jeans' picture of a divine spirit in the form of an omniscient mathematician.

10. For example, in the early twentieth century Wilhelm Ostwald turned energetics, the generalized science of thermodynamics, into a philosophical system (monism) with which he intended to replace Christianity and other theistic religions. He wanted to 'establish in place of the traditional ethics dependent on revelation a rational scientific ethics, based on facts' (quoted in Kragh 2008, p. 129). On Ostwald's scientism and quasi-religious monism, see Hakfoort 1992.

11. *New Scientist*, 7 May 2008.

12. The concept of self-organization has been claimed to be the core of another paradigm shift or revolution that fundamentally changes the game of science and extends its domain to, for example, economy, linguistics, and psychology. Pioneered about 1980 by the physical chemist Ilya Prirogine, the physicist Eric Jantsch, the mathematician René Thom, and others, the science of self-organization and complexity has developed into a large and diverse industry. According to Bernd-Olaf Küppers, a German biophysicist, it is 'a paradigm shift ... which may come to change our scientific world-view in a fundamental way' (Küppers 1990, p. 51). For a philosophically and sociologically oriented introduction, see Krohn, Küppers, and Nowotny 1990.

13. Chen and Kuljis 2003 analyse the history of string theory within the perspective of Kuhnian revolutions.

14. On the historical development of philosophies and standards of science from Aristotle to Kuhn, see Losee 1980.

15. Penrose 2006, p. 2762.

16. In the same paper where Dirac predicted the existence of positrons, he also predicted the existence of monopoles. He believed that since the hypothetical monopole turned up as a solution to the equations of quantum mechanics, 'one would be surprised if Nature had made no use of it' (Dirac 1931, p. 71). But whereas positrons exist, apparently magnetic monopoles do not.

17. Barrau 2007.

18. For a detailed exposition of Bayesian methods of science, see Howson and Urbach 1993.

19. Trotta 2008, p. 101.

20. Palonen 2008.

21. Quoted in Matthews 2008, p. 47.

22. The influence of Popper's ideas is demonstrated in Sovacool 2005. See also Kragh 1996, pp. 244–50, for the role of the falsifiability criterion during the cosmological controversy in the 1950s and 1960s.

23. Brumfiel 2006, p. 11.

24. See the analysis in Johansson and Matsubara 2009.

25. Burgess and Quevedo 2007, p. 33. Johansson and Matsubaru 2009 conclude that although the methodology of string theory differs in many ways from ordinary physics, the theory of superstrings does not constitute a break with the basic methodological norm of physics, namely that ultimately it is experiments that decide the truth of a theory.

26. Schellekens 2008, p. 1.

27. Page 2009, who for the inverse fine-structure constant finds a value of about 162, some 18% from the measured value. Recall that Heisenberg in his unified theory got 120, a slightly better agreement, and that Eddington nearly got it right. But of course there is no real connection between the three kinds of theories.

28. Bonometto 2001, p. 233. The writer John Horgan coined the phrase 'ironic science' for the kind of science-like intellectual discourse that 'by raising unanswerable questions, reminds us that all our knowledge is half-knowledge...[but] does not make any significant contributions to science itself'. Ironic science is 'science that is not experimentally testable or resolvable even in principle and therefore is not science in the strict sense at all', Horgan 1997, p. 31 and p. 94. In Horgan 2006, he mentions the anthropic principle and string theory as two prime examples of ironic science.

29. For a critical analysis of these tendencies in modern physics, see Goenner 2001.

BIBLIOGRAPHY

Abbott, B. P. et al. (2009). 'An upper limit of the stochastic gravitational-wave background of cosmological origin', *Nature* **460**, 990–94.

Abdo, A. A. et al. (2009). 'A limit on the variation of the speed of light arising from quantum gravity effects', *Nature* **462**, 331–34.

Abramowicz, Marek and George F. R. Ellis (1989). 'The elusive anthropic principle,' *Nature* **337**, 411–12.

Adams, Mark B. (2000). 'Last Judgment: The visionary biology of J. B. S. Haldane', *Journal of the History of Biology* **33**, 457–91.

Adams, Fred C. and Gregory Laughlin (1997). 'A dying universe: The long-term fate and evolution of astrophysical objects', *Reviews of Modern Physics* **69**, 337–71.

Adelberger, Erich G., B. R. Heckel, and A. E. Nelson (2003). 'Tests of the gravitational inverse-square law', *Annual Review of Nuclear and Particle Science* **53**, 77–121.

Aguirre, Anthony (2001). 'Cold big-bang cosmology as a counterexample to several anthropic arguments', *Physical Review* D **64**, 083508.

Albrecht, Andreas and João Magueijo (1999). 'A time varying speed of light as a solution to cosmological puzzles', *Physical Review* D **59**, 043516.

Ananthaswamy, Anil (2009). 'How to map the universe', *New Scientist* (4 May, online edition, http://www.newscientist.com/article/mg20227061.200-how-to-map-the-multiverse.html).

Anderson, Philip W. (1972). 'More is different', *Science* **177**, 393–96.

Arabatzis, Theodore (2006). *Representing Electrons: A Biographical Approach to Theoretical Entities*. Chicago: University of Chicago Press.

Arrhenius, Svante (1908). *Worlds in the Making: The Evolution of the Universe*. New York: Harper & Brothers.

Ash, David and Peter Hewitt (1994). *The Vortex: Key to Future Science*. Bath: Gateway Books.

Ashtekar, Abhay (2010). 'Classical singularities and quantum space-time', in David Rowe, ed., *Beyond Einstein* (to be published). Boston: Birkhäuser.

Atiyah, Michael (1990). *The Geometry and Physics of Knots*. Cambridge: Cambridge University Press.

Bailey, V. A. (1959). 'The steady-state universe and the deduction of continual creation of matter', *Nature* **184**, 537.

Balashov, Yuri (1990). 'Multifaced anthropic principle', *Comments on Astrophysics* **15**: 1, 19–28.

Balashov, Yuri (1991). 'Resource letter AP-1: The anthropic principle', *American Journal of Physics* **59**, 1069–76.

Balashov, Yuri (1992). 'On the evolution of natural laws', *British Journal for the Philosophy of Science* **43**, 343–70.

Bibliography

Balashov, Yuri (1994). 'Uniformitarianism in cosmology: Background and philosophical implications of the steady-state theory', *Studies in History and Philosophy of Science* **25**, 933–58.

Banks, Tom (1985). 'TCP, quantum gravity, the cosmological constant and all that...', *Nuclear Physics* B **249**, 332–60.

Banks, Tom, Michael Dine, and Elie Gorbatev (2004). 'Is there a string theory landscape?' *International Journal of High Energy Physics* **08**: 058.

Barbour, Julian (2001). *The Discovery of Dynamics*. Oxford: Oxford University Press.

Barnes, E. William (1933). *Scientific Theory and Religion: The World Described by Science and its Spiritual Interpretation*. Cambridge: Cambridge University Press.

Barrau, Aurétien (2007). 'Physics in the universe', *Cern Courier*, 20 November.

Barrow, John D. (1983). 'Anthropic definitions', *Quarterly Journal of the Royal Astronomical Society* **24**, 146–53.

Barrow, John D. (1990). 'The mysterious lore of large numbers', pp. 67–96 in Bruno Bertotti et al., eds, *Modern Cosmology in Retrospect*. Cambridge: Cambridge University Press.

Barrow, John D. (1992). *Theories of Everything: The Quest for Ultimate Understanding*. London: Vintage.

Barrow, John D. (1998). *Impossibility*. Oxford: Oxford University Press.

Barrow, John D. (2002). *The Constants of Nature: From Alpha to Omega*. London: Jonathan Cape.

Barrow, John D. (2007). *New Theories of Everything*. Oxford: Oxford University Press.

Barrow, John D. and Frank J. Tipler (1978). 'Eternity is unstable', *Nature* **276**, 453–59.

Barrow, John D. and Frank J. Tipler (1986). *The Anthropic Cosmological Principle*. Cambridge: Cambridge University Press.

Bartusiak, Marcia (1993). 'Loops of space', *Discover* Magazine (April), online edition.

Batten, Alan H. (1994). 'A most rare vision: Eddington's thinking on the relation between science and religion', *Quarterly Journal of the Royal Astronomical Society* **35**, 249–70.

Baum, Lauris and Paul H. Frampton (2007). 'Turnaround in cyclic cosmology', *Physical Review Letters* **98**, 071301.

Beck, Christian (2009). 'Axiomatic approach to the cosmological constant', *Physics* A **388**, 3384–90.

Bekenstein, Jacob D. (1986). 'The fine-structure constant: From Eddington's time to our own', pp. 209–224 in Edna Ullmann-Margalit, ed., *The Prism of Science*. Dordrecht: Reidel.

Bernal, John D. (1939). *The Social Function of Science*. London: Routledge & Sons.

Bernstein, Jeremy and Gerald Feinberg (1986). *Cosmological constants: Papers in Modern Cosmology*. New York: Columbia University Press.

Bertola, Francesco and Umberto Curi, eds (1993). *The Anthropic Principle*. Cambridge: Cambridge University Press.

Bettini, Stefano (2004). 'Anthropic reasoning in cosmology: A historical perspective', arXiv: physics/0410144.

Biedenharn, L. C. (1983). 'The "Sommerfeld puzzle" revisited and solved', *Foundations of Physics* **13**, 13–34.

Blanqui, Louis-Auguste (1872). *L'éternité par les astre. Hypothèses astronomique*. Paris: Librarire Germer Baillière.

Blome, Hans Joachim and Wolfgang Priester (1991). 'Big bounce in the very early universe', *Astronomy and Astrophysics* **250**, 43–49.

Bludman, S.A. (1984). 'Thermodynamics and the end of a closed universe', *Nature* **308**, 319–22.

Bojowald, Martin (2005). 'Loop quantum cosmology', pp. 382–414 in Abhay Ashtekar, ed., *100 Years of Relativity. Space-Time Structure: Einstein and Beyond*. New Jersey: World Scientific.

Bojowald, Martin (2007). 'What happened before the big bang?' *Nature Physics* **3**, 523–25.
Bojowald, Martin (2008). 'Follow the bouncing universe', *Scientific American* **299**: 4, 44–51.
Boltzmann, Ludwig (1895). 'On certain questions in the theory of gases', *Nature* **51**, 483–85.
Bondi, Hermann (1948). 'Review of cosmology', *Monthly Notices of the Royal Astronomical Society* **108**, 104–120.
Bondi, Hermann (1952). *Cosmology*. Cambridge: Cambridge University Press.
Bondi, Hermann (1992). 'The philosopher for science', *Nature* **358**, 363.
Bondi, Hermann and Clive W. Kilmister (1959). 'The impact of Logik der Forschung', *British Journal for the Philosophy of Science* **10**, 55–57.
Bondi, Hermann and Thomas Gold (1948). 'The steady-state theory of the expanding universe', *Monthly Notices of the Royal Astronomical Society* **108**, 252–70.
Bondi, Hermann et al., eds (1960). *Rival Theories of Cosmology*. London: Oxford University Press.
Bonnor, William (1964). *The Mystery of the Expanding Universe*. New York: Macmillan.
Bonometto, Silvio A. (2001). 'Modern and post-modern cosmology', in Vicent J. Martínez, Virginia Trimble, and María J. Pons-Bordería, eds, *Historical Development of Modern Cosmology*. San Francisco: Astronomical Society of the Pacific.
Bork, Alfred M. (1964). 'The fourth dimension in nineteenth-century physics', *Isis* **55**, 326–38.
Born, Max (1920). 'Die Brücke zwischen Chemie und Physik', *Die Naturwissenschaften* **8**, 373–82.
Boscovich, Roger J. (1966). *A Theory of Natural Philosophy*. Cambridge, Mass.: MIT Press.
Bostrom, Nick (2002). *Anthropic Bias: Observation Selection Effects in Science and Philosophy*. New York: Routledge.
Bousso, Raphael and Joseph Polchinski (2000). 'Quantization of four-form fluxes and dynamical neutralization of the cosmological constant', *Journal of High Energy Physics* **06**, 006.
Bousso, Raphael and Joseph Polchinski (2004). 'The string theory landscape', *Scientific American* **290** (September), 60–69.
Brain, Robert M., Robert S. Cohen, and Ole Knudsen, eds (2007). *Hans Christian Ørsted and the Romantic Legacy in Science: Ideas, Disciplines, Practices*. Dordrecht: Springer.
Brandenberger, Robert (2008). 'Alternatives to cosmological inflation', *Physics Today* **61** (March), 44–49.
Broad, Charlie D. (1940). 'Discussion of Sir Arthur Eddington's "The Philosophy of Physical Science"', *Philosophy* **15**, 301–12.
Brock, William H. (1997). *Justus von Liebig: The Chemical Gatekeeper*. Cambridge: Cambridge University Press.
Bronstein, Matvei (1933). 'On the expanding universe', *Physikalische Zeitschrift der Sowjetunion* **4**, 114–18.
Brumfiel, Geoff (2006). 'Outrageous fortune', *Nature* **439**, 10–12.
Brush, Stephen G. (1976). *The Kind of Motion We Call Heat: A History of the Kinetic Theory of Gases in the 19th Century*. Amsterdam: North-Holland.
Brush, Stephen G. (2001). 'Is the Earth too old? The impact of geochronology on cosmology, 1929–1952', pp. 157–75 in C. L. E. Lewis and S. J. Knell, eds, *The Age of the Earth: from 4004 BC to AD 2002*. London: Geological Society.
Buchwald, Jed Z. and Andrew Warwick, eds (2001). *Histories of the Electron: The Birth of Microphysics*. Cambridge, Mass.: MIT Press.
Budd, J. and C. D. Hurt (1991). 'Superstring theory: Information transfer in an emerging field', *Scientometrics* **21**, 87–98.

Bueno, Otávio (2005). 'Dirac and the dispensability of mathematics', *Studies in History and Philosophy of Modern Physics* **36**, 465–90.

Bunge, Mario (1959). *Causality and Modern Science*. New Haven: Harvard University Press.

Bunge, Mario (1962). 'Cosmology and magic', *The Monist* **47**, 116–41.

Burbidge, Eleanor M., Geoffrey R. Burbidge, and Fred Hoyle (1963). 'Condensations in the intergalactic medium', *Astrophysical Journal* **138**, 873–88.

Burchfield, Joe D. (1975). *Lord Kelvin and the Age of the Earth*. Chicago: University of Chicago Press.

Burgess, Cliff and Fernando Quevedo (2007). 'The great cosmic roller-coaster ride', *Scientific American* **297** (November), 29–35.

Burniston Brown, G. (1949). 'The philosophies of science of Eddington and Milne', *American Journal of Physics* **17**, 553–58.

Byrne, P. (2007). 'The many worlds of Hugh Everett', *Scientific American* **295** (December), 98–105.

Cahn, Robert N. (1996). 'The eighteen parameters of the standard model in your everyday life', *Reviews of Modern Physics* **68**, 951–59.

Caldwell, Robert R., Rahul Dave, and Paul J. Steinhardt (1998). 'Cosmological imprint of an energy component with general equation of state', *Physical Review Letters* **80**, 1582–85.

Caldwell, Robert R., Marc Kamionkowski, and Nevin N. Weinberg (2003). 'Phantom energy: Dark energy with $w < -1$ causes a cosmic doomsday', *Physical Review Letters* **91**, 071301.

Candelas, Philip, Gary Horowitz, Andrew Strominger, and Edward Witten (1985). 'Vacuum configurations for superstrings', *Nuclear Physics* B **258**, 46–74.

Caneva, Kenneth L. (1997). 'Physics and Naturphilosophie: A reconnaisance', *History of Science* **35**, 35–106.

Čapek, Milič (1961). *The Philosophical Impact of Contemporary Physics*. New York: Van Nostrand.

Capra, Fritjof (1976). *The Tao of Physics*. New York: Bantam Books.

Carr, Bernard J. (1982). 'On the origin, evolution and purpose of the physical universe', *Irish Astronomical Journal* **15**, 237–53.

Carr, Bernard J. and George F.R. Ellis (2008). 'Universe or multiverse?' *Astronomy & Geophysics* **49**, 2.29–2.37.

Carr, Bernard J. and Martin Rees (1979). 'The anthropic principle and the structure of the physical world', *Nature* **278**, 605–12.

Carroll, Sean M. (2006). 'Is our universe natural?' *Nature* **440**, 1132–36.

Carter, Brandon (1983). 'The anthropic principle and its implications for biological evolution', *Philosophical Transactions of the Royal Society* A **310**, 347–1983.

Carter, Brandon (1989). 'The anthropic principle: Self-selection as an adjunct to natural selection', pp. 185–206 in S. K. Biswas, D. C. V. Mallik, and C. V. Vishveshwara, eds, *Cosmic Perspectives*. Cambridge: Cambridge University Press.

Carter, Brandon (1990). 'Large number coincidences and the anthropic principle in cosmology', pp. 121–33 in John Leslie, ed., *Physical Cosmology and Philosophy*. New York: Macmillan.

Carter, Brandon (1993). 'The anthropic selection principle and the ultra-Darwinian synthesis', pp. 33–66 in Francesco Bertola and Umberto Curi, eds, *The Anthropic Principle*. Cambridge: Cambridge University Press.

Carter, Brandon (2006). 'Anthropic principle in cosmology', pp. 173–80 in Jean Claude Pecker and Jayant V. Narlikar, eds, *Current Issues in Cosmology*. Cambridge: Cambridge University Press. (ArXiv:gr-qc/060617).

Carter, Brandon (2007). 'The significance of numerical coincidences in nature', arXiv:0710.3543.

Cartwright, Nancy and Roman Frigg (2007). 'String theory under scrutiny', *Physics World* (3 September, online edition).
Cassirer, Ernst (1956). *Determinism and Indeterminism in Modern Physics: Historical and Systematic Studies of the Problem of Causality*. New Haven: Yale University Press.
Cat, Jordi (1998). 'The physicists' debates on unification in physics at the end of the 20th century', *Historical Studies in the Physical and Biological Sciences* **28**, 253–99.
Chalmers, J. A. and B. Chalmers (1935). 'The expanding universe – an alternative view', *Philosophical Magazine* **19**, 436–46.
Chandrasekhar, Subrahmanyan (1987). *Truth and Beauty. Aesthetics and Motivations in Science*. Chicago: University of Chicago Press.
Chayut, Michael (1991). 'J.J. Thomson: The discovery of the electron and the chemists', *Annals of Science* **48**, 527–44.
Chen, Chaomei and Jasna Kuljis (2003). 'The rising landscape: A visual exploration of superstring revolutions in physics', *Journal of the American Society for Information Science and Technology* **54**, 435–46.
Chew, Geoffrey F. (1961). *S-Matrix Theory of Strong Interactions*. New York: Benjamin.
Chew, Geoffrey F. (1962). 'S-matrix theory of strong interactions without elementary particles', *Reviews of Modern Physics* **34**, 394–401.
Chew, Geoffrey F. (1963). 'The dubious role of the space-time continuum in microscopic physics', *Science Progress* **51**, 528–39.
Chew, Geoffrey F. (1968). ''Bootstrap': A scientific idea?' *Science* **161**, 762–65.
Chew, Geoffrey F. (1970). 'Hadron bootstrap: Triumph or frustration?' *Physics Today* **23** (October), 23–28.
Chew, Geoffrey F. (1971). 'Hadron bootstrap hypothesis', *Physical Review* D **4**, 2330–35.
Chew, Geoffrey (1981). 'Zero-entropy bootstrap and the fine-structure constant', *Physical review Letters* **47**, 764–67.
Chew, Geoffrey (1983). 'Bootstrapping the photon', *Foundations of Physics* **13**, 217–47.
Chew, Geoffrey F., Murray Gell-Mann, and Arthur H. Rosenfeld (1964). 'Strongly interacting particles', *Scientific American* **210** (February), 74–93.
Chown, Marcus (2003). 'Open minds reap rewards', *The Guardian*, 13 March (online edition).
Chown, Marcus (2006). 'My other universe is a Porsche', *New Scientist magazine*, issue 2572 (online edition).
Ćirković, Milan M. (2000). 'The anthropic argument against infinite past and the Eddington-Lemaître universe', *Serbian Astronomical Journal* no. **161**, 33–37.
Ćirković, Milan M. (2002). 'On the first anthropic argument in astrobiology', *Earth, Moon and Planet* **91**, 243–354.
Ćirković, Milan M. (2003a). 'The thermodynamical arrow of time: Reinterpreting the Boltzmann-Schuetz argument', *Foundations of Physics* **33**, 467–90.
Ćirković, Milan M. (2003b). 'Ancient origins of a modern anthropic cosmological argument', *Astronomical and Astrophysical Transactions* **22**, 879–86.
Ćirković, Milan M. (2003c). 'Resource letter: PE-1: Physical eschatology', *American Journal of Physics* **71**, 122–33.
Clausius, Rudolf (1868). 'On the second fundamental theorem of the mechanical theory of heat', *Philosophical Magazine* **35**, 405–19.
Clifford, William K. (1947). *The Ethics of Belief and Other Essays*. eds. Leslie Stephen and Frederick Pollock. London: Watts & Co.

Collins, C. Barry and Stephen W. Hawking (1973). 'Why is the universe isotropic?' *Astrophysical Journal* **180**, 317–34.

Conlon, Joseph P. (2006). 'The string theory landscape: A tale of two hydras', *Contemporary Physics* **47**, 119–29.

Corry, Leo (1999). 'From Mie's electromagnetic theory of matter to Hilbert's foundations of physics', *Studies in History and Philosophy of Modern Physics* **30**, 159–83.

Corry, Leo (2004). *David Hilbert and the Axiomatization of Physics (1898–1918): From Grundlagen der Geometrie to grundlagen der Physik*. Amsterdam: Kluwer Academic.

Craig, William L. (1988). 'Barrow and Tipler on the anthropic principle vs. divine design', *British Journal for Philosophy of Science* **38**, 389–95.

Craig, William L. (2008). 'Time, eternity, and eschatology,' pp. 596–613 in Jerry L. Walls, ed., *The Oxford Handbook of Eschatology*. Oxford: Oxford University Press.

Crowe, Michael J. (1999). *The Extraterrestrial Life Debate, 1750–1900*. Mineola: Dover Publications.

Cunningham, Andrew and Nicholas Jardine, eds (1990). *Romanticism and the Sciences*. Cambridge: Cambridge University Press.

Cushing, James T. (1985). 'Is there just one possible world? Contingency vs the bootstrap', *Studies in History and Philosophy of Science* **15**, 31–48.

Cushing, James T. (1986). 'The importance of Heisenberg's S-matrix program for the theoretical high-energy physics of the 1950s', *Centaurus* **29**, 110–49.

Cushing, James T. (1990). *Theory Construction and Selection in Modern Physics: The S Matrix*. Cambridge: Cambridge University Press.

Darrigol, Olivier (1994). 'The electron theories of Larmor and Lorentz: A comparative study', *Historical Studies of the Physical and Biological Sciences* **24**, 265–336.

Darrigol, Olivier (2005). *Worlds of Flow: A History of Hydrodynamics from the Bernoullis to Prandtl*. Oxford: Oxford University Press.

Davies, Paul C. W. (1973). 'The thermal future of the universe', *Monthly Notices of the Royal Astronomical Society* **161**, 1–5.

Davies, Paul C. W. (1978). 'Cosmic heresy?' *Nature* **273**, 336–37.

Davies, Paul C. W. (1982). *The Accidental Universe*. Cambridge: Cambridge University Press.

Davies, Paul C. W. (1983). 'The anthropic principle', pp. 1–38 in D. Wilkinson, ed., *Progress in Particle and Nuclear Physics* vol. 10. Oxford: Pergamon Press.

Davies, Paul C. W. (2003). 'A brief history of the multiverse', *New York Times*, 17 April (online version).

Davies, Paul C. W. and Julian Brown, eds. (1988). *Superstrings: A Theory of Everything?*. Cambridge: Cambridge University Press.

Davies, Paul C. W., Tamara M. Davis, and Charles H. Lineweaver (2002). 'Black holes constrain varying constants', *Nature* **418**, 602–603.

Dawkins, Richard (2006). *The God Delusion*. London: Bantam Press.

Deakin, Michael, G. J. Troup, and L. B. Grant (1983). 'The anthropic principle in a unique universe', *Physics Letters A* **96**, 5–6.

Deltete, Robert J. (1993). 'What does the anthropic principle explain?' *Perspectives on Science* **1**, 285–305.

Deltete, Robert J. (2005). 'Die Lehre von der Energie: Georg Helm's energetic manifesto', *Centaurus* **47**, 140–62.

Descartes, René (1979). *Le monde, ou traité de la lumière (The World, or Treatise on Light)*. Translated by Michael S. Mahoney. New York: Abaris Books.

Descartes, René (1983). *Principles of Philosophy*. Translated and edited by Valentine R. Miller and Reese P. Miller. Dordrecht: Reidel.

Descartes, René (1996). *Discourse on Method and Meditations on First Philosophy*. Edited by David Weissman. New Haven: Yale University Press.

DeTar, Charleton, J. Finkelstein, and Chung-I Tan, eds (1985). *A Passion for Physics: Essays in Honor of Geoffrey Chew*. Singapore: World Scientific.

Deutsch, David (1986). 'On Wheeler's notion of "law without law"', *Foundations of Physics* **16**, 565–72.

Deutsch, David (1997). *The Fabric of Reality: The Science of Parallel Universes – and Its Implications*. London: Penguin Books.

DeWitt, Bryce S. (1970). 'Quantum mechanics and reality', *Physics Today* **23** (September), 31–35.

Dick, Steven J. (1982). *Plurality of Worlds: The Origin of the Extraterretrial Life Debate from Democritus to Kant*. Cambridge: Cambridge University Press.

Dick, Steven J. (1996). *The Biological Universe: The Twentieth-Century Extraterretrial Life Debate*. Cambridge: Cambridge University Press.

Dicke, Robert H. (1957a). 'Principle of equivalence and the weak interactions', *Reviews of Modern Physics* **29**, 355–62.

Dicke, Robert H. (1957b). 'Gravitation without a principle of equivalence', *Reviews of Modern Physics* **29**, 363–76.

Dicke, Robert H. (1959). 'Gravitation – an enigma', *American Scientist* **47**, 25–40.

Dicke, Robert H. (1961). 'Dirac's cosmology and Mach's principle', *Nature* **192**, 440–41.

Dicke, Robert H. (1962). 'The earth and cosmology', *Science* **138**, 653–64.

Dicke, Robert H. (1964). *The Theoretical Significance of Experimental Relativity*. London: Blackie and Son.

Dicus, Duane A., John R. Letaw, Doris C. Teplitz, and Vigdor L. Teplitz (1982). 'The future of the universe', *Scientific American* **228** (January), 74–85.

Dicus, Duane A., John R. Letaw, Doris C. Teplitz, and Vigdor L. Teplitz (1983). 'Effects of proton decay on the cosmological future', *Astrophysical Journal* **252**, 1–9.

Dingle, Herbert (1937a). 'Modern Aristotelianism', *Nature* **139**, 784–86.

Dingle, Herbert (1937b). 'Deductive and inductive methods in science. A reply', *Nature* **139**, 1011–12.

Dingle, Herbert (1953). 'Science and modern cosmology', *Monthly Notices of the Royal Astronomical Society* **113**, 393–407.

Dirac, Paul A.M. (1929). 'Quantum mechanics of many-electron systems', *Proceedings of the Royal Society* A **123**, 714–33.

Dirac, Paul A.M. (1930). 'The proton', *Nature* **126**, 605–606.

Dirac, Paul A.M. (1931). 'Quantised singularities in the electromagnetic field', *Proceedings of the Royal Society* A **133**, 60–72.

Dirac, Paul A.M. (1937). 'The cosmological constants', *Nature* **139**, 323.

Dirac, Paul A.M. (1939). 'The relation between mathematics and physics', *Proceedings of the Royal Society (Edinburgh)* **59**, 122–29.

Dirac, Paul A.M. (1942). 'The physical interpretation of quantum mechanics', *Proceedings of the Royal Society* A **180**, 1–40.

Dirac, Paul A.M. (1958). *The Principles of Quantum Mechanics*. London: Oxford University Press.

Dirac, Paul A.M. (1961). 'Reply to R. H. Dicke', *Nature* **192**, 441.

Dirac, Paul A.M. (1969). 'Can equations of motion be used?' pp. 1–13 in T. Gudehus, G. Kaiser, and A. Perlmutter, eds, *Fundamental Interactions at high Energy*. New York: Gordon and Breach.

Distler, Jacques, Benjamin Grinstein, Rafael A. Porto, and Ira Z. Rothstein (2007). 'Falsifying models of new physics via WW scattering', *Physical Review Letters* **98**, 041601.

Douglas, Allie V. (1956). *The Life of Arthur Stanley Eddington*. London: Thomas Nelson.

Douglas, Michael R. and Shamit Kachru (2007). 'Flux compactification', *Reviews of Modern Physics* **79**, 733–96.

Drude, Paul (1902). *The Theory of Optics*. London: Longmans, Green & Co.

Duff, Michael J. (2004). 'Comment on time-variation of fundamental constants', arXiv:hep-th/0208093.

Duff, Michael J., Lev B. Okun, and Gabriele Veneziano (2002). 'Trialogue on the number of fundamental constants', *Journal of High Energy Physics* **03**, 023.

Dunbar, D. N. F., R. E. Pixley, W. A. Wenzel, and W. Whaling (1953). 'The 7.68-MeV state in C12', *Physical Review* **92**, 649–50.

Durham, Ian T. (2003). 'Eddington and uncertainty', *Physics in Perspective* **5**, 398–419.

Dyson, Freeman (1953). 'Field theory', *Scientific American* **188** (April), 57–64.

Dyson, Freeman J. (1964). 'Mathematics in the physical sciences', *Scientific American* **211** (September), 128–44.

Dyson, Freeman J. (1971). 'Energy in the universe', *Scientific American* **225** (March), 51–59.

Dyson, Freeman J. (1972). 'The fundamental constants and their time variation', pp. 213–36 in Abdus Salam and Eugene P. Wigner, eds, *Aspects of Quantum Theory*. Cambridge: Cambridge University Press.

Dyson, Freeman J. (1979). 'Time without end: Physics and biology in an open universe', *Reviews of Modern Physics* **51**, 447–60.

Dyson, Freeman J. (2004). *Infinite in All Directions*. New York: Perennial.

Eagles, D. M. (1976). 'A comparison of results of various theories for four fundamental constants of physics', *International Journal of Theoretical Physics* **15**, 265–70.

Earman, John (2001). 'Lambda: The constant that refuses to die', *Archive for History of Exact Sciences* **55**, 189–158.

Eddington, Arthur S. (1918). *Report on the Relativity Theory of Gravitation*. London: Fleetway Press.

Eddington, Arthur S. (1920a). 'The meaning of matter and the laws of nature according to the theory of relativity', *Mind* **120**, 145–58.

Eddington, Arthur S. (1920b). *Space, Time and Gravitation: An Outline of the General Relativity Theory*. Cambridge: Cambridge University Press.

Eddington, Arthur S. (1923). *The Mathematical Theory of Relativity*. Cambridge: Cambridge University Press.

Eddington, Arthur S. (1929). 'The charge of an electron', *Proceedings of the Royal Society* A **122**, 358–69.

Eddington, Arthur S. (1931a). 'On the value of the cosmical constant', *Proceedings of the Royal Society* A **133**, 605–15.

Eddington, Arthur S. (1931b). 'The expansion of the universe', *Monthly Notices of the Royal Astronomical Society* **91**, 412–16.

Eddington, Arthur S. (1931c). 'The end of the world: From the standpoint of mathematical physics', *Nature* **127**, 447–52.

Eddington, Arthur S. (1932). 'The theory of electric charge', *Proceedings of the Royal Society* A **138**, 17–41.

Eddington, Arthur S. (1933). *The Expanding Universe*. Cambridge: Cambridge University Press.

Eddington, Arthur S. (1935a). *New Pathways in Science*. Cambridge: Cambridge University Press.

Eddington, Arthur S. (1935b). 'The speed of recession of the galaxies', *Monthly Notices of the Royal Astronomical Society* **95**, 636–38.

Eddington, Arthur S. (1936). *Relativity Theory of Protons and Electrons*. Cambridge: Cambridge University Press.

Eddington, Arthur S. (1939a). *The Philosophy of Physical Science*. Cambridge: Cambridge University Press.

Eddington, Arthur S. (1939b). 'The cosmological controversy', *Science Progress* **34**, 225–36.

Eddington, Arthur S. (1944). 'The recession-constant of the galaxies', *Monthly Notices of the Royal Astronomical Society* **104**, 200–204.

Eddington, Arthur S. (1946). *Fundamental Theory*. Cambridge: Cambridge University Press.

Edwards, Matthew R. (2002). *Pushing Gravity: New Perspectives on Le Sage's Theory of Gravitation*. Montreal: Apeiron.

Ehrlich, Robert (2006). 'What makes a theory testable, or is intelligent design less scientific than string theory?' *Physics in Perspective* **8**, 83–89.

Einstein, Albert (1931). 'Gravitational and electromagnetic fields', *Science* **74**, 438–39.

Einstein, Albert (1949). *Albert Einstein, Philosopher-Scientist*. Edited by Paul A. Schilpp. New York: Library of Living Philosophers.

Einstein, Albert (1982). *Ideas and Opinions*. New York: Three Rivers Press.

Einstein, Albert (1993). *The Collected Papers of Albert Einstein* vol. 5. Edited by Martin J. Klein, A. J. Kox, and Robert Schulmann. Princeton: Princeton University Press.

Einstein, Albert (1996). *The Collected Papers of Albert Einstein* vol. 6. Edited by A. J. Kox, Martin J. Klein, and Robert Schulmann. Princeton: Princeton University Press.

Einstein, Albert (1998). *The Collected Papers of Albert Einstein* vol. 8, part A. Edited by Robert Schulmann, A. J. Kox, Michel Janssen, and József Illy. Princeton: Princeton University Press.

Eliade, Mircea (1954). *The Myth of Eternal Return*. New York: Pantheon.

Ellis, John (1986). 'The superstring: Theory of everything, or of nothing?' *Nature* **323**, 595–98.

Ellis, George F.R. (1993). 'The theology of the anthropic principle', pp. 363–400 in Robert J. Russell, Nancey Murphy, and C. J. Isham, eds, *Quantum Cosmology and the Laws of Nature: Scientific Perspectives on Divine Action*. Vatican City State: University of Notre Dame Press.

Ellis, George F. R. (1994). 'Piety in the sky', *Nature* **371**, 115.

Ellis, George F. R., ed. (2002). *The Far-Future Universe: Eschatology from a Cosmic Perspective*. Radnor, Pennsylvania: Templeton Foundation Press.

Ellis, George F. R. (2003). 'Einstein not yet displaced', *Nature* **422**, 563.

Ellis, George F. R. (2005). 'Physics, complexity and causality', *Nature* **435**, 743.

Ellis, George F. R. and G. B. Brundrit (1979). 'Life in the infinite universe', *Quarterly Journal of the Royal Astronomical Society* **20**, 37–41.

Ellis, George F. R. and D. H. Coule (1994). 'Life at the end of the universe?' *General Relativity and Gravitation* **26**, 731–39.

Ellis, George F. R. and Jean-Philippe Uzan (2005). '"c" is the speed of light, isn't it?' *American Journal of Physics* **73**, 240–47.

Ellis, George F. R., U. Kirchner, and William R. Stoeger (2004). 'Multiverses and physical cosmology', *Monthly Notices of the Royal Astronomical Society* **347**, 921–36.

Epple, Moritz (1999). *Die Entstehung der Knotentheorie: Kontexte und Konstruktionen einer modernen mathematischen Theorie*. Braunschweig: Vieweg.

Everett, Hugh (1957). '"Relative state" formulation of quantum mechanics', *Reviews of Modern Physics* **29**, 454–62.

Faddeev, Ludwig and Antti J. Niemi (1997). 'Stable knot-like structures in classical field theory', *Nature* **387**, 58–61.
Fadner, W. L. (1988). 'Did Einstein really discover "$E = mc^2$"?' *American Journal of Physics* **56**, 114–22.
Falb, Rudolf (1875). 'Die Welten, Bildung und Untergang', *Sirius* **8**, 193–202.
Faraday, Michael (1844). 'A speculation touching electric conduction and the nature of matter', *Philosophical Magazine* **24**, 136–44.
Farhi, Edward and Alan H. Guth (1987). 'An obstacle to creating a universe in the laboratory', *Physics Letters B* **183**, 149–55.
Farmelo, Graham (2009). *The Strangest Man: The Hidden Life of Paul Dirac, Quantum Genius.* London: Faber and Faber.
Feinberg, Gerald (1972). 'Philosophical implications of contemporary particle physics', pp. 33–46 in Robert G. Colodny, ed., *Paradigms and Paradoxes.* Pittsburgh: University of Pittsburgh Press.
Feuer, Lewis S. (1934). 'On the use of "universe"', *Mind* **43**, 346–48.
Feynman, Richard P. (1985). *QED: The Strange Theory of Light and Matter.* Princeton: Princeton University Press.
Feynman, Richard P. (1992). *The Character of Physical Law.* London: Penguin.
Feynman, Richard P., Robert B. Leighton, and Matthew Sands (1963). *The Feynman Lectures on Physics* vol. 1. Reading, Mass.: Addison-Wesley.
Feynman, Richard P., Robert B. Leighton, and Matthew Sands (1964). *The Feynman Lectures on Physics* vol. 2. Reading, Mass.: Addison-Wesley.
FitzGerald, George F. (1888). 'Presidential address to the section of mathematical and physical sciences', *British Association for the Advancement of Science*, Report, 557–62.
FitzGerald, George F. (1896). 'Helmholtz memorial lecture', *Transactions of the Chemical Society* **69**, 885–912.
FitzGerald, George F. (1897). 'Dissociation of atoms', *Nature* **59**, 103–104.
Folger, Tim (2008). 'Science's alternative to an intelligent creator: The multiverse theory', *Discover* Magazine, December (online version).
Frautschi, Steven (1982). 'Entropy in an expanding universe', *Science* **217**, 593–99.
Freedman, Daniel Z. and Peter van Nieuwenhuizen (1978). 'Supergravity and the unification of the laws of physics', *Scientific America* **238** (February), 126–43.
Freese, Katherine and William H. Kinney (2003). 'The ultimate fate of life in an accelerating universe', *Physics Letters B* **558**, 1–8.
French, Steven (2003). 'Scribbling on the blank sheet: Eddington's structuralist conception of objects', *Studies in History and Philosophy of Modern Physics* **34**, 227–59.
Freundlich, Yehudah (1980). 'Theory evaluation and the bootstrap', *Studies in History and Philosophy of Science* **11**, 267–77.
Friedan, Daniel (2003). 'A tentative theory of large distance physics', *Journal of High Energy Physics* **10**, 063.
Fröman, Per O. (1994). 'Historical background of the tachyon concept', *Archive for History of Exact Sciences* **48**, 373–80.
Gale, George (1974). 'Chew's monadology', *Journal for History of Ideas* **35**, 339–48.
Gale, George (1981). 'The anthropic principle', *Scientific American* **246** (December), 114–22.
Gale, George (1990). 'Cosmological fecundity: Theories of multiple universes', pp. 189–206 in John Leslie, ed., *Physical Cosmology and Philosophy.* New York: Macmillan.
Gale, George (2002). 'Cosmology: Methodological debates in the 1930s and 1940s'. http://plato.stanford.edu/cosmology-30s/

Gale, George and John Urani (1999). 'Milne, Bondi and the 'second way' to cosmology', pp. 343–76 in Hubert Goenner et al., eds, *The Expanding Worlds of General Relativity*. Boston: Birkhäuser.

Galison, Peter L. (1995). 'Theory bound and unbound: Superstrings and experiments', pp. 369–408 in Friedel Weinert, ed., *Laws of Nature. Essays on the Philosophical, Scientific and Historical Dimensions*. Berlin: Walter de Gruyter.

Gamow, George (1949). 'Any physics tomorrow?' *Physics Today* **2** (January), 17–22.

Gamow, George (1951). 'The origin and evolution of the universe', *American Scientist* **39**, 393–407.

Gamow, George (1952). *The Creation of the Universe*. New York: Viking Press.

Gamow, George (1967). 'Electricity, gravity, and cosmology', *Physical Review Letters* **19**, 759–61.

Gamow, George, Dmitrii Ivanenko, and Lev Landau (2002). 'World constants and limiting transition', *Physics of Atomic Nuclei* **65**, 1373–75.

Gardner, Martin (1986). 'WAP, SAP, PAP, & FAP', *New York Review of Books* **33** (8 May), 22–25.

Gardner, Martin (1991). 'Tipler's omega point theory', *Skeptical Inquirer* **15**: 2, 128–32.

Gardner, Martin (2007). 'M for messy', *The New Criterion* **25** (April), 90.

Garriga, Jaume and Alexander Vilenkin (2001). 'Many worlds in one', *Physical Review* D **64**, 043511.

Garriga, Jaume, Viatcheslav F. Mukhanov, Ken D. Olum, and Alexander Vilenkin (2000). 'Eternal inflation, black holes, and the future of civilizations', *International Journal of Theoretical Physics* **39**, 1887–1900.

Gasperini, Maurizio (2008). *The Universe Before the Big Bang: Cosmology and String Theory*. Berlin: Springer.

Gasperini, Maurizio and Gabriele Veneziano (1993). 'Pre-big-bang in string cosmology', *Astroparticle Physics* **1**, 317–39.

Gaukroger, Stephen (1995). *Descartes: An Intellectual Biography*. Oxford: Clarendon Press.

Gaukroger, Stephen (2002). *Descartes' System of Natural Philosophy*. Cambridge: Cambridge University Press.

Gaukroger, Stephen, John Schuster, and John Sutton, eds (2000). *Descartes' Natural Philosophy*. London: Routledge.

Gefter, Amanda (2009). 'Dark flow: Proof of another universe?' *New Scientist* online edition, issue 2692, http://www.newscientist.com/article/mg20126921.900-dark-flow-proof-of-another-universe.html.

Gell-Mann, Murray (1987). 'Particle theory from S-matrix to quarks', pp. 473–98 in Manuel G. Doncel et al., eds, *Symmetries in Physics (1600–1980)*. Barcelona: Seminari d'Historia de les Ciències.

Gell-Mann, Murray (1994). *The Quark and the Jaguar*. London: Little, Brown and Company.

Gheury de Bray, M.E.J. (1936). 'The velocity of light; History of its determination from 1849 to 1933', *Isis* **25**, 437–48.

Ginsparg, Paul and Sheldon Glashow (1986). 'Desperately seeking superstrings', *Physics Today* **39** (May), 7–8.

Goenner, Hubert F. M. (2001). 'The quest for ultimate explanation in physics: reductionism, unity, and meaning', Max Planck Institute for the History of Science, preprint 187.

Goenner, Hubert F. M. (2004). 'On the history of unified field theories', *Living Reviews in Relativity*. http://relativity.livingreviews.org/Articles/lrr-2004-2/

Gold, Thomas (1949). 'Creation of matter in the universe', *Nature* **164**, 1006.

Gold, Thomas (1956). 'Cosmology', *Vistas in Astronomy* **2**, 1721–1726.

Goldberg, Stanley (1970). 'The Abraham theory of the electron: The symbiosis of experiment and theory', *Archive for History of Exact Sciences* **7**, 7–25.

Gonzáles-Diaz, Pedro F. (2010). 'The origin of eternal life in the multiverse', *Journal of Cosmology* **4**, 775–759.

Good, I. J. (1970). 'The proton and neutron masses and a conjecture for the gravitational constant', *Physics Letters* A **33**, 383–84.

Gordin, Michael D. (2004). *A Well-Ordered Thing: Dmitrii Mendeleev and the Shadow of the Periodic Table*. New York: Basic Books.

Gorelik, Gennady and Victor Ya. Frenkel (1994). *Matvei Petrovich Bronstein and Soviet Theoretical Physics in the Thirties*. Basel: Birkhäuser.

Görs, Britta, Nikos Psarros, and Paul Ziche, eds (2005). *Wilhelm Ostwald at the Crossroads between Chemistry, Philosophy and Media Culture*. Leipzig: Leipziger Universitätsverlag.

Gott, J. Richard (1993). 'Implications of the Copernican principle for our future prospects', *Nature* **364**, 315–219.

Gott, J. Richard and Li-Xin Li (1998). 'Can the universe create itself?' *Physical Review* D **58**, 1–43.

Gower, Barry 1973. 'Speculation in physics: The history and practice of Naturphilosophie', *Studies in History and Philosophy of Science* **3**, 301–56.

Grant, Edward (2007). *A History of Natural Philosophy: From the Ancient World to the Nineteenth Century*. Cambridge: Cambridge University Press.

Green, Michael B. (1985). 'Unification of forces and particles in superstring theories', *Nature* **314**, 409–14.

Green, Michael B. and John H. Schwarz (1984). 'Anomaly cancellations in supersymmetric $D = 10$ gauge theory and superstring theory', *Physics Letters* B **149**, 117–22.

Greene, Brian (1999). *The Elegant Universe*. New York: W.W. Norton.

Greene, Brian (2004). *The Fabric of the Cosmos: Space, Time and the Texture of Reality*. London: Allen Lane.

Gribbin, John and Martin Rees (1989). *Cosmic Coincidences: Dark Matter, Mankind, and Anthropic Cosmology*. New York: Bantam Books.

Gross, David J. (2005). 'Where do we stand in fundamental (string) theory?' *Physica Scripta* T **117**, 102–105.

Gross, David J. (2008). 'Einstein and the quest for a unified theory', pp. 287–98 in Peter L. Galison, Gerald Holton, and Silvan S. Schweber, eds, *Einstein for the 21st Century: His Legacy in Science, Art, and Modern Culture*. Princeton: Princeton University Press.

Gross, David J., Jeffrey A. Harvey, Emil Martinec, and Ryan Rohm (1985). 'Heterotic string', *Physical Review Letters* **54**, 502–505.

Guth, Alan H. (1997). *The Inflationary Universe*. Reading, Mass.: Addison-Wesley.

Guth, Alan H. and David Kaiser (2005). 'Inflationary cosmology: Exploring the universe from the smallest to the largest scale', *Science* **307**, 884–90.

Hakfoort, Casper (1992). 'Science deified: Wilhelm Ostwald's energeticist world-view and the history of scientism', *Annals of Science* **49**, 525–44.

Haldane, John B. S. (1928). 'The universe and irreversibility', *Nature* **122**, 808–809.

Hall, P. J. (1983). 'Anthropic explanations in cosmology', *Quarterly Journal of the Royal Astronomical Society* **24**, 443–47.

Hallam, Anthony (1988). *Great Geological Controversies*. Oxford: Oxford University Press.

Halpern, Paul (2004a). 'Nordström, Ehrenfest, and the role of dimensionality in physics', *Physics in Perspective* **6**, 390–400.

Halpern, Paul (2004b). *The Great Beyond: Higher Dimensions, Parallel Universes, and the Extraordinary Search for a Theory of Everything*. Hoboken, N.J.: Wiley.

Halpern, Paul (2007). 'Klein, Einstein, and five-dimensional unification', *Physics in Perspective* **9**, 390–405.

Harder, Allen J. (1974). 'E. A. Milne, scientific revolutions and the growth of knowledge', *Annals of Science* **31**, 351–63.

Harnik, Roni, Graham D. Kribs, and Gilad Perez (2006). 'A universe without weak interactions', *Physical Review* D **74**, 035006.

Harrison, Edward R. (1995). 'The natural selection of universes containing intelligent life', *Quarterly Journal of the Royal Astronomical Society* **36**, 193–203.

Harrison, Edward R. (2000). *Cosmology: The Science of the Universe*. Cambridge: Cambridge University Press.

Hart, Michael H. (1975). 'An explanation for the absence of extraterrestrials on Earth', *Quarterly Journal of the Royal Astronomical Society* **16**, 128–35.

Hawking, Stephen W. (1974). 'The anisotropy of the universe at large times', pp. 283–86 in Malcolm S. Longair, ed., *Confrontation of Cosmological Theories with Observational Data*. Dordrecht: Reidel.

Hawking, Stephen W. (1980). *Is the End in Sight for Theoretical Physics?*. Cambridge: Cambridge University Press.

Heaviside, Oliver (1970). *Electrical Papers* vol. 2. New York: Chelsea Publishing Company.

Hedrick, Reiner (2007). 'The internal and external problems of string theory', *Journal of General Philosophy of Science* **38**, 261–78.

Heimann, Peter M. (1972). 'The Unseen Universe: Physics and the philosophy of nature in Victorian Britain', *British Journal for the History of Science* **6**, 73–79.

Heisenberg, Werner (1925). 'Über quantentheoretische Umdeutung kinematischer und mechanischer Beziehungen', *Zeitschrift für Physik* **33**, 879–93.

Heisenberg, Werner (1943). 'Die "beobachtbaren Grössen" in der Theorie der Elementarteilchenphysik', *Zeitschrift für Physik* **120**, 513–38.

Heisenberg, Werner (1957). 'Quantum theory of fields and elementary particles', *Reviews of Modern Physics* **29**, 269–78.

Heisenberg, Werner (1966). *Introduction to the Unified Field Theory of Elementary Particles*. London: Interscience Publishers.

Heller, Michael (2009). *Ultimate Explanations of the Universe*. Berlin: Springer-Verlag.

Heller, Michael and Marek Szydlowski (1983). 'Tolman's cosmological models', *Astrophysics and Space Science* **90**, 327–35.

Hicks, William M. (1895). 'Presidential address to the section of mathematical and physical science', *British Association for the Advancement of Science* Report, 596–606.

Hilbert, David (1925). 'Über das unendliche', *Mathematische Annalen* **95**, 161–90.

Hirosige, Tetu (1966). 'Electrodynamics before the theory of relativity, 1890–1905', *Japanese Studies in the History of Science* **5**, 1–49.

Hogan, Craig J. (2000). 'Why the universe is just so', *Reviews of Modern Physics* **72**, 1149–61.

Hollands, Stefan and Robert M. Wald (2002). 'An alternative to inflation', *General Relativity and Gravitation* **34**, 2043–55.

Holman, Silas W. (1898). *Matter, Energy, Force and Work. A Plain Presentation of Fundamental Physical Concepts and of the Vortex Atom and Other Theories*. New York: Macmillan.

Hooft, Gerardus 't (1997). *In Search of the Ultimate Building Blocks*. Cambridge: Cambridge University Press.

Horgan, John (1997). *The End of Science*. New York: Broadway Books.

Horgan, John (2006). 'The final frontier', *Discover*, New York (October), 56–62.

Howson, Colin and Peter Urbach (1993). *Scientific Reasoning: The Bayesian Approach*. Chicago: University of Chicago Press.

Hoyle, Fred (1948). 'A new model for the expanding universe', *Monthly Notices of the Royal Astronomical Society* **108**, 372–82.

Hoyle, Fred (1950). *The Nature of the Universe*. New York: Harper & Brothers.

Hoyle, Fred (1954). 'On nuclear reactions occurring in very hot stars. I: The synthesis of elements from carbon to nickel', *Astrophysical Journal, Supplement Series* **1**, 121–46.

Hoyle, Fred (1960). 'A covariant formulation of the law of creation of matter', *Monthly Notices of the Royal Astronomical Society* **120**, 256–62.

Hoyle, Fred (1965a). 'Recent developments in cosmology', *Nature* **208**, 111–14.

Hoyle, Fred (1965b). *Galaxies, Nuclei, and Quasars*. New York: Harper & Row.

Hoyle, Fred (1975). *Astronomy and Cosmology: A Modern Course*. San Francisco: W. H. Freeman and Company.

Hoyle, Fred (1980a). *Steady-State Cosmology Re-visited*. Cardiff: University College Cardiff Press.

Hoyle, Fred (1980b). *The Relation of Biology to Astronomy*. Cardiff: University College Cardiff Press.

Hoyle, Fred (1982). 'The universe: Past and present reflections', *Annual Review of Astronomy and Astrophysics* **20**, 1–35.

Hoyle, Fred (1991). 'Some remarks on cosmology and biology', *Memorie della Societa Astronomica Italiana* **62**, 513–18.

Hoyle, Fred (1993). 'The anthropic and perfect cosmological principles: Similarities and differences', pp. 85–89 in Francesco Bertola and Umberto Curi, eds, *The Anthropic Principle*. Cambridge: Cambridge University Press.

Hoyle, Fred (1994). *Home is Where the Wind Blows: Chapters from a Cosmologist's Life*. Mill Valley, Calif.: University Science Books.

Hoyle, Fred and Jayant V. Narlikar (1962). 'Mach's principle and the creation of matter', *Proceedings of the Royal Society* A **270**, 334–41.

Hoyle, Fred and Jayant V. Narlikar (1963). 'Mach's principle and the creation of matter', *Proceedings of the Royal Society* A **273**, 1–11.

Hoyle, Fred and Jayant V. Narlikar (1964). 'A new theory of gravitation', *Proceedings of the Royal Society* A **282**, 191–207.

Hoyle, Fred, D. N. F. Dunbar, W. A. Wenzel, and W. Whaling (1953). 'A state in C12 predicted from astrophysical evidence', *Physical Review* **92**, 1095.

Hoyle, Fred, Geoffrey Burbidge, and Jayant V. Narlikar (2000). *A Different Approach to Cosmology*. Cambridge: Cambridge University Press.

Hsieh, S.-H. (1982). 'Meaning of a varying gravitational constant', *International Journal of Theoretical Physics* **21**, 673–83.

Hubble, Edwin P. (1936). *The Realm of the Nebulae*. New Haven: Yale University Press.

Huby, Pamela M. (1971). 'Kant or Cantor? That the universe, if real, must be finite in both space and time', *Philosophy* **46**, 120–33.

Hume, David (1980). *Dialogues Concerning Natural Religion*. Ed. by Richard H. Popkin. Indianapolis: Hackett Publishing Company.

Idlis, Grigory M. (1982). 'Four revolutions in astronomy, cosmology and physics', *Acta Historiae Rerum Naturalium Nec Non Technicarum* **18**, 343–68.
Idlis, Grigory M. (2001). 'Universality of space civilizations and indispensible universality in cosmology', *Astronomical and Astrophysical Transactions* **20**, 963–73.
Illy, József (1981a). 'Revolutions in a revolution', *Studies in History and Philosophy of Science* **12**, 173–210.
Illy, József (1981b). 'Lenin, the electromagnetic world view and the theory of relativity', *Acta Historiae Rerum Naturalium Nec Non Technicarum* **14**, 39–45.
Isenkrahe, Caspar (1910). *Energie, Entropie, Weltanfang, Weltende*. Trier: J. Lintz.
Islam, Jamal N. (1977). 'Possible ultimate fate of the universe', *Quarterly Journal of the Royal Astronomical Society* **18**, 3–8.
Islam, Jamal N. (1979). 'The long-term future of the universe', *Vistas in Astronomy* **23**, 265–77.
Islam, Jamal N. (1983). *The Ultimate Fate of the Universe*. Cambridge: Cambridge University Press.
Israel, Werner (1987). 'Dark stars: The evolution of an idea', pp. 199–276 in Stephen Hawking and Werner Israel, eds, *Three Hundred Years of Gravitation*. Cambridge: Cambridge University Press.
Israelit, Mark (1996). 'Nathan Rosen, 1909–1995', *Foundations of Physics Letters* **9**, 105–108.
Israelit, Mark and Nathan Rosen (1989). 'A singularity-free cosmological model in general relativity', *Astrophysical Journal* **375**, 627–34.
Jaffe, Arthur and Frank Quinn (1993). 'Theoretical mathematics: Toward a cultural synthesis of mathematics and theoretical physics', *Bulletin of the American Mathematical Society* **29**, 1–13.
Jaffe, Robert, Alejandro Jenkins, and Itamar Kimchi (2009). 'Quark masses: An environmental impact statement', *Physical Review* D **79**, 065014.
Jaki, Stanley L. (1974). *Science and Creation. From Eternal Cycles to an Oscillating Universe*. Edinburgh: Scottish Academic Press.
Jaki, Stanley L. (1982). 'From scientific cosmology to a created universe', *Irish Astronomical Journal* **15**, 253–62.
Jammer, Max (1999). *Einstein and Religion*. Princeton: Princeton University Press.
Jeans, James (1901). 'On the mechanism of radiation', *Philosophical Magazine* **2**, 421–55.
Jeans, James (1926). 'Recent developments of cosmical physics', *Nature* **118**, 29–40.
Jeans, James (1928a). 'The physics of the universe', *Nature* **122**, 689–700.
Jeans, James (1928b). 'The wider aspects of cosmogony', *Nature* **121**, 463–70.
Jeans, James (1934). 'The new world-picture of modern physics', *Nature* **134**, 355–65.
Jeffreys, Harold (1931). *Scientific Inference*. London: Cambridge University Press.
Jenkins, Alejandro and Gilad Perez (2010). 'Looking for life in the multiverse', *Scientific American* **302** (January), 42–49.
Johansson, Lars-Göran and Keizo Matsubara (2009). 'String theory and general methodology; a reciprocal evaluation', arXiv:0912.3160.
Johnson, Martin (1951). 'The meanings of time and space in philosophies of science', *American Scientist* **39**, 412–21.
Johnson, George (2001). 'New contenders for a theory of everything', *New York Times*, 4 December.
Jones, Eric M. (1985a). '"Where is everybody?" An account of Fermi's question'. Technical Report LA-10311-MS. Los Alamos: Los Alamos National Laboratory.
Jones, Eric M. (1985b). 'Where is everybody?' *Physics Today* **38** (August), 11–13.
Jones, G. O., J. Rotblat, and G. J. Whitrow (1956). *Atoms and the Universe: An Account of Modern Views on the Structure of Matter and the Universe*. New York: Charles Scribner's Sons.

Jordan, Pascual (1952). *Schwerkraft und Weltall*. Braunschweig: Vieweg.

Jungnickel, Christa and Russell McCormmach (1986). *Intellectual Mastery of Nature: Theoretical Physics from Ohm to Einstein* vol. 2. Chicago: University of Chicago Press.

Kaiser, David (2002). 'Nuclear Democracy: Political engagement, pedagogical reform, and particle physics in postwar America', *Isis* **93**, 229–68.

Kaiser, David (2005). *Drawing Theories Apart: The Dispersion of Feynman Diagrams in Postwar Physics*. Chicago: University of Chicago Press.

Kaku, Michio (2005a). 'Testing string theory', *Discover* Magazine (August), online edition.

Kaku, Michio (2005b). *Parallel Worlds. The Science of Alternative Universes and Our Future in the Cosmos*. London: Penguin Books.

Kallosh, Renata, Lev Kofman, and Andrei Linde (2001). 'Pyrotechnic universe', *Physical Review* D **64**, 123523.

Kamminga, Harmke (1982). 'Life from space – a history of panspermia', *Vistas in Astronomy* **26**, 67–86.

Kane, Gordon L., Malcolm J. Perry, and Anna N. Zytkow (2002). 'The beginning of the end of the anthropic principle', *New Astronomy* **7**, 45–53.

Kardashev, Nicolai S. (1964). 'Transmission of information by extraterrestrial civilizations', *Soviet Astronomy* **8**, 217–21.

Kardashev, Nicolai S. (1990). 'Optimistic cosmological model', *Monthly Notices of the Royal Astronomical Society* **243**, 252–56.

Kargon, Robert 1964. 'William Rowan Hamilton, Michael Faraday, and the revival of Boscovichean atomism', *American Journal of Physics* **32**, 792–95.

Karthik, Trishank (2004). 'The theory of everything and the future of life', *International Journal of Astrobiology* **3**, 311–16.

Kauffman, Louis H. (2000). *Knots and Physics*. New York: Macmillan.

Kaufmann, Wather (1901). 'The development of the electron idea', *The Electrician* **48**, 95–97.

Kerzberg, Pierre (1989). *The Invented Universe: The Einstein-De Sitter Controversy (1916–17) and the Rise of Relativistic Cosmology*. Oxford: Clarendon Press.

Khoury, Justin, Burt A. Ovrut, Paul J. Steinhardt, and Neil Turok (2001a). 'Ekpyrotic universe: Colliding branes and the origin of the hot big bang', *Physical Review* D **64**, 123522.

Khoury, Justin, Burt A. Ovrut, Paul J. Steinhardt and Neil Turok (2001b). 'A brief comment on "the pyrotechnic universe"', arXiv:hep-th/0105212.

Kilmister, Clive W. (1994). *Eddington's Search for a Fundamental Theory: A Key to the Universe*. Cambridge: Cambridge University Press.

Kitcher, Philip (1981). 'Explanatory unification', *Philosophy of Science* **48**, 507–31.

Klee, Robert (2002). 'The revenge of Pythagoras: How a mathematical sharp practice undermines the contemporary design argument in astrophysical cosmology', *British Journal for the Philosophy of Science* **53**, 331–54.

Klein, Oskar (1926). 'The atomicity of electricity as a quantum theory law', *Nature* **118**, 516.

Klein, Oskar (1928). 'Zur fünfdimensionalen Darstellung der Reltivitätstheorie', *Zeitschrift für Physik* **46**, 188–208.

Klein, Oskar (1954). 'Some cosmological considerations in connection with the problem of the origin of the elements', pp. 42–51 in P. Ledoux, ed., *Les processus nucléaires dans les astres*. Louvain: Société Royale des Sciences de Liège.

Klotz, A. H. (1969). 'The mesotrons of Eddington', *Nuovo Cimento B* **63**, 309–17.

Kojevnikov, Alexei (1999). 'Freedom, collectivism, and quasiparticles: Social metaphors in quantum physics', *Historical Studies in the Physical and Biological Sciences* **29**, 295–331.

Kojevnikov, Alexei (2004). *Stalin's Great Science. The Times and Adventures of Soviet Physicists*. London: Imperial College Press.

Kolb, Edward W. and Michael S. Turner (1994). *The Early Universe*. Reading, Mass.: Addison-Wesley.

Kragh, Helge (1982). 'Cosmo-physics in the thirties: towards a history of Dirac cosmology', *Historical Studies in the Physical Sciences* **13**, 69–108.

Kragh, Helge (1984). 'Equation with many fathers: The Klein-Gordon equation in 1926', *American Journal of Physics* **52**, 1024–33.

Kragh, Helge (1985). 'The fine structure of hydrogen and the gross structure of the physics community, 1916–26', *Historical Studies in the Physical Sciences* **15**, 67–126.

Kragh, Helge (1986). 'Relativity and atomic theory before quantum mechanics', pp. 1829–35 in R. Ruffini, ed., *Proceedings of the Fourth Marcel Grossmann Meeting on General Relativity*. Amsterdam: Elsevier.

Kragh, Helge (1989a). 'The aether in late nineteenth century chemistry', *Ambix* **36**, 49–65.

Kragh, Helge (1989b). 'Concept and controversy: Jean Becquerel and the positive electron', *Centaurus* **32**, 203–40.

Kragh, Helge (1990). *Dirac: A Scientific Biography*. Cambridge: Cambridge University Press.

Kragh, Helge (1991). 'Cosmonumerology and empiricism: The Dirac-Gamow dialogue', *Quarterly Astronomy* **8**, 109–26.

Kragh, Helge (1995a). 'Arthur March, Werner Heisenberg, and the search for a smallest length', *Revue d'Histoire des Sciences* **48**, 401–34.

Kragh, Helge (1995b). 'Cosmology between the wars: The Nernst-MacMillan alternative', *Journal for the History of Astronomy* **26**, 93–115.

Kragh, Helge (1996). *Cosmology and Controversy. The Historical Development of Two Theories of the Universe*. Princeton: Princeton University Press.

Kragh, Helge (1997). 'The electrical universe: Grand cosmological theory versus mundane experiments', *Perspectives on Science* **5**, 199–231.

Kragh, Helge (1999a). *Quantum Generations: A History of Physics in the Twentieth Century*. Princeton: Princeton University Press.

Kragh, Helge (1999b). 'Steady-state cosmology and general relativity: Reconciliation or conflict', pp. 377–402 in Hubert Goenner et al., eds, *The Expanding Worlds of General Relativity*. Boston: Birkhäuser.

Kragh, Helge (2001a). 'The first subatomic explanations of the periodic system', *Foundations of Chemistry* **3**, 129–43.

Kragh, Helge (2001b). 'The electron, the protyle, and the unity of matter', pp. 195–226 in Jed Z. Buchwald and Andrew Warwick, eds, *Histories of the Electron: The Birth of Microphysics*. Cambridge, MA: MIT Press.

Kragh, Helge (2002). 'The vortex atom: A Victorian theory of everything', *Centaurus* **44**, 32–114.

Kragh, Helge (2003). 'Magic number: A partial history of the fine-structure constant', *Archive for History of Exact Sciences* **57**, 395–431.

Kragh, Helge (2004). *Matter and Spirit in the Universe: Scientific and Religious Preludes to Modern Cosmology*. London: Imperial College Press.

Kragh, Helge (2006). 'Cosmologies with varying speed of light: A historical perspective', *Studies in History and Philosophy of Modern Physics* **37**, 726–37.

Kragh, Helge (2007). *Conceptions of Cosmos: From Myths to the Accelerating Universe*. Oxford: Oxford University Press.

Kragh, Helge (2008). *Entropic Creation. Religious Contexts of Thermodynamics and Cosmology*. Aldershot: Ashgate.

Kragh, Helge (2009a). 'Continual fascination: The oscillating universe in modern cosmology', *Science in Context* 22, 587–612.

Kragh, Helge (2009b). 'Contemporary history of cosmology and the controversy over the multiverse', *Annals of Science* 66, 529–51.

Kragh, Helge (2010a). 'Quasi-steady state and related cosmological models', pp. 141–55 in Rüdiger Vaas, ed., *Beyond the Big Bang*. Berlin: Springer-Verlag.

Kragh, Helge (2010b). 'Cyclic models of the relativistic universe: The early history', in David Rowe, ed., *Beyond Einstein* (forthcoming). Boston: Birkhäuser.

Kragh, Helge and Bruno Carazza (1994). 'From time atoms to space-time quantization: The idea of discrete time, ca 1925–1936', *Studies in History and Philosophy of Science* 25, 437–62.

Kragh, Helge and Bruno Carazza (1995). 'A historical note on the maximum atomic number of chemical elements', *Annales de la Fondation Louis de Broglie* 20, 207–15.

Kragh, Helge and Dominique Lambert (2007). 'The context of discovery: Lemaître and the origin of the primeval-atom universe', *Annals of Science* 64, 445–70.

Krauss, Lawrence M. and Glenn D. Starkman (2000). 'Life, the universe, and nothing: Life and death in an ever-expanding universe', *Astrophysical Journal* 531, 22–30.

Krohn, Wolfgang, Günter Küppers, and Helga Nowotny, eds (1990). *Self-Organization: Portrait of a Scientific Revolution*. Dordrecht: Kluwer Academic.

Küppers, Bernd-Olaf (1990). 'On a fundamental paradigm shift in the natural sciences', pp. 51–63 in Wolfgang Krohn, Günter Küppers, and Helga Nowotny, eds, *Self-Organization: Portrait of a Scientific Revolution*. Dordrecht: Kluwer Academic.

Lanczos, Cornelius (1925). 'Über eine zeitlich periodische Welt und eine neue Behandlung des Problems der Ätherstrahlung', *Zeitschrift für Physik* 32, 56–80.

Landau, Lev D. and Evgeny M. Lifschitz (1974). *Quantum Mechanics*. Oxford: Pergamon Press.

Larmor, Joseph (1900). *Aether and Matter*. Cambridge: Cambridge University Press.

Larmor, Joseph (1927). *Mathematical and Physical Papers* vol. 1. Cambridge: Cambridge University Press.

Larmor, Joseph (1929). *Mathematical and Physical Papers* vol. 2. Cambridge: Cambridge University Press.

Laue, Max von (1921). *Die Relativitätstheorie* vol. 2. Braunschweig: Vieweg & Sohn.

Laughlin, Robert B. and David Pines (2000). 'The theory of everything', *Proceedings of the National Academy of Sciences* 97, 28–31.

LeBon, Gustave (1905). *The Evolution of Matter*. New York: Charles Scribner's Sons.

Leegwater, Arie (1986). 'The development of Wilhelm Ostwald's chemical energetics', *Centaurus* 29, 314–37.

Lehners, Jean-Luc (2008). 'Ekpyrotic and cyclic cosmology', *Physics Report* 465, 223–63.

Lehners, Jean-Luc and Paul J. Steinhardt (2009). 'Dark energy and the return of the phoenix universe', *Physical Review D* 79, 063503.

Leibniz, Gottfried W. (2008). *Theodicy*. Teddington, Middlesex: Echo Library.

Lemonick, Michael D. (2004). 'Before the big bang', *Discover* Magazine (online version), 5 February 2004.

Lepeltier, Thomas (2006). 'Edward Milne's influence on modern cosmology', *Annals of Science* **63**, 471–81.
Lerner, Eric J. (1991). *The Big Bang Never Happened*. New York: Times Books.
Leslie, John (1989). *Universes*. London: Routledge.
Leslie, John, ed. (1990). *Physical Cosmology and Philosophy*. New York: Macmillan.
Lewis, Gilbert N. (1922). 'The chemistry of the stars and the evolution of radioactive substances', *Publications of the Astronomical Society of the Pacifics* **34**, 309–19.
Lightman, Alan and Roberta Brawer (1990). *Origins: The Lives and Worlds of Modern Cosmologists*. Cambridge, Mass.: Harvard University Press.
Linde, Andrei (1983). 'The new inflationary scenario', pp. 205–250 in G. W. Gibbon, S. W. Hawking, and S. Siklos, eds, *The Very Early Universe*. Cambridge: Cambridge University Press.
Linde, Andrei (1986). 'Eternally existing self-reproducing chaotic inflationary universe', *Physics Letters* B **175**, 395–401.
Linde, Andrei (1987). 'Inflation and quantum cosmology', pp. 604–30 in Stephen Hawking and Werner Israel, eds, *Three Hundred Years of Gravitation*. Cambridge: Cambridge University Press.
Linde, Andrei (1988). 'Life after inflation', *Physics Letters* B **211**, 29–31.
Linde, Andrei (1989). 'Life after inflation and the cosmological constant problem', *Physics Letters* B **227**, 352–358.
Linde, Andrei (1989). 'Life after inflation and the cosmological constant problem', *Physics Letters* B **227**, 352–58.
Linde, Andrei (1990). *Inflation and Quantum Cosmology*. Boston: Academic Press.
Linde, Andrei (2003). 'Inflationary theory versus the ekpyrotic/cyclic scenario', pp. 801–35 in Gary W. Gibbons, E. Paul Shellard, and Stuart J. Rankin, eds, *The Future of Theoretical Physics and Cosmology*. Cambridge: Cambridge University Press.
Linde, Andrei (2007). 'The inflationary universe', pp. 127–50 in Bernard Carr, ed., *Universe or Multiverse?*. Cambridge: Cambridge University Press.
Lindley, David (1993). *The End of Physics: The Myth of a Unified Theory*. New York: Basic Books.
Lindsay, Robert B. (1959). 'Entropy consumption and values in physical science', *American Scientist* **47**, 376–85.
Livio, Mario (2000). *The Accelerating Universe: Infinite Expansion, the Cosmological Constant, and the Beauty of the Cosmos*. New York: John Wiley & Sons.
Livio, Mario and Martin J. Rees (2005). 'Anthropic reasoning', *Science* **309**, 1022–23.
Livio, Mario, D. Hollowell, A. Weiss, and J. W. Truran (1989). 'The anthropic significance of the existence of an excited state of 12C', *Nature* **340**, 281–84.
Lodge, Oliver (1883). 'The ether and its functions', *Nature* **34**, 304–306, 328–30.
Lodge, Oliver (1925). *Ether and Reality*. New York: George H. Doran Co.
Loeb, Abraham (2006). 'An observational test for the anthropic origin of the cosmological constant', *Journal of Cosmology and Astroparticle Physics* **05**, 009.
Logunov, A. A., Yu. M. Loskutov, and M. A. Mestvirishvili (1988). 'Relativistic theory of gravitation and its consequences', *Progress of Theoretical Physics* **80**, 1005–23.
Lomonaco, Samuel J. (1995). 'The modern legacies of Thomson's atomic vortex theory in classical electrodynamics', pp. 145–66 in Louis H. Kauffman, ed., *The Interface of Knots and Physics*. American Mathematical Society.
Longair, Malcolm S., ed. (1974). *Confrontation of Cosmological Theories with Observational Data*. Dordrecht: Reidel.
Longair, Malcolm S. (1985). 'The universe – present, past and future', *Observatory* **105**, 171–88.

Lorentz, Hendrik A. (1952). *The Theory of Electrons*. New York: Dover Publications.
Losee, John (1980). *A Historical Introduction to the Philosophy of Science*. Oxford: Oxford University Press.
Luskin, C. (2006). 'The double standard for intelligent design and testability.' http://www.evolutionnews.org (10 August).
MacAlister, Donald (1883). [Review of J. B. Stallo, *The Concepts and Theories of Modern Physics*], *Mind* **8**, 276–84.
MacRobert, Alan (1983). 'Beyond the big bang', *Sky and Telescope* **65** (March), 211–13.
Magueijo, João (2003). 'New varying speed of light theories', *Reports on Progress in Physics* **66**, 2025.
Magueijo, João (2004). *Faster than the Speed of Light. The Story of a Scientific Speculation*. London: Arrow Books.
Magueijo, João (2009). 'Bimetric varying speed of light theories and primordial fluctuations', *Physical Review* D **79**, 043525.
Markley, Robert (1992). 'The irrelevance of reality: Science, ideology, and the postmodern universe', *Genre* **25**, 249–76.
Markovic, Zeljko et al. (1958). *Le symposium international R.J. Boskovich*. Belgrade: Naucno Delo.
Matthews, Robert (2008). 'Some swans are grey', *New Scientist* **198** (10 May), 44–47.
Mattingly, James (2005). 'Is quantum gravity necessary?' pp. 327–38 in A.J. Kox and Jean Eisenstaedt, eds, *The Universe of General Relativity*. Boston: Birkhäuser.
Maxwell, James C. (1965). *The Scientific Papers of James Clerk Maxwell*. Edited by W.D. Niven. New York: Dover Publications.
McAllister, James W. (1996). *Beauty and Revolution in Science*. Ithaca: Cornell University Press.
McCabe, Gordon (2004). 'A critique of cosmological natural selection'. http://philsci-archive.pitt.edu/archive/00001648/
McCormmach, Russell (1966). 'The atomic theory of John William Nicholson', *Archive for History of Exact Sciences* **3**, 160–84.
McCormmach, Russell (1970). 'H.A. Lorentz and the electromagnetic view of nature', *Isis* **61**, 459–97.
McCrea, William H. (1937). 'Physical science and philosophy', *Nature* **139**, 1002.
McCrea, William H. (1950). 'Quantum mechanics and astrophysics', *Nature* **166**, 884–86.
McCrea, William H. (1951). 'Relativity theory and the creation of matter', *Proceedings of the Royal Society* A **206**, 562–75.
McCrea, William H. (1970). 'A philosophy for big-bang cosmology', *Nature* **228**, 21–25.
McMullin, Ernan (1987). 'Scientific controversy and its termination', pp. 49–92 in H. T. Engelhardt and A. L. Caplan, eds, *Scientific Controversies: Case Studies in the Resolution and Closure of Disputes in Science and Technology*. Cambridge: Cambridge University Press.
McMullin, Ernan (1993). 'Indifference principle and anthropic principle in cosmology', *Studies in History and Philosophy of Science* **24**, 359–89.
McVittie, George C. (1940). 'Kinematical relativity', *The Observatory* **63**, 273–81.
McVittie, George C. (1961). 'Rationalism and empiricism in cosmology', *Science* **133**, 1231–36.
McVittie, George C. (1965). *General Relativity and Cosmology*. London: Chapman and Hall.
Mead, C. Alden (1964). 'Possible connection between gravitation and fundamental length', *Physical Review* **135**, B849–62.
Mehra, Jagdish, ed. (1973). *The Physicist's Conception of Nature*. Dordrecht: Reidel.

Mehra, Jagdish (1994). *The Beat of a Different Drum: The Life and Science of Richard Feynman.* New York: Oxford University Press.
Mersini-Houghton, Laura and Richard Holman (2009). ' "Tilting" the universe with the landscape multiverse: The dark flow', *Journal of Cosmology and Astroparticle Physics* **02**, 006.
Michelson, Albert A. (1903). *Light Waves and Their Uses.* Chicago: University of Chicago Press.
Mie, Gustav (1907). *Moleküle, Atome, Weltäther.* Leipzig: Teubner.
Mie, Gustav (1912).'Die Grundlagen einer Theorie der Materie', *Annalen der Physik* **37**, 511–34.
Mie, Gustav (2007). 'Foundations of a theory of matter', pp. 633–97 in Jürgen Renn, ed., *The Genesis of General Relativity* vol. 4. Dordrecht: Springer.
Miller, Arthur I. (1981). *Albert Einstein's Special Theory of Relativity: Emergence (1905) and Early Interpretation (1905–1911).* London: Addison-Wesley.
Miller, Sean and Shveta Verma, eds. (2008). *Riffing on Strings: Creative Writing Inspired by String Theory.* New York: Scriblerus Press.
Milne, Edward A. (1934). 'World-models and the world-picture', *The Observatory* **57**, 24–28.
Milne, Edward A. (1935). *Relativity, Gravitation and World-Structure.* Oxford: Clarendon Press.
Milne, Edward A. (1937). 'On the origin of laws of nature', *Nature* **139**, 997–99.
Milne, Edward A. (1939). 'A possible mode of approach to nuclear dynamics', pp. 208–19 in *New Theories of Physics.* Warsaw: Scientific Collections.
Milne, Edward A. (1940). 'Cosmological theories', *Astrophysical Journal* **91**, 129–58.
Milne, Edward A. (1943). 'The fundamental concepts of natural philosophy', *Proceedings of the Royal Society of Edinburgh A* **62**, 10–24.
Milne, Edward A. (1948). *Kinematic Relativity.* Oxford: Clarendon Press.
Milne, Edward A. (1952a). *Modern Cosmology and the Christian Idea of God.* Oxford: Clarendon Press.
Milne, Edward A. (1952b). *Sir James Jeans. A Biography.* Cambridge: Cambridge University Press.
Minkowski, Hermann (1915). 'Das Relativitätsprinzip', *Annalen der Physik* **47**, 927–38.
Mirowski, Philip (1992). 'Looking for those natural numbers: Dimensionless constants and the idea of natural measurement', *Science in Context* **5**, 165–88.
Moffat, John W. (1993). 'Superluminary universe: A possible solution to the initial value problem in cosmology', *International Journal of Modern Physics D* **2**, 351–65.
Morrison, Mark S. (2007). *Modern Alchemy: Occultism and the Emergence of Atomic Theory.* Oxford: Oxford University Press.
Mosterin, Jesús (2004). 'Anthropic explanations in cosmology', pp. 441–71 in Petr Hájek, Luis Valdés-Villanueva, and Dag Westerståhl, eds, *Proceedings of the 12th International Congress of Logic, Methodology and Philosophy of Science.* Amsterdam: North-Holland. (http://philsci-archive.pitt.edu/archive/00001658/).
Mott, Nevill F. et al. (1959). *Religion and the Scientists.* London: SCM Press.
Mukhanov, Viatscheslav (2007). 'Cosmology and the many worlds interpretation of quantum mechanics', pp. 267–274 in Bernard Carr, ed., *Universe or Multiverse?.* Cambridge: Cambridge University Press.
Mukherji, Visvapriya and S. Kumar Roy (1982). 'Particle physics since 1930: A history of evolving notions of nature's simplicity and uniformity', *American Journal of Physics* **50**, 1100–1103.
Mulligan, Joseph F. (2001). 'Emil Wiechert (1861–1928): Esteemed seismologist, forgotten physicist', *American Journal of Physics* **69**, 277–87.
Munitz, Milton K., ed. (1957a). *Theories of the Universe: From Babylonian Myth to Modern Science.* Glencoe, Ill.: Free Press.

Munitz, Milton K., ed. (1957b). *Space, Time and Creation: Philosophical Aspects of Scientific Cosmology*. Glencoe, Ill.: Free Press.

Mutschler, Hans-Dieter (1995). 'Frank Tipler's physical eschatology', *Zygon* 30, 479–90.

Narlikar, Jayant V. (1973). 'Steady state defended', pp. 69–84 in L. John, ed., *Cosmology Now*. London: BBC.

Narlikar, Jayant V. (1983). 'Cosmologies with variable gravitational constants', *Foundations of Physics* 12, 311–23.

Narlikar, Jayant V. and Geoffrey Burbidge (2008). *Facts and Speculations in Cosmology*. Cambridge: Cambridge University Press.

Narlikar, Jayant V. and T. Padmanabhan (2001). 'Standard cosmologies and alternatives: A critical appraisal', *Annual Review of Astronomy and Astrophysics* 39, 211–48.

Newcomb, Simon (1896). 'The philosophy of hyperspace', *Bulletin of the American Mathematical Society* 4, 187–195.

Newcomb, Simon (1906). *Side-Lights on Astronomy. Essays and Addresses*. New York: Harper and Brothers.

Newton, Isaac (1952). *Opticks*. New York: Dover Publications.

Newton, Isaac (1961). *The Correspondence of Isaac Newton* vol. 1. Edited by H. W. Turnbull. Cambridge: Cambridge University Press.

Nicholson, John W. (1913). 'The physical interpretation of the spectrum of the corona', *The Observatory* 36, 103–12.

Noakes, Richard (2005). 'Ethers, religion and politics in late-Victorian physics: Beyond the Wynne thesis', *History of Science* 43, 415–55.

Nordström, Gunnar (1914). 'Über die Möglichkeit das elektromagnetischer Feld und das Gravitationsfeld zu vereinigen', *Physikalische Zeitschrift* 15, 504–506.

North, John D. (1990). *The Measure of the Universe: A History of Modern Cosmology*. New York: Dover Publications.

Norton, John D. (1992). 'Einstein, Nordstrøm and the early demise of scalar, Lorentz-covariant theories of gravitation', *Archive for History of Exact Sciences* 45, 17–94.

Norton, John D. (2000). '"Nature is the realisation of the simplest conceivable mathematical ideas": Einstein and the canon of mathematical simplicity', *Studies in History and Philosophy of Modern Physics* 31, 135–70.

Nozick, Robert (1981). *Philosophical Explanations*. Cambridge, Mass.: Harvard University Press.

Nussbaumer, Harry and Lydia Bieri (2009). *Discovering the Expanding Universe*. Cambridge: Cambridge University Press.

Olive, Keith A. and Yong-Zhong Qian (2004). 'Were fundamental constants different in the past?' *Physics Today* 57 (October), 40–45.

Öpik, Ernst J. (1954). 'The time scale of our universe', *Irish Astronomical Journal* 3, 89–108.

Ørsted, Hans C. (1809). *Videnskaben om Naturens Almindelige Love*. Copenhagen: Brummer.

Ørsted, Hans C. (1852). *The Soul in Nature*. London: H. G. Bohn.

Ørsted, Hans C. (1998). *Selected Works of Hans Christian Ørsted*, eds. Karen Jelved, Andrew D. Jackson and Ole Knudsen. Princeton: Princeton University Press.

Overbye, Dennis (2002). 'A new view of our universe: Only one of many', *New York Times*, 29 October.

Overbye, Dennis (2003). 'Zillions of universes? Or did ours get lucky?' *New York Times*, 28 October.

Overduin, James, Hans-Joachim Blome, and Josef Hoell (2007). 'Wolfgang Priester: From the big bounce to the Λ–dominated universe', *Naturwissenschaften* **94**, 417–29.

Page, Don N. (2007). 'Predictions and tests of multiverse theories', pp. 411–29 in Bernard Carr, ed., *Universe or Multiverse?*. Cambridge: Cambridge University Press.

Page, Don N. (2009). 'Anthropic estimates of the charge and mass of the proton', *Physics Letters* B **675**, 398–402.

Pagels, Heinz R. (1985). 'A cozy cosmology', *The Sciences* **25** (March-April), 35–38.

Pais, Abraham (1982). *'Subtle is the Lord...': The Science and the Life of Albert Einstein*. Oxford: Oxford University Press.

Palonen, V. (2008). 'Bayesian considerations on the multiverse explanation of cosmic fine-tuning', arXiv:0802.4013.

Pannenberg, Wolfhart (1981). 'Theological questions to scientists', pp. 3–16 in Arthur R. Peacocke, ed., *The Science and Theology in the Twentieth Century*. London: Oriel Press.

Pauli, Wolfgang (1979). *Wolfgang Pauli. Wissenschaftlicher Briefwechsel* vol. 1. Edited by Armin Hermann, Karl von Meyenn, and Viktor F. Weisskopf. New York: Springer-Verlag.

Pauli, Wolfgang (1996). *Wolfgang Pauli. Wissenschaftlicher Briefwechsel* vol. 4, part 2. Edited by Karl von Meyenn. New York: Springer-Verlag.

Peacocke, Arthur R. (2004). *Creation and the World of Science: The Re-Shaping of Belief*. Oxford: Oxford University Press.

Pearson, Karl (1900). *The Grammar of Science*. London: A. & C. Black.

Penrose, Roger (1990). *The Emperor's New Mind: Concerning Computers, Minds, and the Laws of Physics*. London: Vintage.

Penrose, Roger (2006). 'Before the big bang: An outrageous new perspective and its implications for particle physics'. http://accelconf.web.cern.ch/AccelConf/e06/PAPERS/THESPA01.pdf.

Penrose, Roger (2009). 'Black holes, quantum theory and cosmology', *Journal of Physics: Conference Series* **174**, 012001.

Planck, Max (1896). 'Gegen die neuere Energetik', *Annalen der Physik* **57**, 72–78.

Planck, Max (1949). *Scientific Autobiography*. New York: Philosophical Library.

Planck, Max (1960). *A Survey of Physical Theory*. New York: Dover Publications.

Playfair, John (1807). 'Notice de la vie et des ecrits de George Louis Le Sage', *Edinburgh Review* 137–53.

Poincaré, Henri (1913). *Dernières pensées*. Paris: Flammarion.

Poincaré, Henri (1952). *Science and Method*. New York: Dover Publications.

Polkinghorne, John (1985). 'Salesman of ideas', pp. 23–35 in Charleton DeTar, J. Finkelstein, and Chung-I Tan, eds, *A Passion for Physics: Essays in Honor of Geoffrey Chew*. Singapore: World Scientific.

Polkinghorne, John (1996). *Beyond Science: The Wider Human Context*. Cambridge: Cambridge University Press.

Popper, Karl R. (1940). 'Interpretation of nebular red-shifts', *Nature* **145**, 69–70.

Porter, Theodore M. (2004). *Karl Pearson: The Scientific Life in a Statistical Age*. Princeton: Princeton University Press.

Poynting, John H. (1899). 'Presidential address to the section of mathematical and physical sciences', *British Association for the Advancement of Science*, Report, 615–24.

Press, William H. (1986). 'A place for teleology?' *Nature* **320**, 315–16.

Preston, Samuel T. (1879). 'On the possibility of explaining the continuance of life in the universe consistent with the tendency to temperature-equilibrium', *Nature* **19**, 460–62.

Prince, N. H. E. (2005). 'Simulated worlds, physical eschatology, the finite nature hypothesis and the final anthropic principle', *International Journal of Astrobiology* **4**, 203–26.

Pyenson, Lewis (1982). 'Relativity in late Wilhelmian Germany: The appeal to a preestablished harmony between mathematics and physics', *Archive for History of Exact Sciences* **27**, 137–55.

Rabounski, Dmitri (2006). 'Zelmanov's anthropic principle and the infinite relativity principle', *Progress in Physics* **1**, 35–37.

Rademacher, Johann P. (1909). *Der Weltuntergang*. Munich: Münchener Volksschriftenverlag.

Raia, Courtenay G. (2007). 'From ether theory to ether theology: Oliver Lodge and the physics of immortality', *Journal of the History of the Behavioral Sciences* **43**, 19–43.

Rebsdorf, Simon and Helge Kragh (2002). 'Edward Arthur Milne – The relations of mathematics to science', *Studies in History and Philosophy of Modern Physics* **33**, 51–64.

Rechenberg, Helmut (1989). 'The early S-matrix theory and its propagation (1942–1952)', pp. 551–78 in Laurie M. Brown, Max Dresden, and Lillian Hoddeson, eds, *Pions to Quarks: Particle Physics in the 1950s*. Cambridge: Cambridge University Press.

Redhead, Michael (2005). 'Broken bootstraps – the rise and fall of a research programme', *Foundations of Physics* **35**, 561–75.

Rees, Martin J. (1969). 'The collapse of the universe: An eschatological study', *Observatory* **89**, 193–98.

Rees, Martin J. (1998). 'Our universe and others', pp. 51–68 in Hermann Bondi and Miranda Weston-Smith, eds, *The Universe Unfolding*. Oxford: Clarendon Press.

Rees, Martin J. (2003a). 'Our complex cosmos and its future', pp. 16–37 in Gary W. Gibbons, E. Paul Shellard, and Stuart J. Rankin, eds, *The Future of Theoretical Physics and Cosmology*. Cambridge: Cambridge University Press.

Rees, Martin J. (2003b). 'Numerical coincidences and "tuning" in cosmology', *Astrophysics and Space Science* **285**, 375–88.

Rees, Martin J. (2007). 'Cosmology and the multiverse', pp. 57–76 in Bernard Carr, ed., *Universe or Multiverse?*. Cambridge: Cambridge University Press.

Rees, Martin J., Remo Ruffini, and John A. Wheeler (1974). *Black Holes, Gravitational Waves and Cosmology: An Introduction to Current Research*. New York: Gordon and Breach.

Renn, Jürgen, ed. (2007). *The Genesis of General Relativity*. 4 vols. Dordrecht: Springer.

Rescher, Nicholas (1999). *The Limits of Science*. Pittsburgh: University of Pittsburgh Press.

Restivo, Sal P. (1985). *The Social Relations of Physics, Mysticism, and Mathematics*. Chicago: University of Chicago Press.

Reynolds, Osborne (1883). 'Vortex rings', *Nature* **29**, 193–94.

Rice, James (1925). 'On Eddington's natural unit of the field', *Philosophical Magazine* **49**, 1056–57.

Rich, J. (2003). 'Experimental consequences of time variations of the fundamental constants', *American Journal of Physics* **71**, 1043–46.

Richter, Burton (2006). 'Theory in particle physics: Theological speculation versus practical knowledge', *Physics Today* **59** (October), 8–9.

Robertson, Douglas S. (2000). 'Goedel's theorem, the theory of everything, and the future of science and mathematics', *Complexity* **5**: 5, 22–27.

Robotti, Nadia and Massimiliano Badino (2001). 'Max Planck and the "constants of nature"', *Annals of Science* **58**, 137–62.

Robotti, Nadia and Francesca Pastorino (1998). 'Zeeman's discovery and the mass of the electron', *Annals of Science* **55**, 161–83.

Rosenthal-Schneider, Ilse (1980). *Reality and Scientific Truth: Discussions with Einstein, von Laue, and Planck*. Detroit: Wayne State University Press.

Rosevear, N.T. (1982). *Mercury's Perihelion: From Le Verrier to Einstein*. Oxford: Clarendon Press.

Roush, Sherrilyn (2003). 'Copernicus, Kant, and the anthropic cosmological principles', *Studies in History and Philosophy of Modern Physics* **34**, 5–35.

Rovelli, Carlo (1998). 'Strings, loops and others: A critical survey of the present approaches to quantum gravity', pp. 281–331 in N. Dadhich and J. Narlikar, eds, *Gravitation and Relativity at the Turn of the Millenium*. Pune, India: Inter-University Centre for Astronomy and Astrophysics.

Rovelli, Carlo (2003). 'Loop quantum gravity', *Physics World* (November), 1–5.

Rovelli, Carlo (2008). 'Loop quantum gravity', *Living Reviews in Relativity* **11**: 5 (http://www.livingreviews.org/lrr-2008-5).

Rowe, David E. (2001). 'Einstein meets Hilbert: At the crossroads of physics and mathematics', *Physics in Perspective* **3**, 379–424.

Rowlinson, John S. (2002). *Cohesion: A Scientific History of Intermolecular Forces*. Cambridge: Cambridge University Press.

Rowlinson, John S. (2003). 'Le Sage's Essai de chymie méchanique', *Notes and Records of the Royal Society* **57**, 35–45.

Rozental, Iosif L. (1980). 'Physical laws and the numerical values of fundamental constants', *Soviet Physics Uspekhi* **23**, 296–306.

Rozental, Iosif L. (1988). *Big Bang, Big Bounce: How Particles and Fields Drive Cosmic Evolution*. Berlin: Springer-Verlag.

Rueger, Alexander (1992). 'Attitudes towards infinities: Responses to anomalies in quantum electrodynamics, 1927–1947', *Historical Studies in the Physical Sciences* **22**, 309–38.

Rüger, Alexander (1988). 'Atomism from cosmology: Erwin Schrödinger's work on wave mechanics and space-time structure', *Historical Studies in the Physical Sciences* **18**, 377–401.

Rújula, Alvaro De (1986). 'Superstrings and supersymmetry', *Nature* **320**, 678.

Russell, Bertrand (1931). *The Scientific Outlook*. New York: W. W. Norton and Company.

Russell, Colin A. (1971). *The History of Valency*. Oxford: Leicester University Press.

Rutherford, Ernest (1914). 'The structure of the atom', *Scientia* **16**, 337–51.

Sakharov, Andrei D. (1982). 'Many-sheeted models of the Universe', *Soviet Physics Uspekhi* **34**, 404–408.

Sánchez-Ron, José (2005). 'George McVittie, the uncompromising empiricist', pp. 189–222 in A. J. Kox and Jean Eisenstaedt, eds, *The Universe of General Relativity*. Boston: Birkhäuser.

Sauer, Tilman and Ulrich Majer, eds (2009). *David Hilbert's Lectures on the Foundations of Physics 1915–1927*. Dordrecht: Springer.

Schellekens, A. N. (2008). 'The emperor's last clothes? Overlooking the string theory landscape', *Reports on Progress in Physics* **71**, 072201.

Schelling, Friedrich W. J. (1988). *Ideas for a Philosophy of Nature*. Translated by Errol E. Harris and Peter Heath. Cambridge: Cambridge University Press.

Schelling, Friedrich W. J. (2004). *First Outline of a System of the Philosophy of Nature*. Translated by Keith R. Peterson. New York: SUNY Press.

Scherck, Jöel and John H. Schwarz (1974). 'Dual models for non-hadrons', *Nuclear Physics* B **81**, 118–44.

Schlegel, Richard (1962). 'Transfinite numbers and cosmology', *Nature* **193**, 665–66.

Schlegel, Richard (1964). 'The problem of infinite matter in steady-state cosmology', *Philosophy of Science* **32**, 21–31.

Schofield, Robert E. (1970). *Mechanism and Materialism: British Natural Philosophy in an Age of Reason*. Princeton: Princeton University Press.

Schröder, Wilfried and H.-J. Treder (1996). 'Hans Ertel and cosmology', *Foundations of Physics* **26**, 1081–88.

Schrödinger, Erwin (1937). 'World structure', *Nature* **140**, 742–744.

Schuster, Arthur (1898). 'Potential matter – a holiday dream', *Nature* **58**, 367.

Schwarz, John, ed. (1985). *Superstrings. The First 15 Years of Superstring Theory*. Teaneck, NJ: World Scientific.

Schwarz, John H. (1987). 'Superstrings', *Physics Today* **40** (November), 33–40.

Schwarz, John H. (1996). 'Superstring – a brief history', pp. 695–706 in Harvey B. Newman and Thomas Ypsilantis, eds, *History of Original Ideas and Basic Discoveries in Particle Physics*. New York: Plenum Press.

Schwarz, John H. (1998). 'Beyond gauge theories', arXiv:hep-th/9807195.

Schweber, Sylvan S. (1994). *QED and the Men Who Made It: Dyson, Feynman, Schwinger, and Tomonaga*. Princeton: Princeton University Press.

Sciama, Dennis W. (1993a). 'Ist das Universum eigenartig?' pp. 183–94 in G. Börner, J. Ehlers, and H. Meier, eds, *Vom Urknall zum komplexen Universum*. Munich: Piper.

Sciama, Dennis W. (1993b). 'The anthropic principle and the non-uniqueness of the universe', pp. 107–109 in Francesco Bertola and Umberto Curi, eds, *The Anthropic Principle*. Cambridge: Cambridge University Press.

Scoular, Spencer (2007). *First Philosophy: The Theory of Everything*. Boca Raton, Florida: Universal Publishers.

Seife, Charles (2002). 'Eternal-universe idea comes full circle', *Science* **296**, 639.

Seth, Suman (2004). 'Quantum theory and the electromagnetic world-view', *Historical Studies in the Physical and Biological Sciences* **35**, 67–95.

Shapere, Dudley (2000). "Testability and empiricism', pp. 153–64 in Evando Agazzi and Massimo Pauri, eds, *The Reality of the Unobservable*. Dordrecht: Kluwer Academic.

Shapiro, Joel A. (2007). 'Reminiscence on the birth of string theory', arXiv:07113448 (hep-th).

Shermer, Michael (2002). *Why People Believe in Weird Things: Pseudoscience, Superstition, and other Confusions of our Time*. London: Souvenir Press.

Shklovskii, Iosef S. and Carl Sagan (1966). *Intelligent Life in the Universe*. New York: Delta Book.

Siegfried, Robert (1967). 'Boscovich and Davy: Some cautionary remarks', *Isis* **58**, 236–38.

Silliman, Robert H. (1963). 'William Thomson: Smoke rings and nineteenth-century atomism', *Isis* **54**, 461–74.

Silver, Dan (2006). 'Scottish physics and knot theory's odd origins', *American Scientist* **94**, 158–65.

Simmons, Henry T. (1982). 'Redefining the cosmos', *Mosaic Magazine* **13**: 2, 16–22.

Simões, Ana (2002). 'Dirac's claim and the chemists', *Perspectives in Physics* **4**, 253–66.

Sinclair, Steve B. (1987). 'J.J. Thomson and the chemical atom: From ether vortex to atomic decay', *Ambix* **34**, 89–116.

Singh, Jagjit (1970). *Great Ideas and Theories of Modern Cosmology*. New York: Dover Publications.

Singh, Simon (2004). *Big Bang*. London: Fourth Estate.

Sklar, Lawrence (1989). 'Ultimate explanations: Comments on Tipler', pp. 49–55 in Arthur Fine and Jarrett Leplin, eds, *PSA: Proceedings of the Biennial Meeting of the Philosophy of Science Association* vol. 2. Chicago: University of Chicago Press.

Slater, Noel B. (1958). *The Development & Meaning of Eddington's 'Fundamental Theory'*. Cambridge: Cambridge University Press.

Smeenk, Christopher and Christopher Martin (2007). 'Mie's theory of matter and gravitation', pp. 623–32 in Jürgen Renn, ed., *The Genesis of General Relativity* vol. 4. Dordrecht: Springer.

Smolin, Lee (1992). 'Did the universe evolve?' *Classical and Quantum Gravity* **9**, 173–91.

Smolin, Lee (1997). *The Life of Cosmos*. Oxford: Oxford University Press.

Smolin, Lee (2001). *Three Roads to Quantum Gravity*. New York: Basic Books.

Smolin, Lee (2004). 'Atoms of space and time', *Scientific American* **290** (January), 66–75.

Smolin, Lee (2006). *The Trouble with Physics*. London: Penguin Books.

Smolin, Lee (2007). 'Scientific alternatives to the anthropic principle', pp. 323–66 in Bernard Carr, ed., *Universe or Multiverse?*. Cambridge: Cambridge University Press.

Snelders, Harry A.M. (1976). 'A. M. Mayer's experiments with floating magnets and their use in the atomic theories of matter', *Annals of Science* **33**, 67–80.

Soddy, Frederic (1909). *The Interpretation of Radium*. London: John Murray.

Sopka, Katherine R. and Albert E. Moyer, eds. (1986). *Physics for a New Century: Papers Presented at the 1904 St. Louis Congress*. New York: American Institute of Physics.

Sovacool, Benjamin (2005). 'Falsification and demarcation in astronomy and cosmology', *Bulletin of Science, Technology & Society* **25**, 53–62.

Spencer, J. Brookes (1967). 'Boscovich's theory and its relation to Faraday's researches: An analytical approach', *Archive for History of Exact Sciences* **4**, 184–202.

Stachel, John (1993). 'The other Einstein: Einstein contra field theory', *Science in Context* **6**, 275–90.

Stachel, John (1999). 'The early history of quantum gravity (1916–1940)', pp. 525–34 in Bala R. Iyer and Biplab Bjawal, eds, *Black Holes, Gravitational Radiation, and the Universe*. Dordrecht: Kluwer Academic.

Stanley, Matthew (2007). *Practical Mystic: Religion, Science, and A. S. Eddington*. Chicago: University of Chicago Press.

Stapp, Henry P. (1971). 'S-matrix interpretation of quantum theory', *Physical Review* D **3**, 1303–20.

Starkman, Glenn D. and Roberto Trotta (2006). 'Why anthropic reasoning cannot predict Λ', *Physical Review Letters* **97**, 201301.

Steiner, Mark (1998). *The Applicability of Mathematics as a Philosophical Problem*. Cambridge, Mass.: Harvard University Press.

Steinhardt, Paul J. (2004). 'The endless universe: A brief introduction', *Proceedings of the American Philosophical Society* **148**, 464–70.

Steinhardt, Paul J. and Neil Turok (2002a). 'A cyclic model of the universe', *Science* **296**, 1436–39.

Steinhardt, Paul J. and Neil Turok (2002b). 'Cosmic evolution in a cyclic universe', *Physical Review* D **65**, 126003.

Steinhardt, Paul J. and Neil Turok (2004). 'The cyclic model simplified', arXiv:astro-ph/0404480.

Steinhardt, Paul J. and Neil Turok (2006). 'Why the cosmological constant is small and positive', *Science* **312**, 1180–83.

Steinhardt, Paul J. and Neil Turok (2007). *Endless Universe: Beyond the Big Bang*. New York: Doubleday.

Stenger, Victor J. (1995). *The Unconscious Quantum: Metaphysics in Modern Physics and Cosmology*. Amherst, N.Y.: Prometheus Books.

Stern, Alexander W. (1964). 'The third revolution in 20th century physics', *Physics Today* **17** (April), 42–45.

Stewart, Balfour and Peter G. Tait (1881). *The Unseen Universe. Or, Physical Speculations on a Future State*. London: Macmillan.

Stoney, George J. (1881). 'On the physical units of nature', *Philosophical Magazine* **11**, 381–90.

Stoney, George J. (1890). 'On texture in media, and on the non-existence of density in elemental æther', *Philosophical Magazine* **29**, 467–78.
Strominger, Andrew (1986). 'Superstrings with torsion', *Nuclear Physics* B **274**, 253–84.
Suppes, Patrick (1954). 'Descartes and the problem of action at a distance', *Journal of the History of Ideas* **15**, 146–52.
Susskind, Leonard (1997). 'Quark confinement', pp. 233–49 in Lillian Hoddeson et al., eds, *The Rise of the Standard Model: Particle Physics in the 1960s and 1970s*. New York: Cambridge University Press.
Susskind, Leonard (2006). *The Cosmic Landscape: String Theory and the Illusion of Intelligent Design*. New York: Little, Brown and Company.
Susskind, Leonard and Lee Smolin (2004). 'Smolin vs. Susskind: The anthropic principle', *Edge* **145** (August 18). http://edge.org/documents/archive/edfe145.html
Swetman, T. P. (1973). 'Bootstrap – an alternative philosophy', *Physics Education* **8**, 325.
Swinburne, Richard (1990). 'Argument from the fine-tuning of the universe', pp. 154–73 in John Leslie, ed., *Physical Cosmology and Philosophy*. New York: Macmillan.
Swinburne, Richard (1996). *Is There a God?*. Oxford: Oxford University Press.
Tait, Peter G. (1876). *Lectures on Some Recent Advances in Physical Science*. London: Macmillan.
Taubes, Gary (1995). 'A theory of everything takes shape', *Science* **269**, 1511–13.
Taylor, John G. (1993). 'On theories of everything', *Foundations of Physics* **23**, 239–43.
Tegmark, Max (1998). 'Is "the theory of everything" merely the ultimate ensemble theory?' *Annals of Physics* **270**, 1–51.
Tegmark, Max (2003). 'Parallel universes', *Scientific American* **288** (May), 41–51.
Tegmark, Max (2007). 'Many lives in many worlds', *Nature* **448**, 23–24.
Tegmark, Max (2008). 'The mathematical universe', *Foundations of Physics* **38**, 101–50.
Tegmark, Max and Martin J. Rees (1998). 'Why is the cosmic microwave background fluctuation level 10–5?' *Astrophysical Journal* **499**, 526–32.
Tegmark, Max, Alexander Vilenkin, and Levon Pogosian (2005). 'Anthropic predictions for neutrino masses', *Physical Review* D **71**, 103523.
Teller, Edward (1948). 'On the change of physical constants', *Physical Review* **73**, 801–802.
Theckedath, Kumar K. (2003). 'Dialectics and cosmology: The big bang and the steady state theories', *Social Scientist* **31**, 57–84.
Thompson, Silvanus P. (1910). *The Life of William Thomson: Baron Kelvin of Largs*. London: Macmillan.
Thomson, Joseph J. (1883). *A Treatise on the Motion of Vortex Rings*. London: Macmillan.
Thomson, Joseph J. (1892). 'Molecular constitution of bodies, theory of', pp. 410–17 in H. F. Morley and M. M. Pattison Muir, eds, *Watt's Dictionary of Chemistry* vol. 3. London: Macmillan.
Thomson, Joseph J. (1905). 'The structure of the atom', *Proceedings of the Royal Institution* **18**, 1–15.
Thomson, Joseph J. (1907). *The Corpuscular Theory of Matter*. London: Constable & Co.
Thomson, Joseph J. (1908). 'Die Beziehung zwischen Materie und Äther im Lichte der neueren Forschungen auf dem Gebiete der Elektrizität', *Physikalische Zeitschrift* **9**, 543–50.
Thomson, Joseph J. (1909). 'Presidential address', *British Association for the Advancement of Science, Report*, Report, 3–29.
Thomson, William (1867). 'On vortex atoms', *Philosophical Magazine* **34**, 15–24.
Thomson, William (1872). *Reprint of Papers on Electrostatics and Magnetism*. London: Macmillan & Co.
Thomson, William (1891). *Popular Lectures and Addresses* vol. 1. London: Macmillan.

Thomson, William (1902). 'Aepinus atomized', *Philosophical Magazine* **3**, 257–83.

Tipler, Frank J. (1980). 'Extraterrestrial intelligent beings do not exist', *Quarterly Journal of the Royal Astronomical Society* **21**, 267–81.

Tipler, Frank J. (1981). 'A brief history of the extraterrestrial intelligence concept', *Quarterly Journal of the Royal Astronomical Society* **22**, 133–45.

Tipler, Frank J. (1982). 'Anthropic-principle arguments against steady-state cosmological theories', *Observatory* **102**, 36–39.

Tipler, Frank J. (1989). 'The anthropic principle: A primer for philosophers', pp. 27–48 in Arthur Fine and Jarrett Leplin, eds, *PSA: Proceedings of the Biennial Meeting of the Philosophy of Science Association* vol. 2. Chicago: University of Chicago Press.

Tipler, Frank J. (1992). 'The ultimate fate of life in universes which undergo inflation', *Physics Letters* B **286**, 36–43.

Tipler, Frank J. (1994). *The Physics of Immortality: Modern Cosmology, God, and the Resurrection of the Dead*. New York: Doubleday.

Tipler, Frank J. (2003). 'Intelligent life in cosmology.' *International Journal of Astrobiology* **2**, 141–48.

Tipler, Frank J. (2005). 'The structure of the world from pure numbers', *Reports on Progress in Physics* **68**, 897–964.

Tipler, Frank J. (2007). *The Physics of Christianity*. New York: Doubleday.

Tolman, Richard C. (1932). 'Models of the physical universe', *Science* **75**, 367–73.

Tolman, Richard C. (1934a). *Relativity, Thermodynamics and Cosmology*. Oxford: Oxford University Press.

Tolman, Richard C. (1934b). 'Effect on inhomogeneity in cosmological models', *Proceedings of the National Academy of Sciences* **20**, 169–76.

Trimble, Virginia (2009). 'Multiverses of the past', *Astronomische Nachrichten* **330**, 761–69.

Trotta, Roberto (2008). 'Bayes in the sky: Bayesian inference and model selection in cosmology', *Contemporary Physics* **49**, 71–104.

Trotta, Roberto and Glenn D. Starkman (2006). 'What's the trouble with anthropic reasoning?' pp. 323–329 in C. Munoz and G. Yepes, eds, *2nd International Conference on the Dark Side of the Universe*. Madrid: AIP Conference Proceedings, vol. 878.

Turner, Frank M. (1974). *Between Science and Religion: The Reaction to Scientific Naturalism in Late Victorian England*. New Haven: Yale University Press.

Turok, Neil (2002). 'A critical review of inflation', *Classical and Quantum Gravity* **19**, 3449–67.

Urani, John and George Gale (1994). 'E.A. Milne and the origins of modern cosmology: An essential presence', pp. 390–419 in John Earman, Michel Janssen, and John D. Norton, eds, *The Attraction of Gravitation: New Studies in the History of General Relativity*. Boston: Birkhäuser.

Uzan, Jean-Philippe (2003). 'The fundamental constants and their variation: Observational status and theoretical motivations', *Reviews of Modern Physics* **75**, 403–59.

Uzan, Jean-Philippe and Bénédicte Leclercq (2008). *The Natural Laws of the Universe: Understanding Fundamental Constants*. Berlin: Springer-Praxis.

Uzan, Jean-Philippe and Roland Lehoucq (2005). *Les constantes fondamentales*. Paris: Belin.

Van Dongen, Jeroen (2002). *Einstein's Unification: General Relativity and the Quest for Mathematical Naturalness*. Amsterdam: Faculty of Science, University of Amsterdam.

Veltman, Martinus (2003). *Facts and Mysteries in Elementary Particle Physics*. Singapore: World Scientific.

Venable, Francis P. (1904). *The Study of the Atom; or, the Foundations of Chemistry*. Easton, PA: American Chemical Society.

Veneziano, Gabriele (2004). 'The myth of the beginning of time', *Scientific American* **290** (May), 54–65.

Vilenkin, Alexander (2007). 'Anthropic predictions: The case of the cosmological constant', pp. 163–80 in Bernard Carr, ed., *Universe or Multiverse?*. Cambridge: Cambridge University Press.

Vizgin, Vladimir P. (1994). *Unified Field Theories in the First Third of the 20th Century*. Basel: Birkhäuser.

Wallace, Alfred R. (1903). 'Man's place in the universe', *Fortnightly Review* **73** (1 March), 395–411.

Weinberg, Steven (1977). 'The search for unity: Notes for a history of quantum field theory', *Dædalus* **106** (April), 17–35.

Weinberg, Steven (1987a). 'Towards the fimal laws of physics', pp. 61–100 in Richard P. Feynman and Steven Weinberg, *Elementary Particles and the Laws of Physics: The 1986 Dirac Memorial Lectures*. Cambridge: Cambridge University Press.

Weinberg, Steven (1987b). 'Anthropic bound on the cosmological constant', *Physical Review Letters* **59**, 2607–10.

Weinberg, Steven (1989). 'The cosmological constant problem', *Reviews of Modern Physics* **61**, 1–23.

Weinberg, Steven (1992). *Dreams of a Final Theory: The Scientist's Search for the Ultimate Laws of Nature*. New York: Pantheon Books.

Weinberg, Steven (2000). 'Will we have a final theory of everything?' *Time* Magazine, 10 April (online edition).

Weinberg, Steven (2001). *Facing Up: Science and Its Cultural Adversaries*. Cambridge, Mass.: Harvard University Press.

Wesson, Paul S. (2000). 'On the re-emergence of Eddington's philosophy of science', *Observatory* **120**, 59–62.

Weyl, Hermann (1918). 'Reine Infinitesimalgeometrie', *Mathematische Zeitschrift* **2**, 384–411.

Weyl, Hermann (1919). 'Eine neue Erweiterung der Relativitätstheorie', *Annalen der Physik* **59**, 101–133.

Weyl, Hermann (1922). *Space-Time-Matter*. New York: Dover Publications.

Wheeler, John A. (1973). 'From relativity to mutability', pp. 202–49 in Jagdish Mehra, ed., *The Physicist's Conception of Nature*. Dordrecht: Reidel.

Wheeler, John A. (1975). 'The universe as home for man', pp. 261–96 in Owen Gingerich, ed., *The Nature of Scientific Discovery*. Washington, D.C.: Smithsonian Institution Press.

Wheeler, John A. and Wojciech H. Zurek, eds (1983). *Quantum Theory and Measurement*. Princeton: Princeton University Press.

Whitehead, Alfred N. (1933). *Adventures of Ideas*. Cambridge: Cambridge University Press.

Whitrow, Gerald J. (1959). *The Structure and Evolution of the Universe*. New York: Harper.

Whitrow, Gerald J. (1978). 'On the impossibility of an infinite past', *British Journal for the Philosophy of Science* **29**, 39–45.

Whitrow, Gerald J. and Hermann Bondi (1954). 'Is physical cosmology a science?' *British Journal for the Philosophy of Science* **4**, 271–83.

Whyte, Lancelot L. (1954). 'A dimensionless physics?' *British Journal for the Philosophy of Science* **5**, 1–17.

Whyte, Lancelot L. (1960). 'A forerunner of twentieth century physics', *Nature* **186**, 1010–14.

Whyte, Lancelot L. (1961). 'Boscovich's atomism', pp. 102–26 in L.L. Whyte, ed., *Roger Joseph Boscovich: Studies of His Life and Work on the 250th Anniversary of His Birth*. London: George Allen & Unwin.

Wiechert, Emil (1896). 'Die Theorie der Elektrodynamik und die Röntgen'sche Entdeckung', *Schriften der Physikalisch-Ökonomischen Gesellschaft zu Königsberg* **37**, 1–48.

Wigner, Eugene (1960). 'The unreasonable effectiveness of mathematics in the natural sciences', *Communications in Pure and Applied Mathematics* **13**, 1–14.

Wigner, Eugene (1983). 'Remarks on the mind-body question', pp. 168–81 in John A. Wheeler and Wojciech H. Zurek, eds, *Quantum Theory and Measurement*. Princeton: Princeton University Press.

Wilson, David B. (1971). 'The thought of late Victorian physicists: Oliver Lodge's ethereal body', *Victorian Studies* **15**, 29–48.

Wilson, Andrew D. (1993). 'Romantic cosmology', pp. 596–604 in Norriss S. Hetherington, ed., *Encyclopedia of Cosmology*. New York: Garland Publishing.

Wilson, Patrick A. (1994). 'Carter on anthropic principle predictions', *British Journal for Philosophy of Science* **45**, 241–53.

Witten, Edward (1998). 'Magic, mystery, and matrix', *Notices of the AMS* **45**, 1124–29.

Woit, Peter (2006). *Not Even Wrong*. London: Jonathan Cape.

Wood, De Volson (1885). 'The luminiferous æther', *Philosophical Magazine* **20**, 389–417.

Worthing, Mark William (1996). *God, Creation, and Contemporary Physics*. Minneapolis: Fortress Press.

Wünsch, Daniela (2005). 'Einstein, Kaluza, and the fifth dimension', pp. 277–302 in A.J. Kox and Jean Eisenstaedt, eds, *The Universe of General Relativity*. Boston: Birkhäuser.

Wüthrich, Christian (2005). 'To quantize or not to quantize: Fact and folklore in quantum gravity', *Philosophy of Science* **72**, 777–88.

Zachariasen, Fredrik and Charles Zemach (1962). 'Pion resonances', *Physical Review* **128**, 849–58.

Zanstra, Herman (1957). 'On the pulsating or expanding universe and its thermodynamical aspect', *Proceedings of the Royal Dutch Academy of Sciences, Series B* **60**, 285–307.

Zel'dovich, Yakov B. (1981). 'The birth of a closed universe, and the anthropogenic principle', *Soviet Astronomy Letters* **7**, 322–23.

Zelmanov, Abraham L. (2006). *Chronometric Invariants: On Deformations and the Curvature of Accompaying Space*. Rebohoth, NM: American Research Press.

Zinkernagel, Henrik (2006). 'The philosophy behind quantum gravity', *Theoria – International Journal of Theory, History, and Foundations of Science* **21**, 295–312.

INDEX

Note: Indexed items with a page number followed by 'n.' appear in the numbered footnote of the page cited.

Abraham, Max 66, 68–9, 73–4
Adams, Fred 337, 340
Adams, Henry 286 n. 15
Adelberger, Erich 311
aesthetic values 52, 125–6, 153, 198, 204, 208, 307
age of universe 177–8
Aguirre, Anthony 240
Albrecht, Andreas 186–7
Alfvén, Hannes 213
Alpher, Ralph 215 n. 51
Alvarez-Gaumé, Luis 300
analogies, historical 277–8
Anaximander 256
Anaximenes 256
Anderson, Philip 274, 309
anthropic principle 124, 172, 217–48
   Christian 248
   Copernican 338
   final 225, 343, 348
   infinite past 239–40
   name 222, 224, 245
   objections to 243–6
   participatory 225
   precursors 218–20
   predictions 233–41
   religious aspects 247–9
   string theory 266–7, 312–14
   strong 224–5, 231, 239
   weak 224, 236
anthropomorphism 104, 172, 243, 254 n. 97
antimatter 44, 64
a priori physics 29, 31, 81, 98–101, 112–13, 120
Araki, Toshima 138 n. 15
Aristotle 11
Arrhenius, Svante 327
Ashtekar, Abhay 209–10, 316
astrobiology 325, 331–2
astrology 279, 334
astrophysics 88–90, 92, 102
Atiyah, Michael 57 n. 51
atomic theory 20, 24, 30, 80, 83
   see also atoms; vortex atom theory
atoms 15, 17, 37, 61, 141
   Dalton 20
   dynamic 30

electron 63–4
ethereal 38–9
   Larmor 53
   point atoms 20–3
   vortex atoms 35–55
Augustine 194–5
Avogadro's number 95, 170, 326

Bailey, V. A. 123
Banks, Tom 253 n. 74
Barnes, E. William 328
Barrau, Aurétien 279
Barrow, John 228–30, 233, 245
   physical eschatology 335, 337
Baum, Lauris 208–9
Bayes, Thomas 364
Bayesian methodology 364
Becquerel, Henri 84 n. 14
Becquerel, Jean 84 n. 14
Bernal, John D. 92, 350 n. 11
Besso, Michele 317
bioastronomy 331–2
black holes 187, 277–8, 336–7, 342
Blanqui, Louis-Auguste 257, 287 n. 33
Blavatsky, Helena P. 46
Blome, Hans-Joachim 213
Bohm, David 286 n. 25
Bohr, Niels 23, 80, 83, 109, 172, 174, 261
Bojowald, Martin 209–10
Boltzmann, Ludwig 42, 61, 218, 328
   multiverse scenario 258
Bondi, Hermann 109, 118–29, 175, 219, 223
Bonnor, William B. 198
bootstrap hypothesis 149–60
   ideology of 158–60
Born, Max 144, 343
Boskovich, Roger J. 12, 19–26, 30, 257
bounce, cosmic 199, 210, 212–13
Bousso, Raphael 266
Bradley, James 168
Brans, Carl 250 n. 15
Broad, Charlie 101
Bronstein, Matvei 173–4, 292
Brundrit, G. B. 265
Bruno, Giordano 256–7

bubble universes 132, 214 n. 17, 259, 263–4, 342
Bucherer, Adolf 66, 74
Bunge, Mario 130
Burbidge, Geoffrey 134–6

Cahn, Robert 253 n. 81
Calabi, Eugenio 301
Candelas, Philip 301
Cantor, Georg 266
Canuto, Vittorio 180
Capra, Fritjof 159
carbon chauvinism 245
carbon-12 resonance 233–6
Carr, Bernard 228, 231, 279
Cartan, Elie 293
Carter, Brandon 222–7, 236, 245
　falsificationism 241–2, 282
　world ensemble 260–1
　see also anthropic principle
Cartwright, Nancy 314–15
Cassini, Jean Dominique 19
Cauchy, Augustin 160, 266
C-field 131–4
Challis, James 43
Chalmers, B. 181
Chalmers, J. A. 181
Chamberlin, Thomas C. 233
Chandrasekhar, Subrahmanyan 102
Chardin, Teilhard de 344
chemistry 47–50, 62–3
Chew, Geoffrey 142, 149–61, 272
Ćirković, Milan 335
citation data 208, 301, 303–4
Clausius, Rudolf 195
Clifford, William K. 45–6, 175, 223
Cocconi, Giuseppe 329
coincidences, numerical 93–8, 110, 124, 220–1, 347
　see also large number hypothesis
Collingwood, Robin G. 88
Collins, Barry 226, 228
compactification 264, 267, 295, 300–1, 303–4, 313
complexity, see self-organization
constants of nature 110
　concept of 168–74
　Eddington 94–9
　fundamental 169–70
constants, dimensionless 94, 110, 353 n. 91
　see also constants of nature
conventionalism 105–6, 188
Copernican principle 224, 250 n. 23
cosmic strings 311
cosmical number 94–6
cosmobiology 334, 344, 350 n. 11
cosmological constant 90–2
　problem 206, 236–8
　varying 174
cosmological models 90
　cyclic 193–213
　de Sitter 90
　Dirac 94, 110
　Einstein 90
　ekpyrotic 203–5
　electrical 123–4
　Lemaître–Eddington 91, 115 n. 51, 118, 334
　Lemaître–Tolman 259
　Milne 104
　plasma 215 n. 52
　pre-big-bang 210–12
　primeval atom 91
　pyrotechnic 214 n. 25
　steady state 117–37
　see also cyclic cosmologies
cosmological principle 105, 119
　perfect 119–21
cosmology, development of 90–2
cosmophysics 88
cosmythology 111, 129, 244
Coule, D. H. 346
coupling constants 146, 170, 181–2
Craig, William L. 247–8, 335
creationism 254 n. 114, 285
critical density 121
cyclic cosmologies 193–213, 226, 231
　Baum–Frampton 208–9
　conformal 213
　history of 194–9
　Rosen-Israelit 201–2
　Steinhardt-Turok 202–8

Dalton, John 20
dark energy 206–8, 240, 303, 338, 341
dark flow 269, 311
Darwin, Charles G. 109, 112, 233
Darwinism 242, 247, 348–9
Davies, Paul C. W. 187, 228, 231, 239, 277, 284, 336
Davies-Tipler argument 239–40, 363
Davy, Humphry 24, 26–7
Dawkins, Richard 247
deductivism 107–9, 119
Descartes, René 11–18, 35, 97, 107, 224
De Rújula, Alvaro 302
design arguments, see teleology
de Sitter, Willem 90
determinism 22–3, 96, 151, 176
Deutsch, David 288 n. 61
DeWitt, Bryce 223, 262
DeWitt, Cecile 223
Dick, Thomas 326–7

Dicke, Robert 190 n. 25, 200, 219–24, 230
Dingle, Herbert 111–13, 117, 129–30, 244
Dirac, Paul A. M. 64, 86, 145, 250 n. 22, 273
  antiparticles 143
  endless life 340
  large numbers 110–11, 220–1
  reductionism 143, 343
  S-matrix 156
  varying gravitation 175–80
  wave equation 94
doomsday argument 338
Drake, Frank 329–30
Drake equation 330
Drude, Paul 38, 172
Duff, Michael 188
Dürr, Hans-Peter 148
dynamicism 27–30
Dyson, Freeman 145, 160, 182, 223, 226, 246, 357
  endless life 333, 338–43, 345
  universal mind 248–9

Ebertin, Reinhold 334
Eddington, Arthur S. 87, 92–105, 108–13
  constants of nature 94–9, 173
  cyclic universe 197
  end of universe 334
  multiverse 258
  speed of light 105, 185
effectiveness, principle of 230
Ehrlich, Robert 282, 308
Einstein, Albert 72–4, 78–9, 108, 197, 293
  algebraic physics 317
  constants of nature 174
  cosmological constant 90–2
  cyclic model 196
  ethical values 339
  Kaluza-Klein theory 294–5
  mathematics 82, 103
  on Popper 289 n. 87
  rationalism 82–3
ekpyrotic model 203–5
electrodynamics 59–77
  see also electron, theories of
electromagnetic mass 65–7
electromagnetic world view 50, 60–78
electron 49, 65, 77
  early view of 62–4
  inexhaustible 72
  theories of 65–75
elementary particles 64, 78, 98–9, 150
elements, chemical 38, 47–8, 62–3, 68, 89
  maximum number 99–100
Ellis, George 188–9, 244, 248, 344, 346
  infinities 265–6
  multiverse 279, 281, 284

emergent phenomena 274–5
energetics 60–1
energy conservation 120–23, 134, 174
entropy 195–6, 200–1, 218, 258, 266, 327, 329
Epicurus 256
epistemic standards 142, 179, 314–15, 360–6
  anthropic principle 241–4
  multiverse 279
  see also science, criteria of
Ertel, Hans 110
eschatology 335–6 see also physical eschatology
eternal life postulate 225, 348
eternity 23, 328, 337, 345
ether 38–9, 45
  electromagnetic 62–4, 70–1
  squirts 39
ethical values 172, 197, 339
Everett, Hugh 261–2
Everett interpretation, see many-worlds
    interpretation
existence 268, 272, 363
explanations 55, 106, 225
  anthropic 230–1, 242, 246
  multiverse 277, 281
  religious 231
extrapolations 119–20, 335, 359, 367
extraterrestrial life 326–31
  see also pluralism; SETI research

Falb, Rudolf 195–6
falsifiability 128–9, 154, 183, 207, 238, 241–5, 364
  multiverse 281–4
  string theory 307–10
Faraday, Michael 24, 26, 169
Fatio de Duillier, Nicolas 25
fecundity assumption 263
Feinberg, Gerald 163 n. 51, 298
Fermi, Enrico 149, 329
Fermi's paradox 329
Ferrara, Sergio 296
Feuer, Lewis 259
Feynman, Richard P. 145, 167, 181, 272, 356
  on S-matrix 155
  on string theory 306
final theory 70, 296, 312, 355–6
fine-structure constant 83, 85 n. 39, 90, 181
  Chew 157
  Eddington 94, 98
  Heisenberg 148
  Page 367
  varying 182–5, 187
FitzGerald, G. Francis 39, 48, 54, 63
Flammarion, Camille 336
Flint, Henry 190 n. 16
Fock, Vladimir 295

Fontenelle, Bernard de 218
Föppl, August 70
Fowler, Ralph 102
Frampton, Paul 208–9
Frautschi, Steven 149
Freedman, Daniel 296
Freese, Katherine 341
Friedan, Daniel 308
Friedmann, Alexander 196
Frigg, Roman 314–15
fundamental theory, Eddington's 95–100
Fürth, Reinhold 110

Galison, Peter 306
Gamow, George 118
  bouncing universe 212
  constants of nature 173–5, 178–9, 182–3
Gans, Richard 72, 77
Gardner, Martin 225
Garriga, Jaume 265, 342
Gasperini, Maurizio 203, 210–12
gas theory 41, 47
Gell-Mann, Murray 149, 151, 154–5, 225
Georgi, Howard 270
Gheury de Bray, M. 186
Ginsparg, Paul 307
Glashow, Sheldon 157, 270
  string theory 307–8
God 13–17, 45, 107, 194–5, 231, 247–8, 284, 335
  as cosmic singularity 345
Gödel, Kurt 357
Gold, Thomas 118–22, 175, 219, 289 n. 78
Gordon, Walter 294
Goto, Tetsuo 297
Gott, J. Richard 338
gravitation 20, 25, 133
  and atomic structure 161 n. 3
  electrodynamic 67, 72, 77
  mechanical 25
  Newtonian 20–1, 277, 311
  vortex theory 42–4
  see also relativity theory, general
gravitational constant 105–6, 133
  varying 110, 175–80
gravitational waves 207, 212, 293
Green, Michael 300, 305, 322 n. 73
Greene, Brian 255, 305, 310
Gribbin, John 228
Gross, David 245, 300, 309, 314
Gutberlet, Constantin 266
Guth, Alan 137, 186, 348
  anthropic principle 232
  cyclic universe 200
  multiverse 263–5, 267
  string theory 313

Haas, Arthur E. 110, 172–3
hadrons 141, 150–2, 297–9
Haldane, John B. S. 109, 328
Hall, P. J. 230
Halley, Edmund 218
Halpern, Paul 311
Hamilton, William Rowan 24
Harnik, Ronni 238
Harrison, Edward 149, 348–9
Hart, Michael 329
Hartshorne, Charles 252 n. 59
Harvey, Jeffrey 300
Hawking, Stephen 132, 223, 226, 228, 246, 296
Heaviside, Oliver 65
Hegel, Georg W. F. 80
Heisenberg, Werner K. 23, 144–5, 150, 261
  S-matrix 146–7
  unified theory 148, 270
Heitler, Walter 147
Heller, Michael 283
Hellings, R. W. 180
Helm, Georg 60–1
Helmholtz, Hermann von 36, 40
Herschel, William 88, 326
hibernation hypothesis 341
Hicks, William M. 44, 52–3
Hilbert, David 79–81, 265
Hofstadter, Robert 162 n. 41
Hogan, Craig 242
holism 151
Hollands, Stefan 245
Holman, Richard 269
Holman, Silas 51
Hooft, Gerardus t' 157, 320 n. 54, 368 n. 2
Hořava, Petr 203
Horgan, John 369 n. 28
Horowitz, Gary T. 301
Howson, Colin 364
Hoyle, Fred 118, 230
  anthropic principle 235–6
  astrobiology 327
  carbon-12 resonance 233–6
  constants of nature 259
  endless life 339
  steady-state theory 118–37
Hubbard, L. Ron 346
Hubble, Edwin 97, 109
Hubble constant 97–9, 104, 120, 131, 345
Humason, Milton 97
Hume, David 257
humility, principle of 242
Huygens, Christiaan 326
Huxley, Thomas 45, 357
hyperspace 123, 258, 292
hypothetical worlds 253 n. 81, 257

ideology, science and 45, 60–1, 72, 158–60, 343–7
Idlis, Grigory 219–20, 251 n. 46
incompleteness theorem 357
indifference principle 15
infinities, actual 266
   Descartes 17
   multiverse 265–6, 280
   steady-state theory 121–2, 127
inflation models 186–7, 342
   anthropic principle 231–2, 243
   chaotic 263–4
   eternal 263–4
   opposition to 201, 203–8
   and steady-state theory 136–37
intelligent design 279, 285, 307–8
ironic science 369 n. 28
Isenkrahe, Caspar 327, 336
Islam, Jamal 341–2
isotropy, universe 228, 246
Israelit, Mark 201–2, 213
Iwanenko, Dmitrii 173

Jaakkola, Toivo 25
Jaki, Stanley 252 n. 52
Jantsch, Eric 368 n. 12
Jeans, James 64–5, 75, 113, 219, 272, 328
   end of universe 333–4
Jeffreys, Harold 112
Johnson, Martin 109, 125
Jordan, Pascual 109, 113, 177, 182

Kachru, Shamit 303
Kaku, Michio 305, 312
Kallosh, Renata 203, 303
Kaluza, Theodor 293–4, 296
Kaluza–Klein theory 123, 293–6, 299
Kane, Gordon 313
Kant, Immanuel 26, 30
Kardashev, Nicolai 202, 331, 338
Kaufmann, Walter 66, 68–9
Kekulé, Friedrich 48
Kelvin, *see* Thomson, William
Khoury, Justin 203
Kilmister, Clive W. 128
kinematic relativity 102, 109, 119
Kinney, William 341
Kirchner, U. 265
Klein, Oskar 109, 125, 293–5
Knight, Gowin 19, 30
knot theory 40, 55
Kofman, Lev 203
Kolb, Edward 232
Konopinski, Emil 350 n. 11
Kramers, Hendrik A. 109
Krauss, Lawrence 245, 307, 341

Kuhn, Thomas 253 n. 92, 361
Küppers, Bernd-Olaf 368 n. 12

Lakatos, Imre 315
Lanczos, Cornelius 83, 91–2
Landau, Lev 146, 173
landscape theory 244, 267, 304, 313–14
Langevin, Paul 69–70
Laplace, Pierre-Simon 22
   his demon, *see* determinism
large number hypothesis 110, 177, 180, 220–2
Larmor, Joseph 50–1, 53, 62–3, 67
Lasswitz, Kurt 46
Laue, Max von 79
Laughlin, Gregory 337, 340
Laughlin, Robert B. 274–5
laws of nature 15, 31, 52, 81, 96,
   100, 112, 340
   multiverse 269
LeBon, Gustave 72
Lehners, Jean-Luc 207
Leibniz, Gottfried W. 15, 22, 46, 81
   possible worlds 256–7
Lemaître, Georges 91–2, 199
length, smallest 144–5, 173–4, 211, 317
Lenin, Vladimir I. 59, 72
Lenz, Wilhelm 90
Le Sage, Georges-Louis 25–6, 43
Leslie, John 241
Lewis, Gilbert N. 89
Liebig, Justus von 32
life, definition of 342–3, 346
life, endless 338–43
Lifschitz, Evgeny 146
light, speed of 113, 168–9
   varying 185–9
Linde, Andrei 203–4, 206, 263–4, 266–7, 277, 303,
   342, 348
Lindsay, Robert B. 339
Listing, Johann 40
Livio, Mario 235
Lodge, Oliver 45, 47, 52
Loeb, Abraham 238
Logunov, A. A. 201
London, Fritz 295
Longair, Malcolm 232, 334
loop quantum cosmology 209–10
loop quantum gravity 209, 316–20
Lorentz, Hendrik A. 63, 67–71
Lorentz–Einstein theory 73–5
Low, Francis 149
Lowell, Percival 190 n. 11, 327
Lucretius 33 n. 37, 218
Lyell, Charles 233
Lyttleton, Raymond 123

MacAlister, Donald 44
McCrea, William H. 109, 112, 122–4, 184, 199, 215 n. 45
Mach, Ernst 60
Mach's principle 133, 135, 149, 222
MacMillan, William D. 118
McMullin, Ernan 253, n. 91
McTaggart, John 88
McVittie, George C. 111, 125–7, 129, 309
Magueijo, João 186–9
Maldacena, Juan 322 n. 73
Malebranche, Nicolas 14
Mandelstam, Stanley 149
many-worlds interpretation 261–3, 276
Markley, Robert 160
Martinec, Emil 300
Masriera, D. M. 100
materialism 27, 45–6, 61–2
mathematical universe 270–4, 356–7
mathematics, role of 29, 40, 52, 356–7
   Descartes 18
   electron theory 70, 81
   Milne 102–3
   multiverse 270–4
   pre-established harmony 81–2
   vortex theory 52–4
matter creation 120–2, 130, 179
Mattingly, James 323 n. 80
Maxwell, James Clerk 37, 42–3, 168
Maxwell–Lorentz theory 66–70
Mayer, Alfred 49
measure problem 279
Mehra, Jagdish 306
Mendeleev, Dmitrii 38, 48–9, 233
Mersenne, Marin 12
Mersini-Houghton, Laura 268–9
metascientific considerations 125, 197, 339
   see also aesthetical values, simplicity
Meyer, Lothar 48
Michell, John 24
Michelson, Albert 38, 52
microwave background, cosmic 135–6, 207, 211, 240, 311
Mie, Gustav 76–9
Millikan, Robert A. 118
Milne, Edward A. 101–2, 328–9, 336
Minkowski, Hermann 69, 73, 81–2
Misner, Charles 223
Moffat, John 186
monism 61, 368 n. 10
Moore, George E. 339
Morrison, Philip 329
M-theory 203, 270, 303
Mukhanov, Viatcheslav 262, 276, 342
multiverse 23, 230, 244, 255–85

anthropic 264
creationism 285
early ideas 256–9
hierarchy 260
mathematical 270–2
name 259
objections to 276, 279–84
religious aspects 284
Munitz, Milton 130
mysticism 92, 100, 130, 159, 163 n. 52, 359

Nambu, Yoichiro 297
Narlikar, Jayant 99, 131–7
Narlikar, Vishnu 99
natural constants, see constants of nature
natural philosophy, see Naturphilosophie
natural selection, cosmological 349
natural theology 45, 218, 257, 359
Naturphilosophie 12, 26–31 see also Romanticism
Neumann, John von 239
neutrinos 115 n. 35, 144, 240
Neveu, André 298
Newcomb, Simon 258, 292
Newton, Isaac 11, 15, 35
Nicholson, John W. 89
Nielsen, Holger B. 297
Niewenhuizen, Peter van 296
Nordström, Gunnar 293–4
Novikov, Igor 200
Nozick, Robert 263
nuclear democracy 151–2, 160
numerical relations, see coincidences, numerical
numerology 110

observability criterion 150
occultism 46, 57 n. 31, 72
Ockham's razor 26, 276
Olum, Ken 342
omega point theory 344–7
Öpik, Ernst 125
Oppenheimer, J. Robert 144
Oresme, Nicole 24, 256
Origen 194–5
Ørsted, Hans C. 26–32
Ostwald, Wilhelm 61, 351 n. 45, 368 n. 10
Ovrut, Burt 203

Pachner, Jaroslav 199
Padmanabhan, Thanu 136
Page, Don 284, 367
Pagels, Heinz 243–4
Paley, William 254 n. 116
panentheism 335
Pannenberg, Wolfhart 335, 353 n. 77
panspermia hypothesis 327

Pappagiannis, Michael 331
paradigm shifts 275, 360–1, 363, 366, 368 n. 12
　see also epistemic standards
parallel universes, see multiverse
Parmenides 46
Paschen, Friedrich 83
Pauli, Wolfgang 109, 144, 147, 295
Peacocke, Arthur 335
Pearson, Karl 39–40, 44
Peebles, P. James E. 200
Peirce, Charles S. 184
Penrose, Roger 132, 213, 252 n. 58, 317
periodic system 48–9, 68, 80
Perry, Malcolm 313
phantom energy 208, 338
Philoponus, John 138 n. 31
philosophy, role of 125–30, 158, 188, 199–200, 242
　multiverse controversy 280–4
　string theory 314–5
phoenix universe 202, 214 n. 24
　see also cyclic cosmologies
physical eschatology 89, 325, 334–8, 341–3
　early examples 327–9, 333–4
　religious aspects 335
Pines, David 274–5
Planck, Max 61, 71, 73–5
　fundamental units 171–73
Planck density 201, 210
Planck length 171, 295, 299, 317
Platonism 271, 274, 345
Playfair, John 26
plenitude, principle of 256–7, 268, 275, 286 n. 26
pluralism, history of 326–32
Poincaré, Henri 25, 69, 175
Polchinski, Joseph 266, 303, 322 n. 73
Politzer, David 322 n. 73
Polkinghorne, John 155, 248, 252 n. 52
Popper, Karl R. 128–30, 236, 242–3, 307, 364
　Bondi on 128
　multiverse controversy 281–4
　see also falsifiability
positivism 106, 111, 145, 150
postmodernism 160, 189 n. 2, 367
Power, Edwin 223
Poynting, John H. 51, 53
predictions 55, 77, 113, 122, 363
　anthropic 226, 233–41
　multiverse 277–9, 281–3
　omega point theory 345
Press, William 231
pressure, negative 122, 131, 198–9, 202, 213
Preston, Samuel T. 25, 218
Priester, Wolfgang 213
Priestley, Joseph 24
Primack, Joel 208

Prirogine, Ilya 368 n. 12
proton decay 268, 340, 351 n. 37
pseudoscience 326, 331, 346, 348, 364

quantum electrodynamics 142, 144–8
quantum gravity 316
quantum mechanics 93–4, 109, 294
　development of 142–8, 304
　Eddington and 109
　interpretations 261–3
quantum theory 74–5, 83
　cosmology and 90–2
quarks 154–5, 299
quasi-steady-state theory 134–6
quintessence 215 n. 41, 341

Rabi, Isidor 143
Rabounski, Dmitri 220
Rademacher, Johann 336
Ramond, Pierre 298–9
Ramsay, William 63
rationalism 18, 82–3, 100–3, 107–8, 113
Rayleigh, Lord 75
reality 262, 270–4, 345, 357
　see also existence
reductionism 70, 143, 148, 274–5, 305, 343–4, 346
Rees, Martin 200, 226, 228, 240
　eschatological study 335–6
　multiverse 269, 276–80
Regnault, Henri V. 42
relativity theory 24, 72, 93
　general 79–82, 93, 101, 103–4
　special 72–4, 76
religion, physics and 100–1, 139 n. 53, 231, 247–9, 358–60
　see also theology; God
renormalization 145–6
repeatability principle 119
Reynolds, Osborne 54
Rice, James 90
Richter, Burton 280, 307
Ritter, Johann W. 26, 32
Robertson, Howard P. 105
Rohm, Ryan 300
Romanticism 26–32
Rømer, Ole 168
Rosen, Nathan 201–2, 213
Rosenfeld, Léon 261, 292
Rosenthal-Schneider, Ilse 174
Rovelli, Carlo 209, 316–19
Rozental, Iosif L. 170, 200, 230
Ruark, Arthur 190 n. 16
Ruffini, Remo 226
Russell, Bertrand 355, 368 n. 9
Rutherford, Ernest 60, 142

Rydberg, Johannes 63, 169
Ryle, Martin 131

Sagan, Carl 329–30
Sakharov, Andrei 200
Salam, Abdus 157, 270, 295
Salpeter, Edwin 234
Schellekens, A. N. 266, 309, 312
Schelling, Friedrich 12, 26–31
Scherck, Joël 299
Schlegel, Richard 127
Schmidt, Maarten 183
Schönborn, Christoph 289 n. 98
Schramm, David 232
Schrödinger, Erwin 23, 109–10, 294–5
Schuster, Arthur 44
Schwarz, John 298–300, 302, 310, 322 n. 73
Schwarzschild, Karl 66
Schwinger, Julian 145
Sciama, Dennis 229, 275–6, 339
science, criteria of 126, 128–9, 347–9, 360–7
   anthropic principle 241–6
   multiverse 279–85
   string theory 308–10
science, definition of 282–3, 285, 308, 314–15, 360–1, 365
scientism 343–4, 352 n. 65, 359–60
Searle, George 66
Seebeck, Thomas J. 26
self-consistency 151, 153–4, 156, 158, 314
self-organization 29, 368 n. 12
Sen, Amitabha 316
Sen, Ashoke 309–10
SETI research 236, 329–31
Shallis, Michael 252 n. 52
Shapere, Dudley 315
Shapiro, Irving 180
Shklovskii, Iosef 329–31
simplicity 132
   see also Ockham's razor
singularity theorems 132
Sklar, Lawrence 358
slippery slope argument 276–7
S-matrix 146–7, 149–60, 296
Smolin, Lee 209, 237, 247, 281, 314, 348–9
   loop quantum gravity 316–19
Soddy, Frederick 195–6
Sommerfeld, Arnold 69, 77, 82–3, 181, 320 n. 18
Sonnenschmidt, Hermann 336
space, discrete 144, 209, 317, 319
space, many-dimensional 39, 292–6
spectroscopy 37, 41
speculations 28, 117, 279–80, 328–9, 342, 346, 367
speculative physics 26–9, 111
Stapp, Henry 158

Starkman, Glenn 237, 341
steady-state cosmology 117–37, 259
   and new cyclic theory 204–5
Steiner, Mark 254 n. 97
Steinhardt, Paul 202–8, 243–4, 281
Stenger, Victor 288 n. 72
Stern, Alexander 155
Stewart, Balfour 45–6, 55
Stoeger, William 265
Stoney, George J. 39, 62
   natural units 170–2
string theory 210–11, 296–315
   anthropic principle 312–14
   critique of 305–15
   history of 296–305
   loop quantum gravity 316–19
   multiverse 266–7
   Nobel Prize 322 n. 73
   see also cosmological models, pre-big-bang; landscape theory
Strominger, Andrew 301–2
Strum, L. 298
Stueckelberg, Ernst 147
subjectivism, selective 100, 219
Sudarshan, George 298
supergravity 296
superstrings, see string theory
supersymmetry 295–6, 299, 308, 310–11
Suppes, Patrick 18
Susskind, Leonard 241, 247
   Popperianism 283
   string landscape 267–8, 313–14
   string theory 297–8
Swinburne, Richard 248

tachyons 298
Tait, Peter G. 36, 40, 45–6, 54–5
Takeuchi, Tokio 185
Tegmark, Max 240, 283
   multiverse 260, 262
   theory of everything 270–3
teleology 13, 231, 243, 247–9
Teller, Edward 178–9, 350 n. 11
Temple, George J. 87
testing 106–7
   loop quantum gravity 318–19
   mathematical 366
   multiverse 278, 281–3
   non-empirical 126, 310, 312
   string theory 308–12
Thales 11
Theckedath, Kumar 136
theology 45, 108, 243, 248, 284, 307, 329, 346
   see also God
theory of everything 52, 269–75, 355–8

Dirac 111
  objections to 274–5, 358
  string theory 305
thermodynamics 61
  second law 195–6, 258, 333, 339
Thom, René 368 n. 12
Thomson, Joseph J. 47–9, 54, 62–5, 71, 89, 350 n. 6
Thomson, William 24, 36–55, 233
time 195, 204, 333
  electrodynamic arrow 133
  scales of 106, 139 n. 42
  subjective 341, 343
  thermodynamic arrow 214 n. 15
  without end 339
Ting, Samuel 280
Tipler, Frank 57 n. 29, 225, 228–9, 233, 239, 246, 263
  cosmotheology 347
  endless life 342–7
  omega point 344–6
  physical eschatology 335, 337
  SETI 331
  theory of everything 273–4
TOE, see theory of everything
Tolman, Richard C. 90, 108, 259
  cyclic models 196–7
Tomonaga, Sin-Itiro 145
Trivedi, Sandip 303
Trotta, Roberto 237, 364
Turner, Michael 232
Turok, Neil 202–8, 243–4, 281
two-particle paradigm 143
Tyndall, John 45

uncertainty principle 173, 210
unification 27–48, 38, 78, 80, 93, 173, 270
unity of physics 156
  see also unification
universe, accelerating 240, 303
universe, expanding 97, 101, 104, 118, 181
universe, future state of, see physical eschatology
Uzan, Jean-Philippe 189

vacuum 15, 27
  energy density of 92, 236–7
Vafa, Cumrun 303, 322 n. 73
Vallarta, Manuel S. 161 n. 3
Van Vlandern, Thomas 25, 180
Veltman, Martinus 309
Veneziano, Gabriele 203, 210–11, 296–8
Venable, Francis 49

Vilenkin, Alexander 237, 263–5, 342
vortices 15–18, 35, 97
vortex atom theory 35–55
vortex sponges 39

Wald, Robert 245
Walker, Arthur G. 87, 109, 111
Wallace, Alfred R. 218
Wataghin, Gleb 190 n. 16
weakless universe 238
Webb, John 183
Wegener, Alfred 177
Weinberg, Steven 150, 157, 225, 245, 270, 339
  anthropic principle 245
  cosmological constant 236–7, 244
  string theory 301–2, 309, 311–12
Weiss, Christian S. 26, 30
Weyl, Hermann 78–9, 91, 93, 293
Whaling, Ward 234
Wheeler, John A. 146, 191 n. 43, 223–4, 226
Whiston, William 218
Whitehead, Alfred N. 88, 158, 184–5, 335
Whitrow, Gerald J. 87, 109, 126–7
Whittaker, Edmund 93
why-questions 232
Wickramasinghe, Chandra 327
Wiechert, Emil 65–6
Wien, Wilhelm 66–8, 73, 75
Wigner, Eugene 109, 153, 176, 223, 272
Wilczec, Frank 322 n. 73
Witten, Edward 203, 300–5, 310, 322 n. 73
Woit, Peter 308, 314
Wold, P. 185
Wood, De Volson 38
world equations 80
wormholes 276, 341, 352 n. 53
Wynn-Williams, David 332

Yau, Shing Tung 301
Yoneya, Tomiaki 320 n. 19
York, Herbert 350 n. 11
Yukawa, Hideki 145–6

Zachariasen, Fredrik 152
Zanstra, Herman 198–200, 339
Zeeman, Pieter 63
Zel'dovich, Yakov B. 251 n. 46
Zelmanov, Abraham 220
Zemach, Charles 152
Zeno 46
Zöllner, K. Friedrich 353 n. 73
Zytkow, Anna 313

Printed and bound by CPI Group (UK) Ltd, Croydon, CR0 4YY